SCOPE 32

Land Transformation in Agriculture

SCOPE 32

Land Transformation in Agriculture

Edited by
M. G. Wolman
The Johns Hopkins University
and

F. G. A. Fournier
Inspecteur General Honoraire de Recherche de l'ORSTROM

Published on behalf of the
Scientific Committee on Problems of the Environment (SCOPE)
of the
International Council of Scientific Unions (ICSU)
by
JOHN WILEY & SONS
Chichester · New York · Brisbane · Toronto · Singapore

Library of Congress Cataloging-in-Publication Data:

Land transformation in agriculture.

(SCOPE ; 32)
1. Land use, Rural—Environmental aspects.
2. Agricultural pollution. 3. Agricultural conservation.
4. Soil conservation. I. Wolman, M. Gordon (Markley Gordon), 1924– . II. Fournier, F. G. A.
III. International Council of Scientific Unions.
Scientific Committee on Problems of the Environment.
IV. Series: SCOPE (Series) ; 32.
HD111.L255 1987 333.3′35 86-19006
ISBN 0 471 91288 3

British Library Cataloguing in Publication Data:

Land transformation in agriculture.—
(SCOPE ; 32)
1. Land use, Rural—Planning
I. Wolman, M. G. II. Fournier, F. G. A.
III. International Council of Scientific Unions, *Scientific Committee on problems of the Environment* IV. Series
333.76′13 HD108.6

ISBN 0 471 91288 3

Typeset by Acorn Bookwork, Salisbury, Wiltshire
Printed and bound in Great Britain by Biddles of Guildford

International Council of Scientific Unions (ICSU)
Scientific Committee on Problems of the Environment (SCOPE)

SCOPE is one of a number of committees established by a non-governmental group of scientific organizations, the International Council of Scientific Unions (ICSU). The membership of ICSU includes representatives from 68 National Academies of Science, 18 International Unions, and 12 other bodies called Scientific Associates. To cover multidisciplinary activities which include the interests of several unions, ICSU has established 10 scientific committees, of which SCOPE is one. Currently, representatives of 34 member countries and 15 Unions and Scientific Committees participate in the work of SCOPE, which directs particular attention to the needs of developing countries. SCOPE was established in 1969 in response to the environmental concerns emerging at that time; ICSU recognized that many of these concerns required scientific inputs spanning several disciplines and ICSU Unions. SCOPE's first task was to prepare a report on Global Environmental Monitoring (SCOPE 1, 1971) for the UN Stockholm Conference on the Human Environment.

The mandate of SCOPE is to assemble, review and assess the information available on human-induced environmental changes and the effects of these changes on humans; to assess and evaluate the methodologies of measurement of environmental parameters; to provide an intelligence service on current research; and by the recruitment of the best available scientific information and constructive thinking to establish itself as a corpus of informed advice for the benefit of centres of fundamental research and of organizations and agencies operationally engaged in studies of the environment.

SCOPE is governed by a General Assembly, which meets every three years. Between such meetings its activities are directed by the Executive Committee.

R. E. Munn
Editor-in-Chief
SCOPE Publications

Executive Secretary: V. Plocq

Secretariat: 51 Bld de Montmorency
75016 PARIS

SCOPE 1: Global Environmental Monitoring 1971, 68pp (out of print)

SCOPE 2: Man-Made Lakes as Modified Ecosystems, 1972, 76 pp (out of print)

SCOPE 3: Global Environmental Monitoring Systems (GEMS): Action Plan for Phase 1, 1973, 132pp (out of print)

SCOPE 4: Environmental Sciences in Developing Countries, 1974, 72pp (out of print)

Environment and Development, proceedings of SCOPE/UNEP Symposium on Environmental Sciences in Developing Countries, Nairobi, February 11–23, 1974, 4pp

SCOPE 5: Environmental Impact Assessment: Principles and Procedures, Second Edition, 1979, 208pp

SCOPE 6: Environmental Pollutants: Selected Analytical Methods, 1975, 277pp (out of print)

SCOPE 7: Nitrogen, Phosphorus, and Sulphur: Global Cycles, 1975, 192pp (out of print)

SCOPE 8: Risk Assessment of Environmental Hazard, 1978, 132pp

SCOPE 9: Simulation Modelling of Environmental Problems, 1978, 128pp (out of print)

SCOPE 10: Environmental Issues, 1977, 242pp (out of print)

SCOPE 11: Shelter Provision in Developing Countries, 1978, 112pp

SCOPE 12: Principles of Ecotoxicology, 1978, 372pp (out of print)

SCOPE 13: The Global Carbon Cycle, 1979, 491pp

SCOPE 14: Saharan Dust: Mobilization, Transport, Deposition, 1979, 320pp (out of print)

SCOPE 15: Environmental Risk Assessment, 1980, 176pp

SCOPE 16: Carbon Cycle Modelling, 1981, 404pp

SCOPE 17: Some Perspectives of the Major Biogeochemical Cycles, 1981, 175pp

SCOPE 18: The Role of Fire in Northern Circumpolar Ecosystems, 1983, 344pp

SCOPE 19: The Global Biogeochemical Sulphur Cycle, 1983, 495pp

Funds to meet SCOPE expenses are provided by contributions from SCOPE National Committees, an annual subvention from ICSU (and through ICSU, from UNESCO), and annual subvention from the rench Ministère de l'Environment et du Cadre de Vie, contracts with UN Bodies, particularly UNEP, and grants from Foundations and industrial enterprises.

Contents

Preface

This report on Land Transformation in Agriculture is the result of a long effort launched by SCOPE in 1978. At that time, a group of specialists, including Rorke B. Bryan (Canada), F. Fournier (France), S. Hénin (France), V. A. Kovda (USSR), B. Novak (Czechoslovakia), J. Rusek (Czechoslovakia) and M. G. Wolman (USA), defined a basis for a project on land transformation. During a first phase of activity, case studies on the problem prepared by SCOPE National Committees or other groups of scientists were collected. We wish here to acknowledge and to thank all the specialists who prepared the papers and documents presented below by country.

Australia

Effect of wheat cultivation and husbandry practice on soils of the north-west wheat belt

Effect of rice cultivation on grey and brown gilgai of the Murrumbidgee irrigation area

Bulgaria

Influence of reforestation on the ecology of landscape

Czechoslovakia

Effects of changes in soil (land) use as agricultural enterprises pass from 500 to 5000 hectares with intensification of raising of domestic animals

Changes resulting from transformation of small villages to agro-industrial cities

Influence of protected zones (parks and others) of flora and fauna on the environment

Rehabilitation of lands deteriorated by open mining in the lignite basin, North Bohemia

Egypt

Land transformation in the western Mediterranean coastal region

Land transformation in the agro-ecosystems of the Nile Valley and delta after construction of the High Dam

England

Extensive bibliography of land use studies

Rothamstead studies: *ex*. Jenkinson, D. S., and Johnston, A. E. (1976) Soil organic matter in the Hoosfield continuous barley experiment. Rothamsted Rept. 1976, Pt. 2, pp. 87–101

Federal Republic of Germany

Landscape modelling: Ingolstadt

France

Hedges and rural management: effects of hedge removal in Brittany

Effects of spreading pig manure in Brittany

Range and uncultivated areas in mid-altitude mountains: evaluation of land values and changing processes

Effects on evolution of soils of the introduction of conifers

Effects of turnover of straw mulch and grass on loamy soil under long term experimentation

Nitrogen and phosphorus balance in granitic terrain

Drainage of fertilizers from sandy soils and their contribution to pollution

Changes in adventitious vegetation with and without irrigation in different rotations

Effects of cultural techniques in long-established agricultural systems: effect of tillage on soil characteristics under different rotations

Information required on tillage effects to evaluate impacts of rotation

Effects on agricultural lands of electric transmission lines

Effects of underground pipelines on agricultural lands

Impacts of hydroelectric projects on fluvial dynamics, and relation to land management

Impact of tourism and skiing on mountain slopes

Data on the effects of transformation of forest and of prairies in the tropics

Effects of irrigation in Alsace

Effects of pumping underground water on land stability

Effects of pumping in watertables connected to rivers

Ghana

Effective measures for the use and control of land in residential areas

Hungary

The effect of irrigation on the soils of the Hungarian plain

Principles for the development of an environmental monitoring system in the Lake Balaton region

Nutrient adsorption of fertilized soils under field conditions

Mine reclamation with cultivated crops on open pit mines

India

The environment and Cenchrus grazing lands in western India

Indonesia

Seraya Valley Project, Central Java Indonesia (1973–80)

Social and economic aspects of tidal swamp land development in Indonesia

Fifty years of spontaneous and government sponsored immigration in the swamp lands of Kalimantan: past results and future prospects

Israel

Land transformation in the mediterranean north of Israel

A model of land transformation in the Negev during historical times

Côte d'Ivoire

Criteria on the effect of land use in agriculture on landscape

Japan

Nationwide comparative study on large-scale land transformation and related environmental problems in Japan

Netherlands

Effects of intensification of agriculture on nature and landscape

Side-effects of enclosure dams in coastal and delta waters

Urban and traffic effects of development in polder areas

Biological effects of infiltrating river water in coastal sand dune areas for drinking water reserves

New Zealand

Land alone endures: land use and the role of research

Spain

Assessment of land damage and environmental degradation in areas surrounding Madrid

Land evolution (degradation of vegetation) as consequence of soil uses in mountain areas of Andalusia

Influence of persistent and melting snow on high mountain rivers and their effects on erosion

South Africa

Land transformation in South African savanna

Land transformation of the Karoo biome

United States of America

US Dept. Agri. (1980) *RCA: Resource Conservation Act, 1980, Appraisal and Environmental Impact*, 3 Vols, Washington

US National Research Council, National Academy of Sciences (1980) *Report of the Workshop on Soil Transformation and Productivity*, 16–17 October, and associated papers, Washington

US Dept. of Agriculture and Council on Environmental Quality (1981) *National Agricultural Lands Study*

Sheridan, David (1981) *Desertification of the United States*. US Council on Environmental Quality, 142pp

Hawaii: soil erosion by water in the tropics

Changing soil characteristics and productivity of some Kansas wheat growing areas

Sequential changes of land use, soils, and productivity in the southeastern USA

Seepage and salinization associated with changing cropping patterns in the northern Great Plains region, Montana

Evidence of changes in yield and soils from continuous experiments on Morrill plots in Illinois

During the Vth General Assembly of SCOPE, in Ottawa (1982), it was recommended, as a second phase of the project, to concentrate attention on the question of Land Transformation in Agriculture mainly at a global level. A new Scientific Advisory Committee was nominated, composed of the following:

Gordon Wolman, Department of Geography, the Johns Hopkins University, Baltimore, Maryland 21.218, USA; *Chairman*

M. A. Ayyad, Botany Department, Faculty of Science, University of Alexandria, Moharram Bey, Alexandria, Egypt

E. H. Brown, Department of Geography, University College of London, Gower Street, London WC1E 6BT, UK

P. Buringh, Marterlaan 20, 6705, B.A. Wageningen, the Netherlands

F. di Castri, Director, Division of Ecological Sciences, Unesco, 75.700, Paris, France

R. Dudal, Director, Land and Water Division, FAO, Rome, Italy

F. Fournier, Inspecteur Général honoraire de Recherche de l'ORSTOM, 27 Rue Jacques Lemercier, 78.000, Versailles, France

S. Hénin, Directeur honoraire du Departement d'Agronomie de l'INRA, 88 Rue de Jarel, 75015, Paris, France

This report is the final result of the second phase of activity. In its preparation the case studies collected were fully taken into consideration, and some of them are presented here. They form an integral part of the Report strengthening by example observations set forth in the Main Report. Some authors call attention to the fact that the long gestation period of the manuscript may have made obsolete occasional specific observations. The thrust of transformation, however, is clearly evident.

Contributors to the Main Report

L. Bournas	CEMAGREF, Parc de la Tourvoye, 92160 Antony, France
L. Brader	Director, Plant Production and Protection, FAO, Via delle Terme di Caracalla, 00100, Roma, Italia
E. H. Brown	Geography Department, University College, 26 Bedford Way, London WC1H 0AP, UK
P. Buringh	Materlaan 20, 6705 BA Wageningen, the Netherlands
R. Carillon	CEMAGREF, Parc de la Tourvoye, 92160 Antony, France
C. Cheverry	Département du sol, ENSA, 65 route de St. Briex, 35000 Rennes, France
R. Dudal	Faculty of Agricultural Sciences, Irrigation Engineering, University of Leuven, Kardinal Mercierlaan 22, BP 3030, Leuven, Belgique
F. Fournier	Inspecteur Général Honoraire de L'ORSTOM, 27, rue Jacques Lemercier, 78000 Versailles, France
D. Grigg	Department of Geography, University of Sheffield, Sheffield S10 2TN, UK
S. Hénin	Directeur Honoraire du Départment d'Agronomie de l'INRA, 88 rue de Javel, 75015 Paris, France
N. W. Hudson	National College of Agricultural Engineering, Silsoe, Bedford MK4 4DT, UK
R. A. Olson	Department of Agronomy, 279 Plant Sciences, University of Nebraska, Lincoln, 68583 Nebraska, USA
K. Ruddle	National Museum of Ethnology, Senri, Expo Park Suita, Osaka 565, Japan
L. Shanan	18 Hatichon Street, Savyon 56530, Israel
I. G. Simmons	Department of Geography, Sciences Laboratory, South Road, Durham DH1 3LE, UK
G. Spoor	National College of Agricultural Engineering, Silsoe, Bedford MK4 4DT, UK
M. G. Wolman	Department of Geography and Environmental Engineering, The Johns Hopkins University, Baltimore, Maryland 21218, USA

Section I
Main Report

Land Transformation in Agriculture
Edited by M. G. Wolman and F. G. A. Fournier

CHAPTER 1
Introduction to Land Transformation in Agriculture

M. G. WOLMAN AND F. G. A. FOURNIER

Whether couched in Malthusian terms or more current style, concern for the adequacy of land resources for agriculture in the future continues to be expressed by individuals all over the world. These concerns relate to the availability of land for agriculture and to the quality of the land that will be needed forever to produce food and fibre. Increasing numbers of people, and the encroachment of urban, industrial and transportation activities on the agricultural landscape, and the modern techniques of agriculture, are all properly viewed as exerting real or potential pressure on the availability of good land required for future agricultural production. The word 'forever' is not an exaggeration. Trends which show increasing pressures on agricultural land can be projected to demonstrate that the potential to provide food for future world populations is imperiled.

The present volume was stimulated by this worldwide concern for the way in which land used in agriculture is being transformed throughout the world. This concern is a response to increasing reports of damage to the environment and to the ecosystems which comprise it as a result of the changing nature and intensity of a number of human activities. Such reactions, however, may be based on observations of some unusual facts or on environmental changes, the causes of which are often arbitrarily identified in the absence of data or careful analysis. Rational analysis of the factors influencing transformation of the land requires objective data characterizing the transformation, a tie between these observations and their apparent causes, and an understanding of the processes or mechanisms involved in the changes. The emphasis here is upon the land itself, not upon the complex web of social, economic and political factors which may determine what happens to land. Rather, an attempt is made to focus upon how such factors influence or transform the land. The approach is not exhaustive, but illustrative. The intent is to illustrate the kinds of changes taking place on the land, some impacts of significant factors affecting the land, and the nature of the scientific evidence which might permit us to evaluate such impacts, or to determine their extent and potential influence.

The initial overview in Chapter 2 provides a picture of current land use, agricultural production, and the relationship between land availability and potential productivity. Subsequent chapters fall within this initial framework.

While the book deals with the present and the future, historical perspective is essential in considering land transformation and agriculture. First, a historical perspective provides the best possible illustration of the way in which land is transformed, both spatially and qualitatively, through the influence of factors wholly external to agriculture itself. Population change, war, technological revolutions, and a variety of social changes throughout history have resulted in profound transformations of the land. The magnitude of the changes, as well as the rates of transformation, are clearly demonstrated in the history of agricultural development from earliest man through the industrial revolution. Whether one refers to these as lessons or not, history provides superb examples of continuing change, changes mirroring those of today and certain to occur tomorrow. This does not imply that future transformations will be like those of the past or present, but that a myriad of influences will alter how the land is used, what it is used for, and the imprint that such uses will have on the land and soil.

The 'lessons' vantage point is a second contribution of the historical perspective. The same kinds of influences clearly dominate land transformations in agriculture today as they did in the past. Rates of transformation are often accelerated, and analogues from the past may not match precisely those of the present or the future. Nevertheless, without an appreciation of the breadth of factors which influence change today, one can well be misled by assuming that the transformations which take place in and on the land are somehow controlled solely by those who till the land or build upon it. In the last analysis this may be so; yet those who till the land are not always, and are perhaps rarely, independent agents of the world and society in which they live. Historical examples coupled with illustrations throughout the volume from the present scene clearly demonstrate this.

Major technological factors today have a direct influence on the transformation of the land. After clearing the land to make it suitable for producing food and fibre for human beings and feed for domestic animals, improvement of the infrastructure influences land transformation as agricultural products can be more easily transported. In agriculture itself, irrigation, mechanization, conservation practices and wetland reclamation are used, in Chapters 5.I–5.VI, to illustrate the kinds of transformations that are taking place today. Irrigation, while satisfying the need for water, also produces deleterious effects such as waterlogging and salinization. Mechanization has contributed significantly to land transformation. It not only increases the rate at which work can be performed, but also allows various kinds of work to be carried out at appropriate times, thus improving the quality of the work and increasing yields; it also allows new lands to be opened to agriculture. At the same time, mechanization has introduced changes in the physical properties of the soil. Mismanagement of land and cultivation of marginal land have caused severe soil erosion by both water and wind. Thus various soil conservation practices have been developed during the last half century, not only to stop erosion, but also to avoid damaging the land and to improve degraded lands.

Indirect influences of major importance in transforming the land in agriculture, such as the use of fertilizers and plant protection techniques, are described in Chapters 5.I–5.VI. Introduction of inorganic fertilizers, for example, is an important factor causing land transformation as the application of these fertilizers not only increases crop yields, but also affects land use. Intensive application of fertilizers can cause nutrient imbalances in soils, and pollution of soil or groundwater. The control of crop pests and diseases directed toward reducing animal food crop losses currently estimated at $300 billion introduces many changes in land use. Insecticides, fungicides and herbicides have transformed agriculture and land for agriculture. Traditional crop rotation systems could be altered allowing the introduction of monocropping and reduced tillage. Although pesticides have many advantages, there are again shortcomings in side-effects such as the emergence of pesticide-resistant strains of pest organisms, the disturbance of agroecosystems, and pollution of the environment.

These direct and indirect factors influencing land transformation are of fundamental importance throughout the world, and illustrate not only the nature of change, but the kinds of changes associated with such practices. Not every conceivable change is addressed. Thus, forestry and pastural activities are touched upon indirectly. Primary emphasis is given to major factors affecting cropland agriculture.

Last, a key issue in the discussion of transformation of land in agriculture is the establishment of criteria, observable and preferably measurable, that can be used to chart the course of such transformation and to suggest mechanisms for mitigating or reversing long-term deleterious changes. An attempt is made in Chapter 6, using illustrative examples, to describe mechanisms or processes that are associated with changes in land use and that impact the land itself. Knowledge of these processes is fundamental, not only to an understanding of the kinds of changes likely to take place under a given set of conditions, but to the prescription of measures that might be taken to recognize such changes and to alter or reverse them where society wishes to do so. Some processes in the soil, for example, are not well understood. Others are well described and techniques exist for monitoring change. In some cases, scientific and technological information needed to reverse the effects of earlier impacts becomes available after a lapse of years or decades. For others, recognition of incipient and unwanted changes provides a basis for immediate changes in management.

The case studies in Section II of this volume, rather than being separate, form an integral part of the design of the project. The original and current orientation of the study of land transformation under the sponsorship of SCOPE involved the establishment of case studies in diverse regions of the world. These provide a basis for evaluating what is known about land transformation processes and some measure of the potential significance of such processes in different regions of the world. Thus, the case studies, along with the discussion of direct and indirect influences of various practices on the land, provide illustrative examples of criteria which can be established to measure land transformation under different agricultural and other pressures. The case studies reported here are among 30 or 40 such studies from 20 countries identified or prepared for the broader SCOPE Land Transformation Project. Not all of the original studies deal with transformations related to agriculture, and many have been published elsewhere. A list of these studies is available from SCOPE. In addition, the Ministry of Environment of the French Government published a collection of case studies of Land Transformation des Terres (1982), and SCOPE cooperated with COSPAR in a symposium resulting in publication of a volume entitled *Study of Land Transformation Processes from Space and Ground Observations* (Pergamon Press, Vol. 2, No. 8, 1982).

This introduction would be incomplete without several important disclaimers. First, under ideal circumstances it would be desirable to have a complete description of the varied land uses, and their areal extent throughout the world. Thus, it would be desirable to know precisely how much land is in various uses, and the ways in which these uses, and the land devoted to them, are changing in different regions throughout the world. In addition, one would like to know how the quality of the land in each use was changing, as well as the areal extent of such changes in quality. Chapter 2 summarizes some of this information. However, no attempt was made to do a detailed evaluation of the spatial and qualitative changes in lands in all parts of the world. Existing data are insufficient to accomplish this task as data on the spatial distribution of various land uses at relevant scales do not exist for many parts of the world. Even less information is known in detail about the quality of lands associated with these uses. Thus Chapter 2 includes information for relatively large areas within broad classifications of land use and land characteristics.

As the historical discussion suggests, major changes in the land are brought about from distant causes involving social changes in the world, as well as on the local scene. The complex system involving land resources, and cultural, social and political factors, influences what happens to the land. The overview in Chapter 2, as well as the historical perspective, and the conclusions of this volume, address the influence of these complex factors on transformation of the land. However, the focus in this volume is upon the land itself. Thus the emphasis is upon the natural sciences rather than the social sciences. This orientation, or emphasis, is not intended to suggest that the system of resources and man is dominated by processes characterized by the natural sciences. Rather, the emphasis here is simply upon that phase of the complex system involving processes occurring in and on the land itself.

Land Transformation in Agriculture
Edited by M. G. Wolman and F. G. A. Fournier

CHAPTER 2
Agricultural Land Use in Space and Time

P. BURINGH AND R. DUDAL

2.1 THE WORLD LAND AREA AND ITS USE

The land area of our planet is 14.9 billion hectares. This is 29% of the earth's surface, 71% being water. A part of the land area (1.4 billion hectares) is permanently covered by ice. Consequently 13.4 billion hectares is used as arable land, grassland or forests (Table 2.1).

A considerable part of this land area is used for growing food for human beings, feed for domestic and wild animals, biological raw materials for industrial and medical products, cooking, heating etc. Moreover land is used for other purposes such as housing and urban services, industries, mining, communications and recreation. Originally there was only natural vegetation with wild life and some people, who lived by food gathering and hunting, using some wood for shelter and cooking. About 10,000 years ago women began to manipulate nature by cultivating land in order to get more food and to be less dependent on what nature could supply. The number of people was gradually increasing, so more land had to be cultivated for growing food crops. Some natural grazing land was improved in order to get more food for an increasing number of domestic animals, and some cleared forest land was replanted with special tree varieties to provide for timber wood, fuel wood, wood for the paper industry, and so on. More land was needed for non-agricultural purposes, because of the ever-increasing world population, particularly during the last half century, when the population increased exponentially. There were about 165 million people in the year 0, 1 billion in 1820, 2 billion in 1930, 3 billion in 1960, 4 billion in 1976, and there will probably be 5 billion in 1988 and over 6 billion in the year 2000.

The most striking feature of the projected population growth is that the share of world population living in developing countries will increase from the present 72% to 87% in the year 2110, that is 9.1 billion out of the total of 10.5 billion. Within the developing world, differences in fertility levels and a decline of birth rates will entail a marked regional demographic diversity. The

Table 2.1 General land use of the world (FAO Production Yearbook, Vol. 32, 1978)

Category	1961–65 (Mha)	1977 (Mha)
Arable land	1 379	1 462
Grassland	3 054	3 058
Forest land	4 132	4 077
Other land	4 508	4 476
Totals	13 073	13 073

The category 'other land' is mainly land in polar regions, desert land, stony and rocky land in mountains.

stable population of various regions will be reached in different years, ranging from 2030 for Europe to 2110 for Africa. Proportionally, the largest increases are expected in Africa (fivefold) and in southern Asia (threefold)—in the latter case, however, from a much larger 1980 base. Africa and southern Asia together, with 6.3 billion people, will account for over 60% of the world's total population at the time of stabilization (Salas, 1981).

It is evident that the transformation of land with a natural vegetation into arable and grazing land has become an important subject for discussion.

2.2 FUNDAMENTAL ISSUES

Before we begin the discussion of land transformation, some initial comments are needed:

(1) The figures quoted in this chapter are based on the best statistical data. However, these figures are often approximations or estimations, because there are no exact and accurate figures for many countries.
(2) Some figures are over-estimated (for example, because they are published in order to get more foreign aid), while some are under-estimated (data on yields of the main food crops in some countries are at least 10–15% under-estimated, for example).
(3) Figures are often averages (for example, those of yields per hectare). Such averages hardly occur in practice.
(4) The definitions of many words are different in various publications. For example, *desertification* is an expression that sometimes refers to a gradual drying out of land, but sometimes it is also used to refer to all land that is degraded to such an extent that it has become useless for agricultural production (through soil erosion, salinization etc.).
(5) The figures presented here are from various publications (mainly FAO and other international and national organizations). It is, however,

impossible to mention them all, so reference is made only to those articles and books that are relevant and which present more details and references for further studies.

There are four main categories of land use:

Arable land is cultivated and used to grow annual and perennial crops. The main crops are food crops, in particular cereal grains (wheat, maize, rice, etc.). The average yields are rather low in many countries and high in a few countries. This is the result of various modes of production. In many countries the technology applied in growing crops is still low; in some industrialized countries the level of technology is high. It has to be realized that only a rather small proportion of the land area is suitable for growing crops, because of various limiting factors to be dealt with later in this chapter. Moreover, the productive capacity of land suitable for growing crops is very variable, even if the most appropriate technology is applied. Each crop has its specific conditions for optimal growth, and although these conditions can be influenced by various management practices, the ecological conditions remain different all over the world. It is therefore worth while to study land use potentials for various crops.

Grassland is where annual and perennial grasses grow. It may be natural grazing land, where herds of domestic animals or only wild animals are grazing, but it also includes pastures with specific species of grasses that are grown with technological inputs. This indicates a high variability in feed production and use intensity. It is estimated that only a very small proportion (5%) is used rather intensively. Most land in this category produces according to the rules of nature.

Forest land includes all land where trees are growing. Here, too, the intensity of growth is very variable: there are very dense forests with high trees, but also land with only some trees where open spaces are covered by grasses. It is a matter of definition which land is grassland and which is forest land. In the densely populated, industrial countries most forests are artificial, the trees having been planted. However, in many countries forests are the result of natural growth influenced by the action of man, who cuts trees for timber, firewood or for paper factories. There are only a few plantations. Here, too, there is a wide variety in use potential, because there are thousands of tree varieties with different stands and different qualities for various applications.

Non-agricultural land is used for non-agricultural purposes. This category includes land for:

(a) housing and urban services (offices, churches, hospitals);
(b) industry (factories, storage buildings);
(c) mining (open mines, gravel pits, brickworks);

water reservoirs (for electricity, irrigation, drinking water);
sal (urban waste, old cars);
tion (roads, railroads, airports);
parks, sports grounds).

In very remote areas, where only farming families live and where there are only a few simple roads, the non-agricultural use of land is only a small proportion of the land used for agriculture. In densely populated countries the non-agricultural use of land has increased tremendously, particularly during the last few decades. As the world population is rapidly growing this category of land use will increase rapidly in the near future. One-half of the world's population will soon live in towns and big cities that have nuclei in areas with highly productive land; and it is particularly the land around these nuclei that is generally used for non-agricultural purposes.

The history of land use will not be discussed here, except for the last century during which most important transformations have taken place. There have been various stages of agricultural development in different countries. Such stages and the mode of production can still be observed and studied in various parts of the world. There are even now 200 million people applying the simplest modes of production (shifting cultivation). They clear a small area in the forest to grow some food crops for a few years. The natural productivity of this land decreases rapidly, and so the people have to move to another piece of forest land. About 15–20 ha of land are needed to feed one person. On the other hand there is land where three crops per year can be grown, which means that less than one twentieth of a hectare can produce food for one person.

The way man has used and is still using land is not very satisfactory. The productivity and use possibilities of large land areas have declined. Much land has gone out of production because it has been misused. Some land has been eroded, other land has become too saline and has finally been abandoned. There is land that has been degraded to such an extent that its present production capacity is only a fraction of what it could have been. On the other hand man has also developed new techniques to improve the productive capacity of land, for example by irrigation, drainage and reclamation. Unfortunately, though, misuse of land is still increasing and improvement of land is only a small fraction of what it could be. This problem will be dealt with in somewhat more detail in section 2.4.

The subject of agricultural land use cannot be understood without some technical knowledge of growing conditions of crops, including grasses and trees. The main factors determining the suitability and productivity of land are:

(1) *Crop characteristics*. Each crop has various stages of development, and in each stage it needs a specific amount of water and nutrient. It also needs a specific temperature and amount of oxygen.

(2) *Weather conditions*. There are important seasonal variations, particularly in precipitation.
(3) *Soil conditions*. These often vary over short distances. They are important for the development of an optimal root system, because the growing conditions of crops depend on the moisture, nutrient, biological, temperature, air and mineral regimes of a soil.
(4) *Mode of production*. A farmer can to a certain extent manipulate crop production (e.g. by preparing the seedbed, by using manure or chemical fertilizers, by weeding, by plant protection).

The way man can manipulate crop characteristics, weather and soil conditions will be dealt with in section 2.3. There are, however, a number of factors that can hardly be manipulated. The crop production constraints are given in Table 2.2.

Most important is the arable land, which produces almost all our food (92% on a dry-matter basis) except meat, milk etc. (7%) and fish (1%). Cereal grains are the most important crops (79%), followed by tuber crops (7%), sugar and other food crops (6%). The main cereals are wheat, rice and maize. Table 2.3 gives some figures for cereal grain crops and Table 2.4 for root and tuber crops.

Details on land use and crop production are published annually in the FAO Production Yearbooks. Figure 2.1 gives general information on changes in land use since the year 900.

The above technical information is needed to understand that there are

Table 2.2 Crop production constraints of the land area of the world

Constraint	Area (Mha)	Percentage
Ice-covered	1 490	10
Too cold	2 235	15
Too dry	2 533	17
Too steep	2 682	18
Too shallow	1 341	9
Too wet	596	4
Too poor	745	5
Subtotal	11 622	78
Weakly productive	1 937	13
Moderately productive	894	6
Highly productive	447	3
Subtotal	3 278	22
Totals	14 900	100

Table 2.3 World production, area harvested and average yield of cereal crops (data from FAO for the year 1978)

Crop	Area harvested (10^6 ha)	Production (10^6 t)	Average yield (t/ha)
Wheat	266	450	1.9
Rice (in husk)	145	386	2.7
Maize	118	364	3.1
Barley	94	194	2.1
Sorghum	52	69	1.3
Oats	28	51	1.8
Rye	16	32	2.0
Millet	55	36	0.7
Totals	774	1582	

various categories of land use and that there are various factors—such as weather and soil conditions and the ability of the cultivator of the land—that determine how land can be used. Much more important, however, are social, cultural, economic and political factors. They significantly influence not only the way in which land is used, but also how much is produced. Every farmer (all those who cultivate land are called farmers) uses land to provide a living for his family and relatives and members of the social unit to which he belongs.

Long ago almost everyone was directly or indirectly engaged in farming, and fewer than 10% were non-farming families. This can still be observed at the present time in some remote areas. In such a situation basic food crops have to be grown, if land is more suitable for other crops. Since the beginning of the eighteenth century, when industrialization gradually became an important factor in human life in Western Europe and North America, much has changed. More and more people found themselves in non-agricultural sectors for whom food had to be produced, so there was a development of trade and

Table 2.4 World production, area harvested and average yield of root and tuber crops (data from FAO for the year 1978 at 14% water content)

Crop	Area harvested (10^6 ha)	Production (10^6 t)	Average yield (t/ha)
Potatoes	18	55	2.9
Cassava	14	49	3.5
Sweet potatoes	13	34	2.6
Yams and others	4	7	1.8
Totals	49	145	

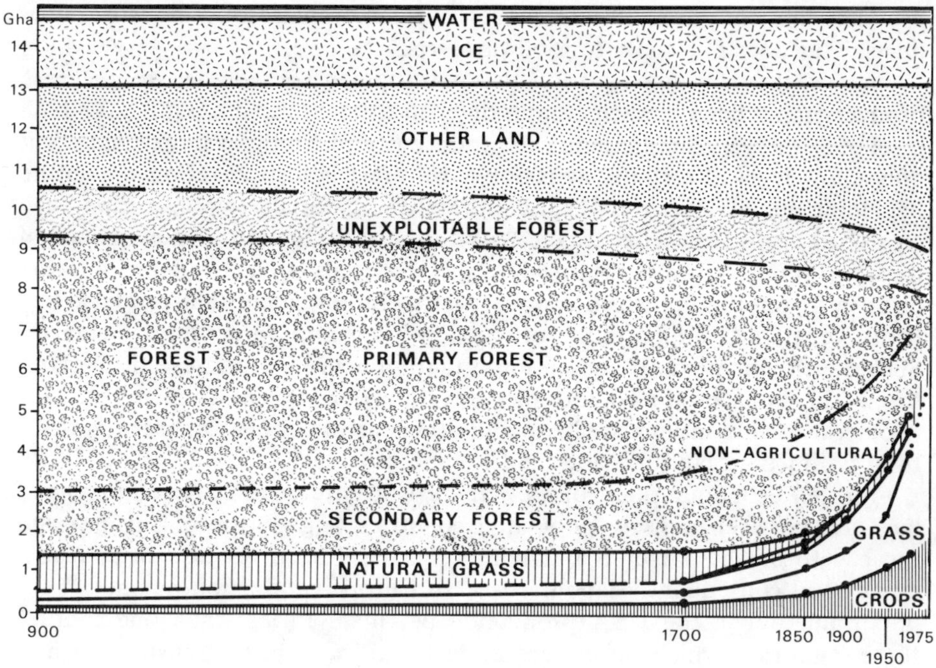

Figure 2.1 Transformation of land in the period 900–1977. Small differences cannot be shown at this scale. It is clear that most important transformations took place during the last 200 years.

of transport of products not only of industrial but also of agricultural products. The development of sciences and of techniques, the education of the people and many other factors have stimulated agricultural production. Most important for agriculture were probably, first, the introduction of ley-farming, which allowed a permanent use of arable land without fallow once in two or three years; later on the introduction of chemical fertilizers, which increased the production per hectare; and then the introduction of mechanization, which not only eased farmwork but also increased the land area for food production, because tractors were replacing horses, a farmer needs approximately one hectare of land to feed one horse.

Mechanization also meant that some tracts of land (e.g. land on rather steep slopes, or stony land) could not be cultivated any more and had to be abandoned. On the other hand, land that previously could not be used for agricultural production became suitable for it; for example, the introduction of the diesel motor made it possible to irrigate land which could not be irrigated before, when all water had to be lifted using simple tools.

As a result of the development of chemical technology some agricultural products like cotton, wool and rubber could be replaced by industrial products. This does not apply to food products.

One of the consequences of the development techniques to increase production has been a much greater diversification in value of land. At a high level of technology similar land can produce five to ten times more than when only a low level of technology is applied. In fact industrialization was in the beginning mainly based on agricultural raw materials (wool, cotton, dye crops, oil crops, flax, rubber, etc.). With the introduction of new means of transportation (railways, steamships) raw materials could be grown at large distances from the industrial centres. Refrigeration meant that products like fruits, vegetables, milk and butter could be transported over large distances. Such developments have had an important influence on land use throughout the world. India produced cotton, Australia wool, Argentina meat, etc., for people in other continents. These and other factors have stimulated agricultural production, land use and land use intensity.

From an economic point of view there is a difference in development of land use between densely populated countries (e.g. Western Europe) and those with much sparser populations (e.g. the USA), because of available labour and consequently labour costs. In Europe production per hectare is important, but in the USA production per man. In less than one century these countries, which once were agricultural, have been transformed into industrialized countries, where agriculture has become a less important sector of the economy.

Using land more intensively means more input of labour or capital or both. In a simple farming system only a few tools are used; but as soon as farming becomes more complicated and various types of technology are introduced, inputs from outside the farm are needed. A farmer needs money to buy these inputs and this money is earned by producing more with lower costs. Farming at the highest level of technology means more dependence on other parts of the economy, and on the availability of energy.

The subject is interesting not only historically or from the economics standpoint, but also from the point of view of land use, because it is evident that all changes in social, cultural, economic and political conditions have influenced the way land is being used. Many developments are hardly possible without the help of a government, particularly in the fields of research, education and extension in agriculture. Moreover, the governments of many countries interfere in the way land is used by land use planning, although this is not always done in favour of agriculture: in many countries this type of planning is done to guide the non-agricultural use of land.

In some countries it is now realized that planning of land use should include the protection of valuable, productive agricultural land, called ‘prime land’. Another important factor is the regulation of prices of the main agricultural products by governments. If, for example, more sugar is needed, an increase in the price of sugar will stimulate the cultivation of sugar cane or beet. The price of the basic food crops is often rather low because governments want cheap food for the many poor people (developing countries) or for the labourers in industry (industrial countries) in order to promote the export of industrial products. In the USA the government provides incentives to farmers *not* to grow cereal crops, because of high surpluses of cereal grains; and in the United Kingdom the other West European countries farmers had to transform some pastural land into arable land during the Second World War.

The foregoing examples demonstrate the influence of government policies on land use. From time immemorial the ruling class has forced farmers to produce specific crops, to increase production, and to use land in a specific way.

It is a generally accepted idea in discussions of foreign aid for developing countries that these countries should at least try to produce the basic food crops they need on land in their own countries. This means that a large part of arable land has to be used for cereal grain crops. It is doubtful whether this is wise, because land—particularly in countries in the tropics—may be more suitable for growing products like cocoa, tea, rubber and other perennial crops, that cover the land surface permanently. If these products could be exchanged for foods produced in countries that are more suitable for growing these crops, much land in the world could be used in a better way.

It may be concluded that indeed it is farmers who use the land, but land use is heavily influenced from outside the agricultural sector, these influences being social, cultural, economic and political. Power may be in the hands of landlords, churches, the 'upper class', local or national governments or international organizations. All of these forces determine whether land is well used or misused today. It is, however, difficult to discover the reasons why land is used in a particular way in a specific area. It is certainly true, though, that much land is not used for purposes for which it is most suitable and that it is not used to its optimal productive capacity. Farmers know the risks of weather conditions, plant diseases, plagues, price fluctuations etc., so they are often more interested in minimizing risks than in maximizing production.

Readers interested in the history of land use and land transformation are referred to the books by Slicher van Bath (1963) and Grigg (1982).

The conclusion is, that all people, and not only those who are engaged in agriculture, are responsible for how land is used. Studies on land transformation in various countries have to be based on exact data. In Western Europe such data are only available of the last one hundred years, in many countries they are not available at all. It is extremely difficult to give a reliable explanation of land transformations, because they are influenced by many, often unknown factors.

Richards, Olson and Rotty (1983) have made a worldwide study on land transformation for the periods 1860–1920 and 1920–78 in order to estimate the releases of carbon dioxide arising from the conversion of land with a natural vegetation to agricultural land. Although they admitted that there were several uncertainties in their estimation, the conclusion was that the net expansion of the cultivated land area was 432 Mha in the period 1860–1920 and 419 Mha in the period 1920–78. In the first period most land was converted in North America (164 Mha) and the USSR (88 Mha). In the second period, Africa (90 Mha), southern Asia (67 Mha), South America (65 Mha) and the USSR (63 Mha) were the most important regions.

2.3 POTENTIAL LAND USE

In most developing countries the mode of agricultural production is still traditional and similar to that in industrialized countries almost two centuries ago. The reasons for this have been explained, and some figures were given in Tables 2.3 and 2.4. The question is now: what is the potential land use?

The word 'potential' has been given many definitions. In this section ecological and technical agricultural management practices will be taken into account, but not the socio-economic conditions: it is necessary to know first what the technical possibilities for agricultural production are.

As cereal grain crops like wheat, rice and maize are the main food crops, most attention will be given to these. Figure 2.2. shows the average wheat yields in selected countries over an extended period (5- or 10-year averages). It is most remarkable that all these countries show an increasing average wheat yield. Moreover there have been and still are important differences. For example, there was an increase in the average wheat yield in the Netherlands from 1 to 2 t/ha in the period 1800 to 1900, and then the increase was rather rapid (excluding the war period). Various factors were responsible for the increased production per hectare. In the period 1800–1900 the main factor was the introduction ley-farming without fallow; after 1900 it was mainly the application of chemical fertilizers followed by the introduction of new wheat varieties, better management as a result of education, extension and research. Finally, biocides and advanced technology, including mechanization, have been applied.

The changes in modes of production causing the increases in the average wheat yields were influenced by industrialization and other factors—in particular the ever-increasing part of the population not working in agriculture who needed food to be produced on the farms. The reasons for increasing agricultural production are still a fruitful research subject for agricultural historians. Hypotheses and explanations that may be correct for one country do not necessarily fit for others.

Differences between the average yields in various countries in the same years are the result of differences in environmental conditions and in modes

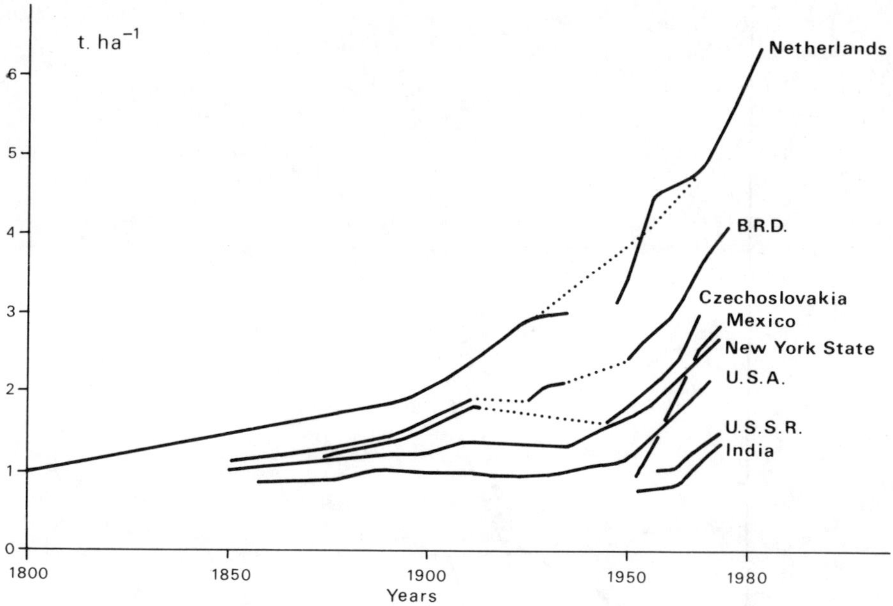

Figure 2.2 Wheat yields in selected countries, 1800–1980

of production. Owing to relatively high precipitation that is well distributed over the growing season, farmers in the Netherlands can apply large quantities of fertilizers in order to get a high production per hectare. In the USA the production per person working in agriculture is more important. The curves for India and Mexico show the influence of the 'green revolution'. Although there are these important differences between countries, this does not mean that farmers getting high yields per hectare earn more than farmers getting much lower yields.

Figure 2.3 shows the average yields, the price and the gross income per hectare of wheat in the Netherlands for the period 1900–80, all based on the year 1900 as index. During this period the yield has increased more than three times and the price more than five times, but the gross income has increased by only about 20%.

The information in Figure 2.2 also indicates three important stages in the Netherland's modes of production (viz. 1800–1900, 1900–50 and 1950–). In other countries these stages may begin and end somewhat later or earlier: they depend in general on the technology applied by farmers.

A similar increase in production per hectare can be observed for other food crops. Figure 2.4 presents data on average rice yields. Here the case of Japan is used to show the increase over a long period, and the present average rice yields of various other countries are indicated on that country's yield curve.

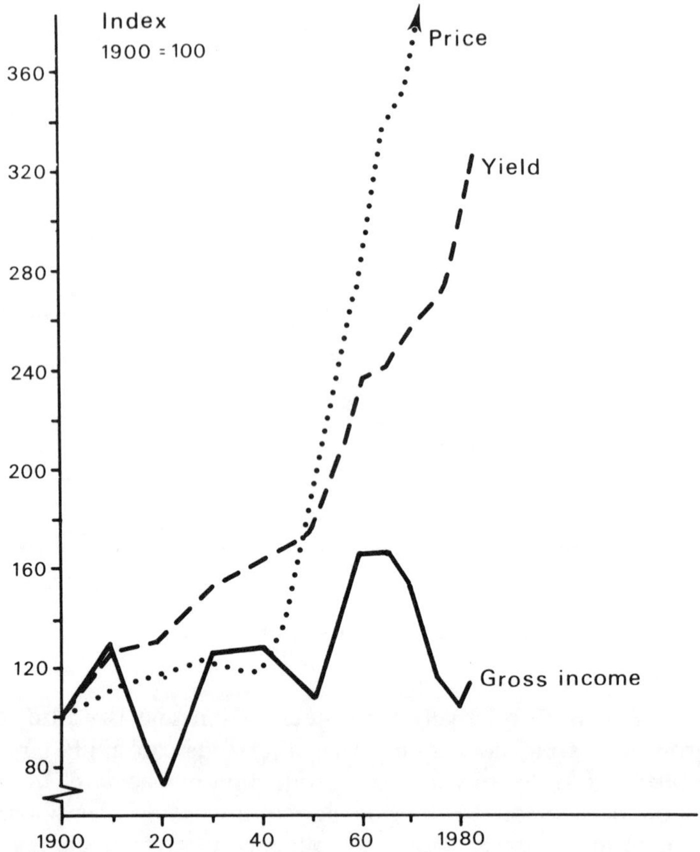

Figure 2.3 Figures for Dutch wheat yields for the period 1900–80, based on 1900 as the index year

Here, too, we see important variations and various stages in the increase of production. Figure 2.4 may suggest that the average rice yields in many countries could be increased to the present, high average yield of Japan. This is not true, because ecological conditions in the various countries are different, although it is clear that production of rice could be considerably increased in many countries.

In order to get an idea of potential land use we also have to know the characteristics of the crop to be grown and of the site (weather and soil conditions) where the crop has to be cultivated. Field experiments have to be carried out. Ratios of the actual average yield and the yield obtained by the best farmers (or a standard yield as obtained in experiments) may be used to get an index for potential land use, and such an index indicates how much the

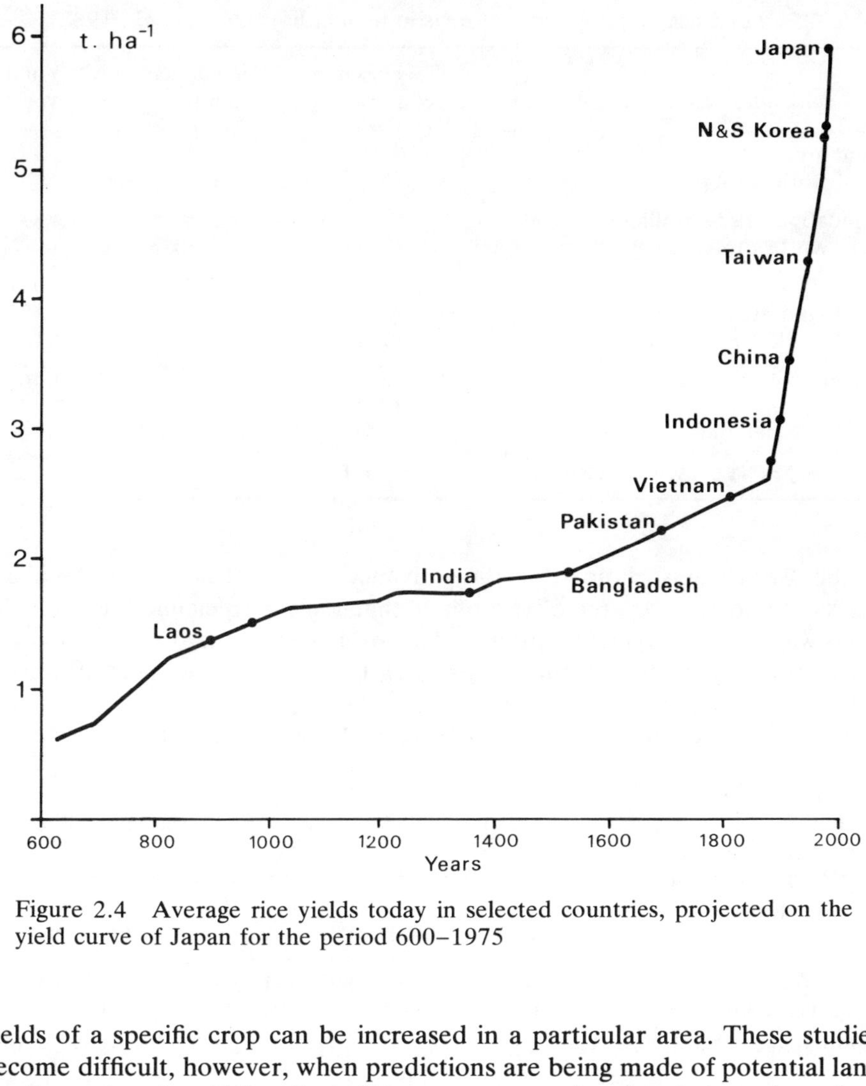

Figure 2.4 Average rice yields today in selected countries, projected on the yield curve of Japan for the period 600–1975

yields of a specific crop can be increased in a particular area. These studies become difficult, however, when predictions are being made of potential land use for regions in which soil conditions are not yet intensively investigated, or in which almost no results of field experiments are available (particularly in young developing countries). In these countries land used for carrying out field experiments is often only representative for a small part of the country, because of variations in weather and soil conditions; in such cases estimates have to be made. In this connection reference can be made to studies carried out by FAO (see, for example, Dudal *et al.*, 1982). Interesting figures on a global scale are presented in Tables 2.5 and 2.6.

Studies of potential land use are not new. In a speech in 1898 the president

Table 2.5 Land use and population (areas in million ha) (Dudal *et al.*, 1982)

	Developing countries	Developed countries	Total world
Land area	7 619	5 773	13 392
% of world's total	(57)	(43)	
Population (1979, millions)	3 117	1 218	4 335
% of world's total	(72)	(28)	
Potentiable cultivable	2 154	877	3 031
% of land area	(28)	(15)	(22)
% of world's potential	(71)	(29)	(100)
Presently cultivated	784	677	1 461
% of potential	(36)	(77)	(48)
% of world's total	(54)	(46)	(100)
Persons per ha presently cultivated	4.0	1.8	3.0

of the British Association for the Advancement of Science, Sir William Cooks, stated that the area of virgin land that could be reclaimed for growing crops was very small, and he predicted a worldwide famine for the year 1930 unless nitrogen fertilizers could be made. In 1923 an article on land utilization in the United States was published by Baker (1923). He concluded that within a few years more agricultural products would have to be imported than could

Table 2.6 Land use and population in developing countries (areas in million ha) (Dudal *et al.*, 1982)

	Africa	S.W. Asia	S.E. Asia	Central Asia	South America	Central America
Land area	2 886	677	897	1 116	1 770	272
% of world's total	(21)	(5)	(6)	(8)	(13)	(2)
Population (1979, millions)	427	153	1 232	947	239	119
% of world's total	(10)	(3)	(28)	(22)	(6)	(3)
Potentially cultivable	789	48	297	127	819	75
% of land area	(27)	(7)	(33)	(11)	(46)	(27)
% of world's total	(26)	(2)	(10)	(4)	(27)	(3)
Presently cultivated	168	69	274	113	124	36
% of potential	(21)	(144)	(92)	(89)	(15)	(49)
% irrigated	(4)	(16)	(24)	(44)	(6)	(18)
Persons per ha presently cultivated	2.5	2.2	4.5	8.4	1.9	3.3

be exported, because of the increasing population in the USA. There are more examples like these.

The main breakthrough was made when the President's Scienee Advisory Committee (PSAC, 1967) published a study based on knowledge of climate and soil conditions of the world. It was stated that the area cultivated was 1406 million hectares, and that the potential area was 3190 Mha (respectively 11% and 24% of the non-ice-covered land area). The conclusion was that approximately nine times more food could be produced.

The first computer model by Meadows *et al.* (1972) showed less hopeful results. In the framework of a model on international relations in agriculture (Linnemann *et al.* 1979) a study was made by Buringh *et al.* (1975) of the maximum food production capacity of the world, when all known, modern agricultural technology was applied on land suitable for crop production (3419 Mha or 26%). In the last study the new Soil Map of the World (FAO/Unesco, 1974–81) could be used as a basis for studying soil conditions. Table 2.7 shows the figures of the American and Dutch studies.

Another study (Buringh and van Heemst, 1977) showed that, even at the present time, not enough food for the present world population could be produced if all suitable land of the world were cultivated and a traditional mode of production applied.

The publications of the American, Dutch and FAO specialists have had two important consequences. First it was clear that the potential cultivatable land area of the world is approximately 23% of the total land area. At present 11% of the land area is cultivated, so there is a reserve of some 12%, now being used as grassland or forest land. Secondly it was made clear that the world food production could be increased some ten times, and that agriculture can produce enough food, when appropriate modern technologies are

Table 2.7 Comparison of American and Dutch calculations on potential agricultural land (PAL)

	American data (1967)					Dutch data (1975)	
	A	Cultivated		PAL		PAL	
	(Mha)	(Mha)	(%)	(Mha)	(%)	(Mha)	(%)
S. America	1 750	77	4.4	681	38.9	596	33.5
Australia	820	32	3.9	153	18.7	199	23.2
Africa	3 010	158	5.2	734	24.4	711	23.5
Asia	4 420	689	15.6	894	20.2	887	20.2
N. America	2 110	239	11.3	465	22.0	627	25.9
Europe	1 040	211	20.2	263	25.3	399	37.9
Totals	13 150	1 406	10.7	3 190	24.2	3 419	26.0

applied. The problem of world hunger and under-nourishment is not attributable to agricultural limitations but is a problem of cultural, socio-economic and political conditions causing poverty. Another conclusion was that it will be better to intensify agricultural production on the presently cultivated land than to reclaim more land.

These conclusions were not entirely new, because they had been expressed earlier by other scientists. However, the studies mentioned above were based on extensive research work. Reference can also be made to more recent studies (e.g. FAO, 1981; Barney, 1981).

It is worth while to undertake global studies, but for many purposes they are too general and do not differentiate enough. Therefore more detailed studies of specific countries have been made. The introduction of computers has facilitated the work. In this connection investigations made by a Dutch group will be mentioned, because of a new approach, developed by de Wit (e.g. de Wit *et al*., 1974). This approach is based on a theoretical calculation of the maximal photosynthetic production of a specific crop, growing on a specific site (known soil and weather conditions) (Centre for World Food Studies, 1980). Use is made of systems analysis and simulation techniques. The growth of a crop is simulated at all stages in a computer model. The system includes the main factors of the systems plant growth, weather conditions, soil conditions and level of agricultural technology—calculation is made, during various stages of development of a crop, of the maximum photosynthetic production taking into account the specific characteristics of a crop, weather conditions before and during the growing season, as well as soil conditions, particularly available water for crop growth, assuming no other limiting factors. This is followed by a calculation in hierarchic sequence of factors that can limit crop production (such as nutrients and weeding). The model is constantly verified, using data from various field experiments. The main advantage of such an approach is that one learns about the real production potential. It is possible to investigate why such a maximal production does not occur in practice. A similar approach has been and is continuing to be developed in countries cooperating in the International Institute for Applied Systems Analysis (Meadows *et al*., 1982). Although significant progress has been made during the past decade, it is evident that much more research work has to be carried out and that models have to be simplified in order to make them suitable for practical application.

We now come back to the information presented in Figures 2.3 and 2.4, because the new calculation procedures provide the possibility of learning the maximum average wheat or rice yields in various countries. This was, for example, calculated for the Netherlands, and the conclusion was that an average yield of approximately 7.5 t/ha will be the maximum average yield, applying the best techniques and using the best wheat varieties, and cultivating the same land for growing wheat as is used at present. New farming techniques, and new wheat varieties that make a more efficient use of the sunlight, may increase wheat production more. If only the best land (some with a production of 10 t/ha) were used to grow wheat, the average would also increase. However, most productive land in the Netherlands is used to grow seed potatoes, sugar beets, onions, etc., and wheat is mainly grown for crop rotation purposes, and to spread the work-load. This example is given here mainly to demonstrate that even the potential yields of crops depend on many factors. Potential land use also depends on how and when the farmer manipulates the growing conditions of a crop. The manipulation technique depends on the experience and knowledge of the farmer, but also on possibilities of input of labour and capital and of marketing prices and facilities to sell the products. A subsistence farmer in a remote area almost without infrastructure has few opportunities to sell the produce which he does not need for his family. On the other hand a farmer in a densely populated, industrialized country has to produce for many people not working in agriculture, who are willing to pay for the agricultural products. This will give him the opportunity to use capital (fertilizers, machines, etc.), particularly when labour is scarce. A comparison of traditional farming and modern farming shows a great difference in potential land use.

Traditional farming is defined here as farming in a traditional way, with only limited input from outside the farm. Yields depend on the natural fertility of soils and the availability of manure, consequently on the presence and number of farm animals, for which feed has to be available from grazing land. In this situation local crop varieties that are not very sensitive to diseases are used. All work in the field is done with simple tools, most of them made by the farmer himself. There may be some animal traction. It is clear that the possibilities for manipulating growing conditions of crops are very limited, that average yields are rather low, and that improvements will take much time.

Modern farming, on the other hand, is defined as farming with high technological input from outside the farm. For example, the farmer buys seeds of proper high-yielding varieties, chemical fertilizers, biocides to kill weeds or to protect crops from diseases. He has machines to prepare seedbeds, to apply fertilizers and biocides, to harvest and to thresh, and tractors to transport the products. Such a mode of production is rather costly and consequently the average yields per hectare have to be much higher than when farming is done in a traditional way; moreover the farmer needs much more knowledge and he has to be educated. Many machines are too costly on small farms.

Under traditional agricultural conditions the best farmers can get a maximum of 1 t/ha of wheat, under modern agricultural conditions probably 5 t/ha on similar land. In the traditional situation the maximum photosynthetic production is approximately 1.4 t/ha, in the modern situation 9 t/ha.

This example demonstrates that potentialities depend strongly on the level of technology applied—and that costs of production are much lower in a traditional mode of production than in a modern mode of production. Many other, similar examples could be given.

There are also possibilities of manipulating crop characteristics, climate and soil conditions. Some examples will be given. Plant breeders can breed new crop varieties that are better adapted to environmental conditions or to new farm management practices. Glasshouses can be built in order to get the most favourable climatic conditions for some expensive crops. Supplementary irrigation can correct a shortage of rain. Drainage can correct too high a precipitation or too high groundwater. Sloping land can be terraced. Land with a less favourable layer below the ploughed layer can be improved by subsoiling or deep-ploughing. Proper management is highly improved by tractors and machines, because work in the field can be done much more easily and quicker, and consequently at the proper time. This short list of manipulation techniques can be extended. However, there are limitations which are partly natural—for example a shallow soil cannot be transformed into a deep soil, for normal crops weather conditions cannot be changed, and often the costs of introducing a new technique are too high. Maybe the most limiting factor is the farmers themselves, because often many are illiterate or have very little education and there are no governmental research stations and extension services.

The introduction of new and better modes of crop production takes time and costs money, a lot of money (FAO, 1981). Most farms are very small, some even smaller than a few hectares. It is difficult to improve production on these farms because of the risks.

The conclusions are that there is a large potential for increasing the area of cultivated land and similarly the potential for large increases in production per hectare when better modes of production are applied. The problem is not technological in that sense. Rather the problems are socio-economic and political.

2.4 INCREASES IN NON-AGRICULTURAL LAND

In all countries much agricultural land is lost and much land is degraded, which means its production capacity decreases. This is partly the result of mismanagement, resulting for example in erosion or salinization, and partly because agricultural land is used for non-agricultural purposes.

The transformation of agricultural land into non-agricultural land is rapidly increasing. Continuous losses and degradation mean that new land has to be reclaimed in order to be able to produce the same quantity of food. Moreover, new land is needed for the increasing population of (80 million people each year). The result is that the reserve of land with potential for growing food crops is rapidly decreasing, a subject to be discussed in section 2.5.

Only a few countries have reliable data on various types of losses and degradation of agricultural land. For most countries the information is rough and figures in the popular press are often exaggerated or copied wrongly, e.g. on desertification. There are almost no figures on the quality of the land that is lost. It makes, for example a great difference whether unproductive land or land with a high productive capacity is lost. A loss of 100 ha of dry, poor grazing land, which can feed some cows (producing 5 kg of meat per hectare per year) is less important than the loss of 100 ha of land where three food crops can be grown annually. Some land that has been eroded may not be suitable for growing crops, but it can probably still be used as grazing land or forest—this is not a total loss.

The example indicates that the subject of losses and degradation is not as simple as is often assumed. Farmers who cultivate their own land will take care of that land in order not to loose the basis of their living and that of the following generations. However, there are some very large farms owned by corporations who are more interested in gaining a profit in a short period than in conserving the land for future use. Farmers who rent their land may be less careful, particularly if they are not sure that they can cultivate the same plots for many years. The same applies to communal land, because no-one feels personally responsible for its misuse. In regions with a high population pressure farmers may reclaim new land that is unsuitable for cultivation (e.g. when it is situated on too steep slopes). Such farmers can hardly be blamed because they do not have any choice. In some countries much land has been misused because the government has introduced land registration for tax purposes, claiming that all land not cultivated is owned by the state. Good grazing land has therefore been ploughed and transformed into useless cropland, because of a too dry climate and severe wind erosion.

A further danger that is not fully recognized is the silting up of water reservoirs that are used mainly for producing electricity and irrigation water for agricultural land. As a consequence of soil erosion in the catchment area, large quantities of silt are continuously deposited in reservoirs. In the end (it may take 50 or 100 years or more) such reservoirs are filled up with silt, which means that large areas of land can no longer be irrigated.

The problem of the non-agricultural use of land has become serious in many countries. Almost all cities are situated in areas with very productive land on which for many centuries food has been produced for the city population. Soon half of the world's population will live in urban areas. In the year 2000 non-agricultural land will amount to approximately 400 Mha, and this will be mainly prime land. In some industrialized countries this problem now gets more attention.

Much land on hills and mountain slopes is seriously eroded and deep gullies are formed when the original forest vegetation has been taken away. A less spectacular type of erosion, sheet erosion, may be even more serious. Sheet erosion often occurs on slightly sloping land from which every year a very thin soil layer (a few millimeters) is eroded. After a few decades the rooting depth of crops may become limited and land productivity decreases.

Besides erosion, reduction in land productivity may also result from certain farming practices. Where an efficient use of fertilizers is not practised chemical degradation may take place through a decrease of nutrient content (leaching of nutrients, carrying off of nutrients with the harvest) and decalcification. Physical degradation may occur through soil surface compaction and the decline of soil structure when inappropriate tillage is applied. Biological degradation may occur through a decrease in biological activity of soils, which is often the result of a decline in organic matter content. The processes that occur in soils are rather complicated, particularly the biological processes that are governed by soil flora and fauna.

A new type of degradation is the result of air pollution that has become a real problem in industrialized countries in recent years. Special attention is currently given to 'acid rains'—more appropriately expressed as 'acid deposition'—where significant amounts of S and N oxides are added to the natural CO_2 components of the air. Olson (1983) considers that likely deleterious effects on soils include reduction of pH, increased activity of various elements with toxicity potential, accelerated mineral weathering, leaching of reserve soil nutrients and undesirable changes in the microflora. Areas with greatest potential for soil damage from deposition are in close proximity and downwind from large coal burning plants, especially on soils which have a low buffering capacity or are inaccessible for corrective measures to be applied. Olson's assessment, however, is that for most of the agricultural regions of the world, the N and S fallout products are more beneficial in supplying needed plant nutrients than they are detrimental. This acidification capacity is insignificant in comparison with that of natural N and S transformation in soils or that induced by conventional fertilizer programmes of farmers. The situation is different, however, for perennial vegetation and forests which may be severely damaged by the cumulative effect of successive acid deposition.

It should be stressed that there is not only loss and degradation of land, there is also reclamation of land and improvement of land productivity, although there are only a fraction of losses and degradation. Examples of reclamation of new land are the drainage of marshes, inland lakes and reclamation of land from the sea in coastal regions. Improvement of productivity can be obtained by drainage, land levelling, terracing etc. Infertile land can be substantially improved through an appropriate application of fertilizers, organic manures, lime or trace elements.

In various countries attempts are now being made to protect land from being eroded and to restore the eroded land. This is also sometimes done with land that has become saline (mainly in arid regions)—that is, irrigated without drainage. Soil conservation is, however, very expensive and it needs the cooperation of many people living in a conservation project area. This means that the national government has to initiate, execute and guide such projects, for which it needs the cooperation of all land owners and land users in a project area, a difficult objective to achieve. The costs are generally too high for farmers, so part of those costs has to be paid by the government and loans have to be provided for the farmers. A further problem is the settlement procedure for newly reclaimed land, which is often the main reason why projects are less successful than anticipated. Farmers in new settlements often do not have enough experience to make use of the new facilities, so the yields are much lower than expected. It is well known from early land reclamation work in various industrialized countries that, very often, the first and even the second generation of farmers are adjudged bankrupt. Such failures can be avoided by creating a special government authority for a project area, responsible not only for initiating a project, but also for the farming for some years until farmers are trained and able to handle the land with supervision and advice provided as needed.

Many specialists have tried to quantify the various losses and types of degradation of land, both on a national and on a global scale. Unfortunately the figures do not indicate the important differences between the areas or regions. In a report for an FAO/UNEP/UNESCO conference almost all the available literature with the most reliable figures was studied, to get at least an idea of the area of land that is lost. Here the general results are given, but more details can be found in Buringh (1982). The base year in this study was 1975 and computations were made for the expected situation in the year 2000. The five main types of land use considered were crop land, grassland, forest land, non-agricultural land and other land. Moreover, four classes were distinguished (high, medium, low and zero), indicating the potential productive capacity of all land not covered by ice.

As a result of erosion, toxification (mainly salinization and alkalization), desertification (drought) and non-agricultural use of land, the potential capacity of land and/or the type of land use can change. As the world population will increase (an average 75 million people each year for the period 1975–2000) a large area of additional crop land will be needed. Taking into account the various losses, changes in and intensivation of land use, and the reclamation of new crop land for the period 1975–2000, the land use per land class is expected to alter considerably (compare Tables 2.8 and 2.9). Although a number of estimates had to be made the figures can be considered to be good working figures.

It turns out that about 200 Mha of agricultural land will be converted into non-agricultural land, 50 Mha will become seriously toxified, and 50 Mha will become desert: a total of some 300 Mha of presently agricultural land in a 25-year period, or an average of 12 Mha per year, or 23 ha per minute. The main reason is the expected loss of agricultural land to non-agricultural use

Table 2.8 Land use (1975) per land class ($\times 10^6$ ha)

	Land class				
Land use	High	Medium	Low	Zero	Totals
Cropland	400	500	600	0	1 500
Grassland	200	300	500	2 000	3 000
Forest land	100	300	400	3 300	4 100
Non-agricultural land	0	0	0	400	400
Other land	0	0	0	4 400	4 400
Totals	700	1 100	1 500	10 100	13 400

(66% or 200 Mha or 8 Mha per year). The total loss of land as a result of mismanagement is about 100 Mha in 25 years or 4 Mha per year (a rather low figure, because it is often assumed to be 5–8 Mha). Not all land is highly productive. When taking into account the four productivity classes mentioned above, the following conclusions are drawn for the 25-year period:

(1) About 4% of all potential productive agricultural land will be lost; however, of all highly productive land 25% will be lost.
(2) The reserve of crop land which is now used as grassland or forest will decrease by 24%; however, the reserve of highly productive land will decrease by 33%.
(3) The total forest area will decrease by 15%. The area of forest on productive agricultural land will decrease by 55%; however, 70% of the forest on highly productive land will disappear.

These figures do not give optimistic assessments of what can be expected in the near future. A comparison of the results with figures presented by various other authors reveals that the figures in this study are rather low. Important,

Table 2.9 Land use (2000) per land class ($\times 10^6$ ha)

	Land class				
Land use	High	Medium	Low	Zero	Totals
Crop land	345	745	710	0	1 800
Grassland	170	320	510	2 000	3 000
Forest land	30	100	230	3 140	3 500
Non-agricultural land	0	0	0	600	600
Other land	0	0	0	4 500	4 500
Totals	545	1 165	1 450	10 240	13 400

however, is the introduction of the four classes of land productive capacity in order to get information on what may be expected to happen with highly productive land. It is repeated once again that the investigation was done on a global scale, that it was based on real data and on assumptions, and it was an important first approximation. A global assessment of land degradation (first phase: northern half of Africa and the Near East) has been made by FAO, UNEP and UNESCO (FAO, 1979) and it was concluded that at the end of this century the developing countries may loose up to 20% of the productive capacity of their land resources. Similar calculations could be made on a country by country basis in order to alert governments.

As can be seen from Table 2.8 the total area of land use in 1975 for non-agricultural purposes is 400 Mha for 4000 million people (average 0.10 ha per person). In the period 1985–2000 the world population is expected to increase by 2000 million people, who will need at least 200 Mha of land for non-agricultural purposes. The average of 0.10 ha per person seems to be realistic (the areas mentioned in various publications generally vary from 0.075 to 0.15 ha per person). In Australia it is 0.5 ha per person. The total annual loss of land to non-agricultural use is 8 Mha. In the USA alone this loss is already 1.2 Mha per year, of which some 37% is highly productive land. Italy is losing 45 000, the United Kingdom 23 000 and France some 65 000 hectares each year, according to the OECD. Table 2.10 gives an idea of the purposes for which the non-agricultural land is used in the Netherlands and in Japan.

A FAO/UNFPA/IISA study (1982) makes a preliminary quantitative assessment of the losses in potential production from unchecked soil erosion if, as at the low input level, no conservation measures are undertaken (Table 2.11). The methodology used for estimating rates of soil loss was a parametric approach using rainfall and wind erosivity indices, soil, topographic, texture

Table 2.10 Examples of annual losses of land to non-agricultural land use in the Netherlands and Japan

	The Netherlands		Japan
	(1950–58 average)	(1977)	(1968–74 average)
Losses (ha)	4 455	6 668	54 873
Land used for (%):			
Housing, urban services	46	42	54
Industries, mining	11	13	13
Communications	27	23	14
Recreation	16	21	19

Table 2.11 Effects of unchecked soil erosion on production potential at low input levels (FAO/UNFPA/IIASA, 1982)

	Africa	Central America	South America	South-east Asia	South-west Asia	Totals
	(% of total production potential)					
Loss in area of rain-fed crop land	16	30	10	36	20	18
Loss in rain-fed crop production	29	44	23	39	35	29
Loss in total potential production[a]	25	25	21	12	5	19

[a]From rain-fed crop land, irrigated land, and the increased area suitable only for grassland.

and land use factors. In addition to quantification of degradation phenomena it was necessary to translate soil losses into decreases in potential productivity. The relationships which have been employed to obtain the figures shown in Table 2.11 are described in the FAO/UNFPA/IIASA (1982) study on potential population supporting capacities of land in the developing world. The loss would in the long run reduce the potential area of rain-fed crop land by 18% and the production of the rain-fed area by 29%. There would be an overall loss of 19% in the combined potential production of rain-fed and crop land, irrigated land, and the increased area suitable only for grassland. Such losses would be particularly serious in Africa and Central and South America. The intermediate input level assumes a 50% reduction in these losses, and the high level their reduction to acceptable proportions.

2.5 RESERVES OF PRODUCTIVE AGRICULTURAL LAND

The present area of crop land is a little more than 1500 Mha, or 11% of the world's land area. Specialists in various countries and the international organization FAO have investigated how much of the land not yet used to grow crops is still available and suitable for food crops. The figures vary from 1700 Mha to 1900 Mha. This means that the reserve of crop land is some 12% of the total area.

Such a figure looks rather optimistic, but there are two factors that should be mentioned. The first is that most of this land has a moderate or low productive capacity, because in most countries the best land has already been reclaimed. Secondly, much newly reclaimed land is needed for the increasing world population, to compensate for the land that is lost to erosion, salinization and degradation, and to compensate for land that is completely lost mainly because of the increase of non-agricultural land use.

In the foregoing section a general idea of these losses was given and special attention was paid to the quality of land for food production. Tables 2.6 and 2.7 give a general idea of the area of potentially productive land in various regions, and Tables 2.12 and 2.13 of the potentials of soil orders and soil units.

In Table 2.12 the world land area is grouped according to soil conditions that are classified according to the US Soil Taxonomy. The total area of potentialy arable land is some 3120 Mha, or 23.7% of the world's land area. Table 2.13 lists the major soil units of the world as presented on the FAO/UNESCO Soil Map of the World. Here the area of potentialy arable land is 3270 Mha, or 24.8%. It should be kept in mind that both tables give approximations because soils in large areas of the world have not yet been fully investigated. Moreover there may be some differences in interpretation. There is often some confusion on the suitability of various soils in tropical regions, particularly the old tropical soils. Various experiments have clearly shown that many deep tropical soils can be permanently cultivated if an appropriate management is applied: the main problem of these soils is not irreversible hardening or laterite formation, as often is believed, but the aluminium toxicity that can be corrected by liming. A poor nutrient status is corrected by appropriate fertilization.

The figures on potentially arable land presented above are valid if modern technology is applied. It is evident that this requires rather large inputs (FAO, 1981). Most reserves of land are located in Africa and South America, where respectively 21% and 15% of the potential agricultural land is currently used (Dudal, 1982). Most specialists, however, argue that reclaiming new land is hardly necessary in many countries, because yields of crops grown on the

Table 2.12 World land area in different soil orders (Mha) (source: USDA Soil Geography Unit, Soil Conservation Service, 1973)

Soil order	Potentially arable	Non-arable but grazeable	Non-arable non-grazeable	Total	Percentage
Alfisols	640	690	400	1 730	13.1
Aridisols	80	250	2 150	2 480	18.8
Entisols	150	290	650	1 090	8.2
Histosols	1	20	100	120	0.9
Inceptisols	230	230	710	1 170	8.9
Mollisols	630	340	160	1 130	8.6
Oxisols	650	350	120	1 120	8.5
Spodosols	100	210	250	560	4.3
Ultisols	270	330	130	730	5.6
Vertisols	140	60	30	230	1.8
Mountain soils	230	910	1 670	2 810	21.3
Totals	3 120	3 680	6 370	13 170	100.0
Percentages	23.7	27.9	48.4	100.0	

Table 2.13 Major soil units of the world

	Totals		Potential crop land	
	Area (Mha)	Proportion (%)	Area (Mha)	Proportion (%)
Acrisols	1 050	8.0	300	9
Andosols	101	0.8	80	2
Cambisols	925	7.0	500	15
Chernozems, Greyzems, Phaeozems	408	3.1	200	6
Ferralsols	1 068	8.1	450	14
Fluvisols	316	2.4	250	8
Gleysols	623	4.7	250	8
Histosols	240	1.8	10	0
Lithosols, Rendzinas, Rankers	2 264	17.2	0	0
Luvisols	922	7.0	650	20
Planosols	120	0.9	20	1
Podzols	478	3.6	130	4
Podzoluvisols	264	2.0	100	3
Regosols, Arenosols	1 330	10.1	30	1
Solonchaks, Solonetz	268	2.0	50	2
Vertisols	311	2.4	150	5
Xerosols, Kastanozems	896	6.8	100	3
Yermosols	1 176	8.9	0	0
Miscellaneous land units	420	3.2	0	0
Totals	13 180		3 270	

presently cultivated land are low and it is not very difficult to increase yields by introducing better management practices. This seems not only less expensive, but it is also better from a point of view of conservation and protection of the environment—and moreover there should be reserves of land for coming generations.

In Section 2.4 it was stated that in the period 1975–2000 some 140 Mha of potentially productive agricultural land will be lost, being 4% of the total of that land area. This figure is rather low in comparison with figures presented by some other authors, who ended up with figures of 5% and 6%. The conclusion that the reserve of land will be enough for a very long time is, however, not correct. First, the figures are given for the whole world, but regionally there are very important differences. For example, Egypt has almost no reserves while Brazil has very large reserves. Secondly, there is the difference in productive capacity of land: although the total loss is only 4%, the loss of highly productive land is 22% in the 25-year period.

The figures presented in section 2.4 also show that the reserves of highly productive crop land (now being used as grazing land or forest) will disappear at a high rate. If no special measures are taken global reserves of highly productive land will be lost in about 75 years. Specialists in the USA have indicated that in that country this category of land will be used up within 25 years. According to our calculations the area of forest land will decrease at a rate of 15% in 25 years (0.6% a year), a figure that is lower than the figure calculated by various foresters. Forest land that is highly productive for growing food crops will decrease by 70% in 25 years. This means that such land will be deforested in about 40 years. Without going into more detail, it should be noted that the figures presented here are low compared to those mentioned in some reports. We have tried not to exaggerate an already very bad situation. One should realize that man has already used land for some 10 000 years and there never has been a shortage, and now it is observed that all reserves will have gone within a few decades or so. In some countries there are no known reserves at the present time.

During the last decade various investigations of land use, land use potentialities, losses and misuse of land, land reserves, etc., have been made. It seems worth while to make this type of investigation in all countries, taking into account the productive capacity of the land. Such an investigation could start with an overall survey made in a short time period, to be improved by more detailed surveys at a later stage.

One conclusion is that the reserves of productive agricultural land are limited (particularly those of highly productive land) and should be protected. As land reserves in many countries are very limited it is worth while to investigate the various problems related to land use and to show on a map where real problems may be expected. Another conclusion is that the cost of production and the use of energy in agriculture will gradually increase, and consequently the costs of food products will increase.

2.6 PROJECTIONS TO MEET FUTURE NEEDS

Projections made at the end of the last century and at the beginning of this century turned out to be wrong, which may be a good reason to distrust projections made in recent years. However, recent projections are based on the availability of much more information. This information is regularly updated and trends can be studied, although these trends can change rapidly because of new technologies. Agricultural production, on the other hand, never changes rapidly: introducing new technologies over large areas and reclaiming land in large projects takes several years. Projections for the next 20 years may therefore have some value, assuming that there will be no disasters.

It is expected that the world population will increase to a total of at least 6000 million in the year 2000. In section 2.4 calculations have already been quoted for:

(1) the area needed for non-agricultural land use;

(2) the area needed to produce food for the people who are annually added to the world population; and
(3) the area needed to compensate for agricultural land that is lost annually.

In addition, the degradation of the productivity of the presently cultivated land should be taken into account, and furthermore one can expect an improvement in the living conditions for at least some of those people whose living conditions are currently rather poor. These two factors are difficult to evaluate. They are therefore neglected for the time being in order to avoid too much speculation. On the other hand an improvement of production technology in agriculture, and consequently an intensification of production, may be expected at least in some areas. In most projections this factor is taken into account.

In recent years various projections have been made—for example, projections of the consumption and production of chemical fertilizers (regularly published by FAO), of the increase of the irrigated land area, of the world labour force and of the world's food production. Although such projections are valuable and should be made, and although the information is often given for individual continents, groups of countries or specific countries, they should be interpreted very carefully. For example, although the figures for a particular country showing transformations in land use over a certain period may be reliable, it is difficult to explain the reasons for these transformations, because one also needs to know changes in population, professions, and cultural, socio-economic and political conditions. Unfortunately all these data may be unreliable for political reasons. Very often this is not admitted and figures presented are merely assumptions instead of real figures. Also, the grouping of figures for statistical purposes is often done in different ways, even in different parts of the same country. Various international organizations, particularly those of the United Nations, produce regularly updated information on various subjects related to land transformation, population, food problems, etc. Usually this information is the best there is, and consequently it is used by most specialists. It is hardly surprising, therefore, that their conclusions are very similar, but not necessarily correct. Moreover, the basic information must come from farmers themselves who, for understandable reasons, may not provide accurate data when it can be used to calculate their income and capital.

There are various projections of the consumption of chemical fertilizers in the near future, but all specialists expect a continuous increase of fertilizer application. Most countries have to import these fertilizers. Storage and transport facilities and financial factors are important.

The irrigated land area, being now over 200 Mha, can be extended to some 400 Mha, but the availability of suitable irrigation water sets a limit. In this connection the increasing need for water for human and animal consumption and the consumption by industries must be mentioned. It takes about 20 years to plan and execute an irrigation project.

Taking into account the possibilities for intensifying the use of currently cropped land and the possibilities for increasing the area of cultivated land, it is easy to calculate that enough food can be produced not only for the present population but also for the population in the year 2000 and far beyond. If the land resources are used in an appropriate way, there should be no problems in growing enough food for a world population of 1200 or 1300 million people, the number of people who are expected to live on earth in the middle of the twenty-second century. From the point of view of agricultural technology the food problems can be solved. However, as explained before, the food problem is not a technical but a cultural, socio-economic and political problem. Moreover the world is not one unit, and there are very important regional differences. Therefore while global projections may suggest optimism, practice is different. It seems to be much more worthwhile to make projections for individual countries. If all countries could make their own projections and those results could be combined, much better global projections could be made. Unfortunately most countries have only a small number of specialists who could make such national projections and they are often occupied with other work.

The World Bank and other international agencies have made studies of several countries, mainly to assist in economic planning, it is well known that this can create problems similar to those a housewife has if some guests are interfering in the household business, but projections to meet the needs of the actual population can best be made by the countries themselves. As agriculture, particularly food production, is a part of the total economy of a country, projections of the needs of the population have to be made on a much wider basis than agriculture alone. Consequently the whole subject of 'projections' is much more complicated than making projects of food, fertilizers, irrigated areas, etc. The needs of the population are more than the needs for food, clothing and housing, although these may have a high priority. In the total economy various options can be recognized. Not all wishes can be fulfilled at once or in the same period: choices have to be made and priorities have to be set, which shows the importance of government policies.

Most countries have economies based on agriculture. Some specialists claim that more attention should be given to agricultural development, while others believe that industries should be developed or that the infrastructure should be improved.

Developing a country is an extremely complicated and difficult task. Furthermore it has to be realized that many problems are interrelated and the economy of a country also depends on the economy of other countries. In order to get an insight into such matters economic models are being developed and intensive use is being made of computers. Such models can be used to show that a certain policy should be followed. However, the outcome depends on the input and programming of the computer. If computer models are used to calculate the possible results of various options, of various types of measures to be taken, policy makers can have an idea of the effects of the planned policies on the economy. This may enable them to take the most appropriate decisions. Economic models used in this way and made by competent, independent scientists may be of great value for administrators and policy makers. Such models do not predict the future, they only try to indicate what may happen under specific conditions.

New light is thrown on the foregoing in a study recently carried out by FAO, in collaboration with the International Institute for Applied Systems Analysis (IIASA), with the support of the United Nationals Fund for Population Activities (UNFPA). The study makes the first detailed assessment of the potential population-supporting capacities of lands in the developing world, in comparison with their actual and projected populations (FAO/UNFPA/IIASA, 1982). The study takes as its starting point the FAO/Unesco Soil Map of the World (FAO, 1971–81) and the FAO Agro-Ecological Zones Project (FAO, 1978–81). It covers 117 developing countries in Africa, Central and South America, and south-east and south-west Asia. East Asia is not included in the study because of insufficient climatic data for China at the time it was prepared.

The methodology developed to assess potential population-supporting capacities is based on an overlay of a specially compiled climatic inventory on to the 1:5 000 000 FAO/Unesco Soil Map. The assessment considers 18 crops, to ascertain maximum potential calorie production, at three levels of inputs. The land resources of countries, expressed in terms of soil and climatic conditions, are matched to the soil and climatic requirements of crops to determine suitability classes. These suitability ratings allow the computation of crop-specific land productivity potentials, which can in turn be converted in calorie–protein production potentials. Then potentials can be compared to country-specific calorie–protein requirements so that the degree of self-sufficiency can be determined for various levels of inputs and different population densities.

The global results of the study confirm that the world as a whole should have enough cultivatable land to feed far more than the maximum population it is ever likely to have to support. These global results, however, presuppose massive movements of food, and probably also population, on a scale that seems hardly feasible. Land resources are very unevenly distributed in relation to population, both between and within the different countries and regions.

It is, therefore, more realistic and instructive to examine the results of the study at lower levels of aggregation. The results for individual countries come closest to political reality, in that greater self-sufficiency in food is being

Table 2.14 Critical countries with population-supporting capacities less than 1975 or projected 2000 populations, at different input levels (FAO/UNFPA/IIASA, 1982)

	Africa	Central America	South America	S.E. Asia	S.W. Asia	Totals
Number of countries	51	21	13	16	16	117
1975						
Total population (millions)	407	106	216	1 118	136	1 983
Critical countries with						
Low inputs	22	11	–	7	15	55
Intermediate inputs	7	4	–	1	11	23
High inputs	2	1	–	1	9	13
2000						
Total population (millions)	828	215	393	1 937	265	3 638
Critical countries with:						
Low inputs	30	14	–	6	15	65
Intermediate inputs	13	7	–	2	15	37
High inputs	4	2	–	1	12	19

sought as a major policy objective at this level. The data by country imply the more realistic assumption of the free movement of food and population within each country.

The study identifies 55 'critical' countries whose potential population-supporting capacities at the low level of inputs are less than their actual 1975 populations (Table 2.14). The countries include 22 out of the regional total of 51 in Africa, 15 out of 16 in south-west Asia, 11 out of 21 in Central America, and seven out of 16 in south-east Asia. Raising the level of inputs has a very marked effect, reducing the total number of critical countries to 23 at the intermediate level and 13 at the high level.

To summarize these country data in another way: 52 out of the 117 countries could, by using all their potentially cultivatable land, feed from domestic production populations of the size projected for the year 2000 at the low level of inputs; 28 could do so only at the intermediate level and 18 only at the high level; and 19 could not do so even at the high level of inputs (Table 2.15).

Irrigation contributes a substantial share of the total production potential, especially in Asia and Central America (Table 2.16). Its share falls as inputs are increased on the rain-fed land, because the irrigated land is assumed to be at present or projected input levels. Irrigation is assessed on the basis of the actual area and production in 1975 and that projected for 2000. If the full irrigable potential could be assessed, its contribution would be still larger.

Table 2.15 Countries classified by lowest input level at which they are to feed from domestic production populations of the size projected for 2000

Input level	Africa	Central America	South America	South-east Asia	South-west Asia	Totals
Even at low level	21	7	13	10	1	52
Only at intermediate level	17	7	–	4	–	28
Only at high level	9	5	–	1	3	18
Not even at high level	4	2	–	1	12	19

Figures are derived from Table 2.14.

It is clear that various projections are needed to construct such models. Projections to be made for the agricultural sector of the economy can be made in a realistic way, at least when they are made for a period of one or two decades.

Little attention has been given here to projections of energy consumption in agriculture, a problem that has attracted much attention in recent years. The introduction of new technologies into agriculture has greatly increased the use of energy, directly and indirectly. The replacement of draught animals by tractors, the introduction of diesel pumps and motor transportation, and particularly the application of large amounts of nitrogen fertilizers, for which production much energy is needed, have increased energy consumption. However, agricultural production is only a modest consumer of energy in comparison with transport or industry. It accounts for only 4.5% of total commercial energy use in the developing countries and 3.5% in the developed

Table 2.16 Contribution of irrigation to total production potential, actual area and production in 1975 and projected in 2000 (FAO/UNFPA/IIASA, 1982)

	Africa	Central America	South America	South-east Asia	South-west Asia
	(% of total production potential)				
1975					
Low inputs	9	38	6	39	80
Intermediate inputs	<5	14	<5	14	49
High inputs	<5	5	<5	8	32
2000					
Low inputs	20	64	13	73	89
Intermediate inputs	5	33	<5	41	66
High inputs	<5	14	<5	28	49

Table 2.17 Commercial energy use in agriculture for 90 developing countries

	Total (Mt oil equiv.)	Fertilizer (%)	Machinery (%)	Irrigation (%)	Pesticides (%)
1980	36	54	31	12	3
2000	178	56	38	4	2

countries, excluding food processing, storage and transport. The use of commercial energy in the developing countries in agricultural production is only about one-quarter of the amount used in farming in the developed countries.

The comparatively limited use of commercial energy in agriculture in developing countries holds down the productivity of land and of labour. If farm yields and earnings are to rise, there must be a steep increase in the use of commercial energy. In order to double production in 90 developing countries, it is estimated (FAO, 1981) that total commercial energy consumption in agriculture should rise from 36 million tonnes of oil equivalent in 1980 to five times as much, or 178 million tonnes, in year 2000 (Table 2.17). Fertilizer accounts for practically 60% of the total increase in energy mostly in the Far East, where land scarcity enforces dependence on raising yields. The next largest increase comes in machinery, particularly in the more land-abundant areas of Latin America.

The intensification of developing-country agriculture through an increased input of commercial energy may seem a paradox in an age of energy shortages and high prices. It is, however, essential if agricultural production is to increase adequately, and in view of agriculture's very modest share in total energy consumption it is also a very reasonable increase. If it should come to a choice of priorities in allocating scarce energy supplies, agriculture must be assured of its supplies at equitable prices.

One more subject should be considered when making projections, and that is the decreasing proportion of the working population engaged in agriculture and consequently the decreased importance and interest being accorded to agriculture. In industrialized countries often less than 7% of the working population is engaged in agriculture, and it is often not realized how much this small group contributes to the economy. Many people take for granted that each day plenty of food of all types is available at rather low prices. They spend some 20 percent of their disposable income for food. However, in less industrialized countries where most people still work in agriculture and where often, more than 50 percent of the income is needed for food, the situation is quite different. Even there, however, the significance of agriculture to the economy is often not realized.

2.7 AN OVERVIEW

The world as a whole has enough land to produce food for present and future populations. However, with the uneven distribution of land resources, populations and agricultural inputs, food production falls short of requirements in a great number of countries. In order to avoid dependency on external supplies, these countries will need to increase their domestic food production.

When planning for a higher degree of self-sufficiency, it is essential that differences in land resource endowment and in crop production potentials be fully appreciated. In some countries, land reserves are such that cultivation can be expanded to meet national requirements, and even beyond. In other areas the limits of cultivatable land have already been, or are about to be, reached and most of the increased production will have to come from the intensification of agriculture on land already cultivated. Certain countries with unfavourable soil and climatic conditions may not have means to meet the food requirements of their populations, even if the level of inputs were to be optimized. Furthermore, other needs also have to be met, such as fibre for clothing, raw materials for housing and industry, lumber and fuel wood, environmental conservation and possibly export crops. Therefore a balance should be established within each country, matching needs with the suitability of the land base for various types of use.

The precarious food situation in a number of developing countries indicates that the mere availability of land is not sufficient to fill the gap between demand and supplies. Incentives need to be created for the farmers to remain on the land and to make it produce. Priority has to be given to rural development in terms of investment, pricing policies, energy allocation, access to inputs, technology transfer, transport and credit, training and research. This intensification of agriculture is crucial in many developing countries where land resources are insufficient, at a low level of inputs, to meet the food needs of their present or future populations. The solution to the world food problem is, in the first instance, self-reliance of every nation in accordance with its production potential. With the identification of critical areas in various parts of the world, it clearly appears that food security will also have to rely on international cooperation.

2.8 REFERENCES

Baker, O. F. (1923) Land utilization in the United States: geographical aspects of the problem. *Geogr. Rev.*, **13**, 1–26.

Barney, G. O. (1981) *The Global 2000 Report to the President: Entering the Twenty-First Century* (Council on Environmental Quality and the Department of State), Washington.

Bivas, M. R., and Bivas, A. K. (1978) Loss of productive soil. *Int. Jrnl. Envir. Studies*, **12**, 189–197.

Buringh, P. (1982). Potentials of world soils for agricultural production. *Trans. 12th Int. Congr. Soil Sc.*, Vol. 1, Plenary session papers, 33–41.
Buringh, P. (1984) Organic carbon in soils of the world. In: G. M. Woodwell, *The Role of Terrestial Vegetation in the Global Carbon Cycle* (SCOPE Publ. 23). John Wiley & Sons, Chichester, New York, 91–109.
Buringh, P., and van Heemst, H. D. J. (1977) *An Estimation of World Food Production based on Labour-oriented Agriculture*. Agr. Univ., Wageningen, the Netherlands.
Buringh, P., van Heemst, H. D. J., and Staring, G. J. (1975) *Computation of the Absolute Maximum Food Production of the World*. Agr. Univ., Wageningen, the Netherlands.
Centre for World Food Studies (1980) *The Model of Physical Crop Production*. Research Report SOW-80-5, Amsterdam-Wageningen, the Netherlands.
Dudal, R. (1981) Land resources and production potential for a growing world population. *Proc. 12th Congr. Int. Potash Inst. Bern*, 277–288.
Dudal, R. (1982) Land degradation in a world perspective. *Jrnl. Soil and Water Conservation*, **37**, (5), 245–249.
Dudal, R., Higgins, G. M., and Kassam, A. H. (1982) Land resources for the world's food production. *Proc. 12th Int. Congr. Soil Sc.*, New Delhi.
FAO (1971–81) FAO/Unesco Soil Map of the World, 1:5 000 000, Vols. I–X. Paris.
FAO (1978–81), *Reports of the Agro-ecological Zones Project*, Vol. 48/1–4, Rome.
FAO (1981) *Agriculture: Toward 2000*. Rome.
FAO/UNEP/Unesco (1979) *Provisional Methodology for Soil Degradation Assessment*. Rome.
FAO/UNFPA/IIASA (1982) *Potential Population-supporting Capacities of Lands in the Developing World*. Rome.
Grigg, D. (1982) *The Dynamics of Agricultural Change*. London.
Hopper, W. D. (1976) The development of agriculture in developing countries. *Scient. American*, **235**, (3), 197–205.
Linnemann *et al*. (1979) A Model of International Relations. Amsterdam.
Meadows, D., *et al*. (1972) *The Limits to Growth*. New York.
Meadows, D., *et al*. (1982) *Groping in the Dark: the First Decade of Global Modelling*. John Wiley & Sons, Chichester, New York.
Olson, R. A. (1983) The impacts of acid deposition on N and S cycling. *Environmental and Experimental Botany*, **23**, (3), 221–223.
President's Science Advisory Committee (PSAC) (1967) *The World Food Problem*, Vol. II. Washington, D.C.
Richards, J. F., Olson, J. S., and Rotty, R. M. (1983) *Development of a Database for Carbon Dioxide Releases Resulting from Conversion of Land to Agricultural Uses*. Oak Ridge Ass. Univ., Oak Ridge, Tennessee.
Salas, R. M. (1981) *The State of the World Population, 1980*. United Nations Fund for Population Activities, New York.
Slicher van Bath, B. H. (1963) *The Agrarian History of Western Europe, AD500–1850*. London.
Wit, C. T. de, and Goudriaan, J. (1974) *Simulation of Ecological Processes*. Simulation Monographs, Pudoc, Wageningen.

Land Transformation in Agriculture
Edited by M. G. Wolman and F. G. A. Fournier

CHAPTER 3
Transformation of the Land in Pre-industrial Time

I. G. SIMMONS

3.1 INTRODUCTION

This chapter aims to provide a perspective upon land transformation before the advent of industrialism, i.e. in the period from approximately 10 000 BC to 1890 AD. It is acknowledged that there may have been instances of human alteration of ecosystems before the Holocene (consider, for example, the antiquity of fire in Africa demonstrated by the evidence from Chesowanja in Kenya (Gowlett *et al*., 1981) and the equivocal but suggestive pollen-stratigraphic analyses from the Hoxnian interglacial in England at Marks Tey and Hoxne itself (West, 1980; Turner, 1976)), but we shall ignore these for present purposes. In spatial terms the spread of different types of land transformation has always been uneven, and by 1890 there were many parts of the world to which even the faintest echoes of the new noises of industrialization had not penetrated, and even today there are people whose way of life is little affected by the industrialized world except perhaps in the possession of iron-based tools. Such people are usually living in remote areas and are usually either hunter-gatherers or pastoralists, as with the Tasaday of the Phillipines, the Panare of Amazonian Venezuela and the Dinka of the Sudd region of the Sudan. These groups are in general disappearing as either their ecosystems or their culture, or both, are subject to transformation by processes emanating from the industrial world. There are also many thousands of agriculturalists whose ways can be described as 'traditional' even though the processes of development are gradually bringing about an infusion of ideas and techniques from the 'modern' world. It can be generally said, of course, that the mid-nineteenth century was a dividing point between those regions of the world which fully adopted an industrial way of life and which have become today's developed countries, and those which did not.

The tenure of the earth by human societies in the period 10 000 BC to 1890 is best illustrated by a series of maps (Figures 3.1–3.5). Figure 3.1 is for *ca* 10 000 BP (8000 BC), when the Pleistocene ice had retreated to positions not too far short of the final postglacial positions. At that time, agriculture was in its incipient stages in south-west and south-east Asia, and so the greater proportion of the populated world was occupied by hunter–gatherers. (Note that practically the whole land surface, with the exception of a few islands, was now populated by *Homo sapiens sapiens*). By 6000 BP/4000 BC (Fig. 3.2), agriculture was firmly established from the Balkans to the Irrawaddy and on the coastal margins of mainland and peninsular east and south-east Asia, with incipient stages like the Saharan Neolithic being visible, and a penumbra of the pastoralism of domesticated animals developing around the Mediterranean–west Asia focus of agriculture. But the world was still mostly the preserve of the hunter–gatherer, though the period through to 4000 BC (Fig. 3.2) shows some erosion of their position with the extension of agriculture into peninsular Europe, savanna Africa, most of east and south-east Asia (with attendant pastoralism in arid, semi-arid and mountainous zones) and the apparently independent nascence of agriculture along the Pacific side of the Americas from Mexico to Peru. But there was still a great deal of land, including coastal margins, where both nomadic and settled huntsmen and fishermen were innocent of the presence of domesticated biota (except the dog) in their lives, though by AD 1 this area was severely diminished (Fig. 3.3), with the Americas, south-east Africa, Australia and

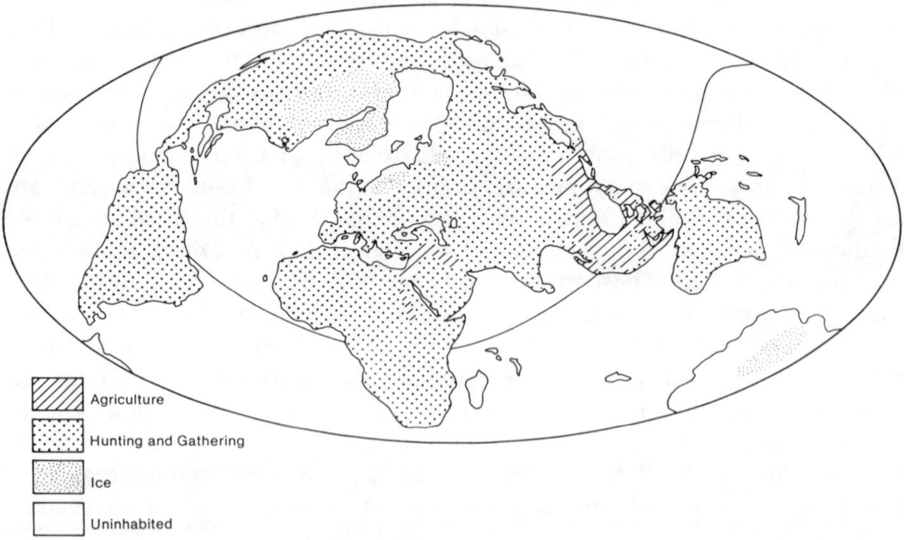

Figure 3.1 Tenure of the earth *ca* 10 000 BP

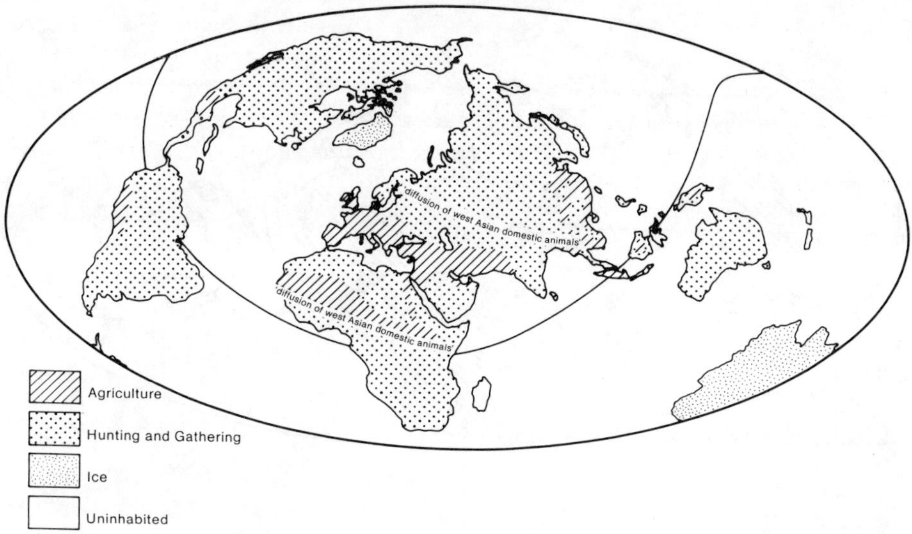

Figure 3.2 Tenure of the earth *ca* 6000 BP/4000 BC

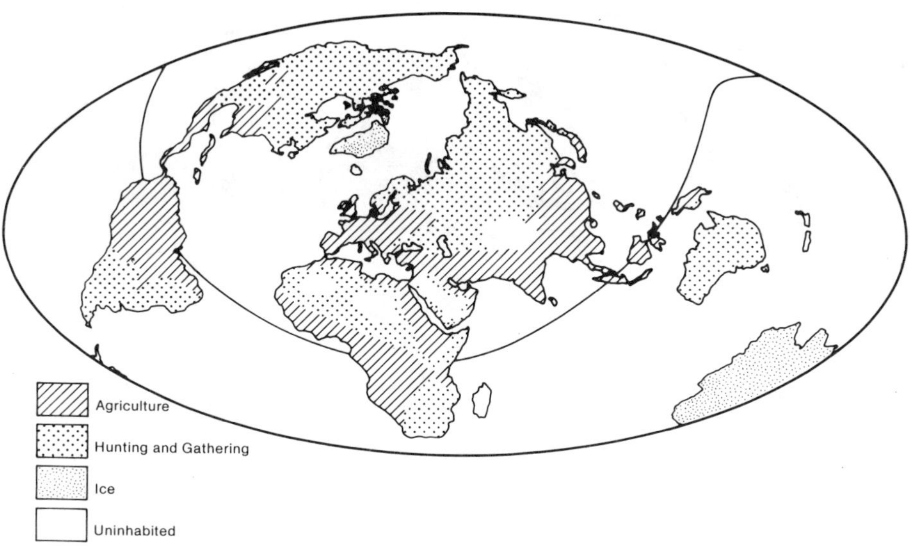

Figure 3.3 Tenure of the earth *ca* AD 1

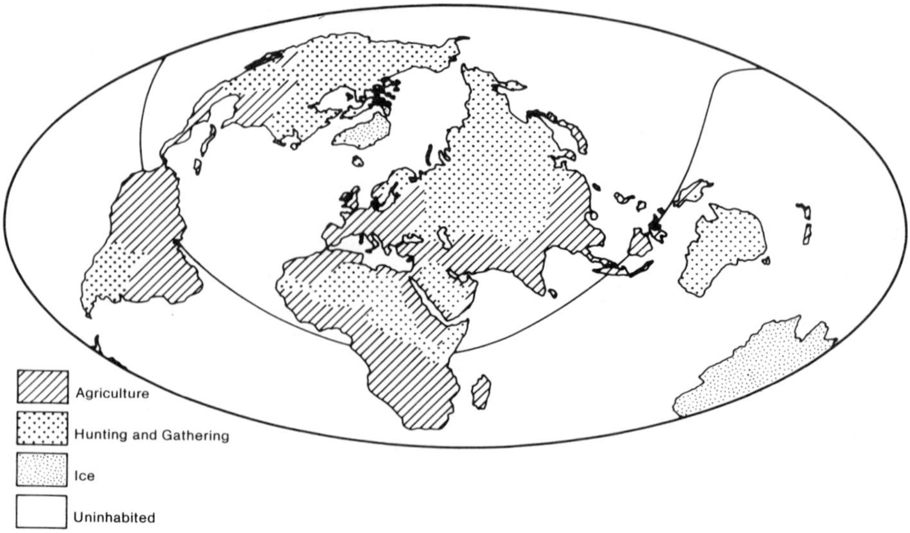

Figure 3.4 Tenure of the earth at 1000 AD

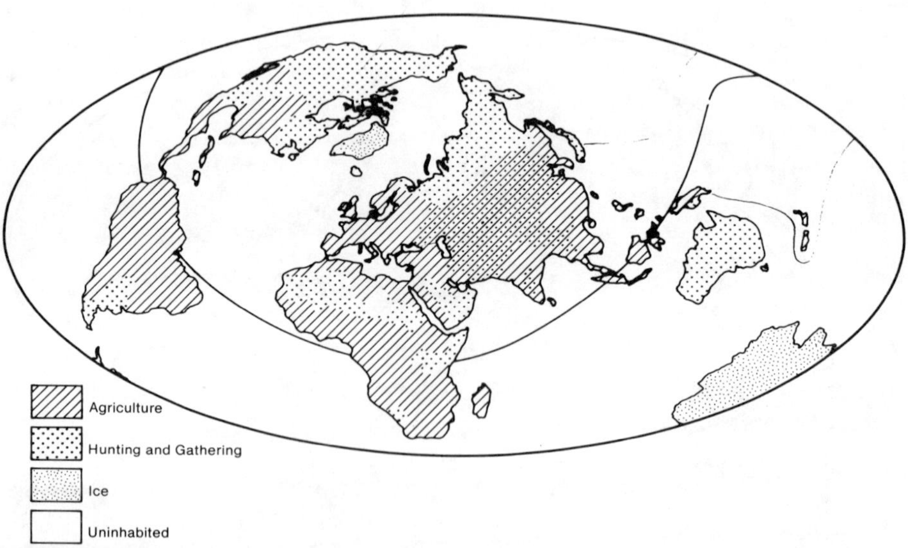

Figure 3.5 Tenure of the earth at 1500 AD

Siberia now being the heartlands of hunter–gatherer activity, including distinctive coastal cultures in many circumpolar regions. Pastoralism, too, now occupied large areas of the interiors of Africa and Asia. This position was consolidated by 1000 AD (Fig. 3.4), with further expansions of indigenous agriculture into both South and North America, so that the grasslands of North and South America, together with Australia, held most of the remaining hunter–gatherer populations, though there were significant numbers of coastal hunters in the far north of the globe. By 1500 AD (Fig. 3.5), the world population of perhaps 350 million numbered only 1% of hunters, compared with virtually 100% at 10 000 BC. European agriculture was starting to be established overseas (e.g. in North America) and pastoralists had replaced hunters in some grasslands. Only Australia remained untouched by agriculture until the coming of European settlers: there was no indigenous development of that mode of subsistence. After 1500 and before 1890, this pattern was again reinforced, with European colonization being one of the major causes of the ousting or assimilation of many hunters (e.g. in North America, Patagonia, South and East Africa) and so progress was made towards the present situation (Fig. 3.10), where hunter–gatherers number about 0.001% of world population; agriculturalists having risen to a peak in proportion in the early nineteenth century are now falling quite rapidly.

This brief account provides us with two pointers. The first is the growing importance of agriculture through most of the period under consideration and its antiquity in certain places where it has had a long period in which to be the agent of land transformation. The second, however, is that hunting and gathering for long periods and in many places held sway as the dominant economic type. So although most attention will be directed at agriculturally induced land transformations, some attention will have to be given to the food-collectors (as distinct from food-producers) to see if they too were the promulgators of ecological change. Immediately, though, we shall consider briefly the technical means by which pre-industrial societies were able to manipulate their environments.

3.2 TECHNOLOGY OF PRE-INDUSTRIAL SOCIETIES

3.2.1 Access to Energy Sources

Numerous writers have pointed out the role of access to energy sources in mediating human relations with the natural environment, and Table 3.1 shows a summary of the sources available at different stages of human history. The greatest turning points were perhaps the control of fire, acquired certainly by 500 000 BP but quite probably before that, and access to the stored energy of the sun in the form of fossil hydrocarbons, the large-scale development of which we call the industrial revolution. The lesson to be gained from

Table 3.1 Energy sources available to different types of culture

Earliest hunter–gatherer	Solar energy via plants and animals
Advanced hunter–gatherer	As above, plus stored solar energy as fire
Early agriculturalist	As above, plus domestic animals (traction, dung)
Advanced agriculture (pre-industrial)	As above, plus wind and water power
Early industrialist	As above, plus power from coal
Advanced industrialist	As above, plus oil, natural gas, hydropower and geothermal power
Contemporary	As above, plus technological methods of trapping solar and tidal power, nuclear power from fission and fusion

Table 3.2 Comparative energy statistics

(A) Historical

Type of society	Energy consumption per capita per day (kcal)	$J \times 10^6$
Hunter–gatherers	2 000	8.4
Agriculturalists	10–12 000	42–50.4
19th c. industrial society	70 000	294
20th c. industrial society	120 000	504

(B) Contemporary

Nation/development stage	Energy consumption per capita per day (1970s)[a] (kcal)	$J \times 10^6$
USA	219 000	919.8
DC average	117 000	491.4
Nepal	300	1.2
LDC average	6 800	28.5
World average	37 400	157

[a]Not food or fuel wood.

this historical perspective is that not only has the spectrum of energy availability increased, but so has the magnitude of it (Table 3.2). Thus industrial societies have under control quantities of energy far in excess of their predecessors and are able, as this volume shows, to effect transformations on hitherto impossible scales, some of which are of an irreversible character. This does not mean, though, that the impact of pre-industrial societies was negligible, for the input of solar, human and animal energy, and fire, channelled through technology, was responsible for widespread ecological changes.

3.2.2 Implements and Agents of Change

By the time of the emergence of *Homo sapiens sapiens* (*ca* 35 000 BC) the control of fire was accomplished, and it had spread with human societies out of Africa into much of Eurasia and was reaching the New World and Australia. This meant that many environments were subject to manipulation by fire, and there is a long list of proven instances of perhistoric and pre-European contact societies using it for manipulation. Indeed, most major biomes are susceptible to fire at some season, with the obvious exception of deserts with little vegetation. Intact, moist tropical forests are also probably unburnable, but secondary forests are vulnerable: in 1983, a fire consumed 3.5×10^6 ha of secondary forest in Kalimantan after a period of abnormally low rainfall (P. Buringh, personal communication). In temperate forests, the ecology of the regular use of fire is such that it increases the value and nutritional quality of food for ungulates by a factor of 10, since it increases the quantity of browse as well as improving the protein content of leaves (via the mechanism of mineralizing litter and thus enhancing the nutrient status of the soils), so opening the way for an increased density of animals either on an absolute basis or in the form of a concentration at a chosen place.

The ability to affect the movement of animals by attracting them to food sources or by making them flee in particular directions (as with some of the North American plains buffalo hunts) must have been a major positive contribution to the energetics of hunting, since it reduces by many times the amount of effort that has to go into harvesting: human energy is replaced by the release of energy stored in dead or otherwise combustible plants. It seems likely, too, that the use of fire reduces uncertainty in the hunt: used as a catalyst of improved forage, this is obvious; in other contexts, it appears that burning to reduce the shrub layer of a forest improves sight-lines and hence kill-frequency; and at the very least if a large area is burned, something edible is almost bound to flee the fire in the direction of a party of hunters (Mellars, 1976). But energy can find its application in fabricated tools as well. Many of them act as concentrators of energy, taking the essentially diffuse energy of, for example, muscles and directing it down a narrower channel with greater force. The blowpipe, the bow and arrow, the spear thrower all do this for hunters, the plough and the hafted axe for agriculturalists; the grinding wheel concentrates the diffuse energy of wind and water for those living off cereal agriculture, and the windmill brings water to (or removes it from) crops.

The energy contained in some chemicals was harnessed in pre-industrial times as well, with gunpowder being used in warfare and in quarrying, which is a distinctly non-reversible land transformation.

All these technologies had two effects: to facilitate the spread of pre-industrial human societies into most parts of the earth, and once they were established, to enable them to deflect the natural flows of energy and matter to their own ends.

3.2.3 Processes of Transformation of Natural Systems

3.2.3.1 Direct Transformations

The application of technology (including fire) to many natural systems results in a perturbation of those systems. Their response is varied and depends *inter alia* upon the intensity of the perturbation (i.e. the quantity of human-directed energy per unit area) and the time over which it is applied. Some systems are extremely resilient, and if the disturbance ceases will undergo succession back to their pre-existing state or something very like it; other ecosystems will not recover and will find a new level of equilibrium in a different condition, even if the hand of man is removed. Any attempt at a complete list of such transformations is bound to include gaps, but we may quote a number of major examples:

(1) fire in the burning of grasslands and dry forests;
(2) forest clearance by axe or fire;
(3) soil manipulation with plough, hoe, digging stick; production of ridges and furrows;
(4) construction of boundaries, like fences, dykes, ditches, bunds;
(5) pasturing of domesticated or semi-domesticated animals;
(6) drainage works and irrigation canals;
(7) continued hunting of wild animals and collection of wild plants;
(8) building of agricultural terraces;
(9) construction and maintenance of domestic and other buildings;
(10) impact of wood gathering for domestic use upon many kinds of vegetation community.

The resulting land transformations, as suggested above, vary greatly. Some hunting communities are both so small in number and peripatetic that they affect the vegetation and animal communities scarcely at all, and their low level of material possessions makes very little impact even at their former dwelling sites. At the other extreme, the irrigation agriculture of the Nile, founded in the early years of agriculture itself, has persisted to the present day and the original reedswamps, wet grasslands and seasonally inundated scrub have never had a chance to come back. Some of these transformations, then, affect only short-cycle components of ecosystems such as plants and animals, but others affect long-cycle constituents like the soils, parts of the hydrological cycle and even the rocks themselves, as in the quarrying which heavily scarred parts of classical Greece (Hughes, 1975).

3.2.3.2 Indirect Transformations

Considered spatially, each impact has an epicentre, whether this be the killing of individual animals, or the clearing and ploughing of a patch of forest. But ecosystems being interactive systems of biotic and abiotic components, the purposeful impact is likely to have unconsidered (and sometimes undesired) consequences elsewhere. To kill a particular male animal may be to disrupt the social and reproductive unit and lower the chances of replacement of that animal; clearing a patch of forest and cultivating the soil may result in the permanent loss of perhaps 1 m of topsoil and its deposition in a river flood plain or in a delta many kilometres away. Because of the silt load, the river basin hydrology may undergo very large-scale changes indeed, with new patterns of flooding, for example, affecting agriculture or urban life in the riverine lowlands.

To exemplify these generalities we shall now consider separately the two major pre-industrial ways of life and give some examples of ways in which they have effected land transformations in various places and at different times. These examples, together with Figures 3.1–3.5, give some idea of temporal and spatial impact, but the treatment will not be comprehensive.

3.3 HUNTER–GATHERERS AND THEIR ENVIRONMENTS

3.3.1 Increasing Access to Energy

The early development of human societies can be seen in some ways in terms of increasing access to energy sources: the development of hand-held tools is carried further by the invention of hafted types, the spear is complemented (probably not until the Upper Palaeolithic of Europe) by the bow and arrow. Indeed the European Upper Palaeolithic marks a sudden burst of technological invention unparalleled since the start of tool making chronicled at Olduvai Gorge (Leakey and Lewin, 1977, 1979) and the control of fire. Access to energy is low compared with today (Tables 3.1 and 3.2), but we have to remember that (a) hunter-gatherers account for 90% of the time since human evolution, and (b) they have occupied most environments, from the ice-margins of the Neanderthals and near-recent Inuit to the near-deserts of the Saharan neolithic and near-recent Bushmen.

Not only are some of these environments fragile because of the extreme physical conditions, but some have been affected by rapid environmental change during, for example, the sudden climatic changes which mark the end of the Pleistocene (Lewin, 1983). At that time there occurred an apparently sudden and irreversible demise of about 200 genera of large warm-blooded animals, a phenomena which has been labelled 'Pleistocene overkill'. There has been, as might be expected, some investigation into whether natural events were entirely responsible, or whether human agency was involved (Martin, 1967). The phenomenon is perhaps most closely observable in North America: here, two-thirds of the mammalian megafauna (species with adults weighing 50 kg) of the end of the Pleistocene (*ca* 11 000 BP) disappeared, including 3 genera of elephants, 6 genera of giant edentates (armadillos, ant-eaters, pangolins and sloths), 15 genera of ungulates, and various giant rodents and carnivores. There is no evidence of such extinctions in earlier periods which might have been expected if previous Pleistocene climatic changes had been the proximate cause, nor any firm evidence of survival of these genera past the 500-year period during which the extinctions appear to have taken place. These data have led to the hypothesis that it was the introduction of *Homo sapiens* as a predator against which these animals had no genetically evolved defence behaviour which was the cause of the extinctions. In North America, the date of introduction of man via the Bering Strait land bridge is usually taken as being about 12 000 years BP, and a simulation of the extinction pattern supposes a wavefront advance by hunters whose population periodcially exploded because of the new and favourable habitat (Martin, 1973; Mosimann and Martin, 1975). At the advancing front, an annual human population growth rate of 3.4% may have given a density of 0.4 persons/km and a forward movement (due to hunting-out of the large mammals) of 16 km/year. Thus the front could sweep from Canada to Mexico in 350 years (Martin, 1973) and extirpate the large mammals through superior predation techniques.

Some support is given to this hypothesis by the effects of the arrival of man into other hitherto unpeopled places. The disappearance of the moa from New Zealand occurred within a few hundred years of the human occupance of the islands; the megafauna of Madagascar disappeared within a similar period after the first human occupation in 1000 AD, including a large terrestrial bird *Aepyornis*, and a pygmy hippopotamus. In Java and Celebes, populations of dwarf elephants did not long survive the coming of man. It would appear, therefore, that when introduced into a new habitat man the hunter is capable of rapidly exterminating flightless birds and large mammals. By contrast, the end of the Pleistocene in the Old World presents a somewhat different picture (Reed, 1970), for in Eurasia only a total of 9 species were made extinct, compared with the New World total of at least 24 genera. The Eurasian losses comprised the woolly mammoth (*Mammuthus primigenius*), woolly rhinoceros (*Coelodonta antiquitatis*), giant Irish elk (*Megaloceros giganteus*), musk-ox (*Ovibos moschatus*), steppe-bison (*Bison priscus*), a buffalo of northern Africa (*Homoioceros antiquus*) and three species of associated carnivores. All except the buffalo are animals of the cold steppe which was widespread outside the ice-sheets of Europe during the last glacial phase and all the herbivores, especially the mammoth and bison, had been hunted for tens of thousands of years yet survived in large numbers until about 12 000–10 800 years BP. But this time-horizon does not mark the introduction of man to Eurasia, not even a particularly great cultural change, so that it appears much more likely that climatic amelioration leading to forest growth of birch and pine was the main cause of the animals' disappearance. At any rate, when their habitats returned briefly in the cold period of 10 800–10 510 years BP, the animals did not. Some other species of the cold steppe proved more adaptable: the reindeer persisted on the tundra, and both the saiga (*Saiga tartarica*) and wild horse (*Equus przewalskii*) took to the grasslands of central Eurasia. A site in New South Wales, Australia, seems comparable with Eurasia since it appears to show 7000 years of coexistence of man and megafauna without any evidence that the human societies were the cause of the eventual extinction of the other animals (Gillespie *et al*., 1978).

As added evidence there has emerged the picture from the North American prairie in the Holocene of the killing of large numbers of animals without any lasting effect upon their populations. We have a picture of the Indians (in the days before their possession of horses) killing in the autumn large numbers of animals in order to provide both an immediate feast and meat for the winter (Frison, 1978). In the open environment of the High Plains the animals had to be trapped in some way, either so that they could be killed at close range or so that they killed themselves. Butchery then took place at or near the kill site, which was then probably abandoned for some time, at least until the rotting remnants of the carcasses had ceased to be noxious: at one site a layer of maggot cases 2.5 cm thick testifies to the fact that not all the meat was eaten. At the Olsen–Chubbuck site in south-east Colorado, dated 8200±500 BC, the hunters stampeded a herd of buffalo into a narrow canyon or *arroyo*, some 2–3 m wide and 2 m deep. The age of the calves suggests this happened unusually early, in late June rather than the autumn. The remains extended some 57 m along the arroyo and about 190 animals were killed (a typical herd would number 200–300 at that time of year). The bone analysis shows that 57% of the animals were mature, 37% immature and 6% juvenile; but no fetal remains were found, either because no gravid animals were killed or because the fetuses were taken away as a particular delicacy. At one time during the years 1670–1740 AD another group of hunters used a spectacular site in central Wyoming, where a bluff 14 m high was used. The vertical part of the cliff is restricted, and the animals had to be carefully controlled: a metre or two either side and they would likely escape down steep but not lethal slopes. In this case, the drive lane suggests the control of the buffalo herds' movements for 1.5–4.5 km before they were finally stampeded over the edge. The bones suggest that this and similar bluffs might be used in rotation: one hundred buffalo stomachs left to rot might not be approachable for a year or two, even to people less easily made queasy than ourselves.

To summarize, the evidence suggests that kill operators were bringing in bison to kill sites from several kilometres round each site, usually but not always in the autumn. Fire was often used in this as in other types of buffalo hunting. So men were reaching out and removing several hundred cows and calves from the breeding population and perhaps doing this in adjacent areas in the same season. Although doubtless plants were eaten, the bison was a staple; and although nobody suggests that the hunters were responsible for anything like the reduction in buffalo numbers that came later, we may surmise that the population dynamics of herds may sometimes have been affected with consequent effects upon vegetation and upon those areas of soil which they physically disturbed, as in wallowing areas.

3.3.1.1 The Later Mesolithic in Upland Britain

As an example of environmental impact in prehistoric times, we shall consider the last hunter–gatherer culture of Britain, groups of people who lived by hunting (most probably concentrating on red deer) and gathering, and probably by fishing. One of the environments they occupied was that of the uplands of England and Wales, and it is in this zone that palaeoecological research points to their role as manipulators of their surroundings. The period of their tenure of these lands in *ca* 8500–5000 BP, ending at *ca* 3500 BC with the arrival of farming (Neolithic) people.

The British uplands present today a largely treeless, peat-clad landscape, vegetated by various kinds of moorland dominated above 300 m by ericaceous shrubs, grasses, sedges and mosses. The land is now used as sheepwalk, grouse moor, watershed and recreation area, and some of it is afforested with exotic conifers. In spite of their wet and cool climate these uplands were largely covered in natural woodland (pine forest by 9600 BP and mixed oak by 5500 BP) during the mid-postglacial period, and it is with the beginning of the disappearance of this forest that this account is largely concerned. Because most of the uplands have extensive deposits of peat and have never been cultivated more than sporadically during historical times, they preserve to a considerable extent both the archaeological and the palaeoenvironmental evidence for the reconstruction of prehistory. They have therefore attracted more attention than some lowland areas (though there are notable exceptions like Star Carr and the Somerset Levels) and the findings given for a particular set of environments must not be taken to extend to other areas for which no comparable evidence is yet available.

Environmentally, the later Mesolithic occurs within the ‘Atlantic’ period of postglacial climate. Rainfall was probably on average 11% above present-day levels, and average temperatures approximately 2°C above those of today. In the uplands, the higher rainfall would have been amplified by years with rainfalls 25% or even 50% above the average, and evaporation hindered by the lapse rate effect of the altitude upon temperatures. Whereas high summer air temperatures in the lowlands in the late Mesolithic might have averaged at 17.5°C at 500 m, the equivalent would have been more like 15.4°C. This oceanicity was, of course, inherent in the position of Britain, but exacerbated by rising sea-level which finally insulated Britain *ca* 7500 BP, i.e. near the beginning of the later Mesolithic. Nevertheless, postglacial vegetation succession had proceeded so far as to cover most of England and Wales with mixed deciduous forest in which oak and probably lime were important trees. The climatically determined tree-line on the uplands seems in places to have been upwards of 700 m, evidence being provided by tree remains in peat at such altitudes.

Yet pollen analysis shows that on some uplands (e.g. southern Pennines, Dartmoor, North York Moors), the tree-line was, during late Mesolithic times, well below the climatically feasible limit; an upper level of forest of 320–290 m was more characteristic. In the Southern Pennines, Tallis (1975) has postulated an upward extension of tree-lines after Mesolithic times (during a period of deteriorating climate), noting that the organic deposits of Mesolithic age in the upland contain not tree remains but charcoal. Similar evidence from the North York Moors has been found (Simmons and Innes, 1981). A number of types of evidence have been described:

(a) 'Clearance phases' in pollen diagrams. These are analogous to the *landnam* phases (*sensu* Iversen, 1949) found in agricultural phases of prehistory in that woodland appears to have been cleared but later regeneration takes place.
(b) Forest recession in pollen diagrams. The pollen indicators of mature forest recede and are replaced by indicators of open ground (e.g. grasses, bracken fern, *Ericaceae*) which stay permanently. Both this type and type (a) sometimes contain small quantities of pollen of aquatic plants at the appropriate horizons, suggesting the nearby presence of water.
(c) Many pollen diagrams contain very large frequencies of the pollen of hazel (*Corylus avellana*). This shrub flowers prolifically when a forest canopy above it is removed, and stump-sprouts after cutting or burning, and will form a scrub. The nuts are nutritious and often turn up in excavations of Mesolithic sites.
(d) Organic deposits in basins show one or more 'inwash stripes' of inorganic matter during the Mesolithic. These stripes interrupt otherwise continuous organic materials and are presumed to have come from open ground yielding silt to the runoff, as would happen at times of deforestation. The

pollen content of the deposits confirms the recession of forest at the time they were deposited. All these phenomena occur during later Mesolithic times and in areas with appropriate artifacts, but it should be stressed that they are not necessarily associated directly with layers of, for example, implements. Charcoal is, however, frequently found in deposits of appropriate age.

None of the evidence, quite naturally, shows any direct evidence for interference with animal populations; while culling of them must have affected numbers there is no way at present of telling whether the overall magnitude and composition of mammal populations would have been affected on anything other than a short-term basis. Pollen analysis does, however, provide evidence for forest recession during late Mesolithic times in places where lithic remains are found, and so these are usually attributed to human activity and called clearances. In lowland England they appear to be commonest on siliceous substrates, often at places which are now heathland such as the Weald of south-east England, and the Breckland in East Anglia. The quantity of silt suggests that sufficient tree cover had been removed to allow the runoff to transport soil material downhill into accumulating rivers (Richards, 1981) and to form spreads of inorganic material over and among the vegetation. In some places the forest limit is either pushed downhill or never attains its climatic potential with acid grassland or shallow peat forms. After the end of the Mesolithic period, deciduous forest colonized the shallow peats and grasslands and in this case the opening was held for hundreds of years. The presence of charcoal in the organic deposits draws attention to the probable role of human communities. Another type of sequence is found on the highest parts of uplands where the angle of slope is low. Forest clearance was followed by acid grassland, then heath vegetation dominated by heather (*Calluna vulgaris*), and then blanket peat (2–3 m deep is common). This process is not reversible in the manner of the other two instances; once the forest has gone it does not return.

In the context of these data, some speculation about possible reasons for environmental manipulation may be made. For example, if in summer the red deer scattered (and roe deer are not herd animals at any time) then they must have been attracted by the combination of abundant browse and water. Culturally, this might coincide with Zipf's principle of least effort: it is easier on the limbs and lungs to attract the beasts than to run after them. There is, as well, the possibility of circumstances in which several millenia of occupation even at a very low density and rate of population growth had filled up the terrain to the point where no 'free.' land existed for any people who could not be supported on a particular territory. Then (a) seasonal movement and (b) burning would be seen as adaptations (i) to increase food supplies by tapping as wide a set of food resources as possible; and (ii) as a form of intensification of land use in which the logical hypothetical sequence would be free hunting: manipulative hunting, herding and, eventually, agriculture.

3.3.1.2 Examples of Transformations by Near-recent Hunter–Gatherers

Near-recent examples of the utility of fire to hunter–gatherers are plentiful, but perhaps the most notable concentration is in the Antipodes. In the Cape York peninsula of Australia, Harris (1977) reports the usefulness of large stands of cycad trees (*Cycas media*) which, in yielding some 131 kcal/m^2/year of edible kernels, equal some cultivated crops. He suggests that these stands are a consequence of firing the area repeatedly, thus encouraging the fire-resistant cycad at the expense of other trees. The cycad *Macrozania*, if fired, produces all its seeds simultaneously and at 7–8 times the unfired quantity. Further into the interior, the aborigines used fire for hunting, land clearance and communications as well as for domestic purposes, with a total of 5000 bush fires a year being estimated for the years of early European occupance. Fire was used to 'clean up' a landscape, it was said (rather as we might—some of us—tidy up a room), and in effect it created a patchy array of habitats with greater floral and faunal diversity than had previously existed, and many of the biota thus encouraged could be eaten with some of the animals easily flushed out by the flames: long-necked turtle and granna lizard are still hunted in this way. At the southern extremity of Australia, the climax forest of Western Tasmania is *Nothofagus* (southern beech) forest, but a high fire frequency converts this to wet sclerophyll forest and thence (if fire is continually applied) via wet scrub to heath or tussock grassland. In the drier north-west of Tasmania the same process has produced large areas of open grassland like the English downlands. The fire enhances the growth and spread of edible plants like the bracken fern *Pteridium esculentum*, which colonized newly burned areas and became a carbohydrate staple; the patchy environment encouraged wallabies, bandicoots and possums, all of which were edible (Jones, 1975).

In New Zealand, there is evidence that at European contact time, the bracken fern was the starchy staple of the Maori and that firing was used regularly to encourage its growth and spread. Burning was extended to the podocarp forests which covered most of the eastern half of the South Island in the period 1000–1400 AD, and much of this forest land became grassland and scrub. Along with the forests and with the first arrival of man, many animal species became extinct: six genera containing 20 species of moa became extinct within 500 years of man's arrival, along with 20 other species of bird, and the extinction of elephant and fur seals except in remote areas. So predation and environmental change brought an analogy to 'Pleistocene overkill' in remote New Zealand.

For a more distant period, we have evidence from south-western Victoria during the period 40 000–7000 BP of the construction of large-scale artificial drainage channels in swamps. This appears to have been a management system for the procurement of eels. The channels were more than harvesting devices, for they retained water in swamps during drought and kept up the supply of eels in what was in fact a margical part of their range. A high density of semi-sedentary hunter–gatherer people (equal to that attained under shifting agriculture in New Guinea) was achieved thanks to a broad-based subsistence base supplemented with the products of these water controls (Lourandos, 1980).

If vegetation cover is affected by subsistence hunters, then we would expect soils likewise to be affected. Following repeated fire, and deforestation, for example, we might expect the loss of soil and perhaps regolith from sloping areas and its downhill and downstream aggregation as a fluviatile deposit. It is very difficult to tie such deposits to hunter–gatherer populations, but we may note in passing recent dating of fluvial sediments in river valleys in southern England and the North York Moors of northern England, which suggests that they were the erosion product of openings in the largely closed deciduous forest of the Mesolithic (terminal hunter) period of those regions. Aboriginal burning in the mountainous areas of Tasmania and eastern Australia is thought to have led to soil loss following large-scale vegetational change.

3.4 PRE-INDUSTRIAL AGRICULTURE AND ITS ENVIRONMENTS

As Figures 3.1–3.5 show, agriculture has not been quite so successful as hunting and gathering in its occupation of the earth's surface. Nevertheless, if we include the pastoralism of domesticated beasts, then not only have the obvious zones of valleys and temperate lowlands been colonized, but mountains, forests and semi-arid lands as well; the barley fields of Zanskar and the cattle herds of Karimojo have been as much part of pre-industrial agriculture as the wheatfields of eighteenth century Europe, the cotton of the American south or the rice terraces of dynastic China.

The access to energy by farmers down the ages has aided the processes of colonization and subsequent adaptation, for they have added to fire and human energy the power of domestic animals as well, and have been able to harness the force of the wind not only in the processing of agricultural products, but also to help with drainage and irrigation. Power generated by falling water, though useful for processing, did not in general lead to land transformation until the development of hydroelectric power.

In general terms, though, we must not under-estimate the role of human energy in pre-industrial agriculture: most of the inputs to farming come from this source, especially where construction is involved, e.g. of wells, fences or terraces. Only when the task is virtually totally repetitive—such as ploughing, lifting water and leading crops—can domestic beasts be of much use in arable farming systems. Fire, too, is still of crucial importance in shifting cultivation and in certain types of pastoralism.

Although it may not show in the landscape, the permanence of transformation requires that the nutrient cycles be maintained: the export from cities of night soil, horse-dung and pigeon manure, the keeping of stalled beasts for their manure output, the use of lime and marl, the gathering and spreading of seashells and seaweed, were all employed at some time to nourish the growing crops.

3.4.1 General Examples of Transformation

The imprint of agriculture has been strong since its apparent evolution in the 9th–6th millenia BC in south-west and south-east Asia. Animal populations, the vegetation, the form of slopes and valleys and the soil cover of those land units, have all been altered, many of them in apparently irreversible ways. The processes of tillage and fallow, of terracing, of irrigation and of drainage have had considerable consequences for such features as the transformation of soil type (like from brown earth to podzol on siliceous substrates in north-west Europe or from skeletal to deep colluvial in the Mediterranean), for the erosion of slopes (as when forests have been cleared in many regions of the world, increasing the sediment yield by perhaps 20 times) and the subsequent aggradation of valleys (which may have been accidental, as in the river valleys of southern England during the Bronze Age, or deliberate where systems of weir terraces were constructed), and the formation of deltas in oceans and lakes where silt from the land surface came to rest. At a smaller scale, the ridging of fields in pre-Columbian South America was a widespread practice (Parsons and Denevan, 1967).

These generalizations can be amplified by a few selected examples of the imprint of agriculture; we may note in passing that even where the agricultural system has gone out of use, the transformations may remain in relict form, as with abandoned terraces, for example, or deserted field systems where the boundaries remain; and soil lost from deforested and subsequently cultivated slopes is unlikely to be regenerated unless a new generation of forest trees reclaims the site and is undisturbed for perhaps tens of years in the tropics, and longer still in the temperate zone.

3.4.2 Regional Examples

3.4.2.1 Prehistoric Central America

In Central America the remains have been found of a complex urban society to be ranked with those of highland South America and of ancient Mesopotamia. Recent work has excavated not only the cities and ceremonial centres of the culture, but has also thrown considerable light on its subsistence base and its environmental relations. The Maya inhabited what is now parts of Guatemala, Mexico (Yucatan), Belize and the fringe of Honduras, mostly in the lowlands but with some extension of occupied area into the upland areas that interrupt the plains. The chronology usually adopted has an emergent phase ('Preclassic Maya') from 2000 BC to 250 AD, a 'Classic Maya' phase from 250 AD to *ca* 1000 AD, followed by the collapse of 800–900 AD (with a sudden decline of population and the desertion of cities); thereafter the Post-classic Maya continued at a lower level of complexity until the advent of the Spaniards in 1520 AD completed the disintegration of the ancient culture.

The original vegetation of the area had a number of elements, the chief of these were:

(1) the swamp or *bajo* dominated by buttressed trees;
(2) an evergreen mesic forest;
(3) monsoon forest on well-drained sites and clay-filled valleys, interspersed with palm forest also on well-drained and gentle slopes;
(4) *aqadas*: small depressions that hold water throughout the year but surrounded by a high forest with a closed canopy;
(5) oak woodland on alluvial clay soils;
(6) an orchard-like savanna vegetation of shrubs with grassland beneath;
(7) an open grassland with trees; and
(8) *campo*, an open grassland on level plains.

Within this matrix the Maya elaborated a subsistence base with a number of lineaments. The Pre-classic period was dominated by swidden agriculture of the 'slash and burn' type, mainly growing maize in the lowlands so that these areas were a mosaic of temporary cornfields (*milpas*) and scrub; the uplands were not occupied. It was once thought that this system was the economic base of the Classic Maya civilization, but scholars now believe that a suite of agricultural techniques occupied the different ecological niches; during the Collapse there was a transition from an agriculture-dominated landscape to one dominated by forest, apparently resulting from an all-embracing catastrophe. In the Post-classic era, the forest reigned with flickers of agriculture until the Maya civilization as such was extinguished by the *Conquistadores*.

There now seems no doubt that the Classic Maya were underpinned economically by a set of subsistence techniques. Of these, the most important spatially was probably swidden agriculture of maize, with sweet potato and manioc as other important crops. Each cultivator probably had several plots to reduce the risk of failure and equally to minimize the effects of soil loss which would have been a constant difficulty on exposed soils. A variation of this technique was the intensive *milpa* practised in areas of year-round rainfall. This involved multi-cropping with a high diversity of crops and high ground coverage to reduce the hazard of soil loss. Even though fallow periods were short, this system was appparently stable. Even the most intensive variant needed some fallow time, but in some lowland areas an artificial rain-forest of trees, vines, roots and seed agriculture was created, growing cassava and yams as well as the basic maize, beans and squash.

Another important element in the diet may have been the nuts of the *ramon* tree (*Brosimium alicastrum*), which may have been harvested from a mono-cultural orchard system or perhaps from a more mixed arboriculture with *guayo* and *aquacate* trees. At any rate, yields of 1245 kg/ha/year of nuts could be obtained and stored underground for long periods, so that these nuts would have been a valuable addition to the subsistence base. It seems, too, that dooryard gardens were a feature of rural areas: these were small fenced areas which experienced much fertilizing and mulching and so yielded a diverse crop of shade trees, fruit, herbs, spices and medicines.

Of greater relevance for our theme, it appears that large-scale alteration of the land surface was within the capability of the Maya, including the creation of raised fields in areas of swamp and river flood plain. These raised fields (*chinampas*) are especially characteristic of the valley of Mexico, but some 21% of the land surface of the Peten area seems to have been devoted to them. Basically, canals are dug and the mud dredged up forms a rich soil. The canal harbours fish and turtles. The crops grown included maize and cotton, but in Maya times, cacao may have been an important product of these areas. Each year the act of keeping free the canals added more rich silt to the fields, and some of the waterways had raised sills at each end to act as traps for silt or fish, or both. (Figure 3.6).

On the upland areas which interrupt the lowlands, large areas were terraced, including one tract of 10×10^3 km^2 in the Rio Bec region. These increased the available agricultural land while avoiding erosion and continual leaching of soils, and there is evidence that soils were carried upslope within the terrace systems and perhaps even from inundated lowland areas. Such terrace systems (Fig. 3.7) were supplemented by simple devices such as rows of stones set into the slope, and by silt trap terraces and weirs set in the valleys where they might accumulate more than 50 cm of soil.

The presence of such large areas of terracing argues for a more or less total land use, since all the available lowland arable would be undertaken, and other land uses besides agriculture must also be considered. Hunting was one of these, although there is very little information on its environmental impact. Another important environmental resource was wood. Each family would need about 15–20 cords a year, with one cord of wood equivalent to 120 trees. Since short fallow cycles would yield insufficient wood of the right size from their regeneration phases, woodlot plantations were probably necessary: either in the form of swidden areas with a long fallow for its wood harvest, or isolated wood-yielding 'plantations' grown within other agricultural systems. Lastly, we may note that the Classic Maya practised water management: they

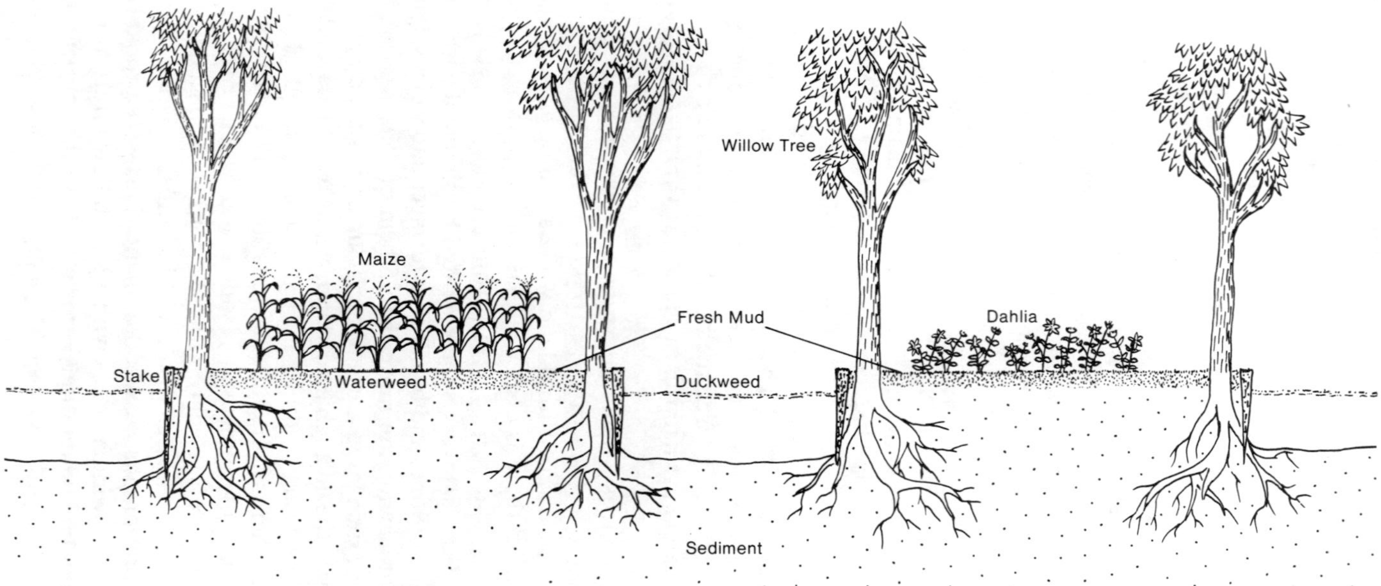

Figure 3.6 Part of a typical *chinampa*

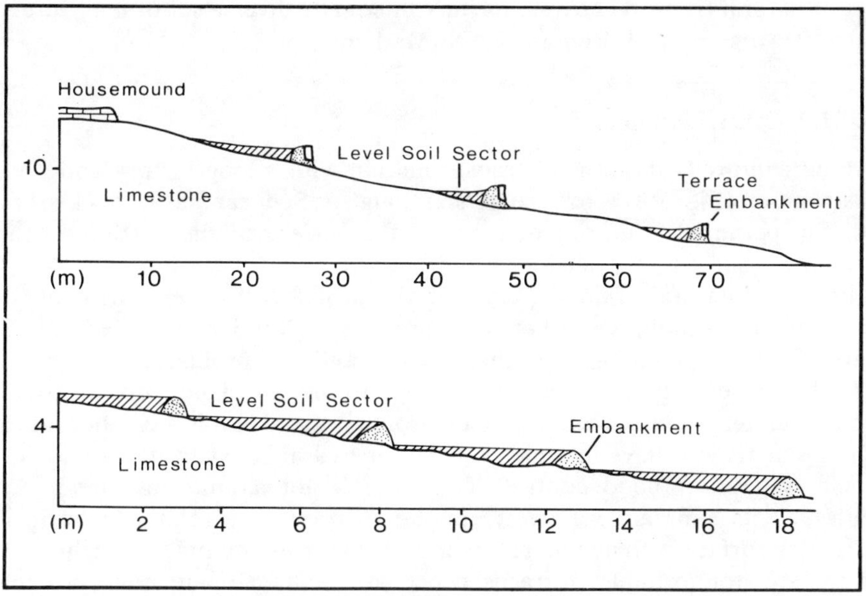

Figure 3.7 Terrace systems

collected water from limestone caves in Yucatan, lined pools with stone, cut chambers into solid rock with stone 'catchments' leading to their mouths, dug wells, and fed moated fortresses by canals as well as erecting water storage systems for cities which covered 17 km^2.

It seems very likely, therefore, that the Maya transformed much of the landscape they inhabited, though presumably patches of virgin savanna, grassland forest remained in the uplands, and there were areas of open water and swamp whose ecology remained undiverted by human acitivity; but in general an airborne observer at the time, were such possible, would have seen a largely cultural, rather than natural, landscape. Beyond this certainty, questions remain: did swidden agriculture in fact precede more intensive forms, or was the core area the lowland production areas which were supplemented by slash-and-burn as the population grew? And could the problem of intensification of agriculture be at the heart of the Maya collapse? Had perhaps the population grown to the extent that the battle between men and loss of soil fertility was lost: that the society could no longer keep up the nutrient cycles; that in effect so much natural cover had been lost that many soils were exposed to leaching and chemical weathering and too little organic matter was going back into them? Further research may illuminate this problem but we can, *interim*, note that no very sophisticated technology (no plough, for example) was needed for a numerous population to transform a set of natural tropical ecosystems into an equally diverse set of man-directed ones (Harris, 1978; Wiseman, 1978; Matheney, 1982).

3.4.2.2 Other Examples

An agriculturally based land transformation with a long history and a wide distribution (Fig. 3.8) is terracing, 'stamping', as Spencer and Hale (1961) put it, 'the permanent imprint of man on the surface of the earth'. As those authors point out, it transforms the natural slope of landforms, alters the patterns of natural drainage, changes the profile and development of soils, counteracts natural cycles of erosion, produces culturally controlled sedimentation and the growth of softer than natural landform profiles. It is interesting, too, that its distribution does not conform to the harshlands of the earth and so is a matter of culture more than environment; and in this it is different from the great river valleys in which the major hydraulic civilizations developed their systems of flood control. Terracing is not a uniform phenomenon, although its aims are relatively simple: to produce a relatively horizontal cropping surface with adequate water at the appropriate time. To achieve this there are, for example, terraces which are built in open, non-entrenched drainage channels and which employ a simple weir to spread water over a surface wider than in the natural condition so as to soak the surface and to cause the accumulation of silt; more common perhaps are rock-embanked contour terraces with sloping fields which start out near a channel from which irrigation water may be diverted to irrigate the terrace; and uppermost in the

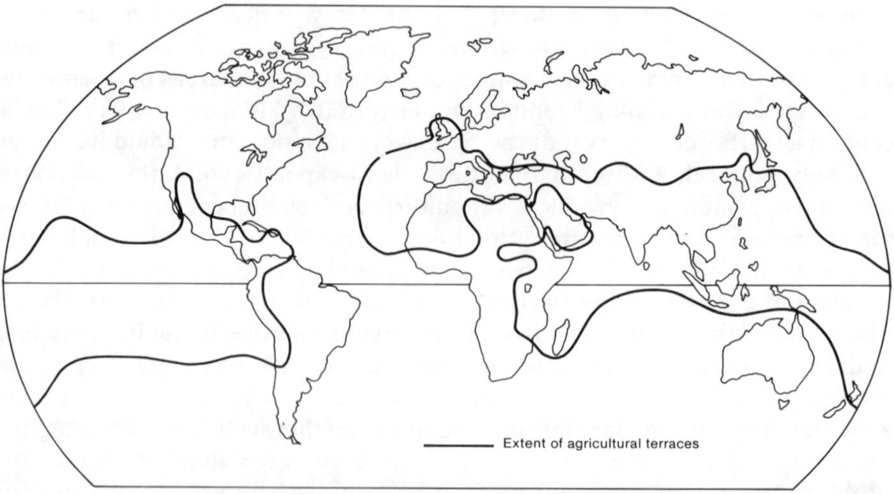

Figure 3.8 The worldwide distribution of agricultural terracing

public conception is the contour terrace laid out laterally across slopes using embankments to impound water on wet fields, using water derived from higher terraces or upland sources.

The origins of terracing are probably in the semi-arid foothill margins of the ancient Near East in the period 9000–5000 BP, with diffusion outwards to other areas. Northern Indochina acted as a secondary centre of invention and evolution out of which spread the complex wet-terrace systems for growing rice, which were then disseminated into southern China, Korea, Japan, the Phillipines and south-east Asia, Sri Lanka and Madagascar, the Himalayan front and the margins of lowland India. Some of the terraces visible in the landscape today are relatively recent (e.g. eighteenth and nineteenth century in China and peninsular India) but yet others are ancient: the much photographed systems of northern Luzon in the Phillipines are thought to be 2000 years old.

We may look in more detail at the growing of wet rice, or *padi*. This is a Malaysian word meaning 'rice on the straw', and has been anglicized to refer to the growth of rice in conditions which require it to be submerged beneath 100–150 mm of slowly moving water for three-quarters of its growing period. Most of it is thus found in flat lands near rivers, but in some places there is an elaborate terracing of slopes which represents a considerable environmental impact. Today, only a relatively small part of the total area of Asia is under wet-rice cultivation, but it supports a large part of the rural population.

Cultivated rice (*Oryza sativa*) was a dryland crop, usually grown in swidden systems, before it became a wetland staple. The early history of *padi* is uncertain, with an apparent reference as early as 5000 BP to the Emperor of China as being the only person allowed to sow *padi*. More solid references refer to wetland rice in the Wei valley of Shansi in 781–771 BC, and in the Yangtse valley of 148 BC we read of areas 'where the land is tilled with fire and hoed with water'. Quite firm dates tie the cultivation of *padi* rice to India, Indonesia and the Phillipines in the second millenium BC, to Sri Lanka in 400 BC and to Egypt in 375 BC. Thereafter the spread included Sicily in Roman times, northern Italy in the fifteenth century and subsequent export by European colonizers to such places as Brazil, the Spanish Americas, South Carolina (1685) and Louisiana (1718), Hawaii (1853) and by German influence to New Guinea and French carriage to New Caledonia.

The classical ecological description of pre-industrial *padi* is by Clifford Geertz (1964) writing about Java in the early 1950s before mechanization had any significant effect. He notes that after 10–20 years of cultivation the yields tend to remain stable almost indefinitely. Geertz suggests that the characteristic nutrient poverty of tropical soils is circumvented through the nutrients brought by water from the higher volcanic slopes on to the terrace; through the fixation of nitrogen by the blue–green algae which proliferate in the warm water; the release of nutrients by the decay of stubble and of other organic matter used to fertilize the fields; and as a *sine qua non*, good community-based control of the water. The working of the waterlogged soil produces 'puddling' so that the soil becomes impervious to moisture and the loss of nutrients by downward moving water is minimized. So the *padi* is a fertile environment: a farmer may add 10–15 cartloads/ha of manure (the water buffalo is a normal concomitant of *padi* farming), ashes, straw, alluvium and perhaps green compost from leguminous crops. Further, there can be 16 t/ha fresh weight of algae which may accumulate 100 kg/ha of nitrogen, and other N_2-fixing organisms also exist in the root zone of the rice plants. Given that the flooded *padi* may also yield animal protein in the form of fish or crustacea, then its virtues as a food-producing system are convincing. As Geertz says of its environmental relations,

> . . . wet rice cultivation is essentially an ingenious device for the agricultural exploitation of a habitat in which heavy reliance on soil processes is impossible and where other means for converting natural energy into food are therefore necessary . . . here we have . . . the fabrication of an aquarium.

In order to achieve this floodable environment and to keep its fertility high, a number of management practices are used. Murton (1980) lists 26 soil management techniques which are applied in wet-rice cultivation in southern Asia, and we may take it that most if not all of them are common in other regions. Animal manure, human manure, oil seed waste, green manure and leaves from trees, ploughing, wet and dry bunding and even soil transfer are prominent in the list. Water management is equally complex: the natural hydrological cycle is manipulated at several points in order to provide the right quantity of water to the fields at the right time, with a correspondingly complex social framework to ensure the equitable distribution of the resource. Again, Murton lists 23 forms of water management, including storage wells and tanks, lifting devices, delivery systems, application practices, and ways of preventing the flooding of *padi* by estuarine or river waters in low-lying areas, especially in the deltas of the major river systems. The greatest environmental impact, however, comes through slope control, with terrace construction, the levelling of fields so that they can be evenly flooded, and the construction of dams in runoff streams to trap any valuable silt which might otherwise escape.

Thus at some stage in history, whole landscapes were realigned so that flows of energy, moisture and nutrients were channelled year in and year out into a relatively small area of cultivated land. As there was relatively little fallowing and not a great deal of suitable wild land, draught animals (mostly water buffalo) were relatively few and so nutrients other than those of domestic stock were very important. Hence the reliance upon night soil, the nutrients brought by the water, and the blue–green algae of the *padi* itself. Where major nutrient leaks in the *padi* system took place (for example, to urban centres in eastern Asia), then the cycle was restored by the use of night soil from the cities.

In summary, we can perhaps say that *padi* rice is not such a radical transformation of an entire landscape as is perhaps commonly thought: it exists in a matrix of less-manipulated ecosystems except in particularly suitable places like riverine plains. But its potential for environmental change extends outwards from the *padi* fields themselves, both into the settlements as providers of fertilizer, into dryland ecosystems which may be able to do the same, and on to surrounding hillsides which supply the vital water (often with important nutrients in solution or suspension) without which the crop could not be grown. Given a stable 'natural' and 'social' ecology, then, *padi* rice is a heavily yielding crop even under traditional conditions. Moreover, the output of most terraces can be increased by attention to detail: seeds are transplanted rather than broadcast, shoots are planted in exactly spaced rows, the terrace is periodically drained and ploughed, and better varieties allow double and even triple cropping. So in pre-industrial times any extra pressure put upon food supplies by population growth could very often be absorbed by applying that labour to the fine tuning of the *padi* system for extra output.

Later than the original Near Eastern domestications of plants and animals, we see the development of a new form of economy, based upon the use of herds of mammals as the major source of food, with camels, cattle, sheep and goats being the most usual. From these animals, the staple food is milk, and the animals represent a food storage and conversion mechanism which gives them a certain buffering against the effects of short-term aridity, especially since mixed herds provide an additional insurance. An environmental feature of pastoralism is that the people make no attempt to improve the pasture of their beasts, and may not practice selective breeding. With some species interbreeding with wild animals was still common in pre-industrial times, as with the camel and reindeer, for example.

This way of using domesticated animals is clearly different from the pattern which developed in Eurasia, where the animals were closely integrated into the agricultural system, providing manure for the fields of the sedentary cultivator in return for a share of the crops. The pastoralist instead exploited a niche marginal to the agricultural zone. This may have been literally marginal in the sense of occupying the interstices of the cultivated zone (odd patches of scrub and grassland, and the edges of field, road and village); or it may have been marginal at an entirely different scale, occupying the semi-arid hill lands peripheral to the great river valleys which were the foci of permanent agriculture. But common to both systems is the characteristic that one area of pasture will not suffice and that the animals must be moved, either from a permanent base or by moving the human settlement along with the animals. The environmental relations of near-recent pastoralists encompass a number of defineable strategies: movement is the most obvious of these, as when they seek to utilize seasonal pastures or leave a drought-stricken area (Spooner, 1973; Swift, 1977; Equipe Ecologie, 1981). Collection of evidence about traditional pastoralism in the Sahel zone suggests that there was little probability of degrading the environment to any marked degree. Pastures were scattered, small-capacity wells the main water source, and animal disease prevalent. Damage to pasture around the wells was limited and the vegetation easily restored during the next wet period. (The same cannot be said of damage to the pastoralists themselves, who seem to have suffered famine at regular intervals since the sixteenth century.) A singular cultural practice which increases environmental impact is the provision of live sheep and goats to the valley of Mina, near Mecca, for sacrifice by pilgrims. The animals needed nearly equal the number of pilgrims and they must be ready for the 10th–12th days of the month of Dhu'l-Hijja; but since the Muslim calendar is lunar, each date falls ten or twelve days earlier in each solar year, so that if the sacrifices fall near the end of the dry season then large numbers of animals have to be kept on pastures least able to sustain them. Grazing lands are therefore denuded of their vegetation, and pasture is expanded at the expense of agricultural land and forest, all in a climate with a long dry season that may end with violent rainstorms (Stewart, 1978).

Pastoralism of a less intensive kind can of course cause a great deal of environmental change when pursued over a long time: in the Mediterranean during classical times, shepherds were renowned as the chief agents of deforestation, using fire and animal herds to increase the areas available to their animals. Stabilization of this and other land-use practices occurred, however, and until the downfall of the Ottoman Empire a kind of equilibrium in the non-cultivated upland ecosystems was achieved, with regular but not excessive burning. Grazing and coppicing produced a mosaic of scrub, woodlands and grasslands, closely interwoven with cultivated land in the form of terraces or small patches of tilled land. The twentieth century, however, has often seen the disintegration of this equilibrium: heavier grazing and cutting of wood for fuel has produced many areas of dense scrub dominated by the unpalatable scherophyll *Pistacia lentiscus* which alternate with grassy patches of inedible low grasses and herbs (Naveh, 1982). Great damage has also been done to the forests on the valley slopes of northern India by pasture of goats and sheep; like most Indian pastoralism this is done from a village base and so impact is greatest nearest the settlement and declines outward. In drier areas of the sub-continent (e.g. in Rajasthan and Baluchistan), semi-nomadism has been practised, with the pastoralists stopping at places within their territory to grow a little wheat, some onions and to tend date palms and fruit trees, where a small stream can provide some irrigation. As in most places, these pastoralists were not self-sufficient and they traded with sedentary agriculturalists: exchanging dates for ghee in Baluchistan, for example (Leshnik and Southeimer, 1975). Only the nomads of the interior of Central Asia appear to have been completely self-contained; in South America the pastoralists integrate their economy with that of lower zones and there is perhaps more transhumance than nomadism; but mobility is high, aided by foods such as semi-dried meat, frost-dehydrated potatoes and parched maize (Webster, 1973).

Ecological change of varying degrees seems, therefore, to have been an inescapable accompaniment to pastoralism, and in many instances fire has been one of the tools used by the herders to increase the amount of forage (by encouraging early sprouting of fodder plants, for example, when other shading plants are burned off) in a pasture, or by extending scrub and grassland at the expense of woodland. Most pastoralism takes place in areas with a substantial dry season (Andean America is a notable exception) and so burning is rarely difficult; the impetus to carry it out in woodlands must have been even greater when forests held larger numbers of the carnivores which were potential predators of domestic herds: the wolf especially, but also the Asiatic lion, and the leopard for example. An example of reaction to this can be seen in the Edicts of As'oka, emanating from India *ca* 247–2 BC, where one law lays down that 'forest fire shall not be lit unnecessarily and with a view to killing living beings', and the prophet Joel encapulates some of the environmental problems of pastoralists in semi-arid climates when he says (1:20), 'the beasts of the field cry also unto thee: for the rivers of waters are dried up, and the fire hath devoured the pastures of the wilderness'. This and the later development of pastoralist economies led F. Fraser Darling (1955) to aver that subsistence pastoralism, if nomadic, might be able to last thousands of years, though not tens of thousands, for the land transformation—of plants, animals, soils and slopes—would be too great for it to remain without being changed into something else.

3.5 CONCLUSIONS

In conclusion, we might try to give a synoptic view of the degree of the making-over of the earth's surface by 1800 AD. General categories which might be proposed are:

(1) urban–industrial areas with intensive agriculture;
(2) areas of arable agriculture with considerable manipulation of soil and vegetation;
(3) areas of largely unmanipulated land with isolated areas of considerable manipulation, e.g. for minerals or forest;
(4) areas with some or little manipulation occupation occupied by hunter–gatherers, pastoralists or shifting cultivators in forests;
(5) virtually uninhabited areas.

An attempt is made at these divisions for the world in Fig. 3.9. If we except Antarctica and contrast categories (1)–(2) with (3)–(5), then we see that the

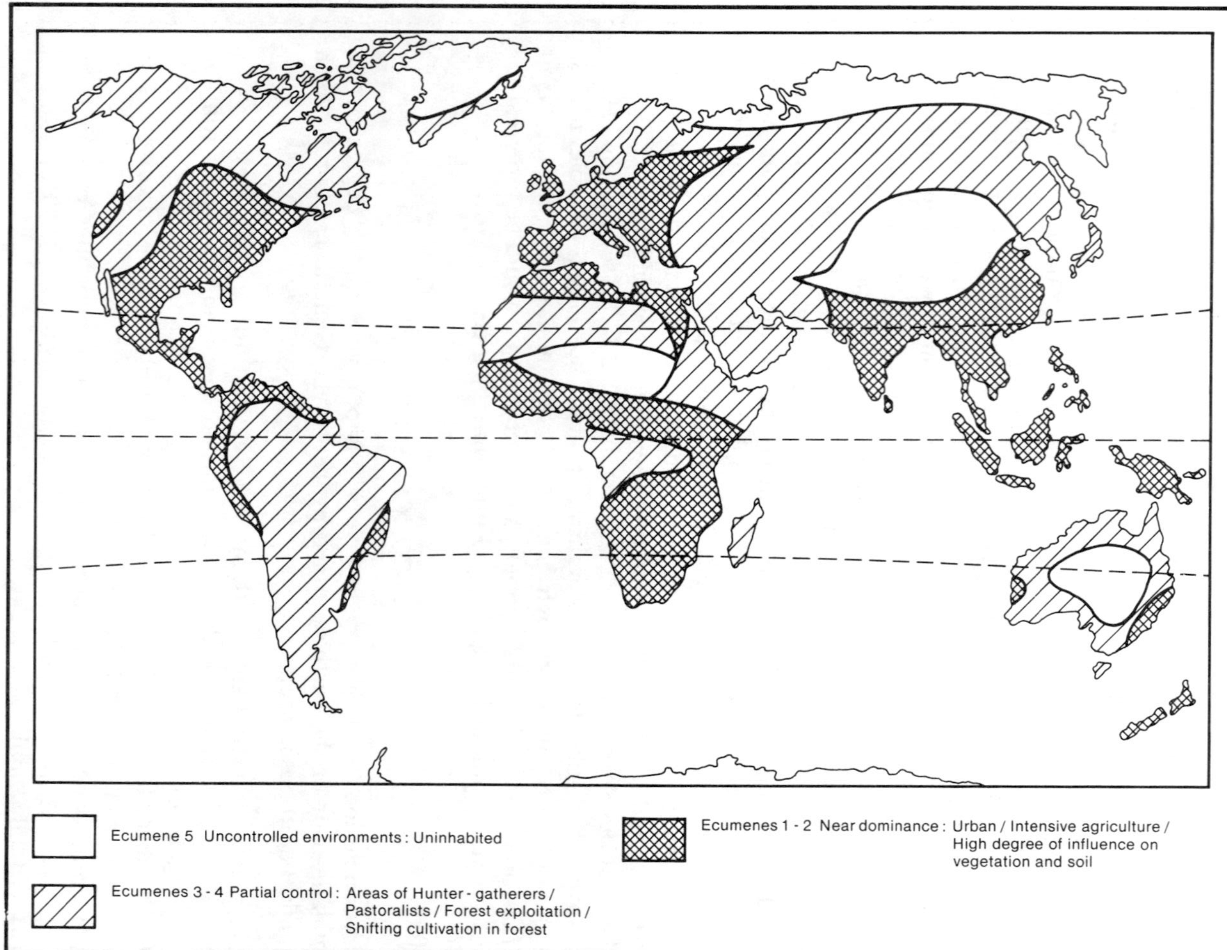

Figure [illegible] Estimation of the making over of the earth's surface by 1800 AD

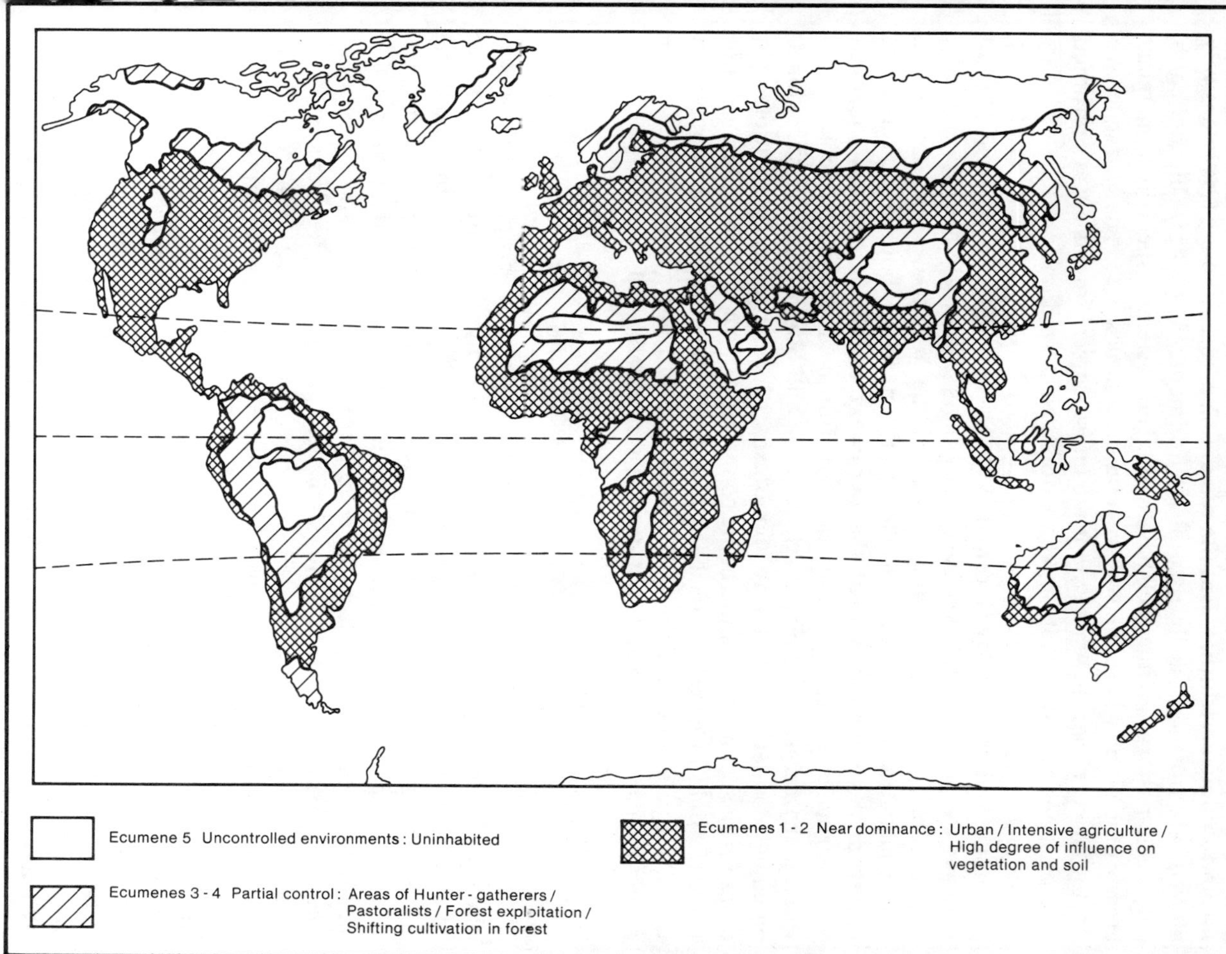

Figure 3.10 Estimation of the making-over of the earth's surface at the present day

highly manipulated areas are still relatively small, though made very large by the addition of the African savannas, considered as products of centuries of fire and shifting agriculture. But the remaining areas, although less intensively altered, are not necessarily natural environments, since fire, shifting, cultivation and pastoralism may well have altered ecosystems, especially in marginal areas of the globe where the soil–vegetation complex is especially fragile. We do not know, either, to what extent the moist tropical forests always returned to a pristine state after shifting cultivators had moved on.

This picture can be compared with Figure 3.10 which is an estimation of the position of the present day (Drew, 1983). As we would expect, areas (1) and (2) have expanded at the expense of the others. The world has experienced a considerable quantity of manipulation as the result of urbanization and industrialization this century (Hooke and Kaim, 1982), both in the centres of industrial development and at the periphery to which they have reached out—particularly during the period of European-dominated colonialism when many rural economies in Africa, Asia and South America were dominated by commercial agriculture and forestry which produced plantation-style crops for the economies of the home countries. So, whereas in 1800 the world pattern would in general look back to a predominantly rural economy with much self-sufficiency (albeit with a great deal of trade as well), by 1890 the major lineaments of today's pattern are well established.

3.6 ONWARD LINKAGES

To choose 1890 as a terminal date for this essay is to accept that there was already present in certain places a great deal of environmental impact, much of it *de facto* irreversible, caused by the early phases of the industrial revolution. In such cradles of industrialism as the English midlands, the economic nexus of coal, iron and textiles produced by deep mines and factories brought about places of heat, smoke and toxic wastes. The smoke and the toxic wastes that were discharged into the atmosphere and into watercourses wrought many changes, both on-site and at a distance. Streams were further diverted and upstream areas dammed to provide impoundments, and habitations for the burgeoning workforces replaced fields and other open areas.

Industrialization found its way into rural areas as well, particularly through the development of agricultural machinery (Gifford, 1981) and the application of chemical fertilizers. But for parts of England, western Europe (especially France, Belgium and Germany) and eastern North America, William Wordsworth's lines of 1814 sum up many of the changes:

. . . From the germ
Of some poor hamlet, rapidly produced
Here a huge town, continuous and compact,

Hiding the face of the earth for leagues—and there,
Where not a habitation stood before,
Abodes of men irregularly massed
Like trees in forests—spread through spacious tracts,
O'er which the smoke of unremitting fires
Hangs permanent and plentiful as wreaths
Of vapour glittering in the morning sun.
And, wheresoe'er the traveller turns his steps,
He sees the barren wilderness erased
Or disappearing;

So from such beginnings came the full force of the nineteenth and twentieth centuries in their alteration of plant and animal genetics, biotic communities, soils, slopes and watercourses and the application of industrialism to agriculture in the form of 'development', whose results are evident in many chapters of this book, and whose results we see—and are concerned about—today. By comparison the impact of human societies in the years before 1890 AD were often less thorough-going, though there are some exceptions, like the irrigation of the Nile, the Yellow River basin, the great rice terraces of south-east Asia, and the stability of agriculture in western Europe, which must rank alongside any later achievements. But they were by no means insignificant: they affected much of the earth's surface, often irreversibly, and perhaps more significantly supported a great variety of cultural groups on a sustainable basis in which a resilient *modus vivendi* between man and nature was achieved. Will we be able to say the same of the transformations of the late nineteenth and twentieth centuries?

3.7 REFERENCES

Drew, D. (1983) *Man–environment Processes*. London, Allen & Unwin.

Equipe Ecologie et Anthropologie des Societes Pastorales (1981) *Pastoral Production and Society*. Cambridge, University Press/Paris, CNRS.

Fraser Darling, F. (1955) Pastoralism in relation to populations of plants and animals. In J. B. Cragg and N. W. Pirie (Eds.), *The Numbers of Men and Animals*. Edinburgh and London, Oliver & Boyd, 121–128.

Frison, G. C. (1978) *Prehistoric Hunters of the High Plains*. New York, Academic Press.

Geertz, C. (1964) *Agricultural Involution: the Processes of Change in Indonesia*. Berkeley and Los Angeles, University of California Press.

Gifford, R. C. (1981) *Agricultural Mechanization in Development: Guidelines for Strategy Formulation*. Rome, FAO Agricultural Services Bulletin 45.

Gillespie, R., Horton, D. R., Ladd, P., Macumber, P. G., Rich, T. H., and Wright, R. V. S. (1978) Lancefield Swamp and the extinction of the Australian megafauna. *Science*, **200**, 1044–48.

Gowlett, J. A. J., Harris, J. W. K., Walton, D., and Wood, B. A. (1981) Early archaeological sites, hominid remains and traces of fire from Chesowanja, Kenya. *Nature* (London), **294**, 125–129.

Harris, D. R. (1977) Subsistence strategies across Torres Strait. In J. Allen, J. Golsen

and R. Jones (Eds.), *Sunda and Sahul. Prehistoric Studies in Southeast Asia, Melanesia and Australia*. New York, Academic Press, 421–463.
Harris, D. R. (1978). The agricultural foundations of lowland Maya civilzation: a critique. In P. D. Harrison and B. L. Turner (Eds.), *Pre-hispanic Maya Agriculture*. Albuqerque, University of New Mexico Press, 301–323.
Hooke, J. M., and Kain, R. J. P. (1982) *Historical Change in the Physical Environment: a Guide to Sources and Techniques*. London, Butterworth.
Hughes, J. D. (1975) *Ecology in Ancient Civilizations*. Albuqerque, University of New Mexico Press.
Iversen, J. (1949) The influence of prehistoric man on vegetation. *Danm Geol. Unders*., Ser. 4, **3**, (6), 1–25.
Jones, R. (1975) The Neolithic, Palaeolithic and the hunting gardeners: man and land in the Antipodes. In R. P. Suggate and M. M. Cresswell (Eds.), *Quaternary Studies*. Wellington, Royal Society of New Zeland, 21–34.
Leakey, R. E., and Lewin, R. (1977) *Origins*. London, Macdonald and Jane's.
Leakey, R. E., and Lewin, R. (1979) *People of the Lake: Man: his Origin, Nature & Future*. London, Collins.
Leshnik, L. S., and Sontheimer, G.-D. (Eds.) (1975) *Pastoralists and Nomads in South Asia*. Wiesbaden, Otto Harrassowitz.
Lewin, R. (1983) Who killed the giant mammals? *Science*, **221**, 1036–1037.
Lourandos, H. (1980) Change or stability? Hydraulics, hunter–gatherers and population in temperate Australia. *World Archaeology*, **11**, 245–264.
Martin, P. E. (1967) Prehistoric overkill. In P. S. Martin and H. E. Wright (Eds.), *Pleistocene Extinctions: the Search for a Cause*. New Haven, Yale University Press, 75–120
Martin, P. E. (1973) The discovery of America. *Science*, **179**, 969–974.
Matheney, R. T. (1982) Ancient lowland and highland Maya water and soil conservation strategies. In K. V. Flannery (Ed.), *Maya Subsistence: Studies in Memory of Denis E. Puleston*. London, Academic Press, 157–178.
Mellars, P. A. (1976) Fire ecology, animal populations and man: a study of some ecological relationships in prehistory. *Proc. Prehist. Soc*., **42**, 15–45.
Mosimann, J. E., and Martin, P. S. (1975) Simulating overkill by Palaeoindians. *American Scientist*, **63**, 304–313.
Murton, B. J. (1980) South Asia. In G. A. Klee (Ed.), *World Systems of Traditional Resource Management*. London, Edward Arnold/New York, Halstead Press, 67–99.
Naveh, Z. (1982) Mediterranean landscape evolution and degradation as multivariate biofunctions: theoretical and practical implications. *Landscape Planning*, **9**, 125–146.
Parsons, J. J., and Denevan, W. M. (1967) Pre-Columbian ridged fields. *Scientific American*, **217**, 92–100.
Reed, C. A. (1970) Extinction of mammalian megafauna in the Old World late Quaternary. *BioScience*, **20**, 284–288.
Richards, K. S. (1981) Evidence of Flandrian valley alluviation in Staindale, North York Moors. *Earth Surface Processes and Landforms*, **6**, 183–186.
Simmons, I. G., and Innes, J. B. (1981) Tree remains in a North York Moors peat profile. *Nature* (London), **294**, 76–78.
Spencer, J. E., and Hale, G. A. (1961) The origin, nature and distribution of agricultural terracing. *Pacific Viewpoint*, **2**, 1–40.
Spooner, B. (1973) *The Cultural Ecology of Pastoral Nomads*. Reading, Mass., Addison-Wesley.
Swift, J. (1977) Sahelian pastoralists: underdevelopment, desertification and famine. *Ann. Rev. Anthropol*., **6**, 457–478.

Tallis, J. H. (1975) Tree remains in southern Pennine peats. *Nature* (London), **256**, 482–484.

Turner, C. (1975) The correlation and duration of Middle Pleistocene periods in Northwest Europe. In K. W. Butzer and G. L. Isaac (Eds.), *After the Australopithecines*. The Hague & Paris, Mouton, 259–308.

Webster, S. (1973) Native pastoralism in the south Andes. *Ethnology*, **12**, 115–133.

West, R. G. (1980) Pleistocene forest history in East Anglia. *New Phytol*., **85**, 571–622.

Wiseman, F. M. (1978) Agricultural and historical ecology of the Maya lowlands. In P. D. Harrison and B. L. Turner (Eds.), *Pre-hispanic Maya Agriculture*. Albuquerque, University of New Mexico Press, 63–115.

Land Transformation in Agriculture
Edited by M. G. Wolman and F. G. A. Fournier

CHAPTER 4
The Industrial Revolution and Land Transformation

DAVID GRIGG

4.1 INTRODUCTION

For most of human history man has been a hunter and gatherer; although his use of fire, and local over-population, may have altered the environment, the consequences of his actions were small compared with those of farmers. Plants and animals were first domesticated some 12–10 000 years ago, and sedentary agriculture slowly became established in most parts of the world.

Agriculture inevitably transforms the land. Crop production requires the removal of the natural vegetation, or its drastic modification. Thus even shifting cultivation at low population densities, with periods of fallow in which forest is allowed to regenerate, causes changes in the species composition and structure of the secondary vegetation. Most other forms of crop production require the almost complete removal of the natural vegetation; in areas of cereal production in eastern England, not only has the original natural vegetation long been removed, but even the hedges and isolated trees that were planted around arable fields after enclosure are now being destroyed. The slow expansion of cultivation has greatly reduced the forests which covered so much of medieval Europe (Figures 4.1 and 4.2), and the woodland of all the great agricultural civilizations has been much reduced. In this century alone the forest area of Afro-Asia and North America has been halved (World Bank, 1982, p. 60).

Arable farming, particularly where the plough is the main means of cultivating, has profoundly altered the soil by both adding and removing plant nutrients, reducing acidity with lime, draining excess soil moisture with underground pipes, removing stones, and changing the soil structure. More fundamental changes have been brought about by the removal of surface water. Long and painstaking drainage schemes, notably in the English fenlands, the Dutch polders and inland lakes, and in the Po valley of northern

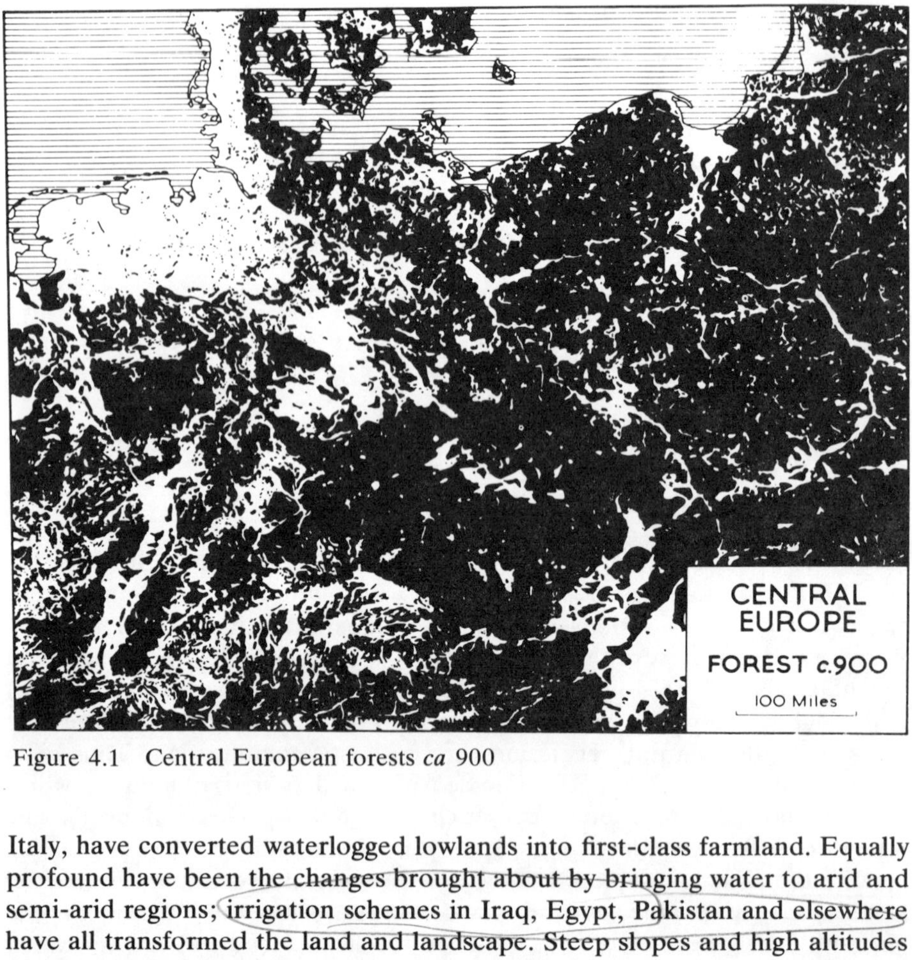

Figure 4.1 Central European forests *ca* 900

Italy, have converted waterlogged lowlands into first-class farmland. Equally profound have been the changes brought about by bringing water to arid and semi-arid regions; irrigation schemes in Iraq, Egypt, Pakistan and elsewhere have all transformed the land and landscape. Steep slopes and high altitudes present great obstacles to the farmer, but terracing has made arable farming possible in many parts of upland Asia and North America.

Agriculture, in short, transforms the environment; other chapters in this book deal in detail with these matters. The purpose of this chapter is not to describe how agriculture changes the environment, but to explain why changes in agriculture have come about.

4.2 MAJOR CAUSES OF AGRICULTURAL CHANGES

At present agriculture exists in a bewildering variety of forms, ranging from the simple shifting cultivation of parts of the Amazon basin to the technologically complex systems of the United States and Western Europe based upon advances in biological science and the rise of modern manufacturing indus-

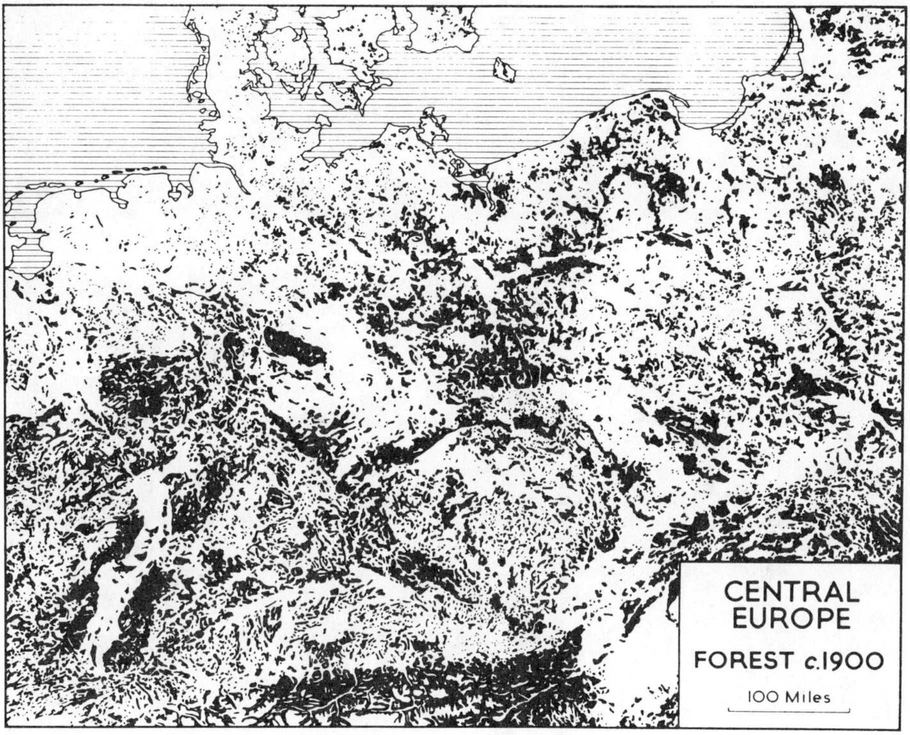

Figure 4.2 Central European forests *ca* 1900

tries. The origins of such farming systems can hardly be reduced to one or two major causes. Most of this chapter will be devoted to a few forces that have been important in causing change over time, particularly in Europe: population growth, urbanization, industrialization, transport changes and the role of science and the state. A smaller section deals with some of the major types of farming and their evolution.

4.2.1 Population Growth

Agriculture must provide food supplies if population growth is to be sustained; conversely population growth requires changes in the way agriculture is carried on, prompting the expansion of the area under cultivation and the more intensive use of land already in cultivation (Boserup, 1965, 1981). Both cause land transformation.

Human beings have occupied the earth for some 250 000 years, and until the last 12 000 years they were few in number and population densities were everywhere very low. The domestication of crops and livestock some 12 000

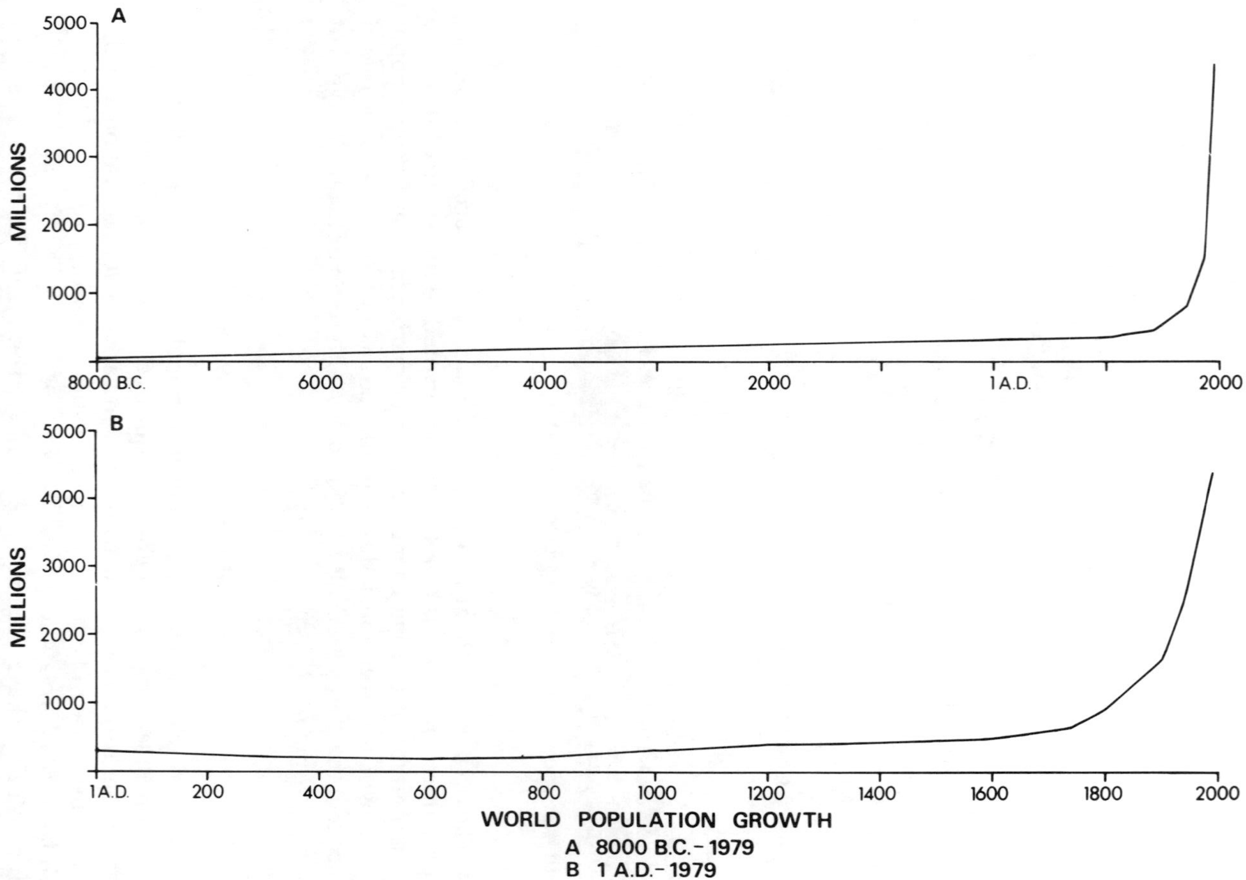

Figure 4.3 World population growths

years ago made higher densities possible and numbers began to increase, albeit very slowly (Figure 4.3). Accurate information on population numbers is not available until the eighteenth century, when the first regular population censuses began to be taken; many countries had no reliable censuses until this century. Nonetheless, most population historians would agree that the rate of world population increase was very low until the eighteenth century. Since then the rate of increase has progressively risen until the last decade (Table 4.1). Until the nineteenth century, increase was not continuous. Periods of sustained increase were separated by periods of stagnation or even decline. Such periods have been identified in Europe. There was a decline in numbers after the fifth century, increase from *ca* 950 to 1300, decline from the mid-fourteenth century to the mid-fifteenth century, increase from then until the early seventeenth century, followed by slow increase or stagnation until the continuous and rapid increase that began in the early eighteenth century and only faded in the 1920s (Helleiner, 1967). China also has a history of growth alternating with decline (Ho, 1959). Little is known of Africa's population history before the nineteenth century, but in Latin America there was a period of sharp decline after the arrival of Europeans in the sixteenth century, and growth may not have commenced again until the eighteenth century (Durand, 1977).

Table 4.1 World population increases (Clark, 1967; United Nations, 1982)

Year	Population (millions)	Average rate of increase since preceding date (% per annum)
8000 BC	5–10	–
AD 14	256	0.04
600	237	−0.01
1000	280	0.04
1340	378	0.1
1600	498	0.1
1650	516	0.1
1700	641	0.4
1750	731	0.3
1800	890	0.4
1850	1171	0.6
1900	1668	0.7
1950	2486	0.8
1960	2982	1.8
1970	3632	2.0
1980	4414	1.8

It was formerly argued that the great modern upsurge in numbers began first in North West Europe in the eighteenth century and was due to a decline in mortality, without initially any fall in the high crude birth rates. The fall in mortality was thought to be due to improvements in the food supply, the control of contagious diseases, and rising standards of living associated with the industrial revolution. This view has been sharply criticized in recent years. It has been shown that advances in medical treatment had little effect on mortality until the twentieth century (McKeown, 1970). It has also been shown that in England a rise in the birth rate was as important as the fall in mortality between 1750 and 1838 (Wrigley and Schofield, 1981); nor was the increase in European areas confined to those states which underwent industrialization, for Ireland, Sweden and Norway had rapid increases in their populations in the late eighteenth century and the first two-thirds of the nineteenth century without experiencing industrialization (Connell, 1950; Drake, 1969; Thomas, 1941). Nor was the increase confined to Europe. J. L. Durand's estimates of world population numbers suggest that the eighteenth century saw a comparable increase in Russia, Asia and Latin America, although not in Africa until the late nineteenth century (Table 4.2, Figure 4.4). Europe, Russia and the areas of European settlement in North America and Australia did increase more rapidly than Afro-Asia in the nineteenth century, so that they made up only a quarter of the world's population in 1750, but a third by the end of the nineteenth century. In this century non-European areas have increased more rapidly than European, particularly since 1950, so that Afro-Asia and Latin America again constitute three-quarters of the world's population as they did in 1750.

Table 4.2 Rates of population increase by major regions, 1750–1979 (% per annum) (Durand, 1977, pp. 253–296; United Nations, 1982)

	1750–99	1800–49	1850–99	1900–49	1950–79	1750–79
Africa	0.01	0.07	0.4	1.0	2.6	0.6
Asia	0.5	0.5	0.3	0.8	2.1	0.7
Latin America	0.8	0.9	1.3	1.6	2.7	1.4
Developing	0.4	0.5	0.3	0.9	2.3	0.7
Europe	0.4	0.6	0.7	0.6	0.7	0.6
(W. Europe)	0.5	0.8	0.5	0.4	0.7	0.6
USSR	0.6	0.6	1.1	0.6	1.3	0.8
N. America	2.5	2.7	2.3	1.4	1.3	2.1
Australia and New Zealand	–	–	1.4	1.8	1.8	2.9
Japan	–	0	0.8	1.3	1.1	0.6
Developed	0.4	0.7	1.0	0.8	1.0	0.75

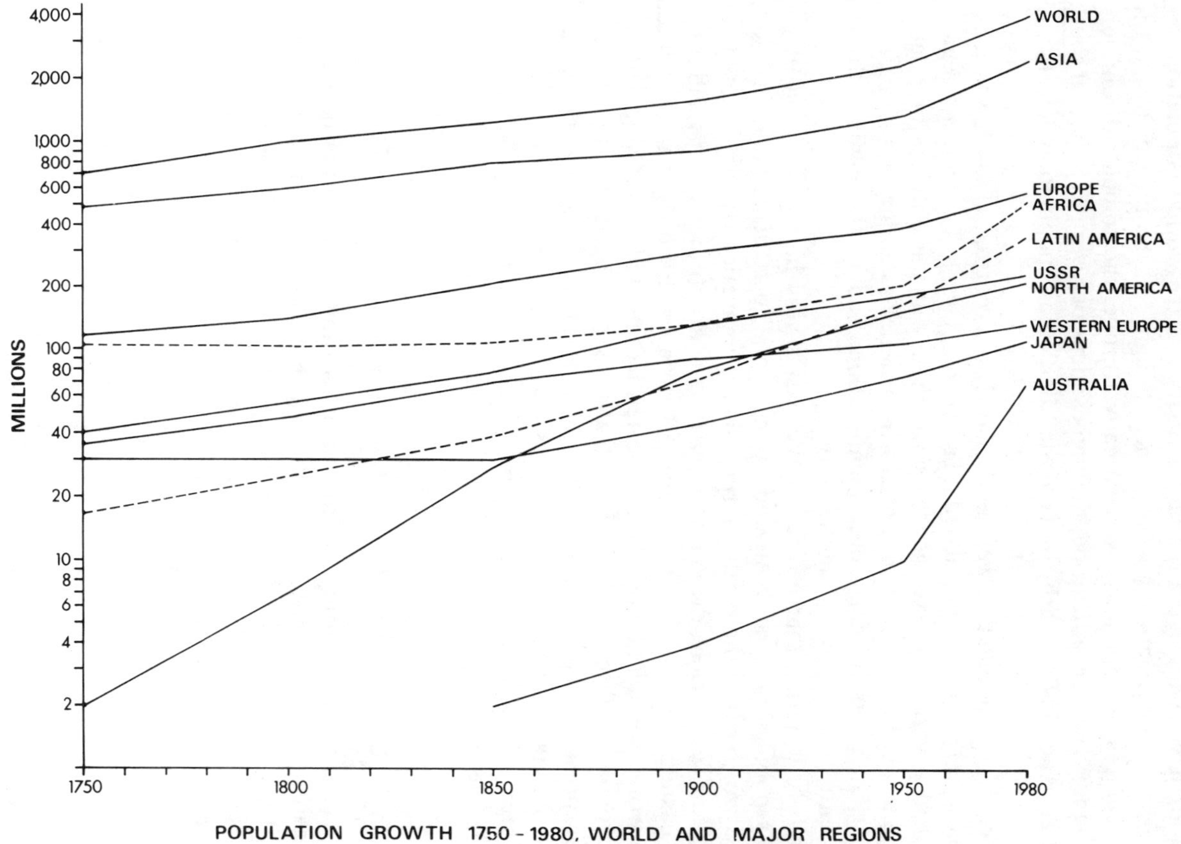

Figure 4.4 Population growths for major regions

4.2.1.1 Population Distribution

Regional variations in mortality and fertility combined with long-distance migration have changed the spatial distribution of the world's population, but not perhaps so dramatically as was once thought. At the time of Christ the major centres of population were already Europe, the Indian subcontinent and China; however, in Europe, the highest densities were to be found south of the Alps and in the nearby parts of the Middle East; population density and numbers were low in northern Europe (United Nations, 1953). In China the great centres of population were in the north and centre; the south was as yet comparatively sparsely populated (Trewartha, 1969). The world's regions have had a variety of experiences since then, but the only major relative change has been the emergence of North America as an important centre of population, a result of European emigration, and in the eighteenth and nineteenth centuries, very high fertility.

It is worth noting that now as in the past a very high proportion of the world's population is concentrated upon a small part of its area. In 1960, nearly two-thirds of the earth's surface had a population density of less than 13 per km^2; conversely, the major population concentrations—east and south Asia, eastern North America, European Russia and Europe—had two-thirds of the world's population upon only one-tenth of its land area (Grigg, 1969b). The proportion of the earth's surface that is used for agricultural purposes is also surprisingly small. In 1980 only 11.1% was used for growing crops and rotational grasses; however, 23.8% was used for grazing livestock, including large areas of prairie, savanna and scrub largely unimproved by the pastoralist (FAO, 1982).

4.2.1.2 The Consequences of Population Growth

Until the nineteenth century the greater proportion of the world's population were dependent upon agriculture for their livelihood; indeed as late as 1900, 75% of the world's labour force were employed in agriculture. Furthermore, the great majority of these were subsistence farmers. Their main aim was to provide food for their families, and comparatively little of their output left the farm. Any increase in numbers left the farmer with two possibilities of increasing the food supply. Either he could intensify output by increasing the yields of his crops, or he could bring land hitherto unused into cultivation. Until the nineteenth century the possibilities of increasing yields significantly over a short period were very limited, and the primary response to population growth was an expansion of the area in crops. There are few reliable statistics on land use at the national level before the nineteenth century; but such as do exist, together with contemporary documents, suggest that the expansion of the arable area more or less matched the growth of population except in periods of particularly rapid growth such as the late thirteenth century in Europe or in eighteenth century China (Grigg, 1980). Over the last 100 years, however, arable expansion has not matched population growth in some regions, and area expansion has been increasingly supplemented with yield increases.

The way in which the extra land has been brought into cultivation has varied from region to region. After the migrating movements of the first millenium after Christ, the main features of European settlement were established, and subsequent increases in cropland came from clearing the woodland between villages when numbers rose, and advancing into areas which were initially thought unsuitable for farming. The high middle ages thus saw much settlement of upland areas in western Europe and the reclamation of poorly drained land. There were few agricultural frontiers in early modern Europe, although the settlement of the plains east of the Danube in Hungary in the eighteenth and nineteenth centuries justify this description. In the first millenium after Christ there were, however, major movements into new land in China and Africa. In AD 2 the most densely populated part of China was the lower Hwang Ho, and 43 million of the 58 million in the Han empire lived north of the Yangtse; south of the Yangtse there were few Han Chinese. Over the next 1000 years the Han Chinese occupied the south and the distinctive wet rice economy emerged. By the fourteenth century the major rice growing areas of China had emerged with double cropping, irrigation and intensive cultivation. Over the same period farming pushed northwards in Japan from its source area around the Inland Sea. In Africa, Ethiopia and the West African savannas were early areas of agricultural development, whilst much of eastern and southern Africa remained occupied by hunters and gatherers. It is believed that the Bantu-speaking peoples, originating in what is now Nigeria, spread slowly east and south, bringing with them cattle, sorghum and other crops and practising shifting cultivation. By the eleventh century they had reached what is now the Transvaal (Grigg, 1974).

These great migrations into sparsely occupied areas, prompted by the slow increase in population, took over 1000 years to achieve. But the upsurge of

Table 4.3 The expansion of crop land, 1860–1960 (million hectares) (Grigg, 1974, p. 261)

	1860	1900	1930	1960
United States	65	128	166	158
Russia	49	113	109	196
Canada	–	8	23	25
Argentina	–	6	24	22
Australia	0.4	3	10	12

population in the eighteenth and nineteenth centuries was accompanied by much more rapid migration and increases in the cultivated area, principally by Europeans. The movement of Europeans to eastern North America, Australasia, Argentina and Uruguay, South Africa, and of Russians from the long occupied forest lands into the grasslands of southern Russia and later east of the Urals, led to remarkable increases in arable area (Table 4.3), and in the early part of this century there was a major Chinese movement north into the grasslands of Manchuria.

4.2.1.3 The Intensification of Output

Prior to the nineteenth century yield increases were slow. Between 1300 and 1800 the yields of wheat and rice possibly doubled in England and China, and were much slower elsewhere. Thus arable expansion was the primary response to population increase. There were other ways of increasing output, however, again prompted by population increase. Most traditional farming systems have a period of fallow in the cropping sequence, to help maintain soil fertility, to allow the removal of weeds, and to control plant diseases. The length of fallow is variable; in some shifting agriculture systems it may be as long as 30 years, which allows the regeneration of a forest cover. But increases in population density lead to a reduction in the area in fallow. In medieval Europe between one-half and one-third of the arable land was in fallow. Although eliminated in much of the Low Countries and parts of England by the seventeenth century, fallow persisted in much of Europe until well into the nineteenth century. Slowly legumes, potatoes and sugar-beet replaced the fallow, the root crops allowing the removal of weeds during growth and fodder for livestock, which in turn provided farmyard manure for maintaining crop yields, whilst clover, beans and peas all added nitrogen to the soil (Chorley, 1981; Slicher van Bath, 1963).

In the rice economies of Asia, the occurrence of blue–green algae in the rice paddies helped maintain yields over long periods, and annual cropping without a fallow was common from an early time; indeed, double cropping became possible in some regions where the growing season was long enough and irrigation efficient. Thus the introduction of rice varieties with a shorter growing season allowed double cropping in Southern China in the eighth century (Ho, 1956). In China as in most other parts of the world increases in yields were only possible by using more manure, by the more careful selection of seed, and above all by the use of more labour in weeding and harvesting. Not until the mid-nineteenth century did industrial inputs such as chemical fertilizers become available to the farmer, and the breeding of higher yielding crop varieties has occurred mainly in this century.

Both area expansion and more intensive cultivation have obviously transformed the land, not always for the better. In the nineteenth and twentieth centuries frontiers of settlement ventured into land too arid for permanent settlement without irrigation, as in the Great Plains of the United States in the 1930s and in Kazakhstan in the USSR in the 1950s. But in the nineteenth and twentieth centuries change in agriculture has been prompted by profit as much as by population increase.

4.2.2 Urbanization and Agricultural Change

Whilst general population growth has been a major cause of both the expansion of the arable area and the intensification of farming, changes in the distribution of population have also been important; and in particular urban growth (the increase in the population of towns) and urbanization (an increase in the percentage of the population living in towns).

Until the modern period yields per hectare and output per man were low in Europe—and elsewhere—and this meant that few societies could support many people who were not occupied in agriculture. Thus towns were comparatively few in number, very large towns were rare (Table 4.4), and only a small proportion of the population were engaged in activities other than agriculture, and the urban population were a small proportion of the total population (Table 4.5). Thus in the late eighteenth century four-fifths of the population of France were engaged in agriculture, and this proportion was only significantly lower in Britain and the Low Countries.

The smallness of Europe's urban population was a function not only of low productivity in agriculture, but also of the high cost of transporting food; overland costs were particularly high, and grain would rarely be moved more than 35 km; costs of movement by water were appreciably lower. Not surprisingly most of Europe's towns were on rivers, and the great cities before the nineteenth century were invariably ports, for the cost of shipping foodstuffs by sea was approximately one-eighth that of overland movement (Grigg, 1982b).

The growth of towns—and particularly either clusters of towns, as in the Low Countries, or very large towns such as Paris, London, Naples or

Table 4.4 Number of towns of particular sizes in Europe, 1500–1800 (J. de Vries, 1981)

	1500	1600	1700	1800
Over 100 000	4	8	11	17
50 000–99 999	7	11	20	32
20 000–49 999	46	73	74	109
10 000–19 999	99	125	116	205

Table 4.5 Proportion of total population of Europe in towns of 10 000 and over (J. de Vries, 1981)

	1300	1500	1600	1700	1800	1850	1890
North and West	4.4	6.0	8.1	13.0	14.7	25.1	45.3
Central	3.6	3.8	4.8	6.7	7.1	18.2	27.8
Mediterranean	8.7	9.7	13.2	11.7	12.9	18.0	20.7
All Europe	5.5	6.1	8.0	9.5	10.6	20.1	31.0

Rome—had a profound influence upon the agriculture of their hinterland. The demand for food in the towns meant that there was a ready market, and the commercialization of agriculture first occurred in these regions. The high value of land meant that small farms predominated and the fallow was soon extinguished, for soil fertility could be maintained by bringing horse manure and other wastes from the town. Farming was intensive, and the products generally high-value goods. These regions were among the first to adopt innovations, had access to capital from the towns, and were generally the advance guard of agricultural improvement. This was notably so in northern Italy in the fourteenth and fifteenth centuries, in the Low Countries in the fifteenth, sixteenth and seventeenth centuries, and in the hinterlands of London and Paris in the eighteenth and nineteenth centuries. Wilhelm Abel has described these small regions (Abel, 1980). And small their influence was. In Figures 4.5 and 4.6 the circles represent not only the relative population size of towns over 10 000, but also the area of farmland which was necessary to feed the towns. In the year 1500 (Figure 4.5) only a small proportion of Europe was directly influenced by towns; even in 1800, on the eve of the industrial revolution, large areas remained beyond the innovatory influence of the cities (Figure 4.6).

4.2.3 Industrialization and Agricultural Change

Since 1800 the developed countries of the world have all gone through the process of industrialization, and this has had dramatic consequences for agriculture.

First industrialization was everywhere accompanied by rapid population growth, which in itself increased the demand for food. But in addition, industrialization led to an increasing proportion of the population living in towns, and thus dependent upon the agricultural population for their food. This encouraged the commercialization of agriculture and the decline of subsistence farming.

Second, industrialization led to slow increases in the incomes of nearly all

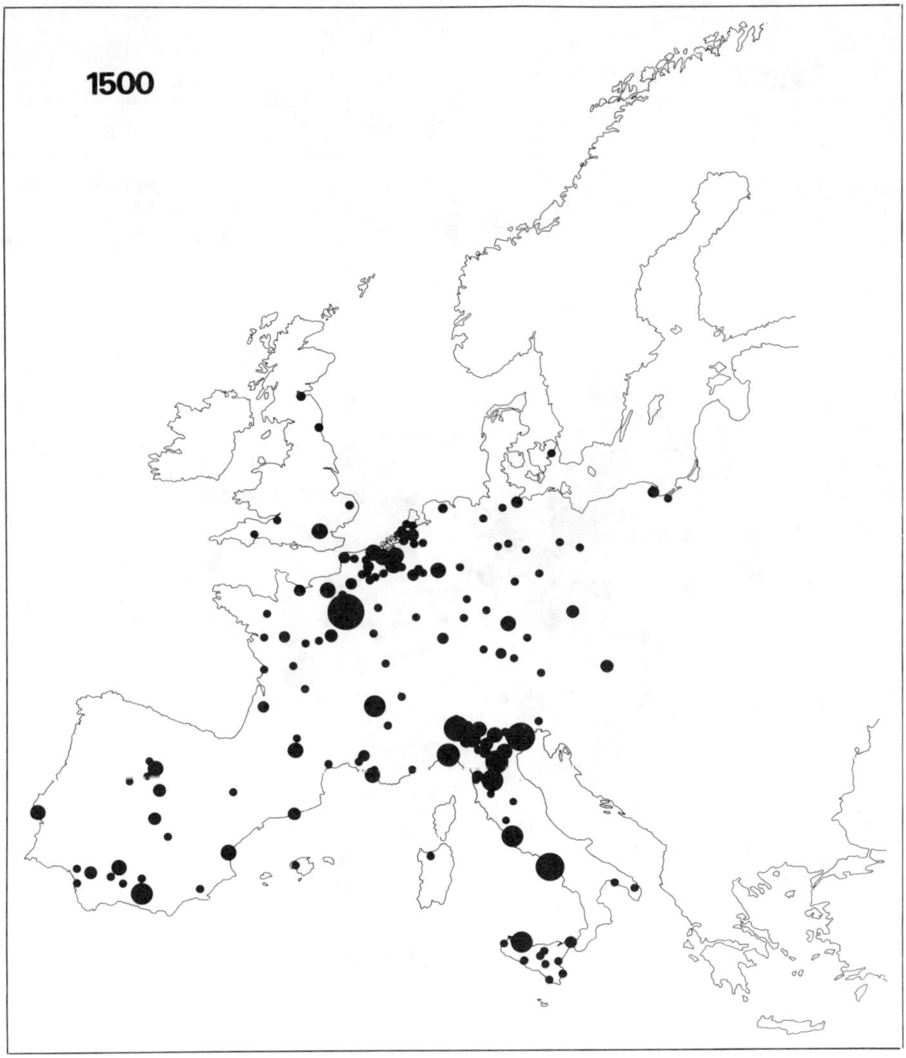

Figure 4.5 The relative population sizes of towns of more than 10 000 people, *ca* 1500

classes. As long as the bulk of the population had wages little above the subsistence level, then most of the diet had to come from cheap foodstuffs (mainly cereals and potatoes). Indeed, as late as 1880, 70% of the French calorific intake came from bread and potatoes (Toutain, 1971). But as incomes rose so the demand for more expensive foodstuffs, and in particular livestock products, increased. From the end of the nineteenth century livestock products became an increasing proportion of the diet in North America

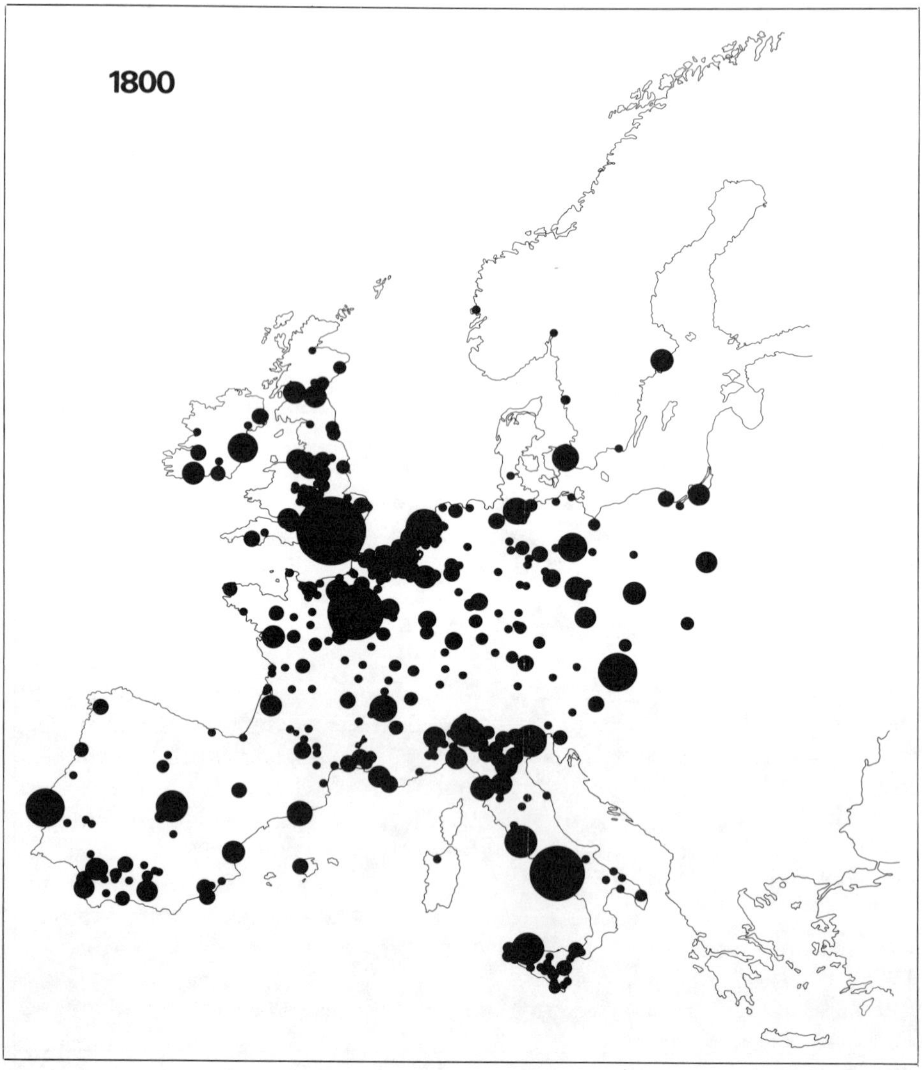

Figure 4.6 The relative population sizes of towns of more than 10 000 people, *ca* 1800

and Western Europe, and as a result farming turned increasingly to the production of livestock products which now make up between 55% and 85% of the value of total farm output in the developed countries. This has not led to any decline in the importance of cereals in the agriculture of these countries. Livestock are fed increasingly upon grain; in the developed countries 70% of all grain is used as livestock feed (Aymard, 1975; Teuteberg, 1975).

Third, the first industrial revolution saw a shift in the raw materials of industry from the organic to the inorganic; even so agricultural products were increasingly needed by manufacturing industry. The development of machinery in the textile industries in Western Europe combined with income increases and population growth led to an almost insatiable demand for cotton and wool, satisfied only by the westward expansion of cotton-growing in the United States, and later more widely in the subtropics, whilst Australia and New Zealand became sheep farms for Bradford. Vegetable oils became of major importance, not only for cooking, but before the exploitation of petroleum, as lubricants, as the basis for paints and for many other industrial purposes. With the development of the electrical engineering industry, and later motor vehicles, rubber moved from a minor curiosity to a major raw material, and in the early twentieth century rubber expanded prodigiously in south-east Asia (Grigg, 1974).

A further consequence of industrialization was that farming ceased to provide all its own inputs. In traditional agriculture farmers got their seed from their own harvest, their manure from their livestock, feed for animals from grass or roots grown on their own farm. Implements were bought locally, generally made by craftsmen, and power came from human and animal muscle. But from the 1830s and 1840s farmers increasingly bought more and more of their inputs from off the farm, much of these from manufacturing industry. In the mid-nineteenth century the theory of Liebig in Germany and the practice of J. B. Lawes in England led to the beginnings of the modern chemical fertilizer industry. At much the same time the availability of cheap iron and the need for more powerful equipment led to the rise of the modern agricultural implements industry which from the 1840s provided iron ploughs, drills, reapers, steam threshing engines, in the 1870s reaper-binders and elevators, in the 1890s the first milking machines, combine harvesters and tractors; much of the initiative now came from the United States, although advances in dairying began in Western Europe (Ojala, 1952; Whetham, 1970; Danhof, 1972).

The dependence of the farmer upon purchased inputs accelerated after the end of the Second World War as pesticides were widely adopted to control disease and herbicides replaced the hoe and harrow; fertilizer consumption has risen dramatically in the developed countries and an extraordinary variety of implements is available to sow and harvest crops. Whereas in the 1930s horses and human muscle provided most of the power on farms in Europe and North America, electricity and the tractor provide nearly all now.

Industrialization has transformed the role of agriculture in the economies of the western world. In the early eighteenth century farmers and farm workers made up three-quarters or more of the labour force in nearly every country. But during the industrialization of the nineteenth century the numbers employed in new manufacturing industries, in transport, and in mining

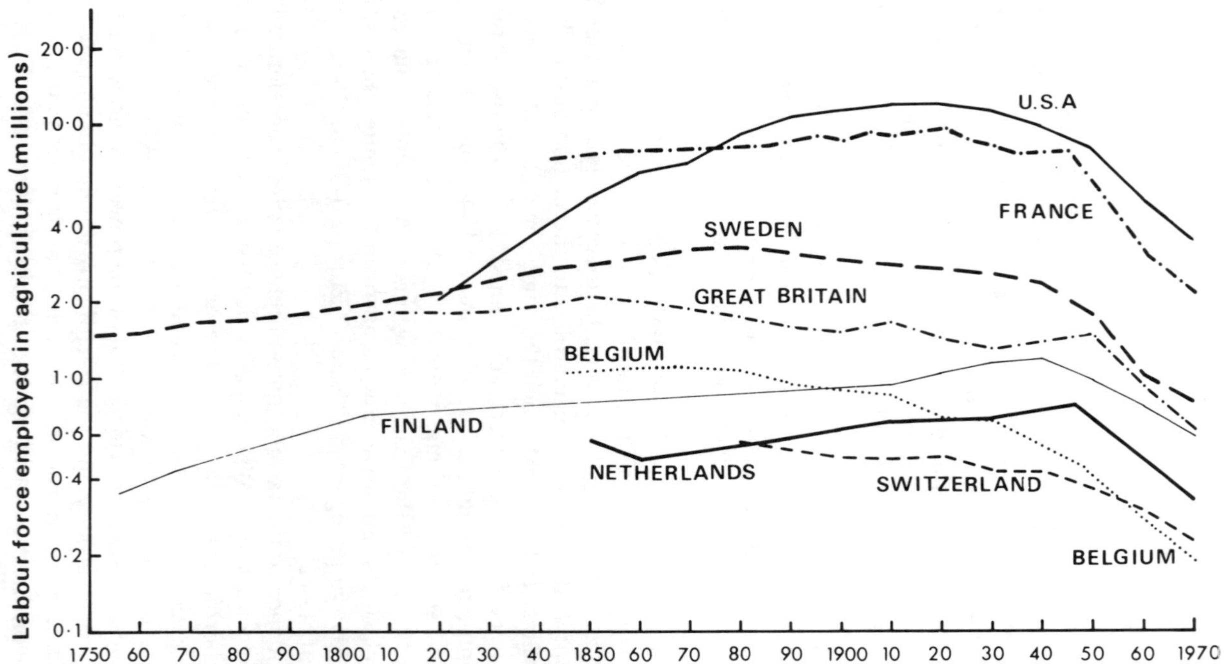

Figure 4.7 Trends in the labour force employed in agriculture

increased far more rapidly than the numbers in agriculture; in the present century this trend has continued as the number in administration and other services has risen. Consequently the proportion of the workforce employed in agriculture has greatly declined. In Sweden, Switzerland, the United Kingdom and the United States it is now no more than 2% (Dovring, 1959; Grigg, 1975). In pre-industrial Europe agriculture provided most of the wealth: now agriculture produces a small proportion of the gross domestic product in most developed countries. In the European Economic Community farming employs only 8% of the workforce and produces but 4% of the gross domestic product.

Although the proportion of the population engaged in agriculture has been declining in Western Europe and North America since the early nineteenth century, the absolute numbers continued to increase for much of the century, for natural increase in the rural areas exceeded emigration to the urban areas. But in the late nineteenth century the labour force began to decline slowly at first, at a dramatic rate after 1945 (Figure 4.7). This has led to pronounced changes in agriculture. Capital, in the form of machinery, has been substituted for labour. Farms have become bigger in order to exploit fully the capabilities of tractors and combine harvesters. Labour productivity has increased rapidly—indeed for the last three decades more rapidly than in manufacturing industry. Whereas on the eve of the industrial revolution it took four men employed in agriculture to feed one man in non-agricultural activities, one man in American agriculture now feeds fifty or sixty in other jobs (Cochrane, 1979; Rasmussen, 1982; FAO, 1972).

4.2.4 Transport and Agricultural Change

For much of human history the transport of agricultural goods over long distances has been impractical because the means of transport have been slow and many foodstuffs are perishable, and because the cost of moving commodities in bulk has been prohibitively high. This had many consequences for farming and the economy in general. It limited the growth of cities because the area needed to feed urban non-agricultural populations was large with the low productivity of agriculture before the nineteenth century; the cost of movement soon equalled the price of the crop transported. It made regional specialization in one or two products for which an area had physical advantages difficult, for each area had to produce as much of its food supply and fibres as possible. It meant that comparatively little agricultural produce entered into trade unless the product had a high value per unit weight (such as wool) or could be moved in the relatively cheap coastal shipping trade. Early regional specialization was thus associated with low-cost movement.

The prime example of this was the Netherlands which, in the sixteenth and seventeenth centuries, became highly urbanized, specialized in dairying and horticulture and imported grain from the Baltic (De Vries, 1974). Maritime freight rates fell almost continuously from the mid-seventeenth century, at between 0.5% and 1% per annum from 1650 to 1770 and at 3.5% per annum from 1814 to 1860. On land the decline in freight rates was slower, although slow improvement of road surfaces, the substitution of the horse for the ox and the use of the four-wheeled waggon showed some reduction in the cost of transport after 1650. It was the spread of the railway after 1820 that accelerated the decline; this was of particular importance in opening up the sparsely populated interiors of Russia, North America and temperate South America. It cost over 100 cents to move a hectolitre of wheat from Chicago to New York in 1873–4, and only 24 cents in 1905. The oceanic freight rate for wheat from New York to Liverpool in 1900 was only a third of what it had been in the 1870s. Combined with advance in the preservation of foodstuffs, and in particular the introduction of refrigeration into steamships in the 1870s, these declines in transport costs allowed the opening up of great new agricultural areas in the Canadian prairies, the American Mid-West, the Argentine pampas, the Russian Steppes and the interior of Australia. In England and the north-eastern United States the railway allowed comparatively remote dairy regions to provide fresh milk to London, Boston and New York. Before the railway few towns obtained fresh milk from more than 35 km. By the 1930s Berlin drew some of its milk from over 700 km (North, 1958, 1968; Morris, 1958; Youngson, 1965; Grigg, 1974, 1982b).

Whereas the initial impact of cheaper transport was upon agricultural produce, the growing dependence of farmers upon purchased inputs was made easier by these advances. In early nineteenth century England canals made possible the movement of coal for threshing machines, pipes for under-drainage and lime and superphosphates; it was the extension of the railways which further cheapened the movement of these goods. But the introduction of the petrol-driven motor vehicle has been even more significant, for fertilizers can come direct to the farm and be more easily moved around the farm and bulk tankers can collect milk. Indeed, much of the change in agriculture over the last 200 years would have been impossible without improvements in transport.

4.2.5 Science, the State and Agricultural Change

Until the nineteenth century most of the advances in agriculture were provided by farmers themselves, who modified implements, selected the better-yielding seeds, experimented in breeding better cattle and acquired new crops. But from the nineteenth century much of the new technologies were dependent upon advances in science; this was particularly true of plant breeding. With the rediscovery of Mendel's ideas on genetics in 1900, plant-breeding stations were established in many countries, and since then state-supported research has produced most of the new crop varieties which account for so much of the increased crop yields of the last 40 years. This is true not only of the new high-yielding rice and wheat varieties developed at experimental stations in Mexico and the Philippines, but also of most new varieties in North America and Western Europe. The growth in chemical fertilizers in the nineteenth century depended upon new knowledge in chemistry, whilst their modern multiplication depended upon the heavy chemical industries. This is even truer of pesticides and herbicides. These advances have required far more specialized knowledge than farmers would possess, and the cost of research and development has been beyond the resources of even groups of farmers. The state has thus been the prime-mover in agricultural research in the developed countries over the last century (Russell 1966; Petersen, 1980).

Nor have the activities of the state been confined to the promotion and finance of research. Until the mid-nineteenth century most European states attempted to ensure their food supplies by protecting their farmers against imports of grains, by controlling the quality of bread, and by forbidding the export of grain when home supplies of grain ran short. From the middle of the nineteenth century free trade spread from Britain to many parts—but not all—of the world. However, from the 1930s Britain, the main food importer—and the United States, the main exporter—both began to defend their farmers against low prices and imports of cheaper foodstuffs. In the United States attempts to both control surplus output and to maintain farm incomes ultimately benefited farm incomes, whilst the federal government bought and stored output that could not be sold. In Britain the import of food continued unimpeded, but until the early 1970s home farmers were paid guaranteed prices, not the market prices. In Western Europe the Common Agricultural Policy has established guaranteed prices for most farm produce and taxed imports from outside the EEC. Whatever the objections to these interventions by the state, farmers have been relatively prosperous in the last 40 years; without this prosperity, they would not have invested in the many farm improvements that have greatly increased output and productivity since 1945 (Fite, 1981; Tracy, 1982).

In the Socialist countries the intervention of the state has been comprehensive. Beginning in the Soviet Union but later followed in Eastern Europe and China, private ownership of land has been abolished and farms have been grouped into collectives or state farms. Many of the decisions about what is to be grown, and with what methods, have been transferred from the farm to planning commissions at both the regional and national levels (Pallot and Shaw, 1981).

The state has also intervened far more than in the past in the developing countries. With the end of colonial rule in the post-war period many newly independent governments have sought to increase the output of food for rapidly growing populations by large-scale agricultural development projects, where both the finance and technical direction has been provided to farmers by the state. Thus in countries of very different political complexions the state has become a major cause of change in farming.

4.2.6 The Diffusion of Technology, Crops and Livestock

By the beginning of the Christian era the *idea* of domestication of plants and livestock had spread to all but the remotest parts of the globe. But initially only indigenous plants were grown, and certain farming methods also remained confined to their regions of origin. Thus the plough probably had its origins in the Near East, and by the beginning of the Christian era was the basis of farming in the great agricultural civilizations of China, India and Europe. In Africa, however, the plough was unknown until brought by Europeans in the late nineteenth century, and still has limited use, except in the Mediterranean north and in South Africa. Nor were the plough or the wheel known in the Americas until after the arrival of Europeans in 1492. Equally marked was the absence of sheep, cattle, pigs, chickens and horses from the New World. However, all were introduced into the Americas by Europeans and were spread rapidly throughout the continent (Isaac, 1970).

Knowledge of the crops indigenous to other regions spread slowly throughout the world before the modern era. Rice is thought to have been domesticated in mainland south-east Asia and spread north into southern China, west into the Indian subcontinent, so that by the time written records are available it was the basis of civilization in most parts of Asia (Chang, 1976). Wheat and barley were first domesticated in the Near East and had been taken or spread into northern Europe long before the Christian era. Other crops remained confined to their hearth lands for very periods. Thus the soy-bean was grown mainly in China until this century; it was taken to the United States in the late nineteenth century and has become a major crop in the last 30 years, whilst its emergence in Brazil has been even more recent (Berlan *et al*., 1977).

The widespread interchange of crops, livestock and techniques accelerated with the discovery of the Americas and the later migration of Europeans to all parts of the world. Crops indigenous to the Americas have profoundly altered the cropping patterns of the Old World. In Europe the potato was known but little grown before the mid-eighteenth century, but became a major food crop by the mid-nineteenth century (Langer, 1975). Maize spread through the warmer southern parts of Europe in the eighteenth century; in the last 30 years the breeding of varieties capable of yielding in cooler, shorter summers has seen its diffusion into northern France and even into southern Britain. African consumption habits were profoundly changed in the slow spread of maize and manioc in the nineteenth and early twentieth centuries, whilst in parts of Asia sweet potatoes and peanuts, as well as maize, have become important supplements to rice and wheat (Geertz, 1963; Jones, 1956; Miracle, 1965; Ho, 1955).

In the last century most of the advances in agricultural technology have come in Western Europe and the United States. Chemical fertilizers were first developed in Britain and Germany, milking machines in Scotland, pesticides in France, the tractor and the combine harvester in the United States. Great improvements in the breeding of livestock and crops have taken place. Farmers in the western world were comparatively slow to adopt these innovations for various reasons; for the most part they were adopted if they were seen to increase output and profit. Thus the rate of adoption has greatly accelerated in the last 40–50 years, during the period in which farmers in Western Europe and North America have been protected from competition, and farmers have had an unrivalled period of prosperity.

In Afro-Asia and Latin America the adoption of innovations was equally slow, but the great increase in population since 1945 has meant an increased demand for food. This has to some extent been met by traditional methods, but there have also been attempts by governments and international agencies to accelerate the adoption of innovations that would raise crop yields. The spread of new high-yielding varieties of wheat and rice clearly illustrate the rise of 'induced innovation'. The research on the breeding of hybrid maize was first undertaken in the United States, on semi-dwarf wheat in the United States and Mexico, and on hybrid rices in the Phillipines by groups of scientists financed by the Ford Foundation of the United States. Since 1965 the new rices have spread rapidly through Asia owing to the intervention of governments who have provided extension workers, who have explained the new farming methods that are necessary, and subsidized the use of fertilizers and irrigation (Dalrymple, 1979; Hayami and Ruttan, 1971).

4.3 MAJOR TYPES OF FARMING AND THEIR EVOLUTION

There have been many attempts to establish a typology of world agriculture; few have been successful, indeed the great diversity of farming and the lack of comprehensive statistical information makes it an almost impossible task. (Grigg, 1969a; Duckham and Masefield, 1970; Kostrowicki, 1964, 1974; Whittesey, 1936). Nor is there space here to discuss all the various types of agriculture. Consideration will be given to three traditional farming systems, and emphasis will be put upon the way in which they maintained soil fertility and how they have adjusted to growing numbers.

4.3.1 Western Europe

The most distinctive features of agriculture in Western Europe before the nineteenth century were, first the use of the ox-drawn plough, and second that most farmers combined the keeping of livestock and the growing of cereal crops. These features had appeared in the Middle ages.

A fallow period of one in two or three years was an essential means of maintaining soil fertility, suppressing weeds and preventing the build-up of plant disease. Sheep and cattle were grazed on grassland held in common, that surounded the arable fields, and were fed on the stubble after harvest. The principal cereal crops varied, but wheat, barley, rye and oats were widely grown, and small amounts of peas and beans added some nitrogen to the soil. This system made the maintenance of yields over long periods possible and, except possibly in the early fourteenth century, there was no sign of soil erosion or falling yields. There was a balance between numbers and the farming system (Duby, 1962; Postan, 1966; Cooter, 1978).

This basic system was modified in different parts of Western Europe between the sixteenth century and the mid-nineteenth century in response both to growing numbers and the extension of commercialized agriculture around the cities. The major change was the greater integration of livestock and arable husbandry. In the medieval system livestock relied upon the grazing of common pasture and woodland for their fodder. However, in the fifteenth century, although mainly from the seventeenth century, part of the arable land began to be allocated to the growth of fodder crops, which included roots, notably the turnip, and clover. Root crops were sown in rows, and weeded during cultivation, which the broadcast cereals could not be; clovers not only provided grazing and hay for livestock, but added significantly to the nitrogen content of the soil. This, combined with the adoption of the potato, an American crop, allowed the considerable increase in food output that sustained the great increase in population between 1700 and 1850, all before industrialization had much effect upon European farming (Slicher van Bath, 1977; Chorley, 1981; Kerridge, 1967; E. Le Roy Ladurie, 1975).

Although the development of West European farms had led to great changes in the physical environment—notably the removal of much of the forest cover and the reclamation of poorly drained areas—it had not led to great damage to the environment. Severe forms of soil erosion, for example, were not found in Western Europe or indeed in those regions to which Europeans had migrated. Farming in eastern North America was still essentially European in origin, as were the types established in south-east Australia, although by the late nineteenth century, intensive agriculture in south-eastern North America was accompanied by severe erosion. But when Europeans moved out of the temperate forest areas where their agriculture had evolved into the drier treeless plains of western North America, Southern Russia and the interior of Australia, in the mid-nineteenth century, they had to discover methods of dealing with low and variable rainfall, and for the first time there were serious problems of soil erosion (Meinig, 1962, 1968; Fite, 1966; Lyaschenko, 1949).

Traditional European farming reached its zenith in the mid-nineteenth century. The oxen had been replaced by the horse, ploughs were increasingly made of metal rather than wood, a variety of implements were available to farmers which had been unknown to their predecessors. Much of the farmers' activity was devoted to feeding livestock to produce not only meat, milk, wool and hides, but the large amounts of manure needed to sustain and increase crop yields; stall-feeding, folding and muck-spreading were a great part of the farming year. This system reached its apogee in Eastern England and the Lothians of Scotland; more labour-intensive systems of equal elegance were found in the Netherlands and north-west Germany. Man and the horse provided the power, livestock returned nutrients to the soil to provide a balanced, self-sustaining farming system (Dovring, 1965; Ernle, 1968).

From the 1850s there were great changes in the implements and methods available to farmers, although it was not until the 1930s that the widespread adoption of these innovations transformed farming in the western world. The differences between European farming in its homeland and in the sparsely populated regions of North America now became apparent. The shortage of labour in North America required improved implements and machines, and these were forthcoming: the reaper replaced the scythe, the reaper–binder shortened stooking, and the combined harvester replaced the threshing machine which earlier had ended the long history of the flail. At the end of the century the first tractors appeared in the United States and were tried in Britain during the First World War. In Europe, advance in farming methods concentrated on increasing, not output per man, but output per hectare and output per animal, although in dairying many other nineteenth century advances also improved output per man. The development of the chemical fertilizer industry, the first crude pesticides and fungicides, and the breeding of new varieties were all well under way in Western Europe before the First World War.

Since then the series of innovations has continued; artificial insemination of livestock; great advances in the control of animal disease, the scientific planning of animal feeding, the discovery of effective herbicides and fungicides in the 1940s, and further advances in plant breeding have transformed the farming of Western Europe and North America (Cochrane, 1979; Staniforth, 1975; Duckham, 1966). These changes have taken place against a background in which three economic changes have been crucial. First, the labour force in agriculture has declined dramatically, demanding the adoption of labour-saving machinery. Second, governments throughout the western world have subsidized inputs and protected producer prices, so that food output has more than doubled in the last 30 years; but demand has not, so that a surplus of agricultural produce has become a constant feature of western agriculture, with the attendant political embarrassments. Thirdly, western diets have become dominated by the consumption of animal products, and in most western countries livestock products account for two-thirds of the value of output (Tracy, 1983).

Thus European agriculture has changed profoundly but slowly over the last thousand years, the underlying forces of change being the growth of population, the rise of urban society, the changing structure of demand brought about by higher incomes, and the technological advances that these changes have induced. At all times the capacity of farmers to change their environment has been enormous, but it has rarely been irreversible. In recent years great concern has been expressed at the capacity of modern machinery, and especially modern chemical inputs to damage the environment. It is perhaps too soon to know whether these changes are as dramatic as sometimes is suggested.

4.3.2 The Rice Economies of Asia

Although rice is very widely grown in Asia it is most closely associated with the densely populated deltas and alluvial valleys of the major rivers. Rice, although rarely a mono-culture, occupies a high proportion of the arable land; ploughs are drawn by zebu cattle or water buffalo; livestock, other than chickens or pigs are uncommon, for land cannot be spared to grow fodder crops, and so great is the pressure on land that grazing land is rare. The crop is frequently weeded, and in some regions is sown initially in nursery beds and then transplanted into the fields. Where there is sufficient water, two crops, although not necessarily both rice, are grown during the year. But the most distinctive feature of these economies is the relationship between the crop and water. Rice—or, more precisely, wet-rice—is sown in soil that is soaked to a muddy state; as the seedling grows, water is let into the small fields or padi in which it is sown, and it is grown with the stalk partly submerged until the water is withdrawn before harvesting. These unique growing conditions explain many features of the distribution of the crop and its ability to sustain high yields over long periods (Grist, 1953; Goor, 1966).

First, the padis contain blue–green algae which fix nitrogen, and play the role that legumes have in European agriculture. Second, the need to keep the water at a constant level on the stalk during growth has meant that rice is ideally grown in deltas and the flat plains of the middle and lower parts of rivers. These areas receive plant nutrients in solution and in suspension from the upper reaches. Third, rice is best grown in areas with a heavy clay subsoil that reduces the seepage of water downwards away from the padis; this also reduces the amount of leaching, and hence loss of nutrients in the soil. Finally, the fact that the fields are covered with water during most of the growing season means that soils are not exposed to high temperatures. This system of farming has proved capable of sustaining high crop yields over very long periods. It is also a system capable of registering small but continuous increases in yields in response to greater labour inputs (Geertz, 1963).

Rice was domesticated somewhere in the south-east Asia between India and South China, but the present distribution and features of the system only emerged in the first millenium after Christ. During this period the Han Chinese slowly moved southwards, and in India, Bengal, which was a frontier region as late as the sixth century AD, was occupied. By the eleventh and twelfth centuries much of the modern pattern was apparent, with high rural densities in southern China, Korea, Japan, much of India, parts of the Phillipines, and Java, but not in the deltas and lower valleys of the Mekong, the Irrawady or the Menam. A variety of means of extending the area under wet rice and increasing yields were practised. Wet rice padis could only be made on flat land, but in areas of great density cultivation was extended up hillsides by terracing, which is common in South China, but rare in the Indian sub-continent. High yields and multiple cropping are only possible if there is water-control and irrigation. The maintenance of water in the padis requires water control, a means of moving water in and out of the fields. This is not the same as irrigation. A majority of Asia's rice is still grown without irrigation and relies either on rainfall or the floods of rivers. Nonetheless the slow improvement of irrigation and water control has offered possibilities of increasing yields by making the supply of water more reliable, and by allowing the cultivation of two crops, often rice and wheat, in a year. The extension of multiple cropping has also been made possible by the selection of rice varieties with a shorter growing season (Grigg, 1974).

Greater labour inputs have also increased crop yields. Transplanting is only practised in the more densely populated areas; the frequency of weeding requires more work. Unlike the traditional European farming system, live-stock manure has played a very small part in increasing or maintaining rice yields, although in southern China night soil has been used and some forms of green-manuring practised. Some extreme forms of labour intensity have been practised; in parts of Java, rice was harvested stalk by stalk with a knife to prevent loss of grain (Krinks, 1978).

Until the middle of the nineteenth century wet-rice cultivation remained little influenced by external forces. Neither urbanization nor industrialization occurred in Asia, and there were few improvements in transport. However, the expansion of European colonialism had two important consequences. First rice became an important export, particularly in French Indochina, Burma, and Thailand, which had remained comparatively sparsely populated. Second, in some parts of Asia the colonial powers introduced modern methods of irrigation, based on the use of massive dams and lined canals; this was most noticeable in British India and in Java (Pelzer, 1945; Randhawa,

1962). However, possibly more important advances were underway in Japan, where the government promoted the improvement of agriculture in the late nineteenth century. Improved varieties of rice were introduced and chemical fertilizers began to be used. This led to the first comparatively rapid increases in rice yields and was the first instance of modernization in the rice economies of Asia. Similar methods were promoted in Korea and Taiwan whilst the latter were Japanese colonies, and between the two world wars rice yields were increasing in these parts of East Asia, but not elsewhere.

The period since 1950 has, of course, seen major attempts to improve farming in Asia, to provide food for a population that has nearly doubled since 1945. The most successful attempt has been the introduction since 1965 of high-yielding varieties of rice and wheat. Both are semi-dwarf varieties with short stems capable of bearing a heavy head of grain without lodging and responding to the application of chemical fertilizers. However, they require for optimum yield adequate moisture throughout the growing season; but only one-third of Asian rice is grown under irrigation. Combined with the use of pesticides the adoption of these varieties has led to substantial increases in crop yields. Although in some parts of Asia the adoption of these yield-increasing techniques has been accompanied by the use of tractors and machinery, Asian agriculture remains firmly linked to its long traditions (Dalrymple, 1979).

4.3.3 Shifting Cultivation and Bush Fallowing

A third way of maintaining soil fertility over long periods is the natural fallow. In most areas, if land is cropped continuously without any means of replacing plant nutrients, crop yields fall steeply and then maintain a low but constant level. One way of restoring the soil fertility without recourse to manure, fertilizers or the special conditions of the wet-rice padi is to crop the land for two or three years, then abandon the land and allow the natural vegetation to colonize the area. The re-establishment of the vegetation cover protects the soil from high temperatures and intense rainfall, and allows humus and plant nutrients to accumulate in the soil. After a period of years, which varies with climate, vegetation and soil type, the plant nutrient level is high enough for the vegetation cover to be partially removed and crops sown again for several years before the land is again abandoned.

Such a method of farming, variously described as shifting cultivation, or bush fallowing, has been widely practised in the past, in nearly all the major climatic and cultural zones. It is still found in parts of Latin America (Watters, 1971) and south-east Asia (Spencer, 1966), but is only the major system in tropical Africa where three-quarters of the arable area is farmed by bush fallowing. Fifty years ago Europeans regarded shifting cultivation as a primitive and wasteful farming system, but it is now realized that it is a successful means of utilizing tropical environments, and its defects are due to very rapid population growth.

There are innumerable ways of practising bush fallowing (Conklin, 1961; Pelzer, 1945; Watters, 1960). However, there are some general features in common. First of all, land to be cropped is cleared from either primary or secondary woodland or grassland; the plots cleared are small, and the vegetation is cleared with the use of axe, machete and fire. Not all the trees are cleared and stumps are frequently left. Ash from the fire provides some plant nutrients; fire makes the surface soil friable, and the presence of stumps accelerates regeneration when the plot is abandoned. Methods of preparing the soil for sowing vary a great deal. In areas with long fallow the soil is hardly disturbed, so only a digging stick is used. In more densely populated areas of Africa, however, soil is piled on mounds or ridges. The distinctive feature of cropping is that a number of crops are planted in the plot; ideally this will include not only cereals but also roots such as yams or cassava, and small shrubs. Together with the larger trees this provides a layered structure that simulates the forest and protects the soil from higher temperatures, and the impact of tropical rainstorms. Intercropping has recently been shown to have other advantages; the total yield of a combination of crops gives a higher yield than single stands of the individual crops, possibly because crops use complementary plant nutrients and the absence of large stands reduces the incidence of plant disease. It also staggers the chronology of harvesting (Morgan, 1969; Papendik *et al.*, 1976).

No fertilizer is used in bush fallowing, and the amount of weeding done varies, but is usually small. When weed growth becomes too exuberant the plot is abandoned. Indeed some have argued that it is this, as much as falling crop yields, that leads to the abandonment of the plot. In all forms of shifting cultivation the plot is abandoned after three or four years cropping, and is colonized by the natural vegetation. The length of the fallow depends upon the type of natural vegetation, but also important is the density of population. In very sparsely populated areas the fallow may last 25 years or more, and the settlement of the cultivators is moved. In more densely populated areas not only is the settlement permanent, but the fallow length has to be reduced; there is not sufficient land to allow long fallows. As the fallow is reduced so alternative means of maintaining soil fertility are sought; in parts of Africa fallows are planted with rapid growing bushes and grasses, some leguminous, and in the very densely populated areas such as northern Nigeria, pastoralists are encouraged to feed their livestock on fallow plots. But in most bush fallowing systems there is an absence of manuring as there is indeed of livestock, the plough is unknown and the complete clearance of natural vegetation is unusual. The chaotic mixture of crops and natural vegetation led Europeans to believe that such farming systems were primitive. It is now recognized that bush fallowing is a satisfactory means of providing food with relatively low labour inputs, and maintaining soil fertility and crop yields over long periods.

What is equally clear, however, is that shifting cultivation can only maintain this ecological efficiency if land is abundant. As fallows are reduced by rising population densities, as has occurred in Africa in the last 50 years, the ability of the natural fallow to restore soil fertility diminishes, and the risks of soil erosion increase.

4.4 LAND TRANSFORMATION AND THE HISTORY OF AGRICULTURAL CHANGE

Throughout the history of agriculture men have been constantly modifying the environment to facilitate the production of crops and livestock. The destruction of vegetation, the salinization of poorly drained irrigated areas, the wind erosion that results when dry areas are farmed with modern machinery, are all damaging to the environment, as are many of the consequences of the use of modern pesticides and chemical fertilizers. It is, however, perhaps unwise to insist that all land transformation by farmers is for the worse, for many of the world's most aesthetically pleasing and ecologically satisfactory landscapes are the result of the long, slow interaction of man and the environment. Other chapters in this book describe the adverse effects of human interference with the environment. It is perhaps as well to recall that a pessimistic view of the advance of agriculture could have been adopted at any time in the last 10 000 years. Land transformation is not new, even though the modern forces for change are far more powerful.

4.5 REFERENCES

Abel, W. (1980) *Agricultural Fluctuations in Europe: from the Thirteenth to the Twentieth Centuries*. Methuen, London.

Aymard, M. (1979) Toward the history of nutrition: some methodological remarks. In Forster, R. and Ranum, O. (Eds.) *Food and Drink in History: Selections from the Annales*. Johns Hopkins University Press, Baltimore, 1–16.

Berlan, J., Bertrand, J., and Lebas, L. (1977) The growth of the American soybean complex. *European Review of Agricultural Economics*, **4**, 395–416.

Boserup, E. (1965) *The Conditions of Agricultural growth: the Economics of Agrarian Change under Population Pressure*. Allen and Unwin, London,

Boserup, E. (1981) *Population and Technology*. Oxford.

Chang, T. (1976) The rice cultures. *Phil. Trans. R. Soc. London*, Part B, **275**, 143–157.

Chorley, G. P. H. (1981) The agricultural revolution in northern Europe, 1750–1880; nitrogen, legumes and crop productivity. *Econ. Hist. R.*, **34**, 71–93.

Clark, C. (1967) *Population Growth and Land Use*. Macmillan, London.

Cochrane, W. W. (1979) *The Development of American Agriculture: an Historical Analysis*. University of Minnesota Press, Minneapolis.

Connell, K. H. (1980) *The Population of Ireland, 1780–1845*. Clarendon Press, Oxford.

Conklin, H. A. (1961) The study of shifting cultivation. *Curr. Anthr.*, **2**, 27–61.

Cooter, W. S. (1978) Ecological dimensions of medieval agrarian systems. *Ag. Hist.*, **52**, 458–477.
Dalrymple, D. G. (1979) The adoption of high yielding grain varieties in developing nations. *Ag. Hist.*, **53**, 704–726.
Danhof, C. (1972) Tools and implements of agriculture. *Ag. Hist.*, **46**, 81–90.
Darby, H. C. (1956) The clearing of the woodland in Europe. In W. L. Thomas, Jn (Ed.), *Man's Role in Changing the Face of the Earth*. Chicago University Press, 183ff.
De Vries, J. (1974) *The Dutch rural economy in the Golden Age, 1500–1700*. Yale University Press, New Haven.
De Vries, J. (1981) Patterns of urbanization in pre-industrial Europe, 1500–1800. In Schmal, H. (Ed.), *Patterns of European Urbanization Since 1500*. Croom Helm, London, 77–110.
Dovring, F. (1959) The share of agriculture in a growing population. *Monthly Bulletin of Agricultural Economics and Statistics*, **8**, 1–11.
Dovring, F. (1965). The transformation of European agriculture. In Habakkuk, H. J., and Postan, M. M. (Eds.), *The Cambridge Economic History of Europe*, Vol. 6, *The Industrial Revolutions and After*, part 2. Cambridge University Press, Cambridge.
Drake, M. (1969) *Population and Society in Norway, 1735–1866*. Cambridge University Press, Cambridge.
Duby, G. (1962) *Rural Economy and Country Life in the Medieval West*. Edward Arnold, London.
Duckham, A. N. (1966) Forty years on: agriculture in retrospect and prospect. *J. Roy. Agri. S.*, **127**, 7–16.
Duckham, A. N., and Masefield, G. B. (1970) *Farming Systems of the World*. Chatto and Windus, London.
Durand, J. D. (1977) Historical estimates of world population: an evaluation. *Population and Development Review*, **3**, 253–296.
Ernle, Lord (1968). *English Farming, Past and Present*. London, Heinemann.
Fite, G. C. (1966) *The Farmers' Frontier 1865–1900*. Holt, Rinehart and Winston, New York.
Fite, G. C. (1981) *American Farmers: the New Minority*. Bloomington, Indiana, Indiana University Press.
Food and Agriculture Organization (1972) *Agricultural Adjustment in Developed Countries*. Rome, FAO.
Food and Agriculture Organization (1982) *Production Year Book*, 35, Rome.
Geertz, C. (1963) *Agricultural Involution: the Process of Ecological Change in Indonesia*. University of California Press, Berkeley.
Goor, G. A. W. van de (1966) *Agriculture with Special Reference to Rice Cultivation in Humid Tropical Deltas: Scientific Problems or the Humid Tropical Zone Deltas and their Implications*, UNESCO, Paris, 305–316.
Grigg, D. B. (1969a) The agricultural regions of the world; review and reflections. *Econ. Geogr*., **45**, 92–132.
Grigg, D. B. (1969b) Degrees of concentration: a note on world population concentration. *Geogr.*, **54**, 325–329.
Grigg, D. B. (1974) *The Agricultural Systems of the World: an Evolutionary Approach*. Cambridge University Press, London.
Grigg, D. B. (1975) The world's agricultural labour force 1800–1970. *Geogr.*, **60**, 194–202.
Grigg, D. B. (1980) *Population Growth and Agricultural Change: an Historical Perspective*. Cambridge University Press, London.
Grigg, D. B. (1982a) Modern population growth in historical perspective. *Geogr.*, **67**, 97–108.

Grigg, D. B. (1982b) *The Dynamics of Agricultural Change: the Historical Experience*. Hutchinson, London.
Grist, D. H. (1953) *Rice*. Longmans, London.
Hayami, Y., and Ruttan, V. W. (1971) *Agricultural Development: an International Perspective*. John Hopkins University Press, Baltimore.
Helleiner, K. F. (1967) The population of Europe from the Black Death to the eve of the vital revolution. In Rich, E. E., and Wilson, C. H. (Eds.), *The Cambridge Economic History of Europe*, Vol. 4, *The Economy of Expanding Europe in the Sixteenth and Seventeenth Centuries*. Cambridge University Press, Cambridge, 1–95.
Ho, P. (1955) The introduction of American food plants into China. *Am. Anthrop*., **57**, 191–201.
Ho, P. (1956) Early ripening rice in Chinese history. *Econ. Hist. R.* **9**, 200–18.
Ho, P. (1959) *Studies in the Population of China, 1368–1953*. Harvard University Press, Cambridge, Mass.
Isaac, E. (1970) *Geography of Domestication*. Prentice Hall, Englewood Cliffs, New Jersey.
Jones, W. O. (1956) Manioc: an example of innovation in African economies. *Econ. Dev. Cu*., **5**, 100–110.
Kerridge, E. (1967) *The Agricultural Revolution*. Allen and Unwin, London.
Kostrowicki, J. (1964) Geographical typology of agriculture, principles and methods. *Geogr. Pol*., **1**, 111–146.
Kostrowicki, J. (1974) *The Typology of World Agriculture: Principles, Methods of Model Types*. International Geographical Union, Warsaw.
Krinks, P. (1978) Rural change in Java: an end to involution? *Geogr*., **63**, 31–36.
Ladurie, E., Le Roy (1975) De la crise ultime à la vrai croissance. In E. Le Roy Ladurie (Ed.), *Histoire de la France Rurale*, Vol. 2, *L'Àge Classique des Paysans, 1340–1789*. Seuil, Paris, 350–624.
Langer, W. (1975) American foods and Europe's population growth, 1750–1850. *Journal of Social History*, 51–66.
Lyaschenko, P. (1949) *History of the National Economy of Russia to the 1917 Revolution*. Macmillan, London.
McKeown, T. (1976) *The Modern Rise of Population*. Edward Arnold, London.
Meinig, D. W. (1962) *On the Margins of the Good Earth: the South Australian Wheat Frontier, 1869–1884*. Rand McNally, Chicago.
Meinig, D. W. (1968) *The Great Colombia Plain: an Historical Geography, 1810–1910*. University of Washington Press, Seattle.
Miracle, M. P. (1965) The introduction and spread of maize in Africa. *J. Afr. Hist*., **6**, 39–55.
Morgan, W. B. (1969) Peasant agriculture in tropical Africa. In Thomas, M. F., and Whittington, G. W. (Eds.), *Environment and Land Use in Africa*. Methuen, London.
Morris, T. N. (1958) Management and preservation of food. In Singer, C., Holmyard, E. J., Hall, A. R. and Williams, T. I. (Eds.), *A History of Technology*, Vol. 5, Oxford University Press, London, 16–48.
North, D. C. (1958) Oceanic freight rates and economic development, 1750–1913. *J.E.H*., **18**, 537–547.
North, D. C. (1968) Sources of productivity change in oceanic shipping 1600–1850. *J. Polit. Ec*., **76**, 953–970.
Ojala, E. M. (1952) *Agriculture and Economic Progress*. Oxford University Press, London.
Pallot, J., and Shaw, D. J. B. (1981) *Planning in the Soviet Union*. Croom Helm, London.

Papendik, R. I., Sanchez, P. A., and Triplett, G. B. (1976) *Multiple Cropping*. American Society of Agronomy, Special Publications No. 27, Madison, Wisconsin.
Pelzer, K. J. (1945) *Pioneer Settlements in the Asiatic Tropics*. American Geographical Society, Special Publication No. 25, New York.
Peterson, T. H. (1980) *Farmers, Bureaucrats and Middlemen: Historical Perspectives on American Agriculture*. Howard University Press, Washington, DC.
Postan, M. M. (1966) Agrarian society in its prime: England. In Postan, M. M. (Ed.), *The Cambridge Economic History of Europe*, Vol. 1, *The Agrarian Life of the Middle Ages*. Cambridge University Press, London.
Randhawa, M. S. (1962) *Agriculture and Animal Husbandry in India*. Indian Council of Agricultural Research, New Delhi.
Rasmussen, W. D. (1982) The mechanization of agriculture. *Scient. Am.*, **247**, 76–89.
Russell, E. J. (1966) *A History of Agricultural Science in Great Britain, 1620–1954*. Allen and Unwin, London.
Slicher van Bath, B. H. (1968) *The Agrarian History of Western Europe, AD 500–1850*. Edward Arnold, London.
Slicher van Bath, B. H. (1977) Agriculture in the rural revolution. In Rich, E. E., and Wilson, C. H. (Eds.), *The Cambridge Economic History of Europe*, Vol. 5, *The Economic Organisation of Early Modern Europe*. Cambridge University Press, Cambridge, 42–133.
Spencer, J. E. (1966) *Shifting Cultivation in South Eastern Asia*. University of California Press Publications in Geography, No. 19, Berkeley.
Staniforth, A. R. (1975) Fifty years of agricultural progress in Great Britain. *Agri. Prog.*, **50**, 1–20.
Teuteberg, H. J. (1975) The general relationship between diet and industrialization. In Forster, E., and R. (Eds.), *European Diet from Pre-industrial to Modern Times*. Harper and Row, London, 61–110.
Thomas, D. S. (1941) *Social and Economic Aspects of Swedish Population Movements, 1750–1933*. Macmillan, New York.
Toutain, J. C. (1971) *Las Consommation Alimentaire en France de 1789 à 1964*. Institute de Science Economique Appliquée, Paris.
Tracy, M. (1982) *Agriculture in Western Europe: Challenge and Response, 1880–1980*. Granada, London.
Trewartha, G. T. (1969) *A Geography of Population: World Patterns*. John Wiley, New York.
United Nations (1953) *The determinants and consequences of population trends*. United Nations, Department of Social Affairs, New York.
United Nations (1982) *Demographic Year Book 1981*. United Nations, New York.
Watters, R. F. (1960) The nature of shifting cultivation: a review of recent research. *Pacific Viewpoint*, **1**, 59–99.
Watters, R. F. (1971) *Shifting Cultivation in Latin America: Food and Agricultural Organization*. Foresty Development Paper No. 17, Rome.
Whetham, E. H. (1970) The mechanization of British farming, 1910–1945, *J. Ag. Econ.*, **21**, 317–331.
Whittlesey, D. (1936) Major agricultural regions of the earth. *A.A.G.*, **26**, 199–240.
World Bank (1982) *World Development Report, 1982*. Washington, DC.
Wrigley, E. A., and Schofield, R. (1981) *The Population History of England, 1541–1871: a Reconstruction*. Edward Arnold, London.
Youngson, A. J. (1965) The opening up of new territories. In Habakkuk, H. J., and Postan, M. M. (Eds.), *The Cambridge Economic History of Europe*, Vol 6, *The Industrial Revolutions and After* part 2. Cambridge University Press, 139–211.

Land Transformation in Agriculture
Edited by M. G. Wolman and F. G. A. Fournier

CHAPTER 5
Agricultural Practices Leading to Land Transformation: Introduction

M. G. WOLMAN AND F. G. A. FOURNIER

The preceding chapters have dealt with the history of agriculture and the forces that influence transformation of the land in agriculture. A major series of essays follows in Chapters 5.I–5.VI. The first four of these chapters describe those factors that directly affect or transform the land itself—including irrigation, mechanization, conservation practices and wetland reclamation. Chapters 5.V and 5.VI then deal with indirect influences—fertilizers and substances used to protect plants. Both the direct and indirect influences may alter the land itself as well as the air and water environment. Each essay attempts to deal with the three media in looking at the way these major forces are transforming land, landscape and environment.

The first type of land transformation was the disturbance of natural vegetation—the clearing of land in order to get land suitable for producing food and fibres for human beings and feed for domestic animals. In the beginning only very simple hand tools consisting of pieces of rock or wood were used. These were improved when iron became available. Transforming land by hand, as was done for thousands of years, was hard and time-consuming. It became somewhat easier when draught animals and appropriate implements became available. Finally, when industry developed, all kinds of special machinery and implements became available to reclaim new land, to improve existing agricultural land and to prepare land for sowing, to do weeding and harvesting. Almost all human and animal work was replaced by mechanical work, including clearing of dense forest to get new arable land or grassland. Human and animal energy were replaced by fossil energy. In the past, much land could not be reclaimed because it was too wet or too dry or because the natural vegetation could hardly be removed. At present, such land can be brought under cultivation, provided knowledge and capital are available.

Improvement of the infrastructure—such as new canals, roads and railways—also influenced land transformation as the agricultural products could be more easily transported to markets. Construction of dams in rivers, of irrigation and drainage canals, and of farm ponds were important for irrigation. When diesel and electric pumps became available much more land, including private land, could be irrigated. In addition to improving productivity, irrigation has been accompanied in places by waterlogging and salinization. These degradative effects and techniques for their amelioration are also considered in the following chapters.

Because mismanagement of land has led to significant soil erosion in a number of areas, conservation and land management are considered as separate land-transforming processes. While the productivity of some lands has been completely lost to erosion, erosion and deposition which accompany it are complex phenomena involving both landscape and river systems. Because modern conservation efforts, like large-scale irrigation and drainage projects, involve many land owners or farmers over large areas, they are also likely to involve governments as well. Careful planning is essential and policies and strategies must be developed to guide governments in these land-transforming activities. These issues are implicit in many of the essays that follow and are emphasized particularly in connection with irrigation and conservation management.

Other examples are presented, dealing with changes in land resulting from crop pests and diseases. Phytophagous organisms have always reduced crop yields or prevented cultivation of specific crops. The value of annual food crop losses is currently estimated to be about $300 billion. Farmers have always tried to minimize such crop losses by adoption of crop rotations. Rotation, for example, is particularly important for controlling organisms such as nematodes which are often crop-specific. Yet epidemics of new diseases have developed, because they were accidentally brought into some regions or countries where they did not exist before. In 1939, following the identification of DDT as an insecticide, a wide variety of synthetic organic pesticides have become available. Insecticides, fungicides and herbicides have changed agriculture and agricultural land use during the last decades, as noted in Chapter 4. Traditional crop-rotation systems have been altered, monocropping and no-tillage practices introduced, and agricultural production increased even as the labour force was reduced. Although pesticides have many advantages, there are again shortcomings. Among them are unforeseen side-effects such as the emergence of pesticide-resistant strains of pest organisms, the disturbance of agro-ecosystems and the pollution of the environment. Use and management of pesticides and other materials, including adoption of integrated pest management systems, markedly influences land and its transformation.

Up to the Second World War, inorganic fertilizers were used in only a few industrial countries. Since then, application has rapidly increased in developed countries and to a more limited extent in developing countries as well. While the application of nitrogen fertilizers is perhaps the single most important factor in worldwide increases in crop yield, land use itself was affected as lands with low nutrient status could be converted into productive crop land. At the same time, intensive application of fertilizers may cause nutrient inbalances in soils, soil and groundwater pollution. Mitigation or avoidance of these effects also transforms the land.

Mechanization in agriculture has contributed significantly to land transformation. As tractors replaced draught animals, much land needed to produce feed for the animals became available for food production. Machinery well-adapted to the situation on the farm reduces the time needed for many tasks, eases the work and frees time for other work. Moreover, various tasks can be carried out at the appropriate time, improving the quality of the work and increasing yields. With the gains, of course, there are losses. For example, a tractor does not produce manure, it makes noise and pollutes the atmosphere. It needs fuel and the capital investment required for mechanization is high. During the last decades more powerful and heavier tractors and machinery have been constructed, leading to increases in weight of the equipment, and in places gradually causing compaction of the soil. In some cases this has changed cropping patterns—for example, the abandonment of root crops on compact, heavy, textured soils.

The following six chapters indicate clearly the advantages and disadvantages of agricultural practices leading to land transformation. Agricultural production depends on land, labour and capital. For traditional agriculture much land and labour and only little capital is needed. In technological agriculture, where more productive crop varieties are used and where inorganic fertilizers, pesticides and mechanization are important, less labour but much capital is needed. Agricultural production must be more intensive and crop yields higher in technological agriculture because of the costs of these inputs. Analyses of historical changes, and of the way current direct and indirect agricultural practices affect the land, clearly demonstrate the dynamic character of forces leading to transformation of the land.

Land Transformation in Agriculture
Edited by M. G. Wolman and F. G. A. Fournier

CHAPTER 5.I
The Impact of Irrigation

L. SHANAN

5.I.1 INTRODUCTION

Irrigation is generally defined as the application of water to the land for the purpose of supplying moisture essential to plant growth. In the process, irrigation projects transform the land in two ways:

(1) by direct modifications of the land surface that occur when canal networks are constructed and land is cleared, shaped and levelled for irrigation;
(2) by indirect in-depth transformations that take place when the water and salt balances in the region are changed following the import of additional quantities of water and salt into the area.

Most of these transformations are permanent and irreversible. This chapter describes the dynamics of these transformations, and by using specific case studies examines the concomitant hazards that develop which may even threaten the existence of the society that created the irrigation project.

Seasonal rainfall fluctuations make rain-fed farming a risky venture. Irrigation reduces some of the uncertainties and so promotes increased production. In desert and arid zones, irrigation enables the economic development of areas which otherwise would be unproductive and uninhabitable. Irrigation development is as old as recorded history. In about 2000 BC, an inscription on the tomb of the Assyrian Queen Semiramis proclaimed that 'I constrained the mighty river to flow according to my will and led its waters to fertilize lands that before had been barren and without inhabitants'. The Bible reports that the prophet Elisha in about 865 BC stated 'thus sayeth the Lord, Make this valley full of ditches . . . Ye shall not see wind, neither shall ye seen rain; yet that valley shall be filled with water, that Ye may drink, both ye and your cattle and your beasts' (Kings II, 3:16–17).

There is evidence of continuous irrigation for thousands of years in the Nile Valley and for comparatively long periods in Syria, Persia, India, Java and

Italy. Egypt claims to have the world's oldest dam built about 5000 years ago to supply drinking water and for irrigation, and basin irrigation introduced at that time still plays a significant role in Egyptian agriculture. Records show that irrigation in China was begun about 4000 years ago, and there are reservoirs in Sri Lanka more than 2000 years old. It is also interesting to note that the potential damages from irrigation have been known since early times. As far back as 2300 BC, the Babylonian Code of Khammurabi provided that 'If anyone opens his irrigation canals to let in water, but is careless and the water floods the fields of his neighbour, he shall measure out grain to the latter in proportion to the yield of the neighbouring field.'

An excellent historic example of extensive land transformations resulting from irrigation development is found in the stoney-gravel limestone desert of the Negev area in Israel. Remnants of these ancient irrigation systems date back from the Israelite period (about 1000 BC) and from the Nabatean–Roman–Byzantine era (300 BC to 600 AD). In the absence of permanent water sources, the ancient farmers developed 'runoff' farm systems that used sporadic flash floods for irrigating. Each farm unit comprised a watershed to collect runoff from the hillsides and a group of terraced fields in the valley bottoms for cultivation. The hillsides were subdivided into small catchments by channels leading the hillside runoff water to the terraced fields. The stone and gravel cover was cleared from the ground surface of the catchments to increase the rates of runoff and placed in geometric patterned heaps of mounds and strips. Gravel mounds and strips, hillside collecting channels and stone-walled terraced fields transformed about 300 000 ha of the barren desert into a productive water-collecting system, and enabled the ancient settlers to cultivate about 15 000 ha and establish a permanent agricultural civilization in a 100 mm annual rainfall region.

5.I.2 THE PRESENT AND FUTURE EXTENT OF IRRIGATION

The area of land irrigated in the world today is close to 250 million hectares, of which about two-thirds is situated in five countries—China, India, Pakistan, the Soviet Union and the USA. With world populations growing concurrently with expectations of higher standards of living, world food and fibre demands will increase. Since arid lands become highly productive with the introduction of irrigation water, and the productivity of rain-fed areas increases considerably with supplementary irrigation, the improvement of existing irrigation projects and the opening up of new irrigated areas will continue to make significant contributions to solving the world's expanding food and fibre needs. In order to ensure the stability and permanency of these projects we cannot afford to overlook the potential hazards intrinsic to irrigation development, and we must be aware of the land transformations that may take place.

5.I.3 LAND TRANSFORMATIONS

Land transformations resulting from irrigation development are in all cases spectacular—especially when seen from the air or recorded in satellite imagery. Aerial photographs showing the boundaries between the desert and the sown in, for example, the Nile Valley in Egypt, the Central Valley in California, or the Rajasthan Canal in India, exhibit dramatic pictures of the contrasting land surfaces and highlight the transformations that have taken place as a result of irrigation making the deserts blossom. These land transformations can be classified into two categories:

(1) surface modifications resulting from implementing various irrigation methods;
(2) in-depth changes caused by disturbing the local or regional salt and water balance.

Most of these transformations are irreversible and permanent. All irrigation projects involve transferring water from one area to another—from a water-surplus region to a water-deficient one. Some major irrigation schemes call for inter-regional water transfers from one basin to another using major dams for inter-annual or inter-seasonal storage. They often require main canals hundreds of kilometres in length to transport the water. Inherent in this irrigation development are the land transformations caused by the construction of these large multi-purpose dams—such as those that have followed the building of the Aswan, Kariba, Tarbella and Boulder Dams. Reservoir sedimentation, steam bed aggradation, flooding of upstream forests and villages are only some of the many problems concomitant to major schemes. This chapter, however, is confined to land transformations that are directly related to the irrigation development.

5.I.3.1 Surface Transformations

Irrigation water is applied to land using one of the four main methods:

(1) surface flooding;
(2) partial wetting of the surface in furrows;
(3) surface sprinkling;
(4) by drip (or 'trickle').

All methods, particularly surface ones, involve considerable land preparation.

The first land transformation follows clearing the land of trees, brush or other vegetation. In the semi-arid zones (for example in the western part of the United States) relatively dense cover of sagebrush, grease-wood, chaparral, scrub oak, pine or mesquite have to be eradicated from proposed irrigated lands, and in the monsoon areas such as Thailand, India, Indonesia etc. tropical and sub-tropical jungle forests must be cut down and eliminated before development can begin. Small brush up to 0.5 m in height can be removed by ploughing, while vigorous vegetative growth usually requires the clearing operations to be carried out with large and powerful tractors. In some cases deep-rooted tree stumps must be blasted away—leaving the land with a bomb-pitted look of a war zone. In other cases stones, boulders and rock outcrops have to be hauled away from the project lands.

The land having been cleared, it next becomes necessary to smooth (or grade the surface) to facilitate uniform water distribution. These works bring about another land transformation. Surface smoothing is usually accomplished by the use of harrows, graders, scrapers, bulldozers and landplanes—but in under-developed countries hand labour and bullock-drawn wooden planing boards are often used for this work. On accurately levelled land, water will be distributed uniformly over the fields; but when the levelling is done poorly and the ground surface remains rough and irregular, some parts of the fields will receive more water than others. The final shape of the land surface will depend on the method of irrigation to be used. In many of the older irrigation systems, where the wild-flooding method is common, water flows from the supply ditches over the fields and is guided only by the slope of the land. This method requires minimum amounts of land levelling (say, up to about 200 cubic metres of earth moving per hectare), but irrigation by this method is generally the least efficient of all methods. More efficient methods of surface flooding call for accurate land shaping and grading and 1500 cubic metres of earth may neen to be moved per hectare. In the border method of irrigation the land is divided into long, narrow strips 5–15 m wide, extending for 100–400 m down the prevailing land slope, and the border strips are separated by levees, 0.1–0.3 m high. Border strips are levelled transversely, so that water will flow at an even depth over the whole width. The longitudinal slope is graded to a design gradient to enable an even percolation of the irrigation water over the entire border. In the level basin method the land is shaped into accurately levelled plots ranging in size from 0.2 to 1.0 ha and surrounded by low bunds. The basin method is suited both to permeable soils which must be covered quickly with water in order to prevent excessive losses near the supply ditches, and to heavy soils with slow percolation rates. It is the method most commonly used for rice ('paddy') irrigation all over the world. Small, level basins are also used for irrigating orchards. Levelling works for constructing either borders or basins transform the original natural land surface into clearly discernable rectangular or square patterned fields.

In the irrigation methods described above, almost the entire land surface is wetted at the time of irrigation. When furrow irrigation is used, the fields are only partially wetted. Furrows (20–30 cm in depth) or corrugations (10–15 cm) are spaced according to the plant rows—one furrow or corrugation being provided for each row. For efficient furrow irrigation the land must be graded accurately and the flow of water into the furrows controlled. The pattern created by this method is neither square or rectangular, but the fields are formed into long, thin, parallel linear strips leading away from the delivery ditches.

In sprinkler irrigation projects, water is applied to the crop in the form of a spray. Unlike the surface methods, sprinkler irrigation requires less land levelling and can be used economically on undulating lands, on soils too shallow to prevent proper levelling, and on soils too porous for good water distribution by surface methods. The land transformation process is partly dependent on whether the sprinkler system is semi-permanent or portable. A particularly dramatic land metamorphosis takes place when central pivot installations are introduced on a large scale. In these schemes, combinations of rotating and fixed sprinklers deliver water from central pivots through continually rotating booms to circular-shaped fields. The fields can have diameters of up to 1 km. These systems when seen from the air or on satellite photographs show up as green circles in a patterned design on a beige desert background.

Drip irrigation systems consist of extensive networks of small-diameter plastic pipes delivering low flows of a few litres per hour to the plants through special emitters. From each emitter, water spreads laterally and vertically in the soil profile. Drip systems require little land levelling. Dissolved salts are left in the soil near the edge of the wetted zone and these salts must be periodically leached out of the root zone by rainfall or by irrigation. The system is adaptable to a wide variety of shallow, sandy or stoney soils as well as undulating and steep topographic conditions. With the world's expanding population pressures, planners will be forced to bring the more difficult areas into production, and we can expect that extensive areas will come under drip irrigation in the future to enable otherwise unirrigable areas to be developed.

5.I.3.2 In-depth Transformations

Since man first began to question the nature of the four basic elements (water, fire, earth and air) he has pondered on the behaviour of water. The early philosophers were puzzled by the continual flow of rivers to the sea and wondered why the level of seas did not continually rise. The early Greek philosophers of the fourth and sixth centuries BC imagined immense underground reservoirs from which the rivers extruded, and it was not until the advent of the early Roman philosophers in the first century AD that the 'evaporation–condensation' phase of the hydrological cycle came to be understood. This lack of appreciation of the hydrological processes contributed to the disintegration of some of the ancient irrigation systems. It was not until the Middle Ages, however, that da Vinci gave the first realistic representation of the water balance equation, when he wrote that in the course of a year, the amount of water that rises from a particular region will be as great as the flows in the rivers and the rainfall that descends in that region.

In the nineteenth century the hydrological cycle was first analysed as a system of storage with a given input producing a specific output. The quantity of water moving through the hydrological cycle in a unit of time could then be evaluated from the water balance equation:

$$I - O = S$$

where '*I*' is the inflow during a given period (surface water, groundwater and precipitation); '*O*' is the outflow during the period (evaporation, transpiration, surface runoff and groundwater discharge from the area); and '*S*' is the change in storage in retention, depression, interception, soil moisture and groundwater levels. Implicit in this approach is the understanding that feedback mechanisms in the system constantly tend to bring about some state of equilibrium.

In a non-irrigated area a stable but changing water balance is developed between the annual rainfall input, ranging from 100 to 1000 mm, the vegetation and the groundwater. With the introduction of irrigation the balance is disturbed. Since irrigation may add the equivalent of, say, 500–1500 mm of water to the annual rainfall, this three- to ten-fold increase in water input will significantly upset the previous prevailing water balance. From its original steady state of equilibrium with system variables oscillating around long-term averages, the additional input forces the entire system over some threshold value into new levels of equilibrium or into meta-stable equilibrium states. The rate of change to the new levels will depend on the pedology, geology, climate and crops grown in the region.

It is important to recognize all the changes that can result from human intervention into the input–output response-process and how the modified hydrological system transforms the land of a particular region. The transformations will depend on both the spatial and the time scales over which the intervention is applied. A few in-depth micro-scale transformations may be observed in relatively short time periods measured perhaps in years, but most macro-scale changes require decades or even centuries to become hazardous threats to the permanency of the system. Some effects may be local and cover only a few hectares, while others may be regional and affect thousands of hectares. In projects where irrigation is supplemental rather than continuous, these effects are generally limited and easier to control. This chapter focuses on the large-scale transformations that develop as dynamic meta-stable processes passing through a number of thresholds before they become problematic and approach a state of irreversibility. Four deleterious land transformations follow irrigation development:

(1) waterlogging;
(2) saline and alkaline soil development;

(3) sea-water intrusion;
(4) land subsidence.

These transformations are now described and illustrated with some typical case studies.

5.1.3.2.1 Waterlogging

Waterlogging is the result of a rise in the groundwater level. An area is generally classed as waterlogged if the water table lies less than 2 m from the ground surface. When waterlogging reaches the root zone of the plants, yields diminish significantly because the roots need a soil–air–water environment to grow and cannot survive in free water. There are four principal causes of waterlogging—leakage from canals, wastage from distribution networks, over-irrigation and lack of suitable drainage facilities. Following the waterlogging process, saline conditions generally develop on the surface.

Many examples of waterlogged areas can be found in the irrigated areas of the world—in India, Pakistan, the USA, Australia, North Africa and China. The case history of Pakistan has been selected because it is a good example of an extensive occurrence of the phenomena and demonstrates the engineering and management measures governments must introduce in order to control regional waterlogging.

There is evidence that irrigation has been practised along the Indus River system in Pakistan from 3000 BC onwards. In the beginning, only narrow strips of land adjoining the river banks were irrigated. With time, irrigation has been extended. Pakistan contains probably the largest irrigation system in the world, and today about 15 million hectares are irrigated, of which more than 75% is served by government canals. The irrigation system comprises the Indus river and its major tributaries, 3 storage reservoirs, 19 barrage/headworks, 43 canal commands and about 90 000 watercourse systems. The total length of the canals is about 56 000 km and there are more than 1.6 million kilometres of watercourses and farm channels.

The physical layout of the canal systems and the methods of delivering water were evolved to fit the pattern of available supplies in the unregulated rivers, and to conform to the legal conditions and water allocations of the Pakistan–India Indus Water Treaty of 1960. The system was planned to bring to maturity the largest possible cropped area with the minimum consumption of water. Consequently, irrigation intensities are generally low (25% in the kharif season and 50% in the rabi season) so that the water can be spread over as large an area as possible. A basic canal design concept underlying the planning has been to achieve equitable water distribution with the minimum of human interference in the management of the system. Overall irrigation efficiencies, however, are relatively low—possibly about 30%—and studies show that 35–45% of the water delivered by the canals never reaches the farmers' fields. Since the natural drainage is inadequate and artificial drainage was not provided in the early project years, extensive over-irrigation combined with excessive leakage from the canals have led to a rise in the water table.

Rates of rise of the water table have been relatively rapid—about 30–40 cm a year. Depending on the initial depth of the water table, areas along the main canals were the first to become waterlogged, but subsequently waterlogging spread to the contiguous areas. The potential need for drainage to prevent this land transformation was realized when large-scale development was initiated, and a network of observation points was established to monitor the changing groundwater conditions. Waterlogging conditions measured in 1978 showed that in the 14.6 million hectares surveyed, the water table was less than 1.8 m below the surface on 13% of the area and between 1.8 m and 3.0 m on 41% of the area. In some areas, the water table is still rising; but in those projects where the government has initiated drainage schemes (see SCARP programs described below), the water table is slowly being controlled and some cases even lowered. Unless these drainage measures continue to be implemented, 50% of the irrigated fields will have to be abandoned in the future and the land will be transformed into unproductive saline waterlogged swamps, which will be useless to Pakistan.

5.I.3.2.2 Salinity—the Second of the 'Twin Menaces'

All irrigation water contain salts and the salt concentrations in soils tend to increase as water evaporates from the surface or is transpired by plants. In order to maintain a favourable root-zone salt balance, more water (called the 'leaching requirement') must be applied to the soil than is used by the evapo-transpiration process. The excess water drains to the groundwater table, and in the absence of drainage it contributes to the development of waterlogging conditions described in the foregoing. Salts accumulate in the upper soil profile as evaporation and transpiration remove water, and highly saline or alkaline conditions develop as more and more water moves upward from the water table by capillary action. This salinization of the upper soil profile increases continuously and when it reaches intolerable levels the land goes out of production.

The rate of salt accumulation depends on the soil surface temperatures, the chemical composition of the ground and surface waters and on the physical and chemical properties of the soil. As the balance between the sodium, magnesium and calcium ions changes in the soil, sodium becomes absorbed on the fine clay particles and the soil properties are modified. Soil swelling, colloidal dispersion and void clogging create impermeable soil profiles and unsatisfactory root aeration conditions, with subsequent significant reductions in production.

In order to prevent the development of these salinity conditions in irrigated areas, two courses of action are possible:

(1) installing surface or subsoil drainage to remove water from the upper 2 or 3 metres;
(2) developing vertical drains (wells) to lower the water table with the use of pumps.

The first method is feasible where suitable outlets for the drains are available. However, this solution may only transfer the problem downstream because when saline drainage water is discharged into local rivers, irrigation projects situated further downstream may be adversely affected. For example, the Colorado River's salinity has been rising during the last few decades as it flows downstream through the States of Nevada, Arizona and California. The current level of salinity at the Imperial Dam is about 900 ppm (parts per million), and is predicted to continue to rise and to reach about 1200 ppm in the year 2000. The cost of the damages due to this high saline content is expected to amount to about $80 million a year and indicates the stake that the irrigators and urban water users in the Colorado Basin have in the measures needed to control the salinity.

The second method of controlling water levels in the ground, pumping from wells, requires a relatively costly energy input to operate, but it has the advantage that a portion of the groundwater pumped can be used for irrigation in conjunction with the surface water. The management of a conjunctive operation, however, can only be done on a large scale and often requires government control of the regional water balance.

Pakistan also provides an example of the salinity problem. As mentioned previously, large-scale irrigation development in Pakistan led to rapid rises in the groundwater table. Groundwater has been a traditional source of irrigation water in the region, and the predominant method of using groundwater was, until the 1970s, the 'Persian-wheel' driven by draft animals. Over 200 000 units are still reported to be in operation. During the last 20 years, modern diesel- and electric-powered tube wells have been introduced in both the public and private sectors.

The water–salt balance in the Indus Basin represents a complex and difficult problem because the average slope of the area is about 1:5000 and there is no natural drainage for disposing of saline effluents. The irrigated plain is underlain by an extensive groundwater aquifer covering about 16 million hectares. One-third of this area has 'fresh' groundwater (1000 ppm TDS—total dissolved salts), one-tenth a moderate salinity content (1000–3000 ppm TDS) and the remainder is highly saline (>3000 ppm TDS). Total recharge of the aquifer is about 50–75 × 10^9 cubic metres per year. At present about 180 000 private tube wells and 12 500 public tube wells jointly pump about 45 × 10^9 cubic metres per year from the groundwater. Without this pumping, the Indus Basin would be a large lake. The Indus

Basin has also been acting as a large salt sink. Average annual river water diverted for irrigation is about 100–120 × 10^9 cubic metres per year, with an initial average salinity of 130 ppm. This means an annual addition of 16 million tonnes of salt to the 16 million hectares of irrigated land, i.e. about 1 tonne of additional salt is added annually to every hectare.

With the recognition of the expanding salinity hazard in the late 1950s, the first large-scale groundwater development and salinity control programme was initiated with the Government Salinity Control and Regulation Project (SCARP I). The plan proposed to control salinity by pumping large quantities of water from the aquifer, spreading if over the land and so leaching down the salts. Parallel with the SCARP I programme, private tube well development took place at an accelerated rate and total private groundwater development today far exceeds the public tube well programme. Because of its success, SCARP I has been followed by a series of five-year SCARP plans and SCARP V is currently in operation. The design of the SCARP projects is based on sophisticated criteria for handling the difficult groundwater and aquifer conditions, such as delineating mixing zones for conjunctive use of different levels of groundwater and surface water, and the programme includes mining certain groundwater areas, combining private and public tube well development, and remodelling canals to increase surface water supplies.

The SCARP programmes are at present under review and an overall strategy is being developed to create a permanent water–salt balance in the Indus Basin by adopting strategies for both prevention and cure. Plans are now being evaluated to dispose of the salts by:

(1) constructing large outfall drains to discharge highly saline effluents to the lower reaches of the Indus River and to the sea during the flood season;
(2) transporting some of the saline water to large evaporating areas in the desert.

During the past 20 years, 30–50% of the water sector's budgetary allocations have been spent on implementing salinity control measures. In the future similar large expenditures will have to continue to be invested in eliminating the long-term salinity hazard of the Indus Basin.

5.I.3.2.3 Seawater Intrusion

Two types of seawater intrusion will be discussed:

(a) subsurface intrusion into coastal aquifers; and
(b) surface flow intrusion into rivers.

(a) Subsurface coastal sea intrusion

Seawater intrusion into coastal aquifers is a problem of worldwide significance. It results when excessive groundwater pumping causes lateral and upward movement of seawater into the coastal wells. Unless controlled, the phenomenon causes significant rises in the salinity levels of the coastal groundwaters and concomitant detrimental effects on the lands being irrigated by these sources. Coastal aquifers discharge their unexploited groundwater to the sea. In many countries these aquifers serve as major water supply sources for the intensively developed agricultural areas and densely settled municipal areas in the coastal plains—such as in California, Florida and Israel. Exploitation of a coastal aquifer can be maintained indefinitely without causing hazardous consequences provided the maximum 'safe yield' of the aquifer is not exceeded. Simulation models for the behaviour of coastal aquifers have been developed in many countries to evaluate potential yields and are useful tools in predicting the behaviour of the groundwater system under alternative abstraction options. These models, although using idealized conditions and approximate assumptions, give results sufficiently accurate to enable the planner to establish the rate of groundwater development most suited to the local or regional conditions. Since lateral and vertical flow rates of groundwater are extremely slow, a critical lowering of the water table near the coast does not bring on immediate seawater intrusion because of the relatively long period required for the salt water to move inland. Thus in interpreting groundwater model results, it is extremely important to give full attention to the time scale which will generally be measured in decades rather than in years.

An excellent sample case history of successful management to date of a coastal aquifer and the prevention of the damaging effects of seawater intrusion on a regional scale is found in Israel. Israel's main water system is a nationwide grid of interconnected waterworks in which surface, ground and reclaimed water are integrated. Israel's water supply is based mainly on two elements:

(1) the waters of the Jordan river and of Lake Kinnereth in the North;
(2) the groundwaters of two principal aquifers—the coastal sandstones and the Yarkon–Turonian limestones.

Other than Lake Kinnereth, there are no favourable surface storage facilities in Israel and the need to provide regulation falls on the two main aquifers. These aquifers make up about 60% of all the nation's freshwater resources, and the coastal aquifer which supplies about 30% of the water consumption of Israel is used as a major long-term regulating reservoir. This aquifer flows towards the Mediterranean, is phreatic with water level depths ranging from a few metres to about 90 m below the ground surface. The aquifer is replenished by percolating rainfall and return flows from irrigation and waste water. The coastal plain overlying the aquifer has a total area of about 2 million hectares and includes the major urban communities of Israel, accounting for about 50% of Israel's population (in 1983 about 3.3 million). The projected population of the coastal plain is expected to double by the year 2010.

The coastal aquifer has an estimated yield of about 160×10^6 cubic metres a year and the limestone aquifer about 200×10^6 cubic metres a year. In order to enable large-scale immigration into Israel and a rapid development of agriculture during the years 1955–64, the coastal aquifer was exploited at an average rate of 330×10^6 cubic metres a year and the limestone aquifer at a rate of 260×10^6 cubic metres a year, causing an annual overdraft amounting to about 230×10^6 cubic metres. Over-pumping of the groundwater led to serious intrusion of seawater into the heavily over-exploited coastal aquifer. By the year 1964, however, the National Jordan conduit had been completed and water could be brought from Lake Kinnereth in the north, to supply the needs of the south. Some of this imported water was injected into wells in the coastal plain in order to repair the saline intrusion damage as rapidly as possible. Since then, using the results of simulation models, groundwater abstraction has been controlled and reduced to rates that ensure a sustained long-term programme of conjunctive use of Jordan water and groundwater.

The combination of a shallow phreatic water table, sandy soils, dense populations, intensive agriculture, and excessive pumping could have led to a potential disaster in Israel. Sound long-term planning using simulation and hybrid computer models, backed up by suitable water laws and competent government management, enabled the full use of the water resources to be achieved without creating critical irreversible and disastrous conditions. Problems still remain. Water quality records show that considerable parts of the aquifer are polluted by nitrates, and unless this is controlled the quality of a major portion of the urban water supply wells in the coastal aquifer will exceed the maximum allowable drinking water standards by the year 2000. On-going studies are evaluating the appropriate engineering techniques needed in the coming decades to ensure that the quality of the aquifer water conforms to these standards.

(b) Seawater intrusion into rivers

As more and more fresh water supplies are diverted or pumped from rivers, the flows reaching the sea decrease; eventually a point is reached when tidal flows begin to move upstream and cause deleterious increases in river salinity. This problem is worldwide, and many downstream riparian areas suffering from serious salt damage will go out of production unless a timely solution is found. The case history of the Chao Phya river in Thailand illustrates the problem.

The Chao Phya provides the major water source for the irrigation in the Central Plain and Delta of Thailand. The average annual supply is about 23×10^9 cubic metres a year, although variations of flow from year to year may be large. Flows reaching the Central Plain are being gradually reduced owing to intensification of rice production in the Northern Regions and expanding irrigation development along some of the upstream tributaries of the Chao Phya. The total crop land in the Central Plain is about 1.4 million hectares in the wet season, and the Royal Irrigation Department (RID) plans to increase production by expanding dry season cultivation from about the present 0.2 million hectares to more than 0.6 million hectares. The additional water for this development must come from storage in upstream reservoirs on the Chao Phya and its tributaries. This will aggravate the present hazardous condition when, during periods of low flow in the dry season (January to May), seawater flows upstream for about 30–40 km from the Chao Phya estuary. The irrigated areas affected today by the saline water intrusion cover about 1500 km^2, and the seawater intrusions also have deleterious effects on the domestic water supplies. The problem is a difficult one to solve because of the conflicting interests between the development targets of the Northern Regions and of the Central Plain. It has been estimated that a minimum flow of about 150 cubic metres a second must be maintained in the Chao Phya river at Bankok to ensure a salinity level of about 1000 ppm (TDS). However, 30 m^3/s of this flow will eventually be required to supply water to the Bankok area, and only 120 m^3/s will remain to flow south of Bankok. Under these conditions the salinity level will rise to 2000 ppm (TDS).

A study made by the RID in 1968 concluded that the value of water released from the upstream dams was greater for salinity control than for power. It is essential to assess accurately the amount of water to be allocated to control seawater in the Chao Phya river and to supply water to all the available irrigable land. Hence priorities for development must be established on a basin-wide plan to take into account irrigation, power, municipal supplies and salinity control requirements.

5.I.3.2.4 Land Subsidence

In recent years it has become apparent that extensive exploitation of groundwater has brought about land subsidence in many localities of the world. A casebook being prepared by UNESCO reports on the occurrence of more than 40 major land subsidence areas in the world and also describes methods for measuring, predicting and controlling the phenomenon.

Subsidence due to excessive groundwater pumping develops under two geological conditions. First, in carbonate rocks overlain by unconsolidated deposits that receive buoyant support from the groundwater, when the water table is lowered that support is lost; unconsolidated material may move downwards into openings in the carbonate rocks, often causing catastrophic collapses. Second, in semi-consolidated sediments of high porosity underlain by sand–gravel aquifers of low compressibility and inter-bedded with clayey aquitards of low permeability and low compressibility, pumpage from the aquifers is accompanied by vertical leakage from the aquitards. Since the compressibility of the clay is one or two orders of magnitude greater than the compressibility of the sand, the total compaction of the clay aquitard is much greater than that of the sand–gravel aquifer; the sediments undergo irreversible compaction as water is squeezed out of the interstices and result in land subsidences. Since the hydraulic conductivity of the clay is several orders of magnitude less than that of sand, the subsidence process is a slow one.

Land subsidence ranges from as little as 20–30 cm to up to 900 cm (in some parts of California), and the areal extent from small areas 10 km^2 in size to extensive zones covering up to 13 500 km^2 (also in California). The principal problems arising from land subsidence are:

(1) differential changes in the gradients of streams, drains and canals;
(2) failure of wells due to rupturing of the well casing;
(3) tidal encroachment in lowland coastal areas;
(4) damage to buildings, roads, railways, power lines and water supply systems.

Land subsidence in Arizona is used here as a typical case history. Arizona's current consumption of water from all sources is about 5.9×10^9 cubic metres a year, of which only about 3.4×10^9 cubic metres are renewable—hence Arizona is mining its groundwater at almost twice the natural replenishment rate, and during the last 30 years Arizona has already overdrawn about 75×10^9 cubic metres. The main culprit for this rate of depletion is agriculture, which uses about 90% of all water consumed in the state. The only way to solve this problem without radically reducing the irrigated areas is to bring in additional water supplies. This is the purpose of the Central Valley Project which, when completed, will include a 480 km canal transporting annually about 1.5×10^9 cubic metres of water from the Colorado River into Arizona. This, however, will eliminate only part of the overdraft. Consequently, the State Legislature has passed the Groundwater Act of 1980 to maintain a long-term balance between groundwater withdrawals and recharge (by natural or artificial means). Arizona has become the first state in the USA to limit the pumping of groundwater. In order to achieve an eventual balance, it is already recognized that agriculture's share of the state's water resources will have to be reduced from 90% to about 70%. This will be accomplished by changing present cropping patterns, improving irrigation efficiencies, reducing waste, introducing more efficient water consumption crops, and initiating a recycling programme for use of sewage effluents.

5.I.4 MODERN PLANNING METHODOLOGIES

Concern with the relationship between land transformation and irrigation is not new. In Mesopotamia, for example, where irrigation was developed on a regional scale in about 2300 BC, the country faced serious salinization difficulties—the earliest and probably most serious land transformation occurring at about 2000 BC when many settlements had to be abandoned. Since that time, irrigation planning has made tremendous strides. Engineers have learned the lessons of earlier civilizations, and today modern technology gives the engineer techniques to analyse the problems and predict the behaviour of the water system on both macro and micro scales. The tools are available to avoid a number of mistakes of the past.

Man's use of the soil has advanced through three stages. In the first stage, an ecological balance was maintained because shifting cultivation, or low irrigation intensities, did not deplete the soil fertility permanently. In the second stage, soil exhaustion occurred because the fertility of the soil was mined without replacing nutrients removed by cultivation, leaching or erosion. In the third stage, man has learned to conserve the productivity of the soil by returning as much or more than he removed, so that permanent cultivation or irrigation has become possible. The objective in irrigation management must be to maintain a dynamic balance in the salt–water input–output relationship and to replace fertility losses with fertilizers and with good farm management. There is no need to assume that technology will remain constant, nor need we prophesy doom for modern irrigation projects.

Historical evidence favours the view that 'necessity is the mother of invention', and new technologies will be developed to meet any potential hazard that may arise—particularly since these problems do not appear suddenly but develop over a decade or two. A good example of present-day regional irrigation planning can be found in the Rio Colorado Basin Project, the plan for which was prepared in the 1970s and included more than 20 irrigation sub-projects covering a total of about 0.8 million hectares. This basin-wide study used multiple social, economic and regional objectives to evaluate various development options in an overall water resource plan, and developed mathematical simulation and optimization models to predict potential problems, future benefits and investment costs. A more recent major study is being carried out in India on a scheme funded by the World Bank—the Sadar Sarovar Project in the State of Gujarat. This multiple-purpose hydro-power and irrigation project will use the waters of the Narmada river to irrigate 2.2 million hectares of land from the year 1990. The Narmada Planning Board established by the State of Gujarat has developed optimization models to examine the benefits from alternative cropping programmes and, because groundwater and salinity problems are expected to develop, surface and groundwater simulation models have been set up to predict the rates of groundwater rise in various zones of the project. The results of these simulation models are already being incorporated into the design, and extensive public and private tube well schemes for conjunctive use of ground and surface waters are being anticipated so as to ensure the permanency of the project by maintaining a satisfactory salt–water balance in the region.

5.I.5 STABILITY AND PERMANENCE

Surface and in-depth land transformation always follow in the wake of irrigation development. The surface modifications take place during the construction period, while the in-depth changes appear only after the introduction of water into an area. While surface modifications can be regarded as beneficial to man, the in-depth changes carry inherent potentially harmful features—waterlogging, salinity, seawater intrusion and land subsidence. These hazards must be controlled in time lest their deleterious effects become malignant and reach dangerous and irreversible levels. Good management is therefore a key ingredient of irrigation development. Management must continually monitor the water and salt balances so as to predict the rates of build-up of the hazards. Finances must be made available in time to introduce the measures needed to control the changes so that they do not reach injurious states, placing the project in jeopardy.

The main lessons to be learned from the history of irrigation are that governments must recognize these potential hazards, in the early project planning stages, and establish competent operational management structures to detect and heed early-warning signals that precede problematic situations. There are no perfect solutions, and choices will always have to be made between various policy options, trading off losses in production and farmer's income in order to achieve long-term stability and permanence. Good planning, the quintessence of engineering, is essential to ensure lasting survival of an irrigation project and the society dependent on it. In the past, emphasis in planning has been placed on the need to meet plant water demands to attain maximum yields (per unit of land). Greater weight should now be given to the management of supplies, with the aim of maximizing economic returns on water (rather than land) and minimizing long-term hazards. In carrying out the economic analysis of a project, the investments required to prevent long-term project failure must be included in the stream of costs, or a concomitant reduction in the stream of benefits must be made to take into account a probable reduction in yields as the hazards advance.

5.I.6 FURTHER READING

Iftikhar, A. S., and Nasri, S. N. (1982) A new approach to waterlogging and salinity control. *Proceedings of National Seminar on New Socio-Economic Order for Pakistan*, Lahore, 1–23.

Al-Layla Anis, M. (1978) Effects of salinity on agriculture in Iraq. *Irr. and Drainage Div., ASCE*, **104**, (IR 2), 195–207.

Evenari, M., Shanan, L., and Tadmor, N. (1982) *The Challenge of a Desert* (2nd ed.). Harvard University Press, Cambridge, 1–436.

Framji, K. K., Garg, B. C. and Luthra, S. D. L. (1982) *Irrigation and Drainage in the World—a Global Review* (two vols, 3rd edn.). ICID, S. P. Dhawan Central Electric Press, New Delhi.

Freeze, R. A., and Cherry, J. H. (1979) *Groundwater*. Prentice Hall, New Jersey.
Gabaly, M. M. (1977) *Problems and Effects of Irrigation in the Near East Region*. FAO Regional Office of Near East, Cairo.
Garde, R. J. (1978) Irrigation in ancient Messopotami. *ICID Bulletin*, **27**, (2), 11–22.
Hansen, V. E., Israelson, O. W., and Stringham, G. E. (1979) *Irrigation Principles and Practices* (4th edn.). John Wiley, New York.
Lieftinick, P., Sadove, A. R., and Creyke, T. C. (1969) *Water and Power Resources of Pakistan* (3 vols). World Bank Publication, The Johns Hopkins Press, Baltimore.
Major, D. C., and Lenton, R. L. (1979) *Applied Water Resources System Planning*. Prentice-Hall, New Jersey.
Michael, A. M. (1978) Salt problems in irrigated agriculture, In *Irrigation*, Chap. 9. Vikas Publishing House Pvt Ltd, New Delhi.
Morrison, A. (1983) Arizona's water strategy: bring more in and restrict its use. *Civil Eng.* (ASCE), April, 46–49.
Poland, F. (1981) Subsidence in the USA due to groundwater withdrawal. *J. Irrig. and Drainage Div., ASCE*, **107**, (IR2), 115–135.
Shuval, H. I. (ed.) (1980) *Water Management Under Conditions of Scarcity*. Academic Press, New York.
UNDP (1971) *A Review of the Water Resources and Other Potential Developments of Northern Thailand*. Final Report, IPD, Project THA–33, UNDP New York, 1–71.
UNDP and IBRD (1981) *Master Water Plan Project, Arab Republic of Egypt*. Main Report UNDP/EGY/73/024, Ministry of Irrigation Cairo, 1–170.
Valentine, V. E. (1974) Impacts of Colorado River salinity. *J. Irr. and Drainage Div., ASCE*, **100**, (IR4), 495–510.
Worthington, E. (Ed.) (1977) *Arid Land Irrigation in Developing Countries*. Pergamon Press, Oxford, 463.

Land Transformation in Agriculture
Edited by M. G. Wolman and F. G. A. Fournier
Published by John Wiley & Sons Ltd

CHAPTER 5.II
The Impact of Mechanization

G. SPOOR, R. CARILLON, L. BOURNAS AND E. H. BROWN

5.II.1 INTRODUCTION

Agriculture is a series of organized processes carried out by man on his natural environment, particularly affecting the soil and vegetation, for the purpose of systematically producing foodstuffs for substenance and raw materials such as fibres. The only other activities competing with agriculture in this respect are the gathering of wild plants, fishing and hunting. From the Mesolithic era when the world's inhabitants numbered just a few millions, agriculture has developed the earth's potential to produce foodstuffs such that, despite localized endemic or episodic famine, the world's population has been able to reach its current level of 4.5 billion over the course of a dozen millennia supported by agricultural production. Continuing rapid and sweeping progress in agriculture will be required in most countries in order to cope with the demographic growth expected in the coming decades (a world population of 5.6–6 billion people forecast for the year 2000, and 10—15 billions for the second half of the first century).

Agriculture is, above all, a process of coordinating the preparation, care, maintenance and harvesting of crops, but it also involves considerable handling and transport operations. It is a difficult activity, for in place of natural equilibria, it substitutes others which, if long-term production is to be maintained, must not have irreversibly destructive effects upon the land.

Agriculture by definition involves the use of various implements (many of which must be mobile or transportable), and the application of directed force. Equipment must be suited to a particular treatment of the soil, crops and animals, in order to respect biological imperatives and protect the new equilibria that are essential to the entire agricultural environment.

Much progress in agriculture has only been made possible through improvements in implements and the means of propelling them. This has involved a change from primitive manual farming to farming using draught animals, the machines for which were perfected relatively slowly, followed by a rapid switch to mechanized farming.

In France, for example, it was only in 1850 that this latter stage began to be apparent; 130 years later 99% of farming is mechanized, involving the use of 1 400 000 tractors and 150 000 self-propelled machines. Agricultural production has more than doubled over the last 30 years, a result achieved over a slightly smaller surface area by a workforce only three-quarters its former strength. There has been consequently a significant shift of farm labour (5 to 6 millions out of a total of 20 millions) into non-agricultural occupations which, until the oil shortages of the early 1970s, allowed for exceptionally rapid growth in the overall wealth produced in the country, and consequently a very significant rise in the standard of living. These changes have also met the demands met by the concurrent improvements in the average Frenchman's diet, the demographic growth of the country, and have created agricultural surpluses that have made large-scale exporting possible. Precisely the same changes are evident in agriculture in the United States, as noted earlier.

Throughout the world, agriculture has followed broadly the same sequence of development and it is possible to recognize stages in the development and introduction of new agricultural implements (Table 5.II.1). At present there are, worldwide, somewhat over 20 million farm tractors in service on 1.5

Table 5.II.1 The development of farm equipment and changes in food supply

Approximate dates	Era or essential characteristics of the era	'Agricultural' implements[a]	Means of acquiring food supply
−1 000 000	TERTIARY↑ QUARTERNARY↓	No implements (in the sense of tools specifically chosen, and possibly adapted, crafted and saved for identical and repeated use)	Gathering (?)
−600 000	Lower Paleolithic	Cut stones { blunt & handled (1 side) ↓ (2 sides) sharp & thrown }	*Hunting* *Fishing* and more methods of acquiring food supply (fruits, grains, leaves, roots, etc.)
−150 000	Middle Paleolithic		
−40 000			
−20 000 . . .	Upper Paleolithic		
−12 000/ −10 000	Mesolithic	Sickles, axes, pickaxes	Ditto, plus forest exploitation, gathering of wild cereals
−7000/−5000	Neolithic	Digging sticks Hoes, grippers, spades Swing ploughs	

Table 5.II.1 (*contd*)

Approximate dates	Era or essential characteristics of the era	'Agricultural' implements[a]	Means of acquiring food supply
−3000/−2500	PREHISTORY↑		
	HISTORY↓	Pickaxes, spades, hand-held hoes	Hunting
		Wheels	
America		Digging sticks/shovels Stone-faced clubs	Fishing *Extensive husbandry*
Sumer, India		Two-handled swing plough with seeder	
Egypt, Greece Rome		One-handled Mediterranean-type swing plough without seeder	*Crops*
Mesopotamia	Antiquity	Transport vehicles	
Mesopotamia		Transport sleds	
Egypt		Blocking hoes and forks	
		Bill hooks, beating rods	
Rome		Harrows Water-mills Screw presses Swinging yokes	
−270		Two-handled wheeled gallows	
	Founding of the Roman Empire	Plough	
Gaul		Ear harvester (disappeared in the 5th century)	
Beginning of the Christian era 395	Roman Empire	Disymmetrical use of the swing plough	*Crops*
		Two-piece flails	Hunting
	Barbarian invasions	Advances in metallurgy	Husbandry
8th century		Beginning of the development of the plough	Fishing
	Middle Ages	Development of the water-mill	
		Mounting coulters on swing ploughs and ploughs	
10th century		Harrows, rollers	
1453			
	Renaissance	Plough supports Metal spades Windmills with rotating roofs	*Crops* Hunting Husbandry Fishing
1600			
	Pre-industrial transition period	First prototypes of seeders	*Crops* Husbandry Hunting Fishing
1789			
	Political unrest		*Crops*
1800–1820			

Table 5.II.1 (*contd*)

Approximate dates	Era or essential characteristics of the era	'Agricultural' implements[a]	Means of acquiring food supply
	Development of transportation and communication Industrial expansion	Widespread use of scythes Numerous industrial inventions applied or specific to agriculture: cultivators, crossbill rollers, hay-makers, mowers, windrowing corn-mowers, binders, combined-harvesters, threshers, fixed-part cubing machines, seeders, fertilizer spreaders, reversible mouldboard ploughs, multiple ploughs, steam-powered ploughing, self-propelled threshers, horse-powered generators and inclined horse-powered generators Progressive replacement of cast iron, iron, and steel for wood in the construction of existing soil working machines.	*Husbandry* Fishing Hunting
1860–1890	--------------	First development of advanced animal-powered equipment for other than soil working	*Crops* *Husbandry* Fishing
1914–1918	1st World War		
1918		Second development of advanced animal-powered machines. Beginning of motorization	*Crops* *Husbandry* Fishing
1929	-------------- Depression	First development of combine-harvesting	*Crops* *Husbandry* Fishing
1939–1945–1950	2nd World War		
1945–1950		Development of motorized machines and mechanized farming	*Crops* *Husbandry* (Fishing)
1973	1st oil shortage		
(1982)		Application of data-processing techniques to some farm equipment	*Crops* *Husbandry* (Fishing)

[a]Hunting and fishing have been considered as 'agricultural' activities, utilizing natural resources in prehistoric times; implements for these activities have therefore been included in the agricultural implements section for this period only.

billion hectares of farmland. This equipment is presently concentrated in the industrialized countries, where barely 10% of the world's active farm workers produce close to one-half the world's food on one-fifth of the world's usable farmland, thanks largely to the efficiency of the equipment and amendments at their disposal. It is almost inevitable that mechanization will in the future have as great an impact upon the land in the other four-fifths of the world if its growing population is to be fed and their rising expectations are to be met. That the area of fertile land cannot easily be extended, and that in most countries such land is also in demand for other purposes, will in the future necessitate increased agricultural productivity from less fertile lands, involving the extended use of farm machinery.

A change from the use of draught animals to tractors as the source of motive power in agriculture is brought about because of perceived economic and human benefits (Table 5.II.2), but there are consequent changes to the

Table 5.II.2 Technical comparison between draught animals and tractors

Draught animals	Tractors
Must be cared for every day	Require maintenance only in terms of the amount of work they perform
Must be fed every day	Only consume energy when they work
Require large areas to be devoted to growing their food	Consume petroleum products which must be obtained outside the farm (and even the country)
Tire after 8–10 hours of work, including the breaks needed during the working day	Can work long, solid hours without breaks or interruptions if the operating personnel work in shifts
Work at one, slow speed—that of man	Work at a choice of several speeds, up to about 30 km/h
Leave discontinuous, shallow marks in the soil	Leave large, continuous tracks in which soil compaction may be deleterious
Can turn in place under certain circumstances (except with some types of hitches)	Have turning radii of several metres, causing wider furrows
Most often require the carter or ploughman to walk behind the hitch (10 km in single-share ploughing per workday)	Allow the driver to sit in increasingly comfortable seats on a heavy vehicle
Can be driven from a long distance but with a reaction time that is sometimes quite long	Require the driver to be in place, but execute his commands immediately
Can make major efforts, but for a short time	Can only get out of difficult spots if high-powered equipment is used
Produce manure	Do not provide manure but spread it quickly and efficiently, especially artificial fertilizers

Table 5.II.2 (*contd*)

Draught animals	Tractors
Are silent and non-polluting	Are noisy and polluting
Even when riding the hitch the carter is subjected to slow bumps; nevertheless, he is not subjected to high-frequency vibrations	Create high-frequency vibrations which the driver feels if they are not adequately absorbed
Are dangerous (fatal accidents possible)	Are dangerous (fatal accidents possible)
Replacements can be made relatively cheaply by breeding	Replacements involve high capital costs
When hauling, can only deliver low continuous effort (75 kg per draught horse) using conventional hitches of 1 to 4 horses (beyond that number, the animals are difficult to hitch and lead, and their output is worse still)	When hauling, can deliver high continuous effort (several tonnes, for example) without difficulty if the engine chosen is sufficiently powerful, if the work speed is not too great, and if traction is satisfactory
When the mechanism is in motion, it is necessary to have a drive wheel operating on the ground; the efficiency of this system is limited by the low traction of the hitch and its low flexibility (does not operate at rest, jerky starts, necessity of working over dry ground, etc.)	When the mechanism is in motion, all operations are possible thanks to the power take-off, and the power transmitted can be very high if tractor class and power are adequate; tractors have made total mechanization of farm work possible
Draught animals cannot work at night	The tractor has headlights
Draught animals can work over loose, wet earth	The tractor's efficiency is limited by its traction and it is inadvisable to insist upon working on earth where the modern tractor has no traction
Draught animals can supply energy in place by means of a rotary device; the energy is limited to the continuous effort of one or perhaps two animals which must take breaks	The tractor can provide energy when stationary without difficulty, with no other limitation than engine power, and without having to take breaks
In the fields, the working part of the machines can only be lowered by the force of gravity and raised again by means of a ratchet wheel	In the fields, lifting and hydraulic depth adjustment of the implements are possible; electric controls are also used
Cannot carry tools	Can carry heavy implements
Do not require special knowledge of mechanics	Require considerable mechanical knowledge, which is new for farmers
Correspond to a long agrarian tradition	Are new for farmers
Hardly allow for progress in work productivity, once animal-powered harvesting equipment, representing the second phase of the mechanization, has been acquired	Only require the use of new, more powerful mechanical equipment to offer new direct possibilities and substantial progress in the productivity of work

soil itself which may or may not in the long run be beneficial. These fall into four main groups:

(1) factors which facilitate an extension of the area under cultivation;
(2) those which lead to compaction of the soil;
(3) factors enabling cultivation to greater depths;
(4) those which encourage changes in the chemical as well as the physical properties of the soil.

The principal items of equipment enabling this intensification of agriculture and consequent transformation are of five principal types. First there are multi-purpose tractors, increasingly of the heavier four-wheeled-drive kind. Secondly there are implements for working the soil—including the plough, still the most widely used of farm implements, but also including deep-working equipment such as chisels, spading machines and sub-soil ploughs as well as a wide selection of cultivators and harrows employed in the preparation of the seedbed. Thirdly there is a very wide range of machines involved in the spreading of wet and dry fertilizers and pest control chemicals. The fourth group of machines are those involved in the harvesting process: combine harvesters for grain crops, hay cutting and turning and silage equipment for grass. Root crops such as potatoes and sugar-beet can also be harvested using mechanical means, often self-propelled, as can vegetable crops such as peas and beans and even grapes and cotton. A fifth type comprises the multifarious pieces of farm machinery which use electricity as a direct source of power.

5.II.2 SOIL MANAGEMENT

Transformation of the land to create a good soil physical environment is a prime requirement for high levels of crop production, and this is synonymous with good soil structure and subsurface drainage. Such a condition provides an appropriate pore size distribution, optimizing available water and aeration status, and helps to provide adequate temperatures with minimum impedance to root development. Soil management through land transformation is concerned with maintaining and, it is hoped, improving this environment; but in the process, two opposing sets of forces are at work, one group tending to improve structure and the other causing deterioration.

Soil structure improvement arises largely through the related activities of plant roots, soil organisms and natural weathering. Roots, through their growth, exudate and breakdown products and coupled with organism activity, are responsible for the formation and the stabilization of the larger conducting pores and structural units throughout the soil profile. Mechanical operations can assist in exposing soils to the weathering agents and in the formation of the larger pores but can do little for their direct stabilization.

The forces causing soil structure deterioration, and thus increasing the negative aspects of transformation, are related to those activities which encourage the oxidation of organic matter, aggregate destruction through mechanical or chemical means, and restrictions to root development through impeding layers and general compaction. The deleterious activities are frequently associated with many farming operations, including soil tillage and harvesting, traffic, poor drainage and irrigation, as well as rainfall impact erosion of bared soil surfaces.

5.II.3 PAST AND CURRENT APPROACHES TO SOIL HANDLING

Experience in Britain exemplifies the changing influences of the mechanical management of soil and consequent soil transformation.

5.II.3.1 Soil Handling in Britain before 1960

The initial method of cultivation based on a digging stick or hoe was shallow, extremely precise, and limited in area, with soil only being disturbed where absolutely necessary—largely for the control of weeds. The introduction of animals as a source of power, and the mouldboard plough for weed burial, changed this precision tillage into a form of more extensive and complete tillage. Adequate burial could only be achieved through deeper working at 50–100 mm depth and complete soil surface inversion. The deeper tillage resulted in the formation of clods, and hence the need for subsequent operations for seedbed preparation.

The ideas and practices developed by Jethro Tull in the eighteenth century led to a considerable increase in the amount of complete tillage, based on the premise that production increased with increased soil working. Soil preparation came to take the form of ploughing and cross-ploughings, followed by multiple harrowing and often inter-row hoeing. These practices continued well into the twentieth century.

Interest in deeper tillage to depths of 500–600 mm, usually linked with subsurface drainage, increased in the nineteenth century, but the practice was never widespread and ceased abruptly as agriculture entered the depression at the end of the century. Protagonists of these practices attributed the improved crop response as much to better drainage as to greater rooting depth.

The introduction of the tractor in the twentieth century did not change soil handling methods, although ploughing depths tended to increase. Yield responses to deeper working were very variable and, where positive, were largely attributed to better weed control Russell (1956).

During the whole of this period, from the introduction of the mouldboard plough by the Romans to the late 1960s, a system of complete uniform tillage was practised with effectively random passes of animals, tractors and implements across the fields. Crops were established largely in the spring, allowing plenty of time for soil weathering following autumn ploughing. Although tillage was frequently excessive, the major benefits came from weed control, and weathering played a major role in conditioning and restructuring structurally damaged soil. The power available for soil movement was low, and hence soil conditions themselves played a large part in dictating whether soil working was possible. Soil loadings over the period were relatively low, and the results of studies of comparative tillage Russell (1956) and of compaction (Soane, 1970; Eriksson *et al*., 1974) show that most of the soil damage resulting from the farming operations was confined within the plough layer. This damage was alleviated annually through ploughing and weathering action.

5.II.3.2 Soil Handling in Britain after 1960

During the 1960s the rate of technological change started to increase rapidly, the changes being initiated by economic factors and research. The major changes which influenced soil handling practices were:

(1) large increases in yield potential of crop varieties;
(2) increased fertilizer, pesticide and herbicide usage;
(3) the introduction of paraquat;
(4) reduction in farm labour forces;
(5) the increased power available;
(6) a change from farmyard manure to slurry;
(7) closer links between producers and processors.

These changes stimulated a rapid increase in yield in many crops and in the quantity of materials to be handled. There were also increases in crop inputs (e.g. fertilizers and herbicides) and in the timeliness of their application; a move towards autumn-drilled crops and earlier and shorter establishment periods; and increased stocking rates and a longer grazing season. These changes in turn resulted in:

(1) harvesting times dictated increasingly by crop growth stage rather than soil condition;
(2) reductions in effective soil weathering time between crops;
(3) the increased importance of soil physical conditions influencing output as fertility levels increased.

The response of the farm machinery industry to all of these changing factors in the agricultural scene was to increase machine capacity and power further, which led to an increase in the weight of the equipment (Table 5.II.3). The tractive performance of power units was improved by making maximum use of weight, and the additional weight, particularly on implements, was supported on high-pressure, high-ply-rating, relatively small-diameter and narrow tyres. The industry also introduced specialist processor-inspired machines, frequently designed with little thought for soil conditions.

Table 5.II.3 Sizes of farm equipment

	Power (kW)	Mass (tonnes)	
		1970	1983
Wheeled tractors	26	2	
	60	3	
	150		10
	200		12
Combine-harvesters (full tank)		6	15
Trailers		5	15
Manure spreaders		4	10
Potato harvesters (tankers)			10
Self-propelled slurry tankers			25
Ministry of Transport maximum allowable axle load (UK)		11 tonnes	

The net effect of all these changes on the soil and its response can be most clearly seen by following the developments and experiences in the direct drilling of cereals on the well-suited soils which started in the early 1960s. Initially the results were very encouraging, with equivalent and sometimes higher yields than from traditional tillage systems (Davies and Cannell, 1975). Continuing improvements were noted in pore size distribution, organic matter levels, surface tilth, soil structure and soil support capacity (Russell *et al.*, 1975). After a period, however, compaction symptoms started to appear, coinciding with significant increases in machine contact loads and pressures. The soil, even in its undisturbed state, was now no longer capable of supporting the surface loads without unfavourable deformation.

An increase in soil problems was also experienced over the same period in the root-crop growing areas, resulting in increasing abandonment of these crops on the heavier soils. Whilst soil damage was not the only reason for a cropping change, these soils were the first to suffer from the more punishing loads, with farmers finding increasing difficulty in rectifying the structural problems created. Similar compaction problems have also arisen in intensive grassland areas, where stocking densities, and the weights of forage- and slurry-handling equipment, have all increased.

This period has therefore seen a very large increase in soil loading and a move from initially satisfactory zero tillage to a situation where there is increasing interest at both farm and research level in the possibilities of deep tillage for the future.

Similar developments in tillage and machinery practices have taken place in many parts of the world, the major difference being the time scale, with certain areas still practising precision tillage and others strip tillage with animal-drawn equipment.

5.II.4 FUTURE SOIL MANAGEMENT OPTIONS

The brief review of past and current soil-handling practices highlights the emphasis placed on management measures to overcome problems, rather than attempt to prevent the problems arising in the first place. The options open for the future are therefore either to continue this reclamation management approach, or to move towards preventative management.

In the tillage/mechanization area the damaged soil zones have until recent years been confined within the top 300 mm of soil. This has allowed the annual reclamation process of tillage plus weathering to alleviate most problems. Any continuation, however, of the current machinery trends with increasing loads will increase the depth and degree of compaction damage, making reclamation increasingly more difficult and expensive. Figure 5.II.1, taken from Eriksson *et al*. (1974), shows the increase in vertical soil deformation at depth that can be expected from higher surface loadings, even at the same contact pressures. The situation has already arisen on many farms, as shown in Table 5.II.3, where the axle loadings on a material—namely soil—which the civil engineer removes immediately as being unsuitable for a road, exceed the permissible loads on a carefully prepared road pavement.

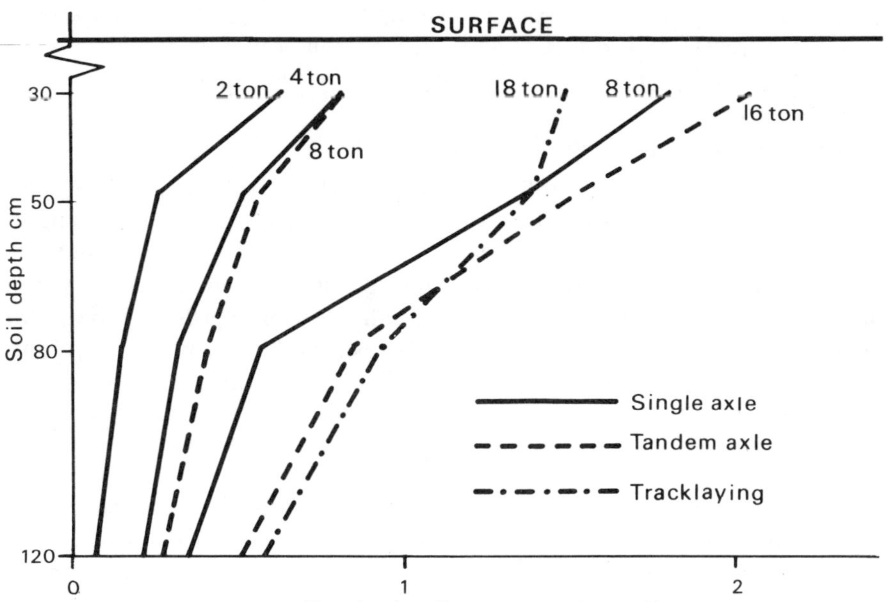

Figure 5.II.1 Vertical soil movement under loaded wheels and tracks (Danfors, 1974)

The results from continuing Swedish work (Hakansson, 1980) on the natural recovery of deeper compacted layers suggest that compaction effects could last for decades. Virtually no alleviation of compaction within a layer at 350–450 mm depth had occurred over a 2-year period, despite the soil freezing annually to depths well below this. The principles and techniques for satisfactory soil loosening at depth are established to enable deeper reclamation to be executed (see Spoor and Godwin, 1978), but loosening alone does not immediately rectify structural damage. The continuation, therefore, of this reclamation form of management in the future must be seriously questioned, for even without deep compaction problems, the annual ritual of loosening soil for the pleasure of recompacting it again makes little sense if it could be avoided. Any move towards a system of preventative management will require a reduction in the magnitude of the punishing loads and pressures applied to the soil. Such a change will almost certainly involve a cost, and hence it is worth while considering the potential benefits which may be achieved from preventative management to weigh against this cost.

5.II.5 POTENTIAL BENEFITS FROM PREVENTATIVE MANAGEMENT

The possible benefits arising from control over the level of soil compaction and traffic can be categorized as follows:

(1) avoidance of yield depressions resulting from incomplete reclamation;
(2) yield increases as a result of an improved soil environment;
(3) reduced establishment costs and increased system capacity;
(4) increased working period;
(5) opportunity for precision tillage.

The yield depression of the following crop resulting from incomplete reclamation can vary widely, depending upon the level of damage remaining and the subsequent weather conditions. Figure 5.II.2, taken from Hakansson (1980), shows equilibrium yield depressions resulting from medium-size farm vehicle compaction every autumn prior to ploughing over a 10-year period in Sweden. The compaction effect was mainly restricted to the plough layer, and an elapsed time of 5 years was required after the experiment to alleviate all the damage done. Continuing yield reductions are also being recorded following compaction damage to 500 mm depth created through one concentrated wheeling treatment with a 16 t tandem axle trailer (see Figure 5.II.3).

Whilst it is recognized that there is an optimum degree of compaction for crop production, deep loosening experiments without subsequent recompaction by wheels give some indication of the potential for yield improvements (see Table 5.II.4 and Stephens (1855), McEwen and Johnson (1979), Rowse and Stone (1980) and Gooderham (1976)).

Table 5.II.5, derived from Patterson *et al.* (1980), shows the influence of

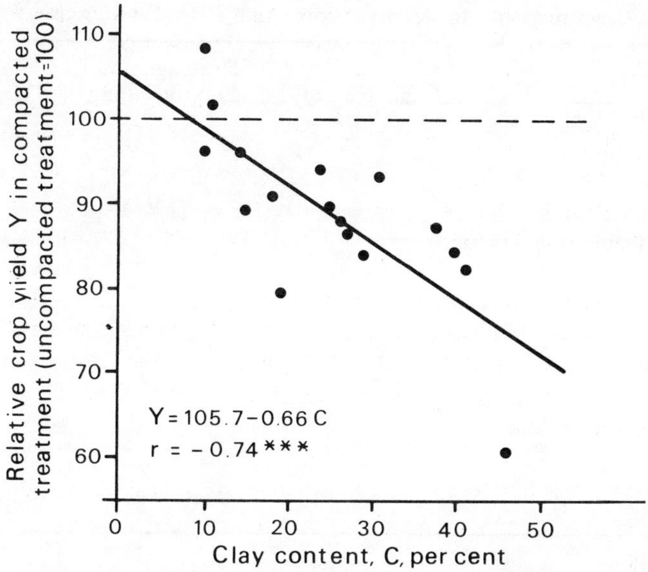

Figure 5.II.2 Yield depression from farm-vehicle compaction during previous harvest (Hakansson, 1980)

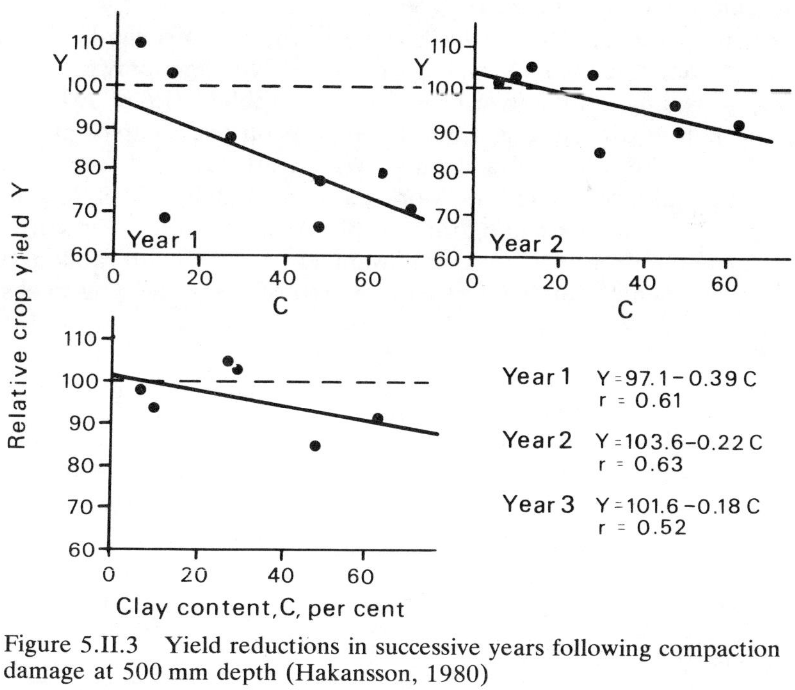

Figure 5.II.3 Yield reductions in successive years following compaction damage at 500 mm depth (Hakansson, 1980)

Table 5.II.4 Crop response to deep-soil loosening (% yield increase)

	Drainage only	Drainage and deep loosening	Deep loosening
Yester 1850 (clay)			
cereals	35	95	
turnips		100	
Rothamsted (sandy loam)			
cereals			20–25
sugar-beet			11
potatoes			0
Wye (silt loam)			
cereals			15
Wellesbourne (sandy loam)			
potatoes			0
broad-beans			25–95
red-beet			−3–20

working depth on the cost and system capacity when establishing winter cereals. The significant penalties from deeper soil working to overcome deeper compaction can be clearly seen.

Spoor and Godwin (1978, 1979) have shown that soils can be effectively loosened at moisture contents well in excess of those commonly considered as limiting, providing confining stresses are low. Table 5.II.6 shows brittle or loosening failure occurring in triaxial compression test samples at moisture contents well above the plastic limit. With traffic control, therefore, the opportunity exists for increasing work days and timeliness, extending crop growing seasons, and extending root-crop production to heavier soils.

The current approach to crop establishment executed with random tractor passes is completely dominated by the action of the wheels or tracks. The

Table 5.II.5 Relationship between cultivation working depth and cost of establishing cereals (Patterson *et al*., 1980)

Cultivation depth (mm)	Relative cost	Relative output (time/unit area)
0	1	1.0
100	0.9	2.0
150	1.3	3.5
200	1.8	4.0
400	2.8	7.0

Table 5.II.6 Effect of soil moisture on type of soil failure at low confining stress (% samples failing in brittle manner)

Soil	Moisture tension (bars) 0.5	0.16	0.50	1.00	Moisture tension at plastic limit
Wicken Series Clay	0		37	90	2
Fladbury Series clay	12	50	100	100	9

tractor wheels continually modify the soil condition produced by the implements, tending to increase the degree of compaction at each pass. This could well be a major reason for the lack of significant differences between implement and no-tillage treatments in many tillage experiments. Control over the dominating wheel opens up the possibility of returning to precision tillage. Particular attention can then be paid to providing optimum but different conditions in various parts of the field and the soil profile for the seed, the roots, wheel support and traction and the conservation of water and soil. Figure 5.II.4 illustrates, from the work of Prestt at Silsoe College, the

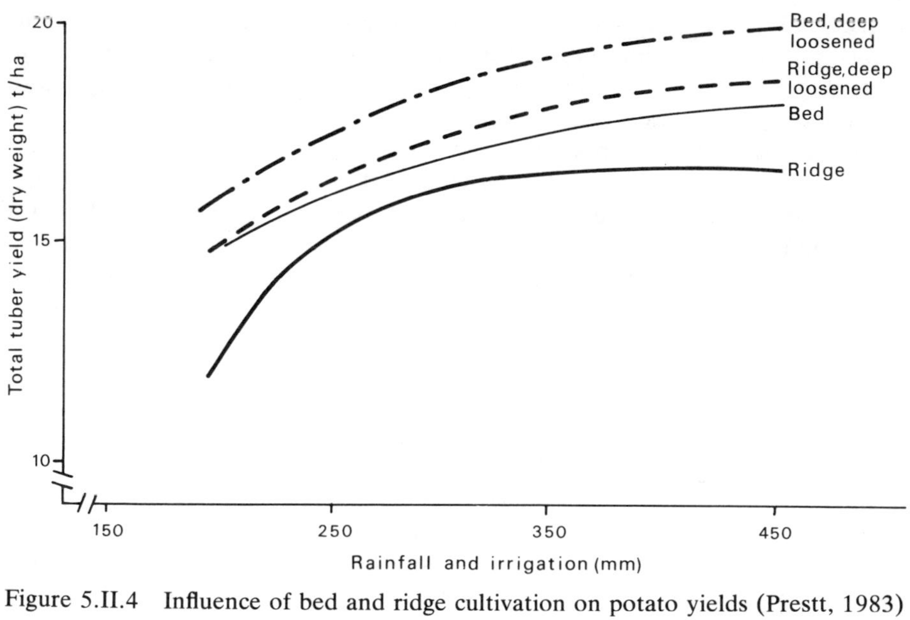

Figure 5.II.4 Influence of bed and ridge cultivation on potato yields (Prestt, 1983)

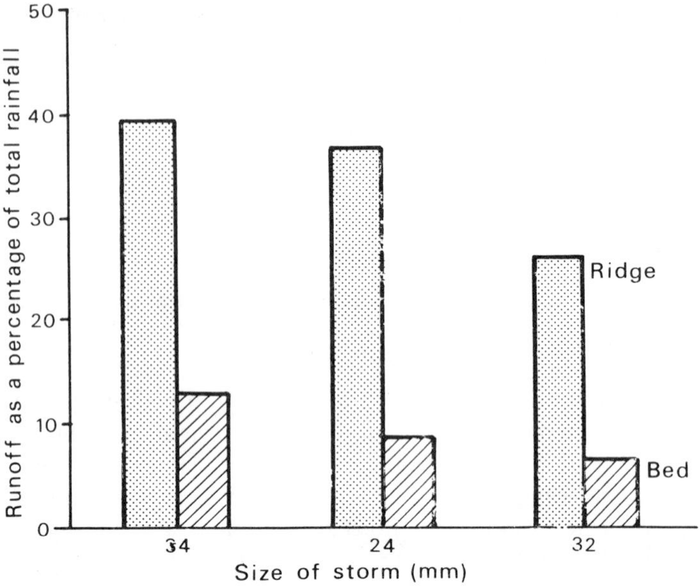

Figure 5.II.5 Influence of bed and ridge cultivation on rainfall runoff

type of yield improvement possible from providing a potato growing area within a bed system rather than in traditional ridges. The bed surface profile was formed to encourage water movement into the growing area, rather than runoff into the compacted traffic support area in the furrows. Differences in percentage runoff between the bed and ridge treatments are shown in Figure 5.II.5.

Larger-scale soil conservation measures rarely eliminate soil loss and are often not feasible in smallholder agriculture. Preventative management can, through significantly reducing soil loss, maintain soil fertility and hence production capacity.

These potential benefits, coupled with the increasing difficulty and cost of satisfactory deeper soil reclamation, force us into the situation where much more effort must be put into preventative rather than reclamation soil management.

5.II.6 A PREVENTATIVE MANAGEMENT APPROACH

In tillage mechanization, preventative management must involve a reduction in the damaging effects of traffic, the modification of husbandry techniques and greater attention to subsurface drainage. Similarly, drainage and husbandry methods must be given greater emphasis in irrigated and erosion prone areas. Interest is increasing in the possibilities of preventative management, but doubts exist as to its feasibility. Possible ways forward are considered under the following categories:

(1) reducing soil compaction potential of traffic;
(2) field control of wheels and tracks;
(3) soil preparation;
(4) drainage;
(5) cropping.

The two major factors influencing the compaction potential of traffic are the contact pressures and loads. Whilst it may be unrealistic to expect significant weight reductions in future, unnecessary weight can be shed and consideration given to restricting the magnitude of future increases. The suggestion made by Hakansson (1979) that axle loads of agricultural vehicles should be restricted to keep compaction damage within the top 400 mm must be considered very seriously.

The level to which farmers can reduce inflation pressures with current equipment and tyres is limited. Very low ground pressures (5–10 kPa) can be achieved on light, low traction vehicles using existing tyres. This is not possible, however, with the current very-high-pressure equipment such as trailers, combines and other harvesters, nor on vehicles with high traction requirements. The use of dual wheels can help, but their application is frequently limited by width restrictions and maximum allowable bending moment and torque levels in existing axles. Many of these problems can be resolved at acceptable cost at the design stage—hence the need for farmers to encourage manufacturers into action to minimize the compaction potential of their products. Crop transport operations are frequently the most punishing to the soil, and the continuing development of interlinked field and road container systems offer much promise for the future. In the meantime, farmers can benefit from making best use of the tyre equipment available, ensuring operation of their current units at maximum allowable tyre deflections and by achieving adequate traction through minimum rather than maximum added weight.

Tramline and bed systems for the control of through crop traffic after establishment and before harvest have proved very successful. Their full potential for complete traffic control through all operations has yet to be realized in most situations.

Situations requiring soil disturbance may be significantly reduced in future if good traffic control can be established. Where soil disturbance is still necessary, however, there are two major information gaps constraining the future development of soil preparation techniques. These are the lack of information on the actual soil environment required, and of objective tests to define the soil environment. These aspects were less important in the past when almost complete reliance was placed on the successful combined action of mouldboard ploughing and weathering to provide the required soil conditions. Future research programmes need to be directed towards providing this information in a form which would enable a more precise approach to be taken to tillage in the field. This requires a clear definition of the necessary environments for the seed, roots, and soil, and water conservation so that appropriate conditions can be produced between the controlled traffic zones.

In the immediate future in situations where traffic control cannot be achieved, a reversal in the sequence of tillage operations should be considered so that soil loosening, if required, becomes one of the last operations rather than the first, to minimize the risk of excessive recompaction. Great scope exists for the better use and selection of tillage equipment to avoid the creation of new problems when overcoming the original one. There is similar scope for maximizing the use of undisturbed soil strips during multi-pass operations, to maximize traction and minimize rolling resistance and compaction.

The long-term maintenance and improvement of soil structure is mainly dependent on root and organism activity together with natural weathering action. This improvement can be best achieved through minimum soil disturbance by cultivation and surface traffic, and good subsurface drainage below. Subsurface drainage is therefore one of the most effective and necessary aids to preventative soil management, contributing by increasing the soil resistance to compaction as well as through improved aeration and salinity control. Drainage problems are most severe and solutions most expensive in the finer textured soils, where close drain spacings are required. Appropriate soil loosening measures and mole drainage have much to offer in these situations, and possibilities should be explored for widening the use of these techniques in conjunction with other subsurface systems.

Soil erosion on erodible soils can only be prevented by protecting the surface soil layers and maximizing infiltration rates and minimizing runoff. Much more attention needs to be paid to the field conditions if soil losses are to be avoided, or reduced, in many areas. Field conditions more resistant to soil loss can be achieved through appropriate crop selection and zero or precision tillage, coupled where possible with the use of herbicides. Trends in recent years have been both towards reducing the number of farm enterprises and increasing farm size, with the demise of the crop rotation. In many situations farm size has now reached a point where more enterprises and crops could be contemplated and investigations into the possibilities of a return to the preventative management rotational systems would be well justified.

To summarize, the major research and development role for the farmer in any move towards preventative management lies in the area of field operation, organization and husbandry. The manufacturers' part is in providing equipment of appropriate width and performance to match the field systems and to reduce the compaction potential of their equipment. The major priority for the research and extension worker is the definition of soil environmental requirements and the development of objective tests, as well as the promotion of field drainage and preventative husbandry measures.

5.II.7 CONCLUSIONS

Changes in mechanization and soil-handling practices in recent years have placed increasing pressure on the soil environment, with soil structural problems tending to become more severe and deep-seated.

The previous approaches to soil management, based primarily on reclamation methods are rapidly becoming inappropriate; hence the need for a change towards more preventative management. Major benefits can be gained from following a preventative management approach, but they will only be attainable following a combined coordinated attack on the outstanding problems by farmers, manufacturers and researchers together. The slogan for the future must be 'prevention is better than cure'.

5.II.8 REFERENCES

Davies, B. D., and Cannell, R. Q. (1975) Review of experiments on reduced cultivation and direct drilling in the United Kingdom 1957–1974. *Outlook on Agriculture*, **8** (special issue), 216–220.

Eriksson, J., Hakansson, I., and Danfors, B. (1974) *The Effect of Soil Compaction on Soil Structure and Crop Yields*. Bulletin 354, Swedish Institute of Agricultural Engineering, Uppsala.

Food and Agriculture Organisation (1971) *Assessing Soil Degradation*. Soils Bulletin, 34, FAO, Rome.

Godderham, P. T. (1976) The effect on soil conditions of mechanised cultivation at high moisture content and loosening by hand digging. *J. Agric. Sci. Camb.*, **86**, 567–571.

Hakansson, I. (1979) *Experiments with Soil Compaction at High Axle Load*. Soil Investigations 1–2 Years after the Experimental Compaction. Swedish University Agr. Sci. Div. Soil Management, Report 57 (with English summary).

Hakansson, I. (1980) Swedish experiments with crop response to wheel-induced soil compaction. *Proceedings of Working Group on Soil Compaction by Vehicles of High Axle Load*. Uppsala, 38–40.

McEwen, J., and Johnston, A. E. (1979) The effects of subsoiling and deep incorporation of P and K fertilisers on the yield and nutrient uptake of barley, potatoes, wheat and sugar beet grown in rotation. *J. Agric. Sci. Camb.*, **92**, 695–702.

Patterson, E. E., Chamen, W. C. T., and Richardson, C. D. (1980) Long-term experiments with tillage systems to improve the economy of cultivations for cereals. *J. Agric. Engng. Res.*, **25**, 1–35.

Prestt, A. J. (1983) Private communication, Silsoe College.

Rowse, H. R., and Stone, D. A. (1980) Deep cultivation of a sandy clay loam. I: Effects on growth, yield and nutrient content of potatoes, broad beans, summer cabbage and red beet in 1977. *Soil Tillage Res.*, **1**, 57–68.

Russell, E. W. (1956) Effect of deep ploughing and subsoiling on crop yields. *J. Agric. Sci.*, **48**, 129.
Russell, R. S., Cannell, R. Q., and Goss, M. J. (1975) Effects of direct drilling on soil conditions and root growth. *Outlook on Agriculture*, **8** (special issue), 227–232.
Soane, B. D. (1970) The effects of traffic and implements on soil compaction. *J. Proc. Inst. Agric. Eng.*, **25**, 115–126.
Spoor, G., and Godwin, R. J. (1978) An experimental investigation into the deep loosening of soil by rigid tines. *J. Agric. Engng. Res.*, **23**, 243–258.
Spoor, G., and Godwin, R. J. (1979) Soil deformation and shear strength characteristics of some clay soils at different moisture contents. *J. Soil Sci.*, **30**, 483–498.
Stephens, H. (1855) *The Yester Deep Land-culture*. Blackwood, Edinburgh.

Note: A major portion of this chapter was previously published by Gordon Spoor in the Agricultural Engineer, 1983, Vol. 38, No. 4, pp. 95–100, entitled 'Soil management or reclamation?'.

Land Transformation in Agriculture
Edited by M. G. Wolman and F. G. A. Fournier

CHAPTER 5.III
Limiting Degradation Caused by Soil Erosion

N. W. HUDSON

There is a world need for more production of food and other agricultural products. This chapter discusses how land degradation resulting from soil erosion can be avoided or reduced.

5.III.1 REVIEW OF THE PRESENT SITUATION

5.III.1.1. Pressures Leading to Unwise Land Use

Increasing pressure on the land as a result of rising populations is a world-wide problem. The long-term effect can be clearly seen in India, where a high density of population on all the arable land over a long period has resulted in the land being severely degraded, and the fertility reduced to the point where it has stabilized at a very low level of production. In the Deccan Plateau of Central India on fertile soils derived from basalt, there are millions of hectares where cultivation began many centuries ago, and the soil loss can be measured in metres rather than millimetres.

Land hunger and the need for food tempts us to take cultivation into marginal or unsuitable land. A notable example is the steady climb of cultivation into increasingly unsuitable mountain slopes. This is particularly true in the humid tropics, and alarming destruction is occurring as a result of the spread of small-scale peasant farming into land which cannot possibly support this use in the long term. Examples are the spread of cultivation up the mountain slopes in central Java, in the Machakos and Aberdare Districts of Kenya, the lower Himalayas of India and Nepal, and the Andes in South America.

There are other examples of land which is vulnerable to damage if used too intensively.

Soils rich in organic matter (peat or muck soils) can have a high initial fertility reduced very quickly if the rate of oxidation is increased by cultivation, whereas they could remain stable almost indefinitely under pasture or woodland. Similarly, the highly-organic soils accumulated under forest can degrade rapidly when cultivated. The limitation may be purely physical—such as a shallow depth of soil which is adequate for grazing but not for cropping; or the limitation which makes the land vulnerable to degradation could be climate, such as low or erratic rainfall. The man-made dustbowl of America in the 1930s is an example of unwise and unsuitable land use leading to damage, and the situation is being repeated in the Sahel region of northern Africa where the combination of unreliable rainfall and over-stocking is leading to irreversible damage.

Pressure on the land can also lead to unwise speculative use. There are many examples where flood plains are being used for farming, or even for housing, in spite of a known history of floods. In the centre of Dehra Dun in India, shacks are built and crops planted on land where the return period of complete inundation by flooding is only about 5 years.

Some forms of land tenure may lead to unwise use. Communal ownership can lead to mismanagement, particularly over-grazing by cattle, or the over-enthusiastic removal of firewood. Share cropping or short-term tenancy may encourage damaging exploitation.

Economic pressures may result in unwise land use, and this is not limited to developing countries. Shrader (1975) gives a clear picture of this occurring in the Mid-West of the USA as a result of economic forces in the 1970s. Up to 1973 the area of erodible land under cultivation was declining because it was not profitable to farm it. Then large increases in the price of maize, wheat, and soy-bean brought large amounts of low-quality land into production. The Soil Conservation Service estimated that of the 3.8 million hectares brought into production in 1974, nearly one-half was land subject to severe erosion.

In addition to personal or family needs, national requirements can result in conflict between the long-term objective of maintaining the country's land resource base, and the more immediate requirements of self-sufficiency. Another national dilemma is the choice between growing cash crops to earn foreign currency or using the land for food production. The understandable desire of national governments to increase food production can lead to excessive pressure on the land (Dudal, 1981).

5.III.1.2 Need for Planning the Use of Land Resources

There is a wide variation in the extent to which countries plan the use of their resources. The sensible approach is to undertake an inventory of the land resources available, and then use a systematic planning method to make the best use of the resources. Some case studies of planned implementation of agricultural development schemes which include soil conservation are discussed in section 5.III.3. In countries which do not yet use a systematic classification and planning approach, a lack of suitable existing methods is not valid as an excuse. There are many good schemes which can serve as a base, perhaps with some local modifications.

Where pressure on the land is not severe it may be possible to make assumptions, such as that the land will not be cultivated if it is steeper than some specified limit. In this situation the USDA land capability classification system is extremely useful, possibly with some local variations such as have been made when using this system in Zimbabwe, Zambia, Malawi, Israel, the Philippines, and some Australian states. In other cases the US capability classification has been merely used as a starting point and its concepts applied in a different manner. For example, the system used in Thailand is similar to the USDA but also includes classification of wetlands for paddy, while in Bangladesh the suitability for paddy dominates. The basic principle is simple but important: land is ranked according to how high it comes on the scale of potential productivity which can be achieved without excessive risk of erosion, and then suitable soil conservation practices are recommended for each class of land and for each form of land use. The US system and its derivatives have soil erosion as the main limitation, but other constraints can be included, such as wetness or salinity. Other objectives can also be included, such as suitability for forestry, or—as in the Malaysian approach—the mining potential. The major difficulty of the USDA system is that it excludes steep land, and so it is not appropriate in much of the humid tropics. However, for this situation there is an alternative classification designed by Sheng of FAO, which gives recommendations for land use and soil conservation practices for any slopes (Sheng, 1975). When neither of these systems is suitable as a starting point for developing a local method, there is the FAO Framework for Land Evaluation which merely defines the principles that serve as the framework, allowing an infinite number of classification or suitability systems to be built upon the framework (FAO, 1976).

It is interesting that a quantitative and systematic inventory of actual and potential land capability at regional or district scale shows up the cases where over-intensive use is causing degradation, but it usually also shows that other areas are being under-utilized. The Agro-Ecological studies by FAO (1978) combine climatic data with the world soil map to provide a large-scale inventory of agricultural capacity. These studies are revealing that in some countries there is unexploited potential, particularly in Africa.

5.III.2 APPROACHES TO THE CONTROL OF SOIL EROSION

In the world today there is a wide understanding of the need to conserve soil, and there is a wealth of knowledge and experience of the methods that can be used. There are some weak points in our knowledge, but these are less important than the gaps between what could be done and what is happening in practice. This section analyses some of the difficulties and constraints, not to emphasize their importance but to show how they can be overcome.

5.III.2.1 Policies and Strategies

5.III.2.1.1 Political Policy for Soil Conservation

Most politicians and political parties pay lip service to the idea of good husbandry and the conservation of natural resources. In practice, soil conservation programmes are less attractive to the voters than programmes to build roads or schools, so the situation in many countries is that plans are made for the conservation of natural resources but there is not the political will to translate them into effective action.

There can be a conflict between the short-term objective of national self-sufficiency and the long-term objective of maintaining the country's land resources (Dudal, 1981). The understandable desire of national governments to increase food production can lead to excessive pressure on the land.

Many developing countries have a historical legacy of land reserved for the use of some agency or power, whether it be a colonial regime or a feudal system or a sovereign state. Although the reservation was often to preserve the income or power of the ruling élite, there are also many examples where the land was deliberately witheld from development because it was ecologically unsuitable. A corollary to tolerating encroachment is the policy of allocating land to non-landowners. In India the declared policy is 'land for the landless', but the policy is not realistic because the land has been intensively farmed and settled for centuries and there is no reserve of arable land that can be allocated.

5.III.2.1.1 Strategies for Soil Conservation

There are several possible approaches to the task of putting soil conservation into practice. The simplest is the concept that soil conservation is such 'a good thing' that it is only necessary to show people what to do and they will start to do it. In this scenario, once the technical knowledge is available, all that is required is education and extension. This has been actively practised in the USA for several decades. Even there, success is not universal and this approach has limited application in developing countries. The conservation message must be right for the time and place, and important changes have taken place in the last 50 years. For example, ploughing and cultivating on the contour is now generally accepted as common sense, but it was a revolutionary idea in the 1930s. Also, the image of soil conservation in the 1930s was dominated by engineering works, whereas today conservation is much wider and assumes that engineering works are just one of the tools used to support sound land use and good husbandry.

Another approach is to use financial incentives to encourage the implementation of conservation works, for example by payment of subsidies for the construction of approved works. Some care is needed to avoid the situation of wealthy farmers on good land being subsidized in their conservation programmes, while more urgently needed works on marginal land are not built because the farmers cannot afford to build them, even with a government subsidy.

The combination of persuasion, education, and financial incentives is sometimes referred to as the 'green ticket'. This is in contrast to the 'red ticket' which is based on penalties. These might be the withholding of benefits until certain conservation practices have been adopted, or the use of legislation to prohibit undesirable practices, or to enforce those considered desirable. There is now sufficient evidence for us to reject the red ticket strategy as impractical and unworkable, although recent legislation in the United States combines withholding of benefits with other financial incentives. In general, to be effective any strategy for soil conservation must have popular approval—that is, it must be a bottom-up movement, not a top-down instruction.

5.III.2.1.3 Social Constraints

The social system may impose stresses on the land resource or constraints on the application of soil conservation. In many countries there is the cultural ethic that everyone has an automatic right to own land, particularly in situations where there has in the past been both sufficient land for everyone and little alternative employment other than in agriculture. However, with a constantly expanding population, this ethic results in increasing pressure on the land, with the danger of over-use.

Another cultural practice which can lead to inefficient land use is the fragmentation which results from dividing an inheritance of land when it passes to the next generation. Extreme examples of this are seen in Indian villages where a family may own many small parcels of land, each too small for efficient cultivation.

It is a fact of life that humans are gregarious and do not like to move from their accustomed environment. An example comes from Indonesia. Java is extremely crowded, and erosion in the mountains is very serious; but the outer islands have vast areas of undeveloped land, much of it with good soil and good rainfall. The Government's transmigration plan aimed to reduce land degradation in Java by a large-scale transfer of population to the other islands, but it has been disappointingly ineffective.

5.III.2.1.4 Paying for Soil Conservation

A major difficulty in justifying expenditure on soil conservation works is that payment has to be made today, but the benefits accrue in the future. Also, it is difficult to use conventional accounting procedures (such as cost–benefit or internal rate of return) to justify soil conservation works, because it is difficult to assess the future value of the soil that is saved today. Another difficulty is deciding who benefits and who should pay, particularly in the case of off-site damage. An example comes from India, where over-grazing and over-cultivation in the foothills of the Himalayas results in flooding and interference with irrigation and hydroelectric power 1000 km down the River Ganges. The peasant farmer in the hills may be the basic cause of the problem, but he cannot afford to do anything about it. However, the money to meet the cost of conservation works must come from somewhere, so the question is how to distribute the costs. In this case the increased cost, or the cost of corrective measures, has to be passed on to the consumer or be borne by the state. The division of costs will vary in different countries. In a wealthy, developed country it is reasonable that there should be a significant contribution by the landowners or the land users. But in countries where the farmers are poor, their ability to contribute towards the cost is low or non-existent. In this case the answer has to be that it is in the long-term interests of the state that the soil should be conserved, so the state should pay. Governments and development agencies recognize that some programmes are desirable even if they do not show an immediate cash profit; for example, improving the nation's health or strengthening its education. Soil conservation has to become equally accepted as an investment for the long-term benefit of the nation.

5.III.2.2 Tactics and Techniques

A most important principle of soil conservation tactics is that the construction of earth-moving protection works can never be a satisfactory solution in isolation. The first requirement must be appropriate land use, using some form of classification to identify the restrictions and the dangers. Equally important is that mechanical conservation works must be accompanied by good farming practice. This is little chance of someone inventing a magic conservation solution to keep erosion down to a tolerable level when the land is used to grow clean-weeded maize on 40° slopes in El Salvador, or bananas on 45° slopes in Venezuela, or upland rice on 60° slopes in Northern India.

Soil erosion is influenced by the combination of many separate factors—soil, climate, topography, land use, and farming practice. It is therefore site-specific, and individual control measures that are appropriate and successful in one set of circumstances cannot automatically be transferred. Each situation requires a package of practiccs which together make up a farming system which meets the local needs.

Some particular gaps in our technical knowledge can be identified. The first is for information on physical conservation measures suitable for land steeper than 10°. There have been some interesting field trials on bench terracing and hillside ditches during FAO projects by Sheng (1975) in Jamaica, El Salvador and Thailand, and by John and van der Goot (1976) in Java. However, our knowledge and experience could be improved, particularly on the difficult question of how to get the surface runoff down steep slopes without expensive structures.

The second need is for information on the reduction of erosion by crop management alone, in situations where physical conservation works are not possible. Two situations illustrate this point. In both El Salvador and in Java, peasant farmers cultivate holdings of 1 or 2 ha on 30° slopes with insecure tenancy, and without either the ability or the incentive to construct bench terraces or hillside ditches. The question is, 'how much can erosion be reduced by contour planting, mulching, multiple cropping, or other agronomic techniques?'.

5.III.2.2.1 Mechanical Soil Conservation Works

The description 'mechanical works' is not very satisfactory, but it is the accepted terminology for conservation works which involve modifying the land surface by levelling, the excavation of ditches, or the building of banks.

One of the most successful and widely used methods is the classical North American system, developed and refined by the Soil Conservation Service of the Department of Agriculture. There are three components of the protection of a piece of arable land. At the upper edge of the land, a diversion drain intercepts any surface runoff which would otherwise flow down from higher ground on to the arable land. This is the first line of defence and vital to the protection system since all the structures lower down are designed on the assumption that it will divert all the runoff from outside the arable land. The second component is graded channel terraces. There are many varieties, but the basic principle is an excavated channel, with the soil used for a bank on the downhill side, and the channel on a gentle grade so that runoff from the arable land is drained away at non-erosive velocities. The third component is a disposal channel into which the water can be discharged from the diversion drain and from the graded channel terraces. Natural watercourses may be used, but an excavated channel planted with grass is more usual. Each of these three components is a channel designed according to well-established design procedures.

This system is only effective on gentle slopes, usually up to about 8° (12% or 15%), and on steeper slopes some form of terracing is usually employed. There are many types of terrace for particular situations; their description and design is documented in standard textbooks, such as that by Hudson (1981). Bench terracing is the practice of converting a steep slope into a series of steps with horizontal or nearly horizontal ledges, and vertical or almost vertical walls between the steps. A great deal of earth-moving is required. Bench terraces may be constructed on the contour to minimize runoff, or with a slight gradient like a graded channel terrace. When used for cultivated crops it is desirable for each bench to be as wide as possible, and this is determined by the steepness of the slope and the soil depth. When regular cultivation is not required, as for tree crops, smaller terraces will be equally effective and need less earth-moving. These are usually called step terraces and are used for fruit trees, tea and coffee.

If the soil is too shallow, or the land too steep, terracing the whole land surface may not be possible. In this case the land may be developed for tree crops by the use of intermittent terraces, sometimes called orchard terraces. These are small terraces each carrying one line of trees. The land between the terraces is planted to a close-growing cover crop.

Another technique for slopes too steep for bench terracing is hillside ditches. There are many variations, the common feature being a ditch dug on the contour to catch soil and water. A variation, used successfully on tea plantations in Sri Lanka, is 'lock and spill drains'. These are ditches with an uneven level in the bottom of the ditch, so that some water is held and can infiltrate, but in heavy rain the runoff spills towards the outlet, which is usually a stone-lined channel.

On gentle slopes (usually up to 2% or 3%) it is usually possible to control erosion by carrying out all cultivation on the contour (contour cultivation), or by growing alternative crops in strips along the contour (strip cropping), or by dividing up the arable land with strips of grass laid out on the contour (grass strips).

5.III.2.3 Implementation

In many developing countries the implementation of soil conservation policies is inhibited by lack of money; in others the constraint is a combination of two interconnected problems—the absence of a soil conservation department, and a shortage of trained personnel. Because soil conservation involves so many aspects of land use, it is not always obvious which arm of government is the most appropriate to be responsible for soil conservation. Most commonly it is the Ministry or Department of Agriculture; but the lead may rest elsewhere, for example with Forestry in Indonesia, the Bureau of Soils in the Philippines, the Prime Minister's Office in Malaysia, the Department of Primary Industries in Queensland, Australia.

The multidisciplinary nature of soil conservation, and doubt about where the responsibility should be placed, has in several cases led to the creation of a special national organization, such as the Presidential Soil Conservation Commission in Kenya and the Presidentially sponsored Greening Programme in Indonesia. Even more common is the creation of a special cross-departmental agency to carry out agricultural development and soil conservation on a project basis. Several examples are discussed in the case studies in section 5.III.3.

The fact that a nation which has severe erosion problems has not developed a Department of Soil Conservation may also be the result of a shortage of suitably trained specialists. As an example, 25 years ago, the Conservation Service of one small British colony in Africa was recruiting and training 30 graduates every year, and a similar pattern existed in the African colonial territories of France, Belgium and Portugal. Today it is doubtful whether 30 graduates a year go into soil conservation in the whole continent of Africa. (This is in no way an apology or justification for colonial rule, but merely a factual demonstration.) A study of the needs and opportunities for conservation training for developing tropical countries defined six separate levels and types of training, ranging from the experience-sharing workshop for senior administrators to practical training for field operators (Hudson, 1982). Although some training facilities exist for each of these six categories, the only one which is adequately catered for is graduate study for teachers and research workers through advanced university degrees which concentrate on theory and research. Training programmes for conservation practitioners which concentrate on 'how to do it' are few and quite inadequate for the task. The shortage of trained personnel is therefore at least partly the result of the absence of suitable training facilities.

5.III.3 CASE STUDIES

A selection of examples is chosen here to illustrate different aspects of agricultural development which include land surface modification. The example from China is bench terracing to reduce soil erosion in the loess plateau, and illustrates land surface modification on a massive scale. The example from Java also illustrates bench terracing, but incorporated into a programme of general agricultural development through watershed projects. The example from Swaziland is another project approach to general rural development in which land selection and physical conservation works are combined with agricultural and non-agricultural facets of development. Kenya brings us an interesting example of a grass-roots soil conservation movement arising from the ashes of a colonial regime. From Australia we have, first, an example of the integration of good farming practice with soil conservation procedures of grazing land, and then the combination of good farming and soil conservation on arable land.

5.III.3.1 Bench Terracing in China

The biggest demonstration of land surface modification is the building of bench terraces in China. The Yellow River has been known as 'China's Sorrow' for thousands of years because of the damage caused by its floods. To quote Lowdermilk (1953): 'Since the days of Ta-Yu, nearly 4000 years ago, the battle of floods with this tremendous river has been lost and won time and again'. The problem of flood control is enormously complicated by the very high silt loads which have raised the river bed above the surrounding surface, so that it requires an unceasing campaign of building dykes to contain it. The main source of the silt is the erodible deep loess soils in the middle reaches of the Yellow River. The loess plateau is 580 000 km^2, of which 430 000 km^2 are severely eroded. The range of soil loss is 2000–20 000 t/km^2, and the average for the whole of the plateau is 10 000 t/km^2 (Robinson, 1979).

Conventional attempts at control of the floods have never been more than temporarily successful, and so the strategy of the Yellow River Conservancy Commission is to control the rates of erosion, and hence the sediment load in the river. The construction of bench terraces has been carried out for centuries, but the scale is now greatly magnified as a result of regional and national conservation programmes. Much of the construction is carried out by hand labour, but some interesting special-purpose machines have also been developed to assist in the prodigious amount of earth-moving required. These include soil-throwers using both rotating discs and moving belts to move large quantities of earth sideways in a manner rather reminiscent of the American Whirlwind Terracer of the 1940s. Other purpose-designed equipment includes specially shaped dozer blades and large reversible ploughs, as described by Fang Zhengsan *et al*. (1981). The programme of terracing is also linked to a general programme of development, including the planting of grass and trees on steeper land which is withdrawn from cultivation, together with gully control, check dams, and trapping of silt in basins for subsequent cultivation. The introduction of more grass also encourages the development of livestock in an area which was previously predominantly used for cereal production.

5.III.3.2 Watershed Management in Java

The population of Indonesia is not high in relation to the total land area; but the distribution is extremely uneven, with very high concentrations on just a few of the islands, particularly Java. As a result, the pressure on the land in Java is resulting in severe deterioration, and the government has developed several programmes to relieve the situation. One of these is the Transmigration Programme which aims to physically redistribute the population more evenly by encouraging settlement in the outer largely unpopulated islands. A second approach is the 'Re-Greening Programme' (Perencanaan Penghijauan), which aims to encourage the establishment of several types of terrace on arable land, the establishment of improved pastures on steeper land, and the restoration of trees and shrubs on land taken out of cultivation. The third approach, which shows the most promise, is the programme of 'Watershed management and development'. This programme envisages a multi-disciplinary approach to erosion control and agricultural development in selected watersheds, with intensive programmes in selected sub-watersheds.

An interesting example is the Citanduy II Project in Central Java which is funded jointly by the Government of Indonesia and USAID, and derives its technology from an earlier FAO Project in Upper Solo (Barreau and Djati, 1985).

There are two important concepts in the Citanduy project. The first involves the idea of expanding and spreading from a nucleus or catalyst. In the sub-watersheds the catalysts are a number of model farms which are a demonstration of a 'package of technology' appropriate to the particular combination of soil, climate, and social factors—such as availability of labour and availability of credit. The components of the technology are hand-built bench terraces with forage species planted on the riser, and improved agronomic practices. It is claimed that the result is to generate three times the income of conventional farming methods, and certainly the practices are being widely adopted by the land owners close to each of the established model farms.

The second principle is that the project will only provide seed for improved varieties of cash crops, planting materials for establishment of grass and trees, the supply of fertilizer and agro-chemicals, and ultimately credit for the purchase of livestock, after the farmers have built bench terraces according to specification. (While not a 'red ticket', this is a carrot and stick approach.)

This improved land use at the village level is reinforced by large-scale works in the watershed, such as impounding reservoirs, and re-afforestation on land too steep for cultivation. Although the Citanduy project has a good technical base, it is not without organizational difficulties. These result mainly from the common problem that government departments responsible for extension, forestry, soil conservation and public works are usually structured according to the divisions of local government (regions, districts and subdistricts), whose boundaries seldom coincide with the natural topography or with the boundaries of watersheds. The solution adopted in Citanduy is to set up a separate multi-disciplinary project to manage all aspects of development within the project. This may achieve that objective but leads, as in Citanduy, to duplication and parallel organizations, with dilution of effort and rivalry. The problem of carrying out development using the watershed approach could only be overcome if there were a massive reorganization of the country's internal administration so that the boundaries of regions and districts were based on the boundaries of watersheds. Few, if any, developing countries have felt able to grasp this nettle. There is a tendency today for development agencies to channel more effort through existing government departments, rather than set up special projects.

5.III.3.3 Integrated Rural Development in Swaziland

This is another example of the concept of pilot projects which are expected to spread to larger areas, using the idea of multi-disciplinary, integrated development. In this case the main components of the scheme were:

(1) the application of a system of capability classification to define land areas suitable for arable, and areas suitable for grazing, together with the reallocation of use where necessary;
(2) to establish roads, fences, and homesteads along the boundary between the arable and the grazing land;
(3) to provide all the inputs to improve production of crops and livestock, including strong extension support, supply lines for inputs, and market outlets for the products;
(4) to improve mechanical soil conservation measures to halt soil erosion on arable lands, and to reduce erosion on grazing lands by a reduction of the level of stocking.

The programme started with four small areas, each of about 5000 ha, and has since been extended to cover about 50% of the Swazi Nation Land. (This is land held in trust by the King for the benefit of the nation as a whole, and comprises all land which is not allocated to private ownership for large-scale irrigation, or forestry schemes, or privately owned farmland.)

The chosen organizational structure was a special programme within the Ministry of Agriculture to administer directly most of the required activities, and to liaise with appropriate departments such as Health or Education for activities which could not be handled within the Ministry of Agriculture. There is the same problem as in the Citanduy Project in Java, in that the establishment of a special organization has led to duplication, and conflict with the Departments whose structure is geographically different.

The project has shown some notable successes, some unanticipated changes and some failures. Although not initially defined as a major objective of the programme, it has certainly improved the general standard of living of the rural population, at least in terms of convenience, by improvements in rural roads, schools and water supplies. The objective of smaller numbers of cattle in better conditions has not been achieved because the policy of fencing the grazing areas has led to increased concentration of stock inside the fenced areas, and the soil erosion from over-grazing has in many cases become worse as a result of the fencing.

An unexpected result was that hybrid varieties of maize were quickly adopted and led to increased yield per hectare; but at the same time the area planted was reduced and so there was no *net* increase in the national production.

The main land changes were that reallocation of land, based on capability classification, paved the way for improved soil conservation. The latter, however, was not successful because the proposed practices were borrowed from another set of conditions and were not suitable for Swaziland. The traditional soil conservation method, introduced about 50 years ago, is to leave grass strips on the contour across all arable land using a very simple specification of a 2 m width of grass at a vertical interval of 2 m. Given the fairly low erosivity of the rainfall, and well-structured resistant soils, this was a very appropriate method, although some strips were not correctly aligned because of a shortage of surveyors. The project called for the replacement of grass strips by the classic North American style of mechanical protection—graded channel terraces leading to grassed waterways. This method, in fact, is unsuitable for two reasons:

(1) The individual arable holdings are usually 0.5–2 ha, and so do not lend themselves to this kind of layout.
(2) It could have been foreseen that, since cultivation rights are allocated on a seasonal basis by the tribal chief, there is no interest in the maintenance of the terraces and waterways, which are seen as impositions of the government.

The lessons of this example are (*a*) that the principle of integrated rural development which combines land selection, soil conservation and improved agricultural practices is sound, but (*b*) there are difficulties in setting up an organization to carry out such development, and (*c*) it is essential that the technology being applied is appropriate to the local situation.

5.III.3.4 Soil Conservation Policies in Kenya

Kenya provides an interesting example of the development of soil conservation policy. Up to independence in 1963, an autocratic soil conservation programme was imposed by the colonial government, consisting of alienation of land for use by different ethnic communities, regulation of land use, and mechanical protection works, mainly contour banks. The policy was acceptable on the large 'white settler' farms, but was enforced in the African areas with at best a grudging acceptance and at worst outright opposition. With independence, all ideas of soil conservation were swept away by the government, as an unwanted relic of colonialism.

However, after some years a grass-roots interest in soil conservation began to emerge, encouraged by good leadership and a strong sense of national unity based on the 'harambee' philosophy, meaning 'let us all pull together'. An important step was the adoption of soil conservation by the 'Mwethya' groups, which are voluntary self-help groups which form each year to carry out communal work during the dry season, and then disband when the rains come, to go about their individual farming activities. The leadership of the groups comes from the women, but men are persuaded or coerced to join for the heavier tasks. The range of communal activities includes bush clearing, water carrying, management of grass or tree nurseries, and ploughing arable land on each farm in turn. The construction of conservation works has now been included, and there has emerged the concept of a new type of terrace called 'Fanya Juu', literally meaning 'throwing upwards'. This consists of excavating a ditch on the contour, with the soil placed on the uphill side to form a bank or 'bund'. The steeply sloping riser and the bank are planted with grass, and there is a self-terracing effect as the grass and the bank retain soil which is washed down the cultivated inter-terrace strip.

This soil conservation movement was given a strong boost when the President established a Permanent Presidential Commission on Soil Conservation and Afforestation, and gave his personal support to a national conservation programme. At the same time the increasingly serious problem of soil erosion in Kenya attracted the attention of national and international aid agencies, who set up many development projects which include an element of soil conservation. The Swedish International Development Agency, SIDA, has been particularly active, with programmes strongly directed towards training for the technical personnel concerned with soil conservation. The net result is that Kenya has today one of the most promising soil conservation programmes in Africa.

5.III.3.5 Soil Conservation on Grazing Land in Australia

The Eppalock Catchment Soil Conservation Project was carried out by the Soil Conservation Authority of the State of Victoria between 1960 and 1975. The object was to reclaim badly eroded land and reduce erosion on productive grazing land in the catchment of the Eppalock reservoir, which was to be greatly increased in size but was estimated to have an unacceptable rate of sedimentation.

The soil conservation works were carried out and paid for by SCA, and consisted of gully control structures and tree planting, silt traps, fencing-off eroded areas, diversion banks, and ripping land suffering from tunnel erosion. The improved pasture management included the aerial application of fertilizer, which carried a state government subsidy, and 'chisel seeding', which is the application of seed and fertilizer in separate narrow chisel furrows. There was also sub-division fencing, and improvement of water supplies for the livestock. Extension advice was provided by both SCA and the Department of Agriculture.

The cost of the dam enlargement was A$6 million, and the cost of soil conservation works A$100 000 per year for 10 years—a total of A$1 million. The main area for treatment was 72 000 ha, but there was another fringe area of 17 000 ha where the improved pasture management practices were adopted by the farmers.

Throughout the project, data were carefully recorded so that a detailed analysis could be made. This showed that the overall economics are good, even after discounting distributional effects, such as the lowering of the wool price in other areas as a result of the increased production in Eppalock, and possibly a similar effect on butter-fat prices as a result of increased production. The internal rate of return is calculated at 25.4%, with a benefit–cost ratio of 2.0 using an 8% discount rate. The report concluded that the project was 'a very sound investment for the community as a whole'—and an excellent investment for the landowners in particular (Australian Soil Conservation Study, 1978).

5.III.3.6 Soil Conservation on Arable Land in Australia

A detailed cost–benefit analysis was also made of a soil conservation project on arable land in Allora Shire on the east Darling Downs in south-east Queensland. The programme was carried out under the Soil Conservation Act of 1965, which allows the designation of 'areas of soil erosion hazard', in which the government takes the initiative of erosion prevention and control through Statutory Conservation Programmes. This project started in 1973 and involved both the landholders and the local authorities.

The problem was extensive sheet, rill and gully erosion on uplands of sloping basalt, causing salt deposition and flooding in the lowlands. In 1972 it was estimated that 75% of the area (i.e. 18 400 ha) required intensive soil conservation measures, and that measures had been applied on 42%, but that much of the area needed more and better soil conservation work. Another 18% (4200 ha) was estimated to need minor land protection practices such as strip cropping and contour cultivation. Wheat yields had been reduced by erosion, but the amount was difficult to quantify because of the effect of increased fertilizer application and better varieties. Other unquantified damages included the necessity to clear road drains and culverts, and the silting up of stream channels.

The main costs were for mechanical protection works (e.g. waterways, diversion banks and contour banks). The work was done by the landholders, who bore 71% of the total project costs and received a 25% subsidy because they were in an area of soil erosion hazard. Other contributors to the costs were the soil conservation branch at 23% (mainly providing extension services) and the Queensland Government at 6% (mainly providing the 25% subsidy on conservation works).

The direct benefit is the increased production arising from the lower rate of decline in productivity. The most significant agronomic practice introduced is 'opportunity cropping'—slipping in an extra crop whenever there is adequate moisture. This leads to an average crop intensity of three crops in two years. Other benefits include lower maintenance costs of roads and rivers as a result of reduced flooding and silt deposition. Intangible benefits include increased commercial activity in nearby towns—for example, supplying the special tillage equipment used for stubble mulching in the opportunity cropping.

A financial and economic analysis of the project shows a benefit–cost ratio of 1.5 at a discount rate of 8%, and an internal rate of return over a 30-year period of 12.4%, or 13.1% over 40 years, not including the uncosted benefits. The project was therefore a sound investment for both landholders and the nation.

Under conventional cropping systems, mechanical protection would break even with costs over 30 years in the absence of subsidy, and this would not be acceptable to landowners. However, with the benefit of opportunity cropping included, the internal rate of return to landowners before tax is 14.9% over 30 years, or 11.2% over 20 years. If the costs of erosion control are not subsidized a long payback period is required, but with a 25% subsidy they become an attractive investment.

5.III.4 CONCLUSIONS

The conservation of the present land-resource base is essential because it would not be good management to make major efforts to develop new resources while the existing ones are being depleted by erosion. Estimates by FAO suggest that the increased area of land required to feed the world population in the year 2000 will be 200 million hectares, but that the rate of loss of useful land through erosion is 5–7 million hectares a year. It is therefore necessary to give equal attention to the preservation of the existing land resource at the same time as expanding it and improving it.

The approach to soil conservation should be to liken it to preventative medicine, not to curative treatment. Soil conservation practices and techniques cannot be expected to prevent erosion if the land use is fundamentally unsound. Physical soil conservation must be one component of an integrated approach to land development through good husbandry. The package will include the evaluation of land so that its potential can be assessed, and the application of appropriate physical soil conservation measures, closely combined with erosion control through good husbandry.

5.III.5 REFERENCES

Australian Soil Conservation Study (1978) *Economic Evaluation of the Eppalock Catchment Soil Conservation Project*. Report No. 9, Commonwealth and State Government Collaborative Soil Conservation Study 1975–77. Australian Government Publishing Service.

Australian Soil Conservation Study (1978) *Economic Evaluation of a Soil Conservation Project in Allora Shire, Queensland*. Report No. 10, Commonwealth and State Government Collaborative Soil Conservation Study 1975–77. Australian Government Publishing Service.

Barreau, E. M. and Djati, K. (1985) The Citanduy project in Java: Toward a new approach to watershed stabilization. In: *Soil Erosion and Conservation*, ed. by S. A. El-Swaify, W. C. Moldenhauer, and A. Lo. Soil Conservation Society of America.

Dudal, R. (1981) An evaluation of conservation needs. In: *Soil Conservation: Problems and Prospects*, edited by R. P. C. Morgan. John Wiley, Chichester.

Fang Zhengsan, Zhou Piehua, Liu Qiande, Lie Baihe, Ren Letian, and Zhang Hangxiong (1981) Terraces in the Loess Plateau of China. In: *Soil Conservation: Problems and Prospects*, edited by R. P. C. Morgan, John Wiley, Chichester.

FAO (1976) *A Framework for Land Evaluation*. Soils Bulletin 32, FAO, Rome.

FAO (1978) *Report on the Agro-ecological zones project*. Vol. 1: *Methodology and Results for Africa*. Vol. 2: *Results from South-West Asia*. Vol. 3: *Latin America 1984*. Vol. 4: *Results from South-East Asia*. World Soil Resources Report 48, FAO, Rome.

Hudson, N. W. (1981) *Soil Conservation* (2nd edn.). Batsford, London.

Hudson, N. W. (1982) Soil conservation research and training requirements in developing tropical countries. In: *Soil Erosion and Conservation in the Tropics* (Chap. 7). Special Publication No. 43, American Society of Agronomy, Madison.

John, B. C., and van der Goot, W. (1976) *Soil Conservation Measures in the Upper Solo Basin*. INS/72/006 Termination Field Document No. 6, FAO, Rome.

Lowdermilk, W. C. (1953) *Conquest of the Land Through Seven Thousand Years*. Agricultural Information Bulletin No. 99. USDA, SCS.

Robinson, A. R. (1979) *Erosion and Sedimentation Problems and Solutions in the People's Republic of China*. Paper 79-2058 presented at ASAE Summer Meeting, Winnipeg, Canada, June 1979.

Sheng, T. C. (1975) *Forestry Development and Watershed Management in the Upland Regions: Watershed Management in Jamaica*. Technical Report 12 (FO:DP/JAM/67/505), FAO, Rome.

Shrader, W. D. (1975) What is happening to our soil resources: losses to erosion. In: *Land Use: Persuasion or Regulation?* Proceedings of 29th Annual Meeting of Soil Conservation Society of America, August 1974.

Land Transformation in Agriculture
Edited by M. G. Wolman and F. G. A. Fournier

CHAPTER 5.IV
The Impact of Wetland Reclamation

K. RUDDLE

5.IV.1 INTRODUCTION

The complex ecosystems that characterize coastal south-east Asia are areas of intimate interaction between aquatic and terrestrial ecosystems, and are zones of high inherent biological productivity and complexity. They are among the most complex, diverse and biologically productive ecosystems in the biosphere, and among the least understood and most fragile. Even more than their temperate zone counterparts, tropical and sub-tropical coastal ecosystems are liable to severe damage from unsuitable development schemes.

Despite the predominance of both freshwater and marine aquatic environments in the tidally influenced wetlands of humid, tropical coastal zones, forest and grassland habitats provide man with a wide range of foodstuffs and economically important raw materials, and levée and beach-ridge soils, for example, provide fertile areas for agriculture. The outstanding natural characteristic of wetland ecological systems is an all-pervasive interaction between terrestrial and aquatic components, a fundamental characteristic clearly reflected in many of the resource systems traditionally employed to exploit wetland renewable natural resources. Such systems have long been used, usually on a relatively small scale, but not inevitably so, to effect agricultural transformations that yield foodstuffs and complementary materials on a sustained basis and with very little environmental repercussion (Ruddle, 1982). Many, but not all, such systems are thoroughly grounded in generations of empirical biological and ecological knowledge, and not uncommonly they employ sophisticated, low-energy technologies of research transformation (Ruddle and Grandstaff, 1978). On the other hand, policy-makers and planners generally foresee more specialized uses for reclaimed wetlands in terms of a combination of large-scale, heavily capitalized agriculture, aquaculture, animal husbandry and forestry; as sources of freshwater for irrigation, industrial and domestic uses; or as sites for ports, industry and human settlements. Some of these destroy the natural ecosystem and others are highly disruptive, although not inherently so if properly conceived and implemented.

Since a worldwide survey of the complex and varied transformations that have been effected in coastal wetlands is beyond the limited scope of one chapter, the coverage here is limited to two contrasting case studies from south-east Asia. The first is from Sumatra, Indonesia, where wetland reclamation for the Transmigration Scheme is being attempted by rapid and large-scale environmental change. Serious problems have arisen with this strategy. It is contrasted, in the same region, with traditional and successful, yet small-scale, wetland agricultural transformations.

The second case exemplifies the quite different results that have been obtained during the two millenia-old and still continuing agricultural transformation of former tidal wetlands of the Zhujiang (Pearl River) Delta of southern China, where for many centuries the resource development policy has been one of gradualism, an adaptation of the natural order to 'turn the worst into the best', to take advantage of changing market demand and ecological opportunity. In this densely populated region no sudden, fundamental method has been applied to radically alter the wetland environment. The result has been the gradual evolution of one of the world's largest and most successful systems of integrated agriculture–aquaculture.

It should be noted that the notion of continental wetlands now includes coral reefs and other nearshore marine ecosystems. Coral reefs are excluded from this treatment, perhaps arbitrarily, as are mangroves. Aquaculture has also been left out, since the widespread brackish water systems based on the reclamation of rapidly accreting shorelines are not essentially agricultural transformations; small-scale integrated farming systems are still not an important feature of the tropical agricultural landscape; and in the case of ricefield fisheries, wetlands were reclaimed for ricefields and fish derived from them has always been a secondary consideration (Ruddle, 1980, 1982).

5.IV.2 AGRICULTURAL TRANSFORMATION OF INDONESIAN COASTAL WETLANDS

Indonesia accounts for 14% of the world's coastline. Some Indonesian coasts, like the north coast of Java, have population densities in excess of 3000 persons/km^2 and are among the most densely populated areas on earth, whereas others, such as the tidal wetlands of southern Sumatra, have only 10–20 persons/km^2. Further, with a population of some 142 million (1980), which is growing at a rate of 2% a year, Indonesia is the fifth most densely populated country in the world.

The predominant agricultural transformation over the course of many centuries and at the hands of the local inhabitants on Java, Bali and Madura, together with those coastal tracts on the Outer Islands lacking wetlands, has been the conversion of pre-existing ecosystems into irrigated rice fields, either for mono-culture or in association with coconut plantations or coconuts interplanted with mixed secondary crops. Many local variants occur within this necessarily generalized picture, but none refutes it.

'Spontaneous migrants', principally Buginese from southern Sulawesi and Javanese and Sundanese from Java, who have migrated mainly as pioneer agriculturalists to thc tidal wetlands of Sumatra and Kalimantan, have also followed the same agricultural practices, i.e. mostly rice and coconut farming or secondary crop–coconut agriculture. Few such migrants settle in coastal zones that lack tidal wetlands, probably because pioneering in the swamps yields higher economic returns. In contrast, most government-sponsored transmigrants are obliged to adopt the mono-culture of irrigated rice, although some have also taken up the cultivation of secondary crops combined with coconuts, partly in response to the poor economic returns from rice.

5.IV.2.1 Traditional Technologies for Wetland Transformation

Sophisticated, low-energy techniques for successful agricultural transformation of peat soils of tidal swamp areas in central and southern Kalimantan (Borneo) have been developed and expanded over many decades by Banjarese, Buginese and migrants from Java. All three groups employ systems based almost entirely on the Banjarese model.

Over some five decades Banjarese agriculturalists have employed and developed their system of sustained agricultural production in former tidal swamps (Collier, 1980). At Samuda Kecil Village, on the River Metaya of central Kalimantan, Banjarese pioneer agriculturalists select a small tributary stream and, by hand, deepen and widen it from a confluence with a larger stream to its source. At the source they 'dig a small canal which follows the natural drainage system' (meaning ambiguous).

The area destined for cultivation is then progressively opened and cleared in 1 ha lots. First, the small stream is canalized for some 50 m, then the trees in an adjacent area of 425 m^2 are felled during the July dry season (since in the wet the area is inundated to a depth of 1 m), sun-dried for three months, and finally cleared by burning either once or twice during September and October. In December and January seedlings propagated in a seedbed are transplanted to the new field. Since only a digging stick is used, the peat layer is hardly disturbed. At the same time residual stumps and logs are burned again. Finally, the only time that first- and second-year fields are weeded is during the transplantation period. More intense weeding is required in third-year fields and consumes 14–20 man-days per hectare per year of labour input. Weed infestation becomes progressively worse, and fourth-year fields, for example, require 30 man-days per year of weeding input for each hectare. Weed infestation and associated problems are the prime reasons for eventually converting rice fields to other forms of cropping. Only one crop *per annum* of local rice varieties adapted to the prevailing swampy conditions is cultivated. The crop is harvested in May.

After the second year of rice cultivation a drainage ditch 1 × 0.5 × 425 m is excavated across the centre of the field. Again, only hand tools are used.

During the third and fourth years of field use, coconut seedlings are interplanted with rice, at a rate of 135/ha, on small mounds made of the earth removed from the shallow drainage ditches that separate the rows of coconuts. The individual mounds must be built up and the ditches deepened biennially until the fifth year, when they are formed into continuous ridges. Coconuts, the principal crop, are then inter-planted with coffee, bananas and vegetables. Rice is cultivated in the shallow ditches between the rows for three more years.

Three years after inter-planting with coconuts, shade impedes rice cultivation, so that the village headman permits the clearance of an additional 2 ha of swamp forest, thereby initiating another 7–8 years of the agricultural transformation process. The Banjarese consider a 6 ha farm to be ideal, with 4 ha under coconuts and associated secondary crops and 2 ha planted to rice, given the time required for their system of swamp forest agriculture to reach maturity, together with the ability to recruit labour.

Under the Banjarese system, rice fields are converted to coconut and secondary crop production because the increasing labour costs required to weed older rice fields, in combination, perhaps, with lower yields, is not economically justifiable. This is particularly so when coconuts produce a higher rate of return.

Over several generations the Buginese from southern Sulawesi have developed a method similar to that of the Banjarese—from whom they might have learned it—for agriculturally transforming tidal swamps. The Buginese, who began migrating spontaneously to the swamps of Kalimantan and southern Sumatra early this century, cultivate mostly coconuts and other commercial crops, and plant rice for subsistence only, until their coconuts produce an income or until rice yields begin to decline on the newly cleared land.

Buginese pioneer settlers carefully select cultivable areas in swamps by judging the quality of the natural vegetation and noting the presence of indicator species. The soil is then checked; peats are cleared for coconuts inter-planted with sweet potatoes and soy-beans, among other catch crops, whereas rice is planted where peat is either absent or only a few centimetres thick (Collier, 1978; Collier *et al*., 1981).

These settlers obtain land from the local clan (*marga*), and clear it by cooperative labour. First, a primary canal is gradually excavated at right-angles to a river and leading towards the centre of a peat dome. Its length depends on the size of the area to be developed, but the depth is about 2.5 m and the width some 3 m. As canal excavation proceeds trees are progressively felled and economically valuable species sold for subsistence. Land is progressively cleared along the canal, and narrow strips, separated by secondary canals some 450 m apart, are divided into 2 ha plots and distributed among the pioneers.

After several years, increasing weed and pest infestation combine with declining soil fertility to reduce yields, such that rice cultivation is no longer economically feasible. New fields are then opened up to repeat the land reclamation process and the older fields converted to coconut plantations.

Successful pioneer swamp agriculturalists in Kalimantan also include 'spontaneous migrants' from Java. Their success stems from having directly copied the Banjarese system, although in their case final results are delayed since before their own fields become productive they must work for others as farm labourers to ensure family subsistence.

5.IV.2.2 Wetland Transformation Via the Transmigration Scheme

As a consequence of increasing land shortage combined with an ever-growing demand for goods and services, together with national and regional policies, many coastal zones of Indonesia that were hitherto perceived as wastelands, or at best as marginal areas, are becoming the foci of economic and social development schemes. Indonesia has a land surface of approximately 200 million hectares, of which only 17 million hectares (8.5%) is under permanent cultivation. Further, 65% of all Indonesians are concentrated on the agriculturally rich zones of volcanic and alluvial soils of the islands of Java, Madura and Bali, which together comprise a mere 7% of the national territory. This extreme concentration of population has been increasingly exacerbated by the better economic diversification and social infrastructure of those islands. Thus it is now widely maintained that the major development problem facing the nation is the concurrent reduction of pressure on the resources of Java, Madura and Bali and the development of those on the Outer Islands, a linkage that will be effected by differential capital investment and through implementation of the transmigration policy.

Transmigration, a basic tenet of Indonesian development policy, is an ambitious scheme to relocate large numbers of people. Massive infrastructural investments are now being made to reclaim coastal wetland tracts formerly regarded by planners as useless, to absorb the transmigrants.

Although initially seen as a means of reducing population pressure and the associated detriments of landlessness, lack of economic opportunity and the degradation of renewable natural resources and the environment on the three overcrowded islands, the continued in-migration to Java prompted a revision of the fundamental objectives of transmigration, which is now principally viewed as the promotion of regional development to the sparsely populated

Outer Islands. In addition to satisfying a strategic need for national defence, this would also serve, along with the BIMAS (Agricultural Intensification) Program, to both augment the national supply of foodstuffs and raw materials, as well as providing land for the landless. Transmigration schemes are also used to absorb populations displaced by either natural disasters or by development projects as well as communities that must be resettled to reduce pressures on damaged or fragile watersheds, particularly those upstream of critically important agricultural zones or densely settled areas.

Both government-sponsored schemes to geographically redistribute population, as well as 'spontaneous migration', are officially termed 'transmigration'. The history of both types is already well documented and merits no further reiteration here (see, for example, Peltzer, 1945; Hardjono, 1977; Collier, 1978; Hanson, 1980) (Table 5.IV.1).

In the Outer Islands an estimated 40 million hectares of potentially productive agricultural land, a large percentage of which is tidal wetland, is considered to be available for development, particularly for irrigated rice (IBRD, 1974). The definition, and therefore the estimates of the area of Indonesian tidal wetlands, vary considerably, and accurate assessment of their

Table 5.IV.1 Summary of agricultural planning for development of coastal wetlands off Java (1880–1976) (Hanson, 1980)

Year	Event
1880–90	Opening of the first in a series of canals in southern Kalimantan for navigation through swamplands. Canals gave Banjarese access to new agricultural lands by digging simple drainage systems emptying into the main canals.
1914–22	Rice shortage during the First World War focused colonial government's attention on outer island land potentials. Some field development started in 1920 in Kalimantan.
1935	Improvement of 1880 original canals in Kalimantan and further settlement in their vicinity.
1936–40	Extensive feasibility studies and soil surveys for massive agricultural development and transmigration to turn Kalimantan swamplands into a 'rice-bowl' in which mechanized agriculture would play a substantial role. First transmigration project initiated in 1937.
1948–52	Kalimantan Polder Plan developed under the direction of H. J. Schophuys and agreed to by the government of Indonesia. Over a 15-year period 840 000 ha of swampland would be employed with proper drainage and pumping systems. Pilot canal development initiated in 1950 on four areas totalling 40 000 ha conducted through regional Polder Corporations directed by government.

Table 5.IV.1 (*contd*)

Year	Event
1953	First 5-year plan was to begin in Polder development but by 1970 only 2 pilot projects totalling 8700 ha were near completion.
1957	Tidal irrigation project to convert 1.5 million ha of Kalimantan and Sumatra coastal swampland to ricefields over a 5-year period announced by Ministry of Public Works. Minister P. M. Noor had formerly been directly associated with the polder plan but shifted emphasis from pumping systems to lower cost open systems in which Public Works would Plan primary and secondary canals in areas under the influence of tidal back-up. A 700 km canal was proposed for Kalimantan.
1960–63	Project supported by government decrees including a 1963 Presidential decree declaring it a vital state project. Actual implementation was limited to minor developments in Kalimantan and, in 1963, the project was reduced in size.
1964–65	Work committee for tidal and polder systems established by Minister of Public Works, resulting in publication of 'Proyek Kanalisasi'. Among other projects proposed was a canal running most of the length of Sumatra.
1967–68	Ministry of Public Works announced a new tidal irrigation project to open 5 250 000 ha of land over a 15-year time span. Considerable exploratory work on possible projects by foreign consultants working for private companies and by FAO.
1968	FAO–Dept. of Agriculture Land and Water Resource Development Project established in south-east Sumatra as an outcome of mission activities. Emphasis on Musi river water basin.
1969–74	Public works tidal irrigation project reduced to 500 000 ha over a 5-year period (REPELITA I 1969–74). Actual area opened over this period was 13 198 ha for 7180 transmigrant families. Three sites (including Delta Upang) were developed in Sumatra and three in Kalimantan. Indonesian universities became involved in planning and design in 1969.
1974	By presidential instruction a plan, developed by the Ministry of Public Works, to open 1 000 000 ha of coastal swampland between 1974–79 (REPELITA II) adopted. Survey activities, training and massive equipment purchases initiated quickly.
1975	Canal excavation started in areas adjacent to REPELITA I pilot projects.
1976	Downward revision of land opening projection to 250 000 ha by 1979.
1975–76	Request to World Bank for design and survey assistance resulted in 3.2 million dollar loan for studies on 300 000 ha at two sites in Sumatra; complemented by 3 million dollar hydrological studies and training from government of the Netherlands.

extent and type is severely hampered by a dearth of scientific knowledge about them. Relatively recently it was estimated that there are some 43.5 million hectares of swampland in the country, excluding that of Irian Jaya (*ibid*.). Excluding the latter, the estimated area thought to have agricultural development potential ranges from 10.5 to 2.2. million hectares (*ibid*.). Further, the area of Indonesian wetland directly or indirectly influenced by tides is also subject to various estimates: Soeriaatmadja (1978) placed it at 5–7 million hectares.

Despite various changes in programme design, the overall scope of the transmigration schemes is large, the Third Five Year Plan (1979–84), for example, aiming to relocate 2.5 million people to some 250 resettlement sites, mostly in Sumatra, Kalimantan and Sulawesi. As one of the major loci for its transmigration projects, since 1968, with the organization of the 'Project for Opening Tidal Swamps for Rice Production', it has become an integral part of government policy to reclaim areas of tidally influenced peat soils as zones for rice production. So important has tidal swampland become in the overall scheme that between 13% and 20% of all transmigrants are now asigned to such areas. However, it is noteworthy that to 1976 some 300 000 ha of tidal swampland has also been converted to agricultural production by spontaneous migrants (DBP, 1978), more than double that developed under government-sponsored schemes to 1980.

5.IV.2.2.1 Agricultural Land Use on Wetland Transmigration Schemes

The geographical location of transmigration schemes situated in tidal wetlands is shown in Figure 5.IV.1. The following description and later analysis of problems encountered is based mostly on one typical such site, that in the Upang Delta of south-eastern Sumatra.

Government-sponsored transmigrants live in official settlements of several hundred families in the centre of a former swamp area previously drained and partially deforested by government contractors. They are given land that has been opened up with an artificial drainage infrastructure excavated by heavy machinery, and the forest has been partially felled, sun-dried and either already removed or made ready for burning by the settlers. Commercially valuable species are first removed. Part of the assigned land remains to be cleared of timber and scrub by the migrants during the slack period of the agricultural cycle. The first rice crop of such newly opened land may be anticipated in either one or two years. Most official transmigrants on the scheme are Javanese and Sundanese, from Java, and a few are Balinese.

During 1970 and 1971 each transmigrant family arriving in the Upang Delta received 2.75 ha of land. Official plans required that 0.25 ha be used for a dooryard garden, 0.75 ha for various dryland crops, and 1.75 ha planted to tidally irrigated rice. This has not been adhered to, since many settlers have

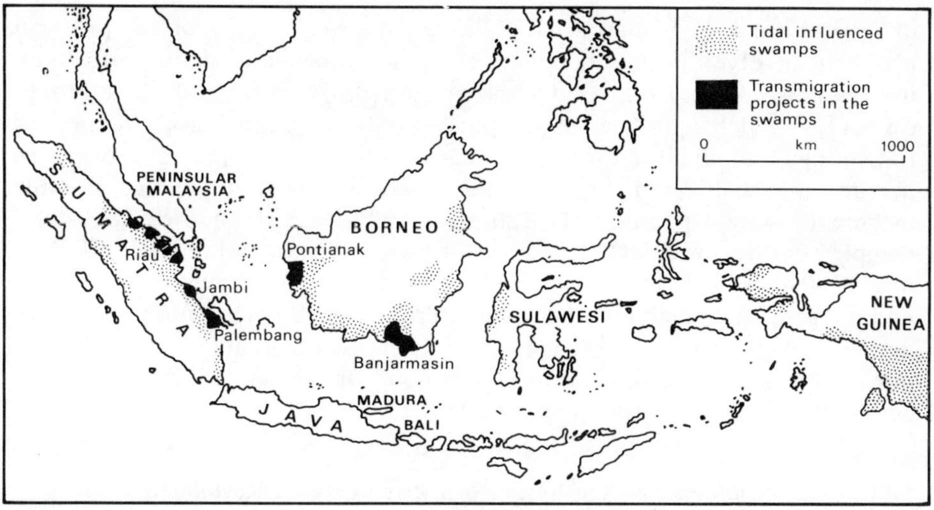

Figure 5.IV.1 Location of official transmigration projects in the tidal wetlands of Indonesia

acquired additional land away from the scheme. Further, since the tidal irrigation system has failed to function as designed, rice production depends almost entirely on rainfall. Hence only a wet-season crop can be taken, and dooryard gardens are almost entirely devoted to rice for household subsistence needs.

With an average of 17.5 quintals/ha (field dry weight), transmigrants obtained higher rice yields in the mid-1970s than either the Buginese (12 quintals/ha) or the local Sumatran population (15.5 quintals/ha) (CNRMES, 1978). This, however, does not indicate the relatively greater success of the transmigration scheme, since both the Buginese and the local population undertake many alternative and complementary economic activities which absorb much of their time and capital, whereas the transmigrants are limited almost exclusively to rice production.

Planners have generally assumed that tidally irrigated rice fields in reclaimed wetlands could be eventually double-cropped. But such optimism is not merited by the evidence. Increasing problems of pest infestation, together with the lessons of the long Buginese and Banjarese experience, make it doubtful whether the agricultural system envisaged for use by transmigrants, based exclusively on irrigated rice, will ever be either ecologically or economically viable. Nor will it permit double-cropping. A further constraint is imposed by water shortages in the dry season, a crop-destroying drought caused by rainfall failure being likely every five years (*ibid.*).

A comparison of local residents, spontaneous migrants and officially sponsored transmigrants cultivating irrigated rice on reclaimed tidal wetland soils in several areas of Sumatra and Kalimantan reveals many of the practical problems involved in the effective reclamation of such areas for agricultural use (Collier, 1980). Compared with rice cultivators on Java, all three groups obtain lower yields per hectare and harvest only one annual crop (as opposed to two on Java), and therefore experience far lower economic rates of return per unit area cultivated. Further, they received (1980) an average monthly income of only Rp12 000 (US$20) from all sources (i.e. rice fields plus complementary activities). This was too low to support a household. By the mid-1980s it is likely that incomes will decline owing to decreasing crop yields and the greater probability of crop loss. Production costs are probably higher in the reclaimed swamp than on Java because of greater weed and pest infestation, and the risk of catastrophic flood or drought is greater than on Java (Hanson and Koesoebiono, 1979).

5.IV.2.2.2 Problems and the Future of Agriculture on Reclaimed Tidal Swamps

An estimated 17–19 million hectares of the Indonesian coastal zone supports lowland peat soils (Pons and Driessen, 1975; Soepraptohardjo and Driessen, 1976). Tidally influenced peat, or tidal swamps, cover about 5 million hectares, mostly as dome-shaped, interfluvial structures, or shallower riparian deposits, approximately 300 000 ha of which have already been developed for rice, coconut and vegetable cultivation by Banjarese or Buginese spontaneous settlers.

Some 5% of the lowland peat soils of Indonesia are chemically rich eutrophic peats that are organically highly productive when managed properly (Pons and Driessen, 1975). Most such peats are already under cultivation (*ibid.*), since for the most part the Banjarese and Buginese have carefully selected them for their spontaneous agricultural settlements. Peat areas that remain available for settlement are mostly the relatively nutrient-poor, interfluvial oligotrophic peats.

The traditional agricultural technology of the Banjarese and Buginese, developed empirically in coastal wetlands over many decades, if not longer, does not appear to be deleterious in eutrophic peat soils; but even the minor perturbations that such technologies can induce in oligotrophic peat cause major and rapid deterioration (Driessen and Suhardjo, 1975). Further, earlier efforts to reclaim oligotrophic peats and areas underlain by potentially acid sulphate soils (estimated to amount to 2 million hectares) in Indonesia (Pons and Driessen, 1975) have a dismal record, several hundred thousand hectares of wasteland having been created as a result (*ibid.*).

As has been demonstrated (Soepraptohardjo and Driessen, 1976), reclamation of infertile tidal swamplands is problematical even where acid sulphate potentials are absent, and such soils remain marginal economically when reclaimed. However, problems can be avoided if soils remain permanently inundated, as under normal swamp forest conditions, and undrained peats can be used to produce sustained yields of sago palms (*Metroxylon* sp.) (Kueh, 1976; Ruddle *et al*., 1978). Rice can also be produced on undrained peats; but results have been disappointing, and poor plant growth and weed infestation need to be overcome by sophisticated and expensive reclamation, drainage and continuous leaching measures (Driessen and Suhardjo, 1975).

In reclaimed tidal swamplands, drainage is the single most important factor required to promote conditions suitable for agriculture. But unless carefully designed, implemented and managed, drainage can lead to high rates of peat subsidence as well as to undesirable and irreversible changes in the reclaimed area. Subsidence in swampland, however, is an inevitable and continuous result of drainage, a consequence that can be controlled but never entirely halted.

Two classes of subsidence have been recognized. 'Initial subsidence', a reorientation and denser packing of peat content (Driessen, 1977), follows the lowering of water tables by drainage and may attain rates in deep peat of 0.6 m a year, as in Sarawak (Tei and Kueh, 1979). 'Continuous subsidence' then follows at lower and more gradual rates: it usually results from the gradual decomposition of peat accelerated by agricultural practices (Driessen, 1977; Chambers, 1979), and at rates determined by them in combination with climatic conditions and biophysical factors inherent in the peat (Driessen, 1977; Chambers, 1979; Tie and Kueh, 1979). Rates of continuous subsistence are generally difficult to estimate, although those of 2–5 cm a year which occur in the Upang Delta Transmigration Project area, in south-eastern Sumatra, would threaten the viability of the project, since little peat would remain by the mid-1980s (Chambers, 1979).

The impact of subsistence on the reclaimed swamp is determined by a combination of factors, the principal of which are such physical elements as shrinkage, settlement and compaction, biochemical oxidation, together with burning and the removal of organic matter during land clearance and subsequent agricultural activities (Driessen, 1977). As a consequence, irreversible changes may occur in the colloidal structure of soils, which affects their capacity to retain water, as well as the degradation of the organic matter into a fine power which is susceptible to accelerated erosion and is of low agricultural value (Tie and Kueh, 1979).

Initial subsidence can be controlled by maintaining water tables as high as crops will tolerate (Driessen, 1977; Tie and Kueh, 1979); hence wet rice is the usual crop planted on peat soils. Reduction of continuous subsistence depends on minimizing the exposure of the reclaimed surface to direct sunlight, usually by cover crops. Rates of peat loss could also be greatly reduced by eliminating burning and by up-grading other agricultural practices.

Lowland tropical peat soils are usually acidic. Typical estimated pH values for those in Indonesia are 3.69–3.55 (Suhardjo and Widjaja-Adhi, 1976). Further, many shallow lowland peat soils are underlain by mineral soils with acid sulphate potential, which on aeration causes acid sulphate conditions and renders them virtually useless for agriculture. In Indonesia many such areas reclaimed by both traditional as well as modern technologies have been abandoned as wastelands (Pons and Driessen, 1975), since once such soils have been formed it is both extremely difficult and often prohibitively expensive to either leach out or neutralize excess acid over many years—an effort that may not be economically justifiable when balanced against the risk of crop failure and low yields from treated lands (Brown, 1972). Even the frequently advocated cultivation of wet rice to reclaim acid sulphate soils is no longer recommended, owing to the difficulties of managing irrigation (Gersie, 1979).

Toxic concentrations of aluminum and iron, together with low nutrient levels, inhibit crop production more than does low pH (Moorman and Pons, 1975). Acid sulphate soils are also phosphate-retentive (Watts, 1969). Management of soils with an acid sulphate potential is directed, via careful management of the water table, either at preventing oxidation and subsequent acid production or at ameliorating conditions by leaching with fresh or brackish water (FAO, 1982). The acidity of shallow peats not underlain by soils with acid sulphate potential can be corrected by the application of lime. Liming must be carefully controlled and requirements are crop-specific (Tie and Kueh, 1979). Moreover, applications that raise the pH values beyond 5.5 have been demonstrated to reduce the availability of trace elements—particularly boron, manganese, phosphorus and zinc (Lucas and Davis, 1961). Draining and ridging for agriculture can result in the formation of acid sulphates. In tropical areas with a pronounced seasonality of rainfall, oxidation can penetrate deep into the soil profile, leading to an annual regeneration of acid sulphates. But depending on the acid sulphate potential as well as on climatic, hydrological and other factors, such acid-tolerant dryland crops as oil palm, coffee, pineapple, cassava and coconuts can be grown successfully on wetland soils reclaimed by draining and ridging, if water tables are carefully managed to prevent acids and salts from reaching root-zone level.

Peats deeper than 2 m in the tropical coastal zone are generally so lacking in nutrients and so difficult to manage that the capital investment required to reclaim them for agriculture is not economically sound (Pons and Driessen, 1975). Since chemical fertilizers are expensive, the surface peat is commonly burned to liberate plant nutrients (Satari, 1979), to raise pH values (Coulter, 1957), to reduce pests and diseases, or to destroy toxic compounds. Opinions vary on the value of this practice. Certainly it is deleterious on shallower peats, where there exists a danger of exposing the underlying soils with an acid sulphate potential.

Any major landscape transformation resulting from radical alteration of biotic and physical assemblages consequent on a change in water–land interactive relationships through wetland reclamation opens broad new niches for a wide variety of organisms. Once wetland soils are drained and chemically treated, for example, conditions for micro-organisms—some beneficial and others not—are enhanced (Tie and Kueh, 1979). Similarly, weeds (especially *Scirpus* sp.) in rice fields, as well as mammalian and insect pests, can reduce crop yields by competition and consumption, or destruction, respectively (Hanson and Koesoebiono, 1979). Further, the alteration of the tidal wetland habitat also creates new niches for the vectors of malaria and other arthopod-borne diseases in humans, since one general aspect of the ecology of human diseases is that their vectors are to be found in man-made or man-modified environments (Ruddle and Manshard, 1981). The large-scale clearance of mangroves, for example, in many parts of the Asian tropics creates conditions suitable to the breeding of malaria vectors (Kunstadter, 1978a). On the other hand, malaria vectors are not endemic in south-east Asian mangroves that are relatively undisturbed or where clearings are small or temporary (Kunstadter, 1978b).

5.IV.2.2.3 Evaluation of Peat Soils for Development

Although the methodology for assessing the biophysical and biochemical capabilities of organic soils in the tropical coast zone is now fairly well established (Soepraptohardjo and Robinson, 1975; Soepraptohardjo and Driessen, 1976; Maas *et al*., 1979)—but not always applied too rigorously in feasibility studies—as is that on farming systems and management requirements, among other of the more important techno-economic considerations, little attempt has been made to assess the sustained investment, capital or manpower requirements needed to maintain agricultural output from reclaimed peat soils. One such attempt, from Sarawak, questions the wisdom of reclaiming tidal swamps by suggesting that the enormous investment costs that underwrite environmentally deleterious methods of reclamation, coupled with the inherent limitations of former swampland soils, are at best difficult to justify economically (Tie and Kueh, 1979).

But such an argument presupposes that the principal reason for developing such lands is to increase national levels of agricultural production. That is not necessarily the case. In Indonesia, for example, the reclamation of tidal wetlands seek to fulfil diverse political objectives, of which increased agricultural production is but one, and not necessarily the most important one at that.

Development plans for both the Central Islands (Java, Bali and Madura) and the Outer Islands have always been linked by the paired concepts of transmigration and irrigated rice production. Yet, ironically, even were it feasible to develop the full extent of Sumatra's tidal wetlands alone by transmigration—some 4 million hectares—the population problem on Java would be far from solved. Further, it is being increasingly demonstrated that tidally irrigated rice production is not suitable for the areas being reclaimed.

Planning for the swampland transmigration schemes in Indonesia is relatively simple in that the only crop envisaged is irrigated rice. But this is flawed since the areas are unsuited to sustained rice mono-culture, as the long experience of the Banjarese and Buginese reveals. Rice production is constrained by several major inherent biophysical problems, among which are edaphic characteristics, flood frequency and height, saltwater intrusion, and drought. Further, planning is complicated by the long-established communities of other ethnic groups in the areas designated for reclamation. These original southern Sumatran inhabitants, together with spontaneous migrants, exploit diverse and complementary resources that are protected and managed by traditional systems of customary rights and administration. The result has been a complex and confusing multiple and uncoordinated allocation of the same resources to different groups and for different purposes.

In conclusion, it is clear that sustaining the agricultural productivity of reclaimed organic soils of former tidal wetlands continues to be problematical, even for the better-adapted traditional practices of the Banjarese and Buginese. The relative success of spontaneous migrants from these two ethnic groups probably derives from their careful selection for reclamation of the shallower and richer peats, as does that of the relatively more successful transmigration schemes.

Monitoring of progress on the wetland transmigration project in southeastern Sumatra reveals a generally decreasing trend in soil fertility, increasing peat management problems and decreasing rice yields from the reclaimed land. Further, rice production via double-cropping is now unlikely owing to a lack of water in the dry season. On the contrary, outright failure of the rice crop is seen as a distinct possibility, owing to rainfall fluctuations (CNRMES, 1978).

At least in its tidal wetland component, the Indonesian transmigration scheme must necessarily be viewed as an expensive experiment, and lamentably one undertaken in advance of an adequate knowledge of the long-term ecological, economic and social costs. Further, the opportunity costs of equal investment foregone in alternative locations has never been properly considered.

5.IV.3 AGRICULTURAL TRANSFORMATION OF WETLANDS BY LARGE-SCALE INTEGRATED FARMING SYSTEMS

5.IV.3.1 The Zhujiang Delta, People's Republic of China

The transformation of tidal and other wetlands through the application of integrated agro-ecosystems, such as the joint operation of agriculture and aquaculture, is an ancient, widespread and enduring practice in south-east Asia. Systems are based on livestock, fowl and fish husbandry in combination with a range of seasonally rotated crops. Small numbers of pigs and ducks, together with fish, provide the household with animal protein and often a small cash income; while aquatic macrophytes, crop residues and kitchen leftovers feed the livestock, the manure of which, together with systematically collected human excrement (at least in the Chinese case), fertilizes the fishpond and eventually the cultivated field. Such diversified and integrated systems have sustained small-scale farm households for centuries. In the more sophisticated variants of such systems nutrients and energy are cycled continuously, little waste results, and an ecological balance is preserved. Whereas many if not most such farming systems are small and usually oriented towards family subsistence, that in the Zhujiang (Pearl River) Delta of southern China is an old-established and highly elaborate integrated system of intensive agriculture combined with poly-culture of carps and other freshwater fish. It

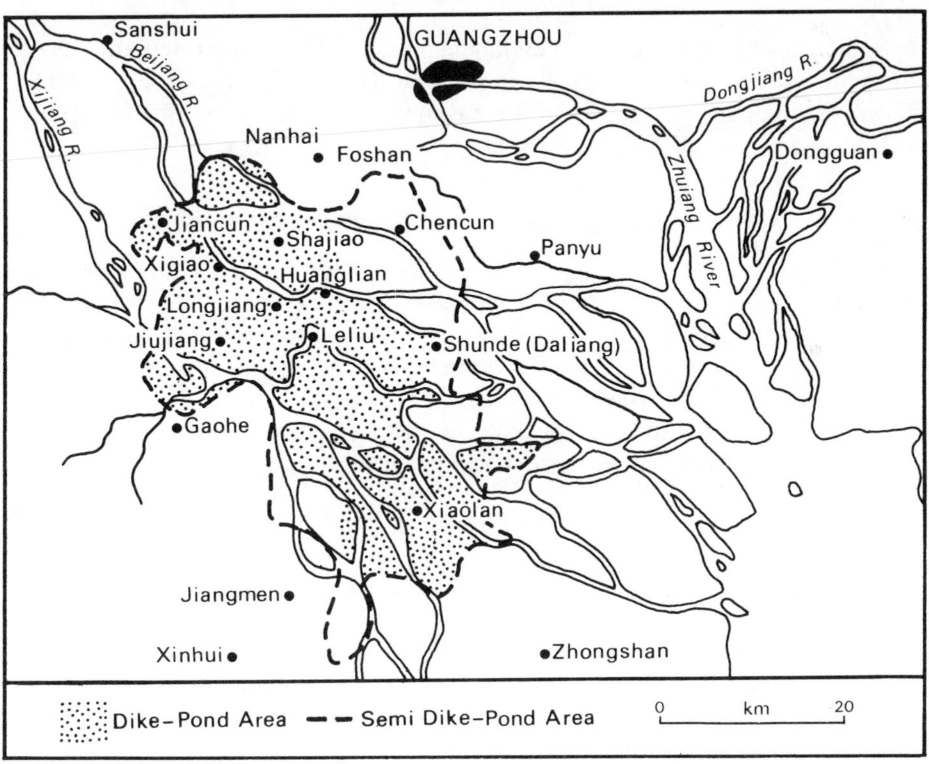

Figure 5.IV.2 Location of the dike–pond area in the Zhujiang Delta (Ruddle *et al.*, 1983)

has evolved over the past two millenia. There, integrated agriculture is operated on a geographical and economic scale unmatched elsewhere in the world (Ruddle *et al*., 1983).

The Zhujiang Delta, a densely populated region of some 12 000 km^2 focusing on the city of Guangzhou (Canton), is the economic and cultural core of south-east China. This delta has been built-up by several large rivers that converge in the region, principally the Zhujiang, Xijiang and the Beijiang (Figure 5.IV.2). Soils have developed from the recent alluvium of former tidal wetlands and are rich in organic matter and only slightly porous. Those in the central delta, for example, were primarily paddy soils with some tracts of meadow-boggy soils; but after centuries of intense cultivation and irrigation under a dike-pond system distinctive classes of silty soils have emerged. The climate is warm, temperate and rainy (Köppen *Cfa*) and is affected by the monsoonal circulation. It is characterized by an annual mean temperature at Shunde, in the central delta, of 22°C and an average annual range of 15°C, with mean temperatures in the warmest months of 28.7°C (July) and 28.4°C (August), and in the coldest months of 13.0°C (January) and 14.7°C (February).

Southern China is subject to damaging climatic hazards, principally Spring drought and monsoon and typhoon flooding (Luo *et al*., 1980). Since 1949, however, their impact has been increasingly controlled by major infrastructural projects (Ruddle *et al*., 1983).

5.IV.3.2 Evolution of the Zhujiang Dike–Pond System

About 1000 years BP, when the coastline in the Zhujiang Delta met the Xijiang, some 60–80 km inland of its present position (Li, 1657), the swampy coastal lowlands supported little agriculture, owing to continual flooding and acid sulphate conditions. Village economies were based mostly on the capture and collection of marine and estuarine resources. At higher elevations, however, fruit cultivation, especially of litchi (*Litchi chinensis*) and longan (*Euphoria longana*), had developed rapidly since its beginning during the Han Dynasty, some 2000 BP.

During the mid-fourteenth century water control began in the wetlands around Jiujiang, when smaller watercourses were dammed and diked to make fish ponds. Ponds were dug to create agricultural land and the excavated soil used to construct dikes. These early artificial ponds were devoted mainly to commerical fry rearing. As a consequence of the region's long experience with fruit cultivation, together with the large market demand throughout China, the first commercial crops grown on the dikes were fruits, particularly litchi and longan. Fruit cultivation may also partly have been a response to declining local water tables caused by the rapid seaward progradation of the delta, and which, when added to the impossibility of retaining water on the dikes, ruled out rice growing. At that time, however, there was apparently little or no conscious organization of an integrated fruit dike–fish pond system via linked flows of materials and energy, although both activities might have been undertaken on the same farm unit.

Mulberry cultivation, silkworm breeding and silk spinning have a history of about 2000 years in the Zhujiang Delta, and were initiated to produce goods for home consumption or to pay taxes. But by the seventh century AD a substantial industry had developed to satisfy the trade that developed rapidly when Guangzhou was opened to international commerce, 1200 years ago. Mulberry growing and silkworm rearing remained separated both geographically and conceptually from fish cultivation, since even by the thirteenth century few dikes were planted to mulberry. By the 1620s, however, mulberry was being widely cultivated on the dikes between the fish ponds, experience having shown that the economic returns from integrated mulberry dike–fish pond systems were greater than those obtained from cultivating fruit. Moreover, pond mud enriched with silkworm excrement and other wastes that had first been utilized to fertilize the pond and feed the fish, was found to be a superior fertilizer for the mulberry bushes than was the raw silkworm excrement applied hitherto, and which when applied to excess damaged the mulberry leaves. With this discovery an integrated dike–pond system that was found to be beneficial to both mulberry and fish, and far better than growing rice, quickly evolved (Zou, 1894). Thus by the 1650s farmers in the area around Jiujiang and Xiqiao began to replace old fruit trees with mulberry and hemp over some 70–80% of the dike area (Li, 1657). This process continued until around 1800, by which time most parts of Shunde County were devoted exclusively to the integrated dike–pond system.

Added impetus to the widespread adoption of this system occurred in the 1750s, when the Qing government closed all ports except Guangzhou to foreign trade, and at the same time limited the export of Hu (Taihu) silk. Thus the demand for Yue (Guangdong) silk increased abruptly and prices soared. Conversion of paddy fields to dike–pond systems continued apace and was extended beyond the areas of tidally influenced wetlands, the transformation to productive use of which was the main objective of the system.

A further boost to silk production occurred after the mechanization of the reeling process, in 1866. With the opening of the first local modern filature the integration of mulberry and fish entered a new phase. Not uncommonly, rice cultivation was abandoned as an increasingly larger area was converted to ponds and dikes. Shunde County, which now accounts for 75% of the Delta's silk production and has the most fish ponds, had converted 6700 ha of paddy fields into integrated mulberry dike–fish ponds farms by the end of the nineteenth century. By the end of the Qing Dynasty (1911) the Zhujiang Delta had some 66 700 ha devoted to this kind of agro-ecosystem; and by 1925, the apogee of Guangdong silk production, 93 000 ha of dikes was planted to mulberry.

Sugar cane has been cultivated in the Zhujiang Delta for about 2000 years, mostly on the higher and drier slopes of the uplands, and since the twelfth century Panyu County has been one of China's important sugar-growing areas. But a relatively poor quality product, with fluctuating yields, Guangdong sugar was grown mostly to satisfy local demand, and the industry was slow to develop. During the 'Great Depression' silk prices fell dramatically, large areas of mulberry dike were left uncultivated, and about one million silk workers were jobless. Seeking a profitable replacement for mulberry, entrepreneurs introduced improved varieties of sugar cane from south-east Asia, and converted abandoned mulberry dikes to sugar-cane growing. When the region's first sugar refinery was established in Shunde County in 1936, the focus of Guangdong sugar cultivation shifted from the uplands to the dike–pond region. Other former mulberry dikes were planted to mixed vegetables or rice, and yet others remained derelict.

Under the Japanese wartime occupation (1938–1945) local silk prices declined precipitously. During that period, too, rice prices increased dramatically, and because little food grain was cultivated in the Zhujiang Delta rice had to be imported from elsewhere in China or from abroad. Unable to purchase expensive rice with the now meagre earnings from their silk, farmers rapidly abandoned the dike–pond system to cultivate rice and other foodstuffs.

Although excavation of fish ponds and dike construction was historically considered to be the best way of transforming formerly marginal and natural hazard-prone wetlands into a productive agricultural region, until 1949 cropping patterns on the dikes were dictated by market prices rather than by the ecological considerations that are the essence of traditional integrated farming systems. But since 1949, although market demand in the guise of fixed production quotas has remained an undeniable force in determining patterns of resource use in the dike–pond region, integrated resource use based on fundamental ecological principles has evolved hand-in-hand with the exigencies of economic dictates. As a consequence, the components and processes of the dike–pond system have become progressively more tightly integrated in terms of complementary inputs and outputs; the system has become more elaborate in terms of the crops and animals raised and in the spatial and temporal patterns of inter-cropping, inter-planting and rotation; and overall system productivity has been raised and energy losses as a result of waste or by-products or niches not exploited, minimized. However, further refinements and improvements that demand a more rigid linking of the various components of the system are still forthcoming.

5.IV.3.2.1 The Fish Pond (Figure 5.IV.3)

Much of the success of freshwater aquaculture in south-east Asia results from the poly-culture of several different yet compatible species in the same body

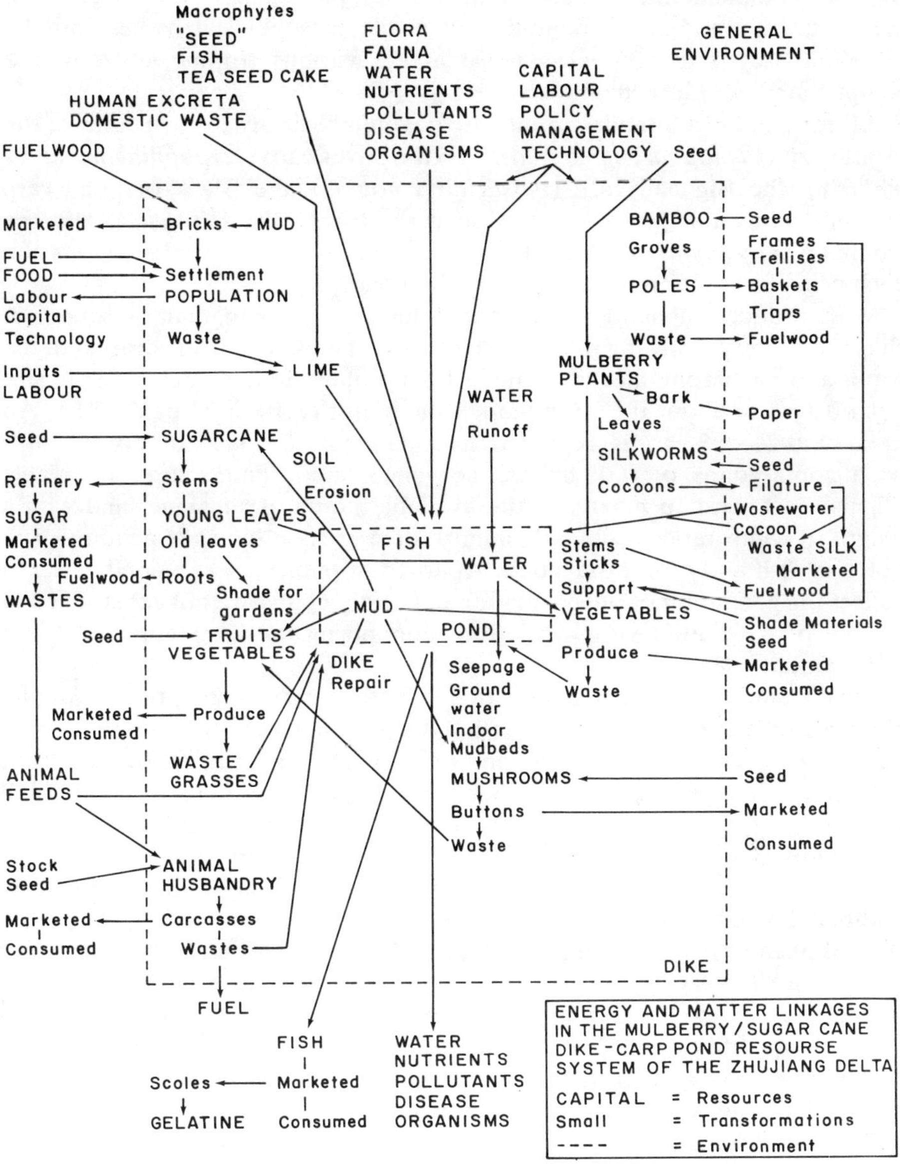

Figure 5.IV.3 Energy and matter linkages in the dike–pond system of the Zhujiang Delta (Ruddle *et al.*, 1983)

of water. This is the most efficient technique for using a limited pond area since the complementarities among the species, particularly in feeding habits, permit the harmonious exploitation of available ecological niches and the optimum use of the foods available at the various trophic levels of the complex aquatic food chain.

All fish ponds in Shunde County are devoted basically to carp culture: the Grass carp (*Ctenopharyngodon idellus*), the Silver carp (*Hypophthalmichthys molitrix*), the Bighead carp (*Aristichthys nobilis*), the Black or Snail carp (*Mylopharyngodon piceus*), the Mud carp (*Cirrhinus moitorella*), and the Common carp (*Cyprinus carpio*). Each species has distinct feeding habits and each occupies a different level within the pond.

Most production ponds in Shunde County are rectangular or square in shape, thereby permitting more efficient natural or mechanical aeration of the water and facilitating mechanization of pond operations. Ponds range in size from 0.2 ha to about 0.5 ha, although the majority are less than 0.5 ha. An area of 0.26–0.33 ha is regarded as ideal, but in zones of porous soils production ponds of 0.66 ha are not uncommon. Ponds are preferrably aligned east–west to maximize the available hours of sunshine, and with a length to width ratio of 6:4 to minimize dike erosion by wave action and to optimize fish and crop production. Apart from public sewage ponds used to collect the communities' faecal matter and urine, ponds are not constructed in the lee of tall buildings or trees, thus minimizing the interception of solar radiation and wind.

Pond water depths range from 2 to 3 m, with a preferred depth of 2.5–3 m. If the water is too shallow strata are not deep enough to optimize poly-culture or to rear large numbers of fish, and over-heating and eutrophication problems can also arise. Conversely, when the water is too deep temperatures and light in the bottom layers are low, bottom waters may become de-oxygenated, and plankton production reduced.

In the Zhujiang Delta the main fish species raised is the Grass carp, combined as environmental conditions and the availability of inputs permit with Bighead, Silver, Black, Mud and Common carp, and sometimes with Bream and Tilapia. Fish cultivation continues to be based on the traditional mixed-age or mixed-size poly-culture, the larger fish being selectively harvested in quantity every 3–4 months, with three or four, and in some places five or six, harvests each year. During the rearing period the pond is selectively restocked with younger individuals; thus in any given pond there are fish at various stages of growth. Although densities and species composition vary within the region, stocking rates of 15 000 per hectare are not uncommon.

Because stocking densities are high, extremely heavy fertilization of ponds and feeding of fish is practised; but since the system is so tightly integrated this presents few if any logistical problems. Only organic fertilizers are applied to both ponds and crops. Of these the most important is pig manure, applied at rates up to 230 t/ha per year. Commonly, pigsties (as well as latrines for humans) are built adjacent to the pond, on the dike, so that manure enters the pond directly. When available, chicken manure is used as fertilizer, at a rate of 15 t/ha per year. 'Fertilized water', which consists of 77% residues of soybean curd and 23% of other fermented wastes from the local food-processing factories, is applied to ponds at a rate of 200 kg per 1 kg of Silver carp per year. Additional pond fertilization is provided by the decomposition of fish excrement and fish food remains.

Fish are fed with organic materials produced on the associated dikes, in some cases after these products have been processed in local factories. Fish do not normally receive high-protein supplements of animal origin. Grass carp is the principal species raised and is fed mainly with grasses, sugar-cane husks, leafy vegetables and aquatic macrophytes grown on the dykes or in proximate water bodies. In general, such rough plant material constitutes 99.6% of fish feed, the balance being fine food consisting of various bean-cake residues and rice bran. The natural fertility of the pond water, enhanced by manuring, promotes the growth of plankton, the principal food of Bighead and Silver carp. Common and Mud carp are detritus feeders, whereas the Black carp feeds on snails. On this basis it has been determined empirically that the crops from 1 ha of dike can provide the feed for fish in the same area of associated ponds.

The principal environmental problems of concern to Shunde fish farmers occur in the pond micro-environment as a result of the extremely high organic loading of the water. Pond fertilization and fish feeding are therefore carefully controlled and water quality strictly monitored, especially to assess the dissolved oxygen content in the hours just prior to sunrise. This is particularly critical in these ponds since they are relatively deep by usual standards. Dissolved oxygen levels are monitored by observing fish behaviour. Optimum levels for ponds in this region are 5 mg/litre; when they decline to 1 mg/litre fish start to surface, and when they fall below 0.2 mg/litre large-scale fish mortality results (Zhong *et al*., 1965). Serious oxygen depletion is indicated when the bottom-dwelling species come to the surface—a rare occurrence in this region, where good management is the norm.

5.IV.3.2.2 The Crop Dike

Since the single most important commodity produced by the dike–pond system is fish, crop patterns on the dikes are geared basically to the feed requirements of the fish and the fertilizer needs of the ponds. The relative emphasis placed on mulberry and sugar cane, the main commerical crops cultivated on the dikes, is also a response to both domestic and international market demands for silk and sugar. Balancing market demand and pond input requirements, and thus deciding on the main dike crop, results in a fluctuating ratio of the different crops.

Four principal types of dike are recognized: mulberry dikes, sugar-cane dikes, fruit dikes and miscellaneous-crop dikes. Compared with those in other regions of integrated farming systems, dikes in the Zhujiang Delta are exceptionally large and varied, both in morphology and function. For the integrated system practised in this region, level-top dikes 6–10 m in width and 0.5–0.7 m in height above the pond surface are considered ideal. Dikes range in width from 6 to 20 m. Cultivation on narrow dikes is inefficient, whereas on the widest it is difficult to provide enough pond mud as fertilizer. Medium-height dikes, 0.5–0.7 m above the pond surface, absorb enough water from the pond to maintain a soil–water content ideal for most crops cultivated. Low dikes—those less than 0.5 m above the pond surface—suffer from waterlogging in the root-zone and are flood-prone in the wet season, whereas dikes higher than 10 m are incapable of absorbing enough pond water.

In the Zhujiang Delta year-round cultivation on the dikes and in the associated waterways presents a complex picture of seasonal and biennial rotation and of inter-cropping. There is no agricultural slack season and the dikes are under continuous cultivation. Cropping is closely connected with the husbandry of mammals, fish and silkworms. The principal commerical crops cultivated are mulberry (*Morus atropurpurea*), sugar cane (*Saccharum officinarum*), various fruits (especially litchi and longan, together with bananas and plantains [*Musa* spp.]) and a wide range of other plants for domestic use, fish feed and for marketing.

In the Zhujiang Delta, 9400 ha of dike is devoted to the cultivation of mulberry, the leaves of which are fed to silkworms (*Bombyx mori*). Mulberry leaves are harvested eight or nine times a year, between March and late-November. In this region 1 ha of dike yields an annual average of 22 500–30 000 kg of leaves, which results in a yield of 1875–2250 kg of silkworm cocoons. Shunde accounts for 90% of the cocoons produced by Guangdong Province (unpublished statistics).

Climatically this region is well-suited to both mulberry cultivation and silkworm rearing. Mulberry grows best at temperatures of 25–30°C, whereas its growth is hindered below 12°C and it cannot survive drought at temperatures greater than 30°C. Thus in this area the mulberry shrubs are pruned in late-November and the dike inter-cropped with various winter vegetables until February of the next year. Rainfall is sufficient for mulberry growth, and without irrigation dikes of medium height absorb enough pond water to maintain adequate moisture at an average depth of 35 cm, the root-zone of the mulberry plant. In drought periods dikes can be irrigated easily from the adjacent ponds.

Temperatures and humidity are important factors in sericulture. Silkworms grow only at temperatures between 15 and 32°C and die in drought periods at temperatures above 40°C. They require a relative humidity of 70–80% for optimal growth. Except in winter, when dry Mongolian air occasionally intrudes, relative humidity in the Zhujiang Delta is always high. From May to September high temperatures and high relative humidity are detrimental to silkworm growth but are mitigated by using bamboo frames constructed for the silkworms in the rearing sheds.

In the Zhujiang Delta mulberry shrubs are pruned after the last harvest of leaves and the dikes are inter-planted with an assortment of vegetable crops. This has the additional advantage of improving mulberry yields. To reduce plant diseases and insect pests the dike is planted alternately with mulberry and sugar cane, which also has been demonstrated locally to increase the yields of both crops as well as to improve soil quality.

With 15 300 ha of dike under sugar cane and an average yield of 75 t/ha, the delta produces 12.5% of Guangdong's sugar and is one of China's leading sugar-producing regions. Sugar cane is well-adapted to annual temperature and rainfall regimes in the region: its greatest water demand occurs in the period June–September, coinciding with the months of highest precipitation, and the second period of high water demand occurs in January–April, when again there is enough water. Mean monthly temperatures normally satisfy the requirements of sugar cane, the growth of which is retarded below 20°C and which stops altogether below 10°C. Frost damage is rare in the Zhujiang Delta. Dikes are planted to sugar in the Autumn and the crop requires 16 months to reach maturity.

In early-Spring, before the mulberrys come into leaf, mulberry dikes are inter-planted with soy-bean (*Glycine max*), mung-bean (*Phaseolus aureus*), taro (*Colocasia antiquorum*), peanut (*Arachis hypogea*), and other vegetable crops, which are harvested in the period May–June. Along the dike edge, bananas, plantains and fruits are cultivated. In Summer and Autumn, at the very edge of the dike, melons and gourds are trained on trellises that hang over the pond, thereby shading the water and preventing excessively high temperatures from endangering the fish. Each November, after the last mulberry leaf harvest, dikes are inter-planted with Chinese cabbage (*Brassica pekinensis*), cabbage (*B. oleracea*), leaf mustard (*B. juncea*), carrot (*Daucus carota*) and radish (*Raphanus sativus*), which provide two harvests, the last being in February. For feeding fish, Elephant grass (*Pennisetum purpureum*), maize (*Zea mays*), sorghum (*Sorghum sp.*) and sweet potato (*Ipomoea batatas*) are planted on all available spare land, including dike slopes, road-sides and along watercourses. To feed fish and pigs aquatic macrophytes such as water lettuce (*Pistia stratiotes*) and water hyacinth (*Eichhornia crassipes*) are cultivated in canals, rivulets and associated water bodies.

The mulberry dike–carp pond represents a highly intensive, integrated system of agriculture and aquaculture in which many outputs of sub-systems constitute inputs for other sub-systems. In this way not only is the medium for the growth of fish and crops provided, but so too is the environment from which their food and fertilizer requirement can be met. The system thus yields for all commodities produced a wider range of products per unit area of land and water than could otherwise be obtained, and lower costs for inputs, which, in the absence of integration, would be imported from outside the system, usually at considerable expense in terms of capital, time and labour. In conventional, semi-intensive aquaculture, for example, under systems not integrated with either animal husbandry or crop production, supplementary feed usually accounts for 50% of the total farm operating budget (Schroeder, 1979) and pond fertilizers and other inputs for the field crop comprise a large item of the budget of a non-integrated farm. In addition to the economic benefits afforded by integrated systems, integration is also beneficial to the environment in that it solves the problems of organic waste disposal in a sanitary manner and obviates *in situ* toxicity and downstream residue problems that arise from the incorrect use of organic fertilizers. In the dike–pond system of the Zhujiang Delta, integration exists exogenously among the pond, dike and general environment (*vide supra*) and endogenously within each of these elements.

Systems such as that of the Zhujiang Delta exhibit many obvious advantages over those that lack integration. Principal among them is that the pond fertilizers and fish needs are produced locally and at low cost, thereby almost eliminating the uncertainties of supply and greatly reducing the costs commonly associated with the use of compounded feeds and inorganic fertilizers. Hence profits from the cultivation of fish are increased by as much as 30–40% in the People's Republic of China (FAO, 1979). By using pond mud as a fertilizer for the crops, soil fertility and crop yields are raised without the expense of inorganic fertilizers. In national terms these savings are considerable when it is realized that an estimated 30% of all agricultural fertilizers used in the PRC are derived from fish ponds (*ibid.*); and, as a consequence of integration, the human consumer is assured a regular and balanced diet as well as a high degree of self-reliance in a wide range of foodstuffs and raw materials.

The different types of agricultural activity involved in complex integrated agro-ecosystems call for a diverse range of skills; and, on the scale practised in the Zhujiang Delta, also requires a large, well-organized labour force. Moreover, integrated systems of such complexity can only be operated on a large scale. Under such systems the economic returns from any one component are not viewed as important. Instead, maximizing the return from the whole system is the objective.

Thus the dike–pond system is ideally suited to the densely populated Zhujiang Delta and to the social and political organization of the PRC. Further, in biological and physical terms the region is well-suited to the assemblage of animals and crops that together form the mulberry dike–carp pond system.

This two millenia-old, ecologically sound system of reclaiming tidal wetlands and transforming them to a flourishing agricultural landscape continues to operate in southern China, with the reclamation of tidal wetlands along the seaward margin of the rapidly prograding Zhujiang Delta. This is a continuous process and is implemented using a field–pond variant of the dike–pond system used in the upper and central deltas. Unlike the dike–pond system, where the ratio of ponds to dikes is 6:4 or 5:5, the field-pond system is dominated by larger fields (replacing the ponds) and relatively few fish ponds, the ratio being 9:1 or 8:2.

In Doumen County, on the seaward edge of the Zhujiang Delta, 50% of the cultivated area is composed of low-lying fields with an average elevation of 0.2–0.7 m above sea-level. It is usually inundated during the rainy and typhoon seasons as well as being subject to salinization during the Winter-season high tides. Such fields are planted with rice and sugar cane, but yields have always been poor. Since 1974 some low-lying fields have been excavated to a depth of 2.5–3 m for use as fish ponds, and the soil removed used to elevate surrounding fields. In addition to providing soil to build up fields, the pond also serves as a drainage basin to lower field water tables. As a consequence, crop yields have risen dramatically and an integrated system of farming has begun to emerge there too (Zhong *et al*., 1982). Gradually the landscape of the inner areas of the delta edge is coming to resemble that of the longer-developed central delta, as an increasingly larger area is converted into fish ponds and the field–pond system evolves slowly but surely toward the dike–pond system.

5.IV.4 THE FUTURE OF WETLAND AGRICULTURAL TRANSFORMATION

To many planners the vast wetlands that dominate large tracts of the coastal zones of many developing nations in the humid tropics represent wastelands that when reclaimed with the appropriate combination of technological inputs supported by a large investment of capital will become a cornucopia of agricultural products, principally rice. This attitude persists despite the failure of many such ambitious reclamation and development schemes to attain planned levels of productivity. Such partial or outright failures, which are not uncommon, do not demonstrate the inherent impossibility of effecting modern transformations in tropical wetlands—rather they highlight the folly of incomplete planning and hurried project design predicated on a narrow knowledge base. They also demonstrate a total failure to evaluate pre-existing and perhaps, in the Western scheme of things, unconventional resource systems that have attained acceptable levels of agricultural productivity and that with careful consideration and planning could be transformed to a modern purpose.

Tropical wetlands are rich in an enormous range of renewable natural resources and their seemingly easily irrigable and largely level surfaces appear to offer ideal locations for extensive rice fields. This belies the potential fragility of such wetlands, the resource base of which can so easily be destroyed by ill-conceived and improperly implemented land transformation schemes.

Given that inherent fragility of wetlands combined with the scanty and uneven state of scientific knowledge regarding the types of transformations to which they should be subjected, large-scale schemes that call for radical environmental alteration are best deferred until such time as planning can be far better informed.

With that in mind, developing nations would be far better advised to consider schemes to rehabilitate for agricultural and other primary production degraded watersheds (such as those reduced to an *Imperata* cover by maladaptive shifting cultivation), as well as seeking ways of extracting higher yields from lands already under cultivation. The failure to consider the opportunity costs of investment in wetland development while not doing so in other areas undoubtedly represents a major squandering of precious development capital.

The focus of large-scale development schemes on the mono-culture of rice is a mistake. The dominant natural characteristic of wetland ecosystems is the all-pervasive and intimate interaction of terrestrial and aquatic components. Almost without exception this characteristic has been reflected in time-honoured resource systems traditionally used by wetland populations, and that yield on a sustained basis a wide range of complementary crops together with harvests of naturally occurring plants and animals. Mono-cultural systems introduced to wetlands are likely to eventually prove maladaptive, whereas properly designed, integrated poly-cultural systems of agriculture–aquaculture–livestock are likely to produce long-term sustained yields.

By definition coastal wetlands are the recipients of the entire up-stream discharge of linked hydrological basins. All too frequently the fundamental implications of this linkage are not considered when wetland development is being planned. The agricultural transformation is planned out of its total context without reference to up-stream uses of resources that can (and invariably do) have deleterious impacts on down-stream resource systems.

The integration of aquaculture and agriculture is an ancient, widespread and enduring practice in south and south-east Asia. Nevertheless, the structure, functions and management of such systems have been little appreciated outside those regions until relatively recently. Almost without exception such traditional and sophisticated resource systems have been overlooked by development planners, perhaps because they have been regarded as small-scale subsistence operations ill-suited to fulfilling the urgent food and raw-material needs of developing nations. As a consequence, few data are available to refute this reasoning, virtually nothing is known of the techniques and technologies used, and data on levels of productivity and farm economy are seriously deficient, if available at all.

Traditional Asian systems of aquaculture and agriculture, particularly those in the People's Republic of China, where such systems have been best developed and applied on the widest geographic and economic scale, have recently aroused major interest among Western scientists and development planners. While this trend is to be applauded, a note of caution is in order. Most integrated farming systems remain in large measure based upon the empirical wisdom of many generations of local farmers. Although Chinese scientists, such as those engaged in the artificial propagation of carps, have recently brought a more analytical approach to bear upon the subject, for the most part the scientific bases for system integration remain to be properly ascertained. This, in turn, has led to the acceptance as proven of many assumptions about integrated systems, a dangerous situation since much basic work remains to be done before integrated resource systems can be transferred with a reasonable assurance of success to other parts of the developing world. Further, virtually all attempts to improve the scientific understanding of traditional integrated systems have concentrated on extremely detailed micro-studies of various biological, physical, technological and economic aspects—such as the nature of the animal waste linkage between livestock and fish from the perspective of fertilizing the pond and feeding the fish—particularly as they concern poly-cultural pond systems. There has been little attempt to relate these detailed studies to the larger-scale relationships such as those between the pond and the dike, or those between the pond–dike system and the general environment within which the integrated resource system functions.

The constraints inhibiting a fuller scientific understanding of integrated systems—and thus their improvement, development and wider dissemination—lie as much in the social as in the natural sciences. The major factors that affect the performance of integrated systems—energy, materials, spatio-temporal considerations and information diversity—are not, as a whole, well-known, and the socio-economic aspects of such systems, which are complex and little understood, are in particularly urgent need of analysis.

The relatively small-scale Banjarese system that has permitted successful transformation of some Indonesian wetlands, together with the large-scale dike–pond system of the Zhujiang Delta of the PRC, when contrasted with the relatively disappointing results of Indonesian wetland transmigration projects, clearly shows the value of both a slower and incremental approach to wetland transformation, as opposed to a rapid and radical change, and the importance of the alternative models exemplified by these non-Western systems. The empirical wisdom—which, granted, in every situation may not be particularly wise—that underpins such systems is of itself a valuable resource that should be distilled and applied to modern techniques and technologies to effect wetland transformation. It has the intrinsic merit of tempering a rampant 'can-doism'. This is not to advocate a species of nativism but rather to suggest an alternative, low-cost path to obtaining the required information to plan adequately for wetland transformation, since it is starkly apparent that budgets will never be sufficient to permit totally comprehensive and fully satisfying scientific pre-development studies of the complex ecosystems that comprise tropical coastal wetlands. And lack of information suitable for planning is probably the largest single constraint on wetland development.

Since successful adaptive strategies in tropical wetlands have largely been based on non-Western concepts, the development of such areas should be based in large part on the adaptation of such systems for larger-scale application and transfer to other regions. The first priority, then, is for information on which to plan for system transformation. This requires sound data on the traditional system to be adapted and its compatibility with proposed innovations. This should be assembled in terms of energy and materials flow, the optimum use of farm (or other minimal unit) space, the temporal synchronization of the various integrated farm activities, information diversity (particularly in management and socio-cultural inputs) and linkages with other off-farm resource systems. Since a wide range of adaptations to wetlands exists throughout the tropics, a large number of crop and animal species should be tested for use in integrated farming systems. Basic research is required on crop (including livestock and fish) biology, seed supply, stocking and planting rates and densities, particularly for candidate species that are hardy and tolerant of the extreme seasonal changes that characterize wetland environments. Research is also urgently required to overcome the principal constraints on integrated systems in wetlands. Of the utmost importance is that on land preparation and soil capabilities, crop requirements, fertilization rates, pest and disease management, weed control and the provision of an adequate and regular water supply. Finally, basic economic research through the cost–benefit analysis of a range of existing systems and proposed introductions and changes is a *sine qua non* for the successful transformation of tropical wetlands.

5.IV.5 REFERENCES

Brown, F. B. (1972) *Land and Water Resources for Development in South East Sumatra*. Department of Agriculture, Jakarta.

Chambers, M. J. (1979) Rates of Peat Loss on the Upang Transmigration Project, South Sumatra. *Proceedings of the Third National Symposium on Aspects of Tidal Swamp Development*, **2**, 765–777. Bogor Agricultural University, Bogor.

CNRMES (Center for Natural Resource Management and Environmental Studies) (1978) *Draft Notes on the Impact of Tidal Irrigation in the Musi-Banuayasin Area*. Bogor Agricultural University, Bogor (mimeo.).

Collier, W. L. (1978) Development problems and conflicts in the coastal zone of Sumatra: swamps are for people. Paper presented at the Agricultural Development Council Inter-Regional Seminar on Resource Use and Management in Coastal Zones of the Asian Humid Tropics, Bangkok.

Collier, W. L. (1980) Resource use in the tidal swamps of central Kalimantan: a case study of Banjaresc and Javanese rice and coconut producers. In: Furtado, J. I. (Ed.), *Tropical Ecology and Development*, Vol. 2, 1047–1064. University of Malaya, Kuala Lumpur.

Collier, W. L., Bochari, R., Supardi, A., Bakir and Rahmadi (1981) Social and economic aspects of rice based cropping systems in the coastal wetlands of Indonesia. Paper presented at the First International Workshop on Research Priorities in Tidal Swamp Rice in Indonesia, Jakarta.

Coulter, J. K. (1957) Development of peat soils in Malaya. *Malayan Agric. J.*, **40**, 188–199.

DBP (Direktorat Bina Produksi) (1981) *Brief Report on the Development of Rice Fields in the Tidal Areas of Kalimantan*. DBP, Jakarta.

Driessen, P. M. (1977) Peat soils, their formation, properties, reclamation and suitability for rice cultivation. Paper presented at the Rice Meeting. International Rice Research Institute, Los Baños.

Driessen, P. M., and Suhardjo, H. (1975) Reclamation and use of Indonesian lowland peats and their effects on soil conditions. Paper presented at the Third ASEAN Soil Conference, Kuala Lumpur.

FAO (1979) *Aquaculture in China*. FAO, Rome.

FAO (1982) *Management and Utilization of Mangroves in Asia and the Pacific*. FAO, Rome.

Gersie, D. (1979) Background of Hydrometric Surveys of P4S as presently Programmed. *Proceedings of the Third National Symposium on Tidal Swampland Development*, 857–868. Bogor Agricultural University, Bogor.

Hanson, J. A. (1980) Transmigration for Marginal Land Development. In: Hansen G. (Ed.), *Agriculture and Rural Change in Indonesia*.

Hanson, J. A., and Koesoebiono (1979) Settling coastal swamplands in Sumatra: a case study for integrated resource management. In: MacAndrews, C., and Chia, L. S. (Eds.), *Developing Economies and the Environment*, 121–175. McGraw-Hill, Singapore.

Hardjono, J. M. (1977) *Transmigration in Indonesia*. Oxford University Press, Kuala Lumpur.

Hart, G. (1976) *Labour Allocation, Income, and Consumption in a Javanese Coastal Village*. Bogor Agricultural University, Bogor.

IBRD (International Bank for Reconstruction and Development) (1974) *Indonesia Agricultural Survey Annex 1*. IBRD, Washington, DC.

Kueh, H. S. (1976) Sago agronomy in the context of Sarawak. Paper presented at the First International Sago Symposium, Kuching.

Kunstadter, P. (1978a) Human settlements, demography and health in Asian and Pacific mangrove forest areas. Paper presented at the UNESCO seminar on Human Uses of the Mangrove Environment and its Management Implications, Dacca.

Kunstadter, P. (1978b) Management of man, malaria and mosquitos in the mangroves: an essay in human ecology (unpublished).

Li, C. X. (1657) *Village Annals of Jiujiang*. Nanhai, publisher not indicated (in Chinese).

Li, H. Z. (1976) *Agriculture in the Zhujiang Delta*. Foshan, publisher not indicated (in Chinese).

Lucas, R. E., and Davis, J. F. (1961) Relationships between pH values of organic soils and availability of 12 plant nutrients. *Soil Science*, **92**, 177–182.

Luo, K. F., *et al*. (1980) *Background to the Utilization of the Dyke–Pond Area of the Zhujiang (Pearl River) Delta, China*. Guangzhou Institute of Geography, Guangzhou (mimeo.).

Maas, E. F., Tie, Y. L., and Lim, C. P. (1979) *Sarawak Land Capability Classification and Evaluation for Agricultural Crops*. Department of Agriculture, Kuching.

Moorman, F. R., and Pons, L. J. (1975) Characteristics of mangrove soils in relation to their agricultural land use and potential. In: Walsh, G. E., Snedaker, S. C., and Teas, H. J. (Eds.), *Proceedings of International Symposium on Biology and Management of Mangroves*, 529–547. University of Florida, Gainesville.

Peltzer, K. (1945) *Pioneer Settlement in the Asiatic Tropics*. American Geographical Society, New York.

Pons, L. J., and Driessen, P. M. (1975) *Waste Land Areas of Oligotrophic Peat and Acid Sulphate Soils in Indonesia* (mimeo.).

Ruddle, K. (1980) A preliminary survey of fish cultivation in ricefields, with special reference to West Java, Indonesia. *Bull. Nat. Mus. Ethnol.*, **3**, 801–822.

Ruddle, K. (1982) Traditional integrated farming systems and rural development: the example of ricefield fisheries in south-east Asia. *Agric. Admin.*, **10**, 1–11.

Ruddle, K., Furtado, J. I., Zhong, G., and Deng, H. (1983) The mulberry dike–carp pond resource system of the Zhujiang (Pearl River) Delta, People's Republic of China: (1) Environmental context and system overview. *Appl. Geog.*, **3**, 45–62.

Ruddle, K., and Grandstaff, T. (1978) The international potential of traditional resource systems in marginal areas. *Tech. Forecast. and Soc. Change*, **11**, 119–131.

Ruddle, K., Johnson, D. V., Townsend, P. K., and Rees, J. D. (1978) *Palm Sago: A Tropical Starch from Marginal Lands*. University Press of Hawaii, Honolulu.

Ruddle, K., and Manshard, W. (1981) *Renewable Resources and the Environment*. Tycooly International Publishing Co. and the UN University, Dublin and Tokyo.

Satari, A. M. (1979) A selective strategy in opening up new tidal areas for aquacultural development. *Proceedings of the Third National Symposium on Tidal Swampland Development*, 42–58. Bogor Agricultural University, Bogor.

Schroeder, G. L. (1979) Fish farming in manure-loaded ponds. Paper presented at the ICLARM–SEARCA Conference on Integrated Agriculture–Aquaculture Farming Systems, Manila.

Soepraptohardjo, M., and Robinson, G. M. (1975) *A Proposed Land Capability System for Agricultural Use in Indonesia*. Bogor Agricultural University, Bogor.

Soepraptohardjo, M., and Driessen, P. M. (1976) *The Lowland Peats of Indonesia: A Challenge for the Future*. Bogor Agricultural University, Bogor.

Soeriaatmadja, R. E. (1978) Plant inventory and vegetational analysis at proposed tidal reclamation project areas in Siak and Aokan, Riau, Sumatra, Indonesia. Paper presented at the Symposium on Mangrove and Estuarine Vegetation in south-east Asia. Agricultural University of Malaya, Selangor.

Suhardjo, H., and Widjaja-Adhi, I. P. J. (1976) *Chemical Characteristics of the Upper 30 cm of Peat Soils from Riau*. Bogor Agricultural University, Bogor.

Tie, Y. L., and Kueh, H. S. (1979) *A Review of Lowland Organic Soils in Sarawak*. Department of Agriculture, Kuching.

Watts, J. C. D. (1969) Phosphate retention in acid sulphate pond muds from the Malacca area. *Malayan Agric. J.*, **47**, 187–202.

Zhong, G., Qin, W., and Huang, F. (1982) *The Ecological Patterns in the Field-Pond System of the Low-Lying Sandy Land of Both Sides of the Zhujiang (Pearl River) Estuary*. Guangzhou Institute of Geography, Guangzhou (mimeo.).
Zhong, L. (1965) *The Biology and Artificial Propagation of Domestic Fishes*. Science Press, Beijing (in Chinese).
Zou, Z. L. (1894) *County Annals of Gaoming*. Gaoming, publisher not indicated (in Chinese).

Land Transformation in Agriculture
Edited by M. G. Wolman and F. G. A. Fournier

CHAPTER 5.V
The Use of Fertilizers and Soil Amendments

R. A. Olson

5.V.1 INTRODUCTION

Soil fertility as it controls the supply of elements and plant nutrition is one of the foremost indirect factors involved in land transformations. The combined provision of moisture and nutrients makes it possible for the desert to bloom in the ultimate transformation. Historical measures for fertility maintenance involving organic inputs have largely been supplanted in developed countries by the growth of a chemical industry that annually supplies over 100×10^6 tonnes of inorganic fertilizer nutrients. Extension of the marketing of these materials into developing countries has only just begun but, as emphasized in the World Food Conference of 1974, must be hastened as rapidly as possible. Many constraints are involved, and technology transfer from developing countries cannot always be effected directly without on-site research.

5.V.2 TRADITIONAL SYSTEMS OF SOIL FERTILITY MAINTENANCE

The importance of soil fertility to agricultural production has been recognized from the beginning of recorded history. The fertile silt deposition from controlled flooding of the Tigris and Nile valley lands made Mesopotamia and Egypt bread-baskets of the ancient world. Agriculturalists of the time recognized the value of these silts to their crops without knowledge of the nutritional elements involved. The Greek historian, Herodotus, in approximately 500 BC, wrote of the very high crop yields achieved in the Tigris valley, and 200 years later Theophrastus described efforts there at holding flood waters on the land as long as possible to maximize silt deposition.

5.V.2.1. Animal and Human Wastes

5.V.2.1.1 Animal Manures

The benefit of animal droppings has undoubtedly been recognized since before the dawn of history. Certainly the very earliest writings associated with agriculture refer to the manuring of vineyards (e.g. Homer in *The Odyssey*). It is significant that Theophrastus stipulated practices in the management of manures that hold today: add bedding to the stalls for holding the urine, and apply the manure abundantly to poor, thin soils but sparingly to rich soils. He classified manures in the following order of decreasing value: human, swine, goat, sheep, cow, oxen and horse, apparently not recognizing the high nutrient quality of fowl manure.

The extreme importance of manure to farming practice throughout medieval times to the recent past is evident in the traditional farmyard of Europe, which fostered the central manurial heap or pit for under-water storage. Likewise in China to this day, animal droppings are not allowed to go to waste. Unquestionably, these animal manures have done much towards maintaining the productivity of European and Chinese soils for thousands of years.

Animal manures that have been stored appropriately during their accumulation can be regarded as well-balanced fertilizers, containing amounts of all known essential plant nutrients. When used continuously on land at modest rates, they are very effective at maintaining most soil nutrients at a fairly high level. Care must be exercised in manure storage, however, to preserve the nitrogen component. Also, manure coming from dairy and beef animals is likely to contain substantial soluble salts (up to 10% of the total dry weight); thus commonly applied annual rate of 25 tonnes manure per hectare may contain several thousand pounds of soluble salt, which can induce a salinity problem in just a few years. Another recognized problem with manure where animals are highly concentrated in feed lots and dairying operations is the contamination of surface waters by runoff from these areas. Fish kills have been observed, and eutrophication accentuated by the organic waste intrusion. This has resulted in regulations in some countries requiring lagoons to catch the runoff waste before it reaches streams or lakes.

The primary shortcoming of animal manure as a fertilizer source, however, is that there is no possible way of producing enough of it to supply the world's soil-fertility needs. Taking out the portion that is used in some countries for heating and cooking purposes, perhaps no more than 10–15% of world fertilizer needs could be met by animal manure.

Manure is bulky, reducing its economic value, and is not aesthetically pleasing to handle, which has accounted for a substantial portion of it being disposed of as a waste product in many of the developed countries. However, it is now being looked at in a different light in view of the escalating cost of the energy that goes into inorganic fertilizer production. At today's fertilizer market prices, the value of the nitrogen, phosphorus and potassium components of an average tonne of barnyard manure is around US$5, without reference to the worth of the micro-nutrients and physical benefits provided by the organic matter.

5.V.2.1.2 Human Sewage

Human wastes have long been used as a soil-fertility restoring measure, exemplified by the complex disposal system established in Athens during the Golden Age of Greece for carrying sewage to adjacent coastal-plain farm fields and vineyards. Productivity of the farmlands of China has been similarly maintained for thousands of years.

There can be no question of the potential value of this nutrient source. Many of the highly developed countries of the world are now putting major efforts into means for returning these 'wastes' to the land rather than disposing of them in streams, lakes, or oceans. It is estimated that the total quantity of US sewage, for example, could supply around 1.5 million tonnes of nitrogen alone, or about 15% of the amount presently marketed as inorganic fertilizer.

Using sewage as a fertilizer is not without problems, however. Disease organisms are rapidly spread, and dysentery is prevalent in those areas of the world where raw sewage is an important fertilizer source. Composting—allowing the heat so generated to kill the organisms—is practised throughout China today, and pasteurizing by other means can also alleviate the hazard. A more difficult problem is that of heavy elements contained in the sewage when industrial wastes are added. Elements like zinc, lead, cadmium, selenium and copper that enter into the sewage system from electroplating and many other manufacturing processes can become long-term soil, plant and consumer toxins. These are elements that are not readily leachable from the soil. The only control possible once the elements have accumulated is to lime the soil to a high pH level and add substantial quantities of another element like phosphorus that inhibits plant uptake of the heavy element. Such measures may prove to be only partially effective and, accordingly, the best solution remains the segregation of industrial wastes from the rest of the city sewage, necessitating a modification of existing sewage systems.

5.V.2.2 Rotations with Legumes

The *Leguminosae* have been recognized for their value in soil-fertility restoration from the earliest historical records. Virgil, the Roman poet, wrote:

> 'Where Vetches, Pulse and Tares have stood
> And Stalks of Lupines grew (a stubborn wood)
> Th' ensuing Season, in return may bear
> The bearded Product of the Golden Year.'

The primary gain from legume production is nitrogen because of the symbiotic nitrogen fixation process in the crop's root nodules. Depending on legume type and thriftiness of growth, residual amounts of nitrogen from the legume in a wide range up to 250 kg/ha have been measured. Such large amounts have been found to benefit yields of non-legumes for two or more years in various climatic regions. Because of the deep-rooting of many legumes there may also be some transport of deep subsoil nutrient reserves to the surface soil as the crop is killed and its residues become part of the surface soil system.

The well-managed legume, when occupying a significant time portion of a rotation sequence, can accommodate much of the nitrogen requirement of the cycle. This obviously fits in best with a livestock farming system that can effectively utilize the associated hay production, which adds a further manure component for soil-fertility maintenance.

Legumes are, however, among the heaviest users of other soil nutrients like phosphorus, potassium and sulphur, necessitating greater inputs of these elements than for production of grain crops. Nor will most of the species tolerate strong soil acidity, which makes liming a required practice in the more humid regions with acid soils. Additionally, legumes have a higher water requirement than non-legumes, making them generally impractical for the drier cropping regions unless irrigation can be practised.

5.V.2.3 Organic Farming

Effective use of manures and legumes has been the nucleus of organic systems of farming in the past and of those espoused presently by 'environmentalists' in the developed countries. Few legitimate arguments can be made against the benefits to be derived where a purely organic system is feasible (USDA, 1980). The additional organic material from the manures contributes to soil structure synthesis, which further assists in erosion control. Requirements for inorganic nitrogen fertilizer are reduced, as well as for most other nutrients if substantial manure treatment is included—a prime example being the pig as a 'fertilizer factory' in China. The most serious limitations to organic farming' worldwide, however, are insufficient amounts of manure and water and the control of incident pests. To provide the necessary animal manure input a livestock system of farming with high meat production is required, and this is a system of low efficiency, compared with food grain production, for providing maximum food calories and protein per capita—a foremost consideration in food-deficit developing countries. Additionally, a liming programme is usually required in humid and sub-humid climatic regions to maintain a higher soil pH level for the *Leguminosae* than is necessary for good grain and fibre crops. Where human sewage constitutes the major organic input, as in China, a high population density is mandatory. Moreover, with both sewage and animal manure there is no control over the timing of when nitrogen is made available for crop use, as can be accomplished with inorganic carriers.

Important as manures are for soil-fertility maintenance in many regions of the world, they serve an equally important role as fuel for cooking and heating in other regions, where their value as fertilizer is lost. In other situations where manures are used first for generating biogas as an energy source, the residue from the process is not greatly reduced in several of its fertility components.

Tidal and continental fish ponds are organic systems representing a quite different type of land transformation. Here rates of sewage and/or pig manure applications are adjusted in accordance with the pond size to supply the nutritional needs of the fish crop. Silt in the pond bottom after drainage is reinforced with nutrients from the fish excrement, and is then either spread on adjacent land or serves as a fertile new land surface.

Again on the plant side of organic farming is the role of blue–green algae in fixing nitrogen in the rice paddy. Amounts in the range 10–20 kg/ha are recognized—not great for the given year's crop but significant over many years' in nitrogen conservancy. Likewise, the Azolla, a nitrogen-fixing fern, living in association with hyacinth in a surface water body or rice in a paddy, is a significant contributor to crop nitrogen economy in China.

The 'natural' connotation given to organic farming systems implies fewer environmental hazards than when inorganic chemicals are employed. In fact, however, quite the converse has commonly been found in carefully documented cases of groundwater nutrient contamination and the eutrophication of surface waters. Seepage of nitrate into wells close to animal impoundments and runoff from fields on which manure had been spread proved to be primary nutrient sources.

Residue of the previous crop as an organic return to the soil cannot be overlooked as a factor in land transformations. It has been regarded variously as a troublemaker in tying up nutrients by microbial immobilization at the expense of the next crop, as a raw material for energy production, and as the foundation for one of the best erosion-control measures available to the farmer. The nutrient content of the stover from a 10 000 kg/ha maize crop is not very high. Nevertheless, adding up the value of all nutrients contained in that stover in relation to the cost of those nutrients in fertilizer gives a value of the order of US$100/ha. In the short term the total quantity is enough to produce the stover for the next crop. Clearly, removal of that residue for alcohol production, or for whatever purpose, takes away that amount of nutrient from the soil, and the effect on soil fertility in the long term will be substantial. Also, its removal will accelerate the rate of decline of organic matter in soil, the regular return of fresh quantities of which is a most important factor in soil structure synthesis and maintenance. Finally, the residue is the key to success of stubble mulch, conservation tillage (reduced tillage) and ecofallow systems aimed at conserving moisture and retarding soil erosion.

5.V.3 ROLE OF INORGANIC FERTILIZERS IN LAND TRANSFORMATIONS

5.V.3.1 Regional and Worldwide Use

The growth of that segment of the chemical industry supplying inorganic fertilizers to agriculture has been quite phenomenal since the Second World War (Table 5.V.1). The total worldwide consumption of the three major nutrients was approximately seven times greater in 1981 than in 1953. Virtually all of the increased production was used in the developed countries through the early 1960s; but in the last 4 or 5 years there has been a levelling off in consumption there, and most of the continued increase has come in the developing countries (Figure 5.V.1).

Of further interest in Table 5.V.1 is the trend in use of the three nutrients relative to each other. Phosphate and potash were the dominant nutrients in the fertilizer trade through the 1950s. With the extensive construction thereafter of efficient nitrogen synthesizing plants, nitrogen consumption has gradually outstripped the other two to the point where it is now double that of phosphate and $2\frac{1}{2}$ times that of potash. These statistics show that fertilizer has become a foremost requisite to economic agriculture, as further exemplified by the fact that food and fertilizer were virtually synonymous terms in deliberations at the World Food Conference in 1974.

Table 5.V.1 World consumption of inorganic fertilizer nutrients (FAO Fertilizer Yearbooks)

	Fertilizer nutrients			
Year	N	P_2O_5 (million tonnes)	K_2O	Totals
1938/39	2.6	3.6	2.8	9.0
1953/54	5.2	6.3	5.7	17.2
1959/60	9.7	9.7	8.6	28.0
1965	16.6	14.0	10.9	41.5
1967	22.1	16.3	12.9	51.3
1969	27.1	18.3	14.5	59.9
1971	31.8	19.8	16.4	68.0
1973	35.8	22.5	18.5	76.8
1975	38.8	23.0	19.5	81.3
1977	46.1	26.6	22.9	95.6
1978/79	53.7	29.9	23.9	107.5
1980/81	60.7	32.8	25.2	118.7
1981/82	60.4	30.9	23.9	115.3

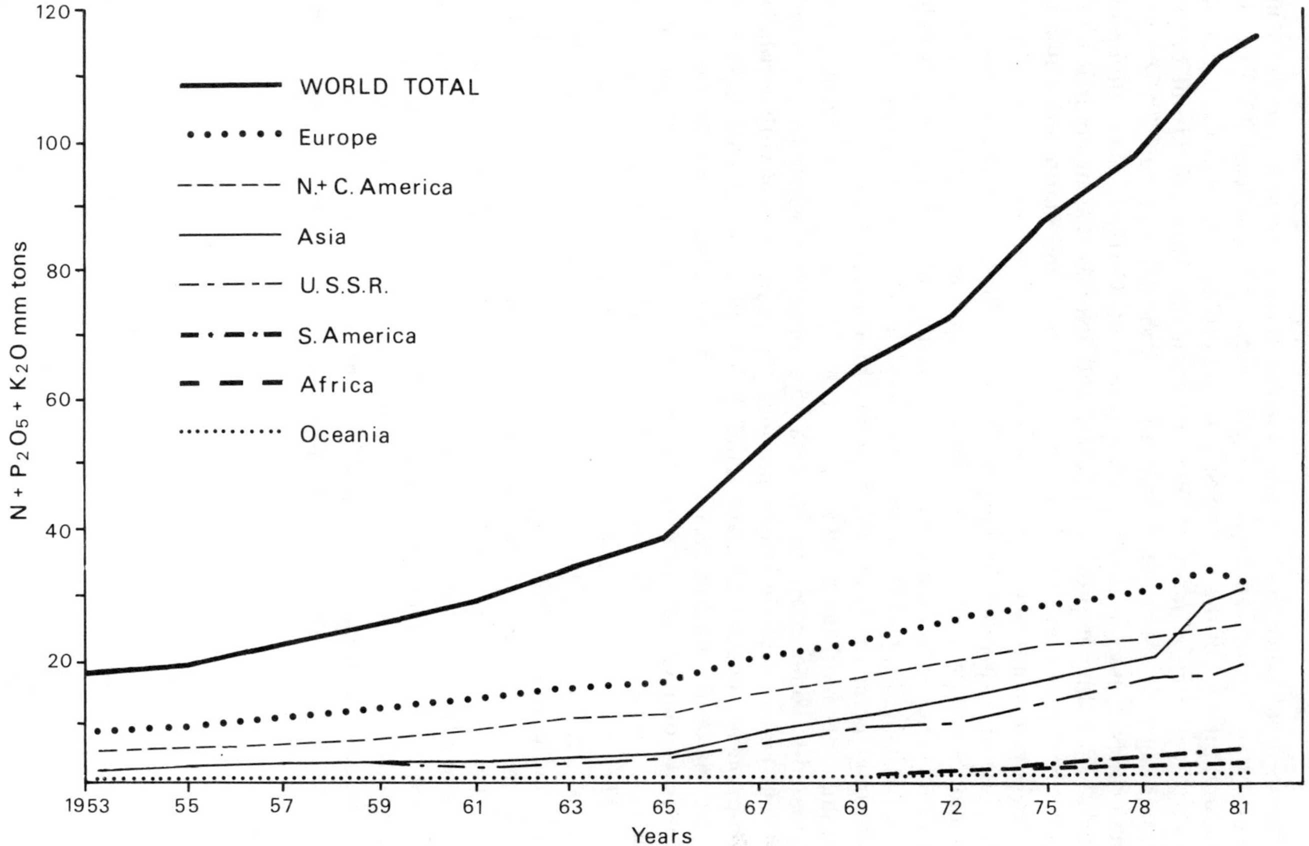

Figure 5.V.1 Growth in world consumption of inorganic fertilizers by regions during 1953–81 (FAO *Fertilizer Yearbooks*)

5.V.3.2 Impact on Agricultural Production

The dominant influence that inorganic fertilizers are having on agriculture in the world today is inescapable. They now account for at least 30% of all agricultural produce of the United States, and probably more of the order of 50% of cereal grain production. Without them the USA would not be an exporter of agricultural products. Indicative are the data of Figure 5.V.2 from the state of Nebraska in the country's grain belt. Certainly many different advances—including improved varieties and hybrids, expanded irrigation practice, better control of pests and diseases, and improved cultural practices—have contributed to the approximate doubling of wheat and quadrupling of corn yields in the past 30 years. These production increases have come about by reason of the fertilizer input in combination with the other factors cited. The data suggest that a levelling off in average state yields has occurred in the past 10 years, awaiting the next major technological breakthrough comparable to the adoption of fertilizer nitrogen.

The impact of inorganic fertilizers on worldwide production of grain crops is expressed to some extent in the data of Table 5.V.2, which lists average wheat and maize yields in major producing countries along with fertilizer consumption per hectare of arable land. When all of the yield data in this table are matched with fertilizer consumed in the respective countries and regions (excluding China and Argentina because of questioned consumption values), a correlation coefficient of +0.71 is found for maize and +0.91 for wheat. This implies that the yield obtained is directly proportional to the amount of fertilizer used. Direct correlations exist, of course, between fertilizer use and advanced technologies in other aspects of agricultural production, but the major role of fertilizer is unmistakable.

An apparently large disparity exists between developed and developing countries in Table 5.V.2 in the amounts of fertilizers consumed. With more than twice the fertilizer rate per arable hectare in the developed countries, maize yields are almost three times as great and wheat 25% larger. (The lesser response of wheat is attributable to the fact that most of the world's wheat is grown in dry regions where yields are limited as much by moisture as by soil fertility.)

5.V.3.3 Projections of Further Needs

The World Food Conference of 1974 made it quite clear that a dependable fertilizer supply was a requisite for meeting the food needs of the developing countries. The amounts required and the cost for their provision would be tremendous. Table 5.V.3 presents one projection of quantities involved and their likely end-effects. If these projections are correct it is obvious that a very serious demographic problem is in the offing for the world. An approximate

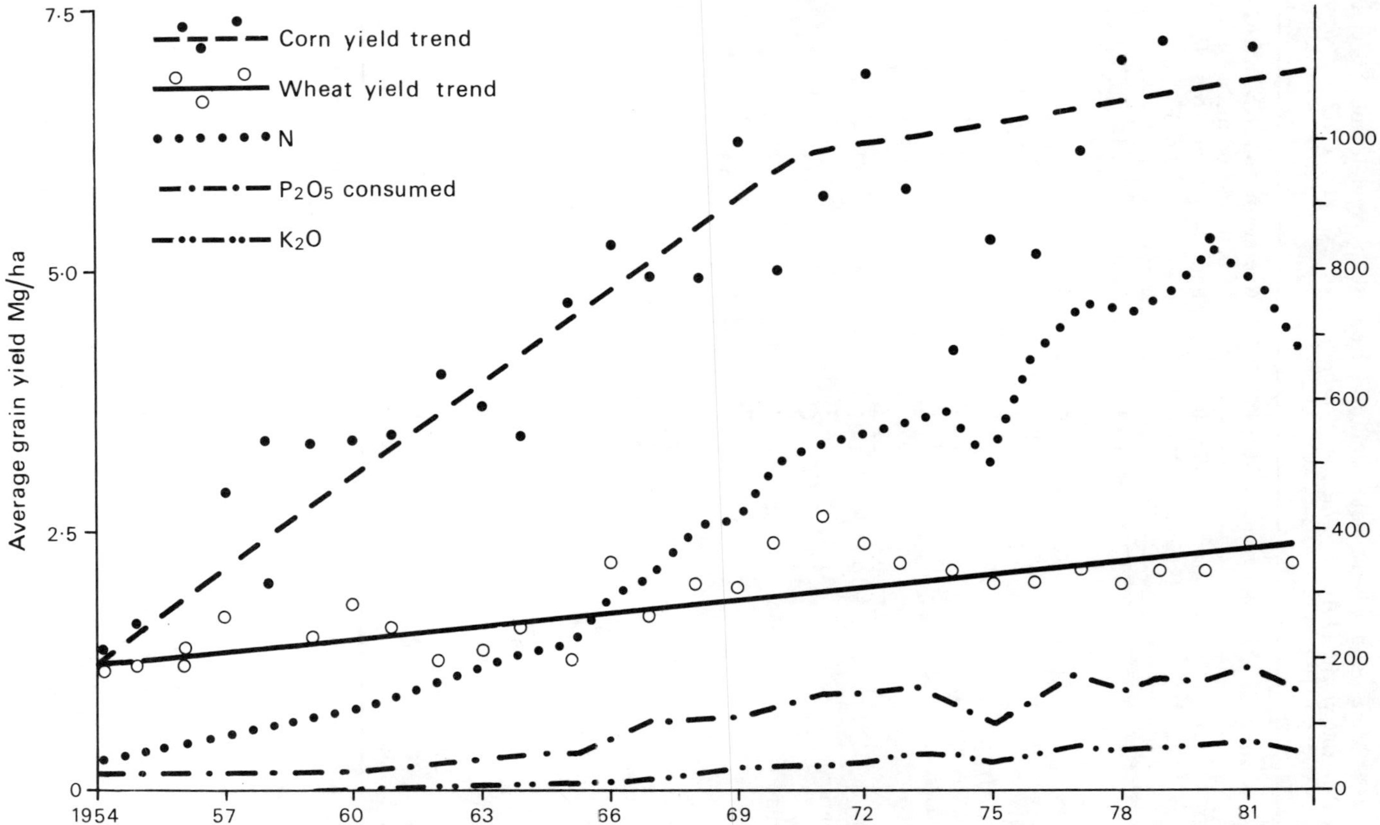

Figure 5.V.2 Relationship between fertilizer N, P_2O_5 and K_2O consumed in Nebraska to state yields of corn (maize) and wheat, 1954–82

Table 5.V.2 Maize and wheat yields and fertilizer use by region and selected countries of the world (*FAO Production Yearbook*, Vol. 36, and *FAO Fertilizer Yearbook*, Vol. 32, 1982)[a]

Region	Average grain yield, 1980–82: Maize (kg/ha × 10^{-2})	Average grain yield, 1980–82: Wheat (kg/ha × 10^{-2})	Average fertilizer N + P_2O_5 + K_2O use on arable land and permanent crops, 1981 (kg/ha)
World[b]	33.2	19.3	78.5
Africa	13.3	11.3	19.5
Algeria		6.8	26.2
Morocco		8.9	23.9
Nigeria	9.1		7.0
South Africa	17.2		90.4
N. and C. America	54.7	22.9	88.4
Canada		19.1	41.9
Mexico	18.3	39.7	66.6
USA	66.0	23.3	102.4
S. America	20.1	14.1	26.9
Argentina	31.3	16.3	2.7
Brazil	17.8	8.9	37.5
Uruguay		13.2	33.3
Asia	23.4	17.3	68.5
China	31.1	20.9	150.1
India	11.3	15.9	33.8
Pakistan		16.2	53.1
Europe	47.8	37.7	225.0
France	57.0	50.7	298.4
Italy		27.0	163.3
Spain		17.3	67.2
UK		59.7	329.6
Romania	35.9	27.7	154.1
Australia		10.3	27.9
USSR	28.1	14.9	82.6
All developed[b]	53.3	21.0	114.8
All developing	19.7	16.8	47.9

[a]Statistics presented for each region and for the two or three major producing countries in each region.
[b]Worldwide (and not just the countries listed).

tripling of fertilizer consumed would increase food grain production by about 50% in developing countries between 1980 and 2000, but with a continuing decline in grain per capita. In short, growth in fertilizer use with its benefits to food supplies cannot keep pace with the productivity of people, and some measure of population control will be essential if widespread starvation is to be arrested in the twenty-first century.

Table 5.V.3 Grain production in developing countries in relation to fertilizer use

Year	Population ($\times 10^6$)	Fertilizer consumption[a] (tonnes $\times 10^6$)	Grain production (tonnes $\times 10^6$)	Fertilizer consumption per capita[b] (kg/capita)	Grain production per capita (kg/capita)
1974	2842	19.0	636	6.7	224
1980	3288	31.0	728	9.4	221
1990	4150	57.0	875	13.7	211
2000	5095	92.0	1021	18.1	200

[a]Weight of nutrient as N + P_2O_5 + K_2O.
[b]Corresponding figure for Europe in 1980 was around 200 kg/ha.

5.V.3.4 Factors Restricting Use

5.V.3.4.1 Raw Materials, Energy, Economics

From the worldwide standpoint, little cause for concern appears to exist with respect to the supply of raw materials for fertilizer production. Known potash mineral reserves can serve world needs for centuries at least, following which the 1.6×10^6 tonnes potash per cubic mile of seawater can become an available source. Phosphate deposits are also extensive throughout the world, with proven reserves of high-grade materials adequate for at least the next two centuries, after which lower-grade materials should become mineable. There is virtually no limit to the amount of nitrogen available for fabrication into fertilizer, with approximately 85 000 tonnes of it in elemental form over every hectare of the earth's surface, besides which it is continuously recycled. The real limitation from the standpoint of nitrogen is the feedstock hydrogen required for its fixation. Natural gas supplies as currently used for nitrogen fertilizer production will eventually be used up; a paradox being the fact that substantially more natural gas is being flared off than presently goes into nitrogen fertilizer synthesis. Where natural gas is not available and coal reserves are, gasification of the coal will yield the necessary hydrogen. Again, the world's reserves of coal are great enough for centuries at least, but once gone hydrogen can be obtained by hydrolysis of water with somewhat greater energy input. Nor is sulphur likely to be in short world supply because of rather large mineral and sour gas deposits of the element, complemented in modern times by the sulphur being captured from coal-fired plants for atmospheric pollution abatement.

Thus, a shortage of the required raw materials is not likely to restrict fertilizer use on a world scale in the foreseeable future. Problems do exist, however, in those developing countries that have none of the mineral reserves mentioned. The highly industrialized societies of the developed countries

Table 5.V.4 Approximate energy required for fertilizer production in the USA in 1973 (Nelson, 1975)

Fertilizer nutrient	Energy input for manufacture ($\times 10^6$ kcal/t of nutrient)
N	11.1
P_2O_5	2.3
K_2O	1.3

without such reserves have their finished products for sale or barter with others having the fertilizer or needed raw materials. As a result, it is especially the substantial number of developing countries without the mineral reserves that are inhibited in their use of fertilizers for enhanced food and fibre production.

An available energy source is essential to fertilizer production, whether from fossil or nuclear fuel, hydroelectric or solar. The latter remains to be developed anywhere on a major scale, so countries without the other sources remain dependent for fertilizer production on those that have them. Of the order of 1200 m^3 of natural gas is required to produce one tonne of ammonia, approximately one-half for its chemical constituents and one-half for the heat required to drive the synthetic process. By far the greatest energy expenditure goes into the production of fertilizer nitrogen among the three major nutrients as expressed in Table 5.V.4. These figures do not include energy for transportation and distribution, although an allowance for handling losses is included. Nitrogen is the most limiting element for crop production wherever agriculture is practised. It has become the largest energy input item, for example, in regions of modern, intensive maize production (Pimental *et al*., 1974).

Ultimately the cost of fertilizer is the factor most restricting its use by the farmer. The work of FAO and other international groups during the 1960s and early 1970s demonstrated substantial economic returns in developing countries from appropriate fertilizer use as expressed by value/cost ratios (FAO, 1981b). Experience had shown in these efforts that the native farmer would readily accept the risk of fertilizer purchase where a value/cost ratio of 2 or more was reasonably assured. The multiplying of fertilizer cost that came with the world energy crisis in 1973/74 has confounded these relationships (Table 5.V.5). There has been some levelling off in costs but at a level twice that of the early 1970s. Gyrations in fertilizer prices as expressed here are neither helpful to the farmer buying fertilizer nor to the industry producing it. International efforts are needed to stabilize prices in order to foster a continued growth in consumption in the developing countries for the needed enhancement of agricultural production.

Table 5.V.5 Fertilizer prices in world markets (*UN Yearbook of Financial Statistics*)

Year	Weighted average cost of all fertilizers/tonne of nutrient[a] (US$)
1966	127
1967	116
1968	97
1969	87
1970	86
1971	85
1972	114
1973	171
1974	520
1975	340
1976	190
1977	200

[a]Index 1967–69 = 100.

5.V.3.4.2 Available Markets

Another important factor restricting use of fertilizer by the farmer in developing countries is the lack of a market in which to sell the added produce effected by the fertilizer input. In remote regions of many developing countries where transportation facilities do not exist, there is little incentive for the farmer to produce more than required for his immediate family. This pinpoints the need for government involvement in the improvement of transport facilities and assistance in establishing local markets for the sale of farm produce. These same markets would serve for merchandizing the various required inputs of improved seeds, pesticides and fertilizers for upgrading the agriculture of the area.

5.V.3.4.3 Interest Rates and Land Tenure

Additional disincentives include the high loan interest rates commonly levied on farmers in developing countries, and the cheap food policies maintained by governments, resulting in low value-to-cost ratios for food products. Nor do land tenure systems in many countries afford farmers the incentive to make improvements in the fertility status of the soil. Where land use rights are controlled by tribes, as in many countries of Africa, or where large land tracts are owned by a few wealthy families, as in Latin America, effective fertilization practice is not likely to be a priority of the farmer.

5.V.3.5 Land Transformations Effected by Fertilizers and Soil Amendments

5.V.3.5.1 Mono-culture

Perhaps the most striking impact of intensive fertilizer use on land use has been fostering the mono-culture of crops. The advent of fertilizer nitrogen along with herbicides for weed control, for example, has made the continuous production of maize on a given tract of land commonplace in the 'Corn Belt' of the USA. Fertilizer nitrogen took the place of the nitrogen previously supplied by a legume crop in rotation. Farmers thereby were better able to specialize in machinery and to acquire expertise in growing that single crop in the most efficient way possible.

Obvious problems with a mono-culture system are the assured build-up of pathogens, weeds adapted to the crop's culture, and insect pests. For the most part, these have not been insurmountable to the able farmer, although costs of the operation have generally increased with time. Two of the most difficult problems are those of pest control and compensating for increasing soil compaction under some climatic and soil textural conditions. Many farmers in the US Corn Belt who practised mono-culture of corn have adjusted to a cropping sequence that periodically includes soy-beans to assist in maintaining soil structure, nitrogen supplies and pest control.

5.V.3.5.2 Transformation of Low-quality Soils to Productive Crop Land

Several examples exist worldwide of land transformation from the low productivity range, bush or sandhills, to productive crop land by the introduction of fertilizers and other appropriate agronomic practices. One of the best examples is found on the Cerrada soils in parts of Brazil and Colombia, SA. The combination of liming and phosphorus fertilization that overcomes the aluminum toxicity of these soils, along with appropriate nitrogen fertilization, has allowed thrifty growth of cereals and soy-beans in place of the existing low-quality range in this otherwise favourable climatic regime. As long as soil pH and erosion are kept within reasonable limits, there is no reason why these large areas should not be major sources of food and fibre for centuries to come.

Vast areas of the humid tropics where either bush fallow has been practised traditionally, or no farming at all, have the potential for transformation to continuous agricultural production with proper fertilization and appropriate pest control measures. The uncultivated portion that is potentially arable totals upwards of 800×10^6 ha, essentially half of the current cultivated land area of the earth (National Research Council, 1972). Temperate region technology will not prove very helpful for long-term management as a general rule because of significant soil and climatic differences. Whereas temperate region soils usually contain 2:1 Si:Al ratio clays and modest to large amounts of organic matter, giving them high buffer capacities, soils in humid tropical regions are generally very low in organic matter and have in their clay fraction primarily iron and aluminum oxides and some 1:1 type clay combined with very low buffer capacity. Fertilization must be practised with extra caution so as not to accentuate acidity or induce imbalances among the various elements of plant nutrition. From the climatic standpoint, the higher temperatures year round in the tropics accelerate chemical and microbiological activities. In most areas there is a wet and a dry season. During the latter, all available soil moisture is removed from the rooting zone by evapo-transpiration. As a result, the next season begins with no stored soil moisture, the opposite of most temperate region farming that allows moisture storage during the autumn, winter and spring months before the next crop is planted. The wet season of the humid tropics usually provides more moisture than can be accommodated by evapo-transpiration, such that leaching, waterlogging or severe erosion are hazards. The combination of these several factors makes the proper management of crops and soils far more complex than in temperate regions, and much on-site research remains to be done to develop techniques appropriate to such regions.

Major changes in the regional and peripheral environments might be expected from the substantial deforestation of the vast Amazon basin were this to occur in the future (currently estimated to be proceeding at a rate in excess of 2.3×10^6 ha/year). An almost certain result would be greater runoff and erosion initially due to reduced soil infiltration rates with severe flooding of the lower Amazon, a river that now contains 15–20% of the global fresh-water supply. Subsequently, with less evapo-transpiration taking place from the region there could be a reduction in annual rainfall, which might also influence the climate in an adverse way in current agricultural regions to the south in Brazil. A concern that has been expressed of a diminution in the oxygen content of the earth's atmosphere by Amazon forest removal is probably not well-founded.

Equally striking as the bush fallow conversion is the transformation of sandy brush and rangelands of semi-arid to arid regions into intensively cropped land with the introduction of centre pivot irrigation and fertilizers, as in the Sandhills of Nebraska and Colorado in the USA, in Saudi Arabia, and in some North African countries. The extreme in this respect is the transformation of desert that supports no more than a few grazing animals per square kilometre to a highly productive agricultural system. The combination of water and fertilizer is required; neither applied alone would be effective.

A foremost problem that fertilizers have created with the transformation of low-quality lands to intensive agricultural production is that of accelerated erosion. Steeply sloping sites with relatively shallow soils when broken from sod or forest may produce sufficiently high yields with fertilization to make cultivation attractive. Once the continuous native cover has been removed, however, these areas are subject to severe water erosion during periods of limited crop canopy. Similarly, very sandy lands of low fertility when transformed from sod or bush to intensive crop production can suffer severe wind erosion when crop or residue cover is limited. Clearly the fault lies not with the fertilizer per se, but rather in the judgement of those who hope to capitalize on the short-term profit potential.

5.V.3.5.3 Substitution of High-value Cash Crops for Food Crops

The accessibility of inexpensive inorganic fertilizers has encouraged the production of high-value cash crops in areas where food grain production might have better served the country's immediate food needs. Examples include cocoa, rubber, coconut, fibre and other plantation crop production encouraged by governments as a means to acquire foreign exchange. The situation has been exacerbated by the provision of surplus grains by exporting countries at below production costs, creating a dependence on the external source. Then, when world food grain supplies are short the country suffers a serious food shortage, a condition experienced in 1973–75.

A contention that every country should organize its resources into a programme of national food security is hardly defensible in the modern 'one world' in which we live. Food grain production is neither ecologically nor economically feasible in much of the highlands of Central America, for example, where coffee thrives. Likewise, where cotton grows well during a hot season that would be depressive to wheat or maize, its fibre should logically exchange with the country possessing an environment favourable for the grains' production. Utilizing such comparative advantages should ideally result in better use and preservation of the land.

5.V.3.5.4 Expanded Area for Crop Production in Dry Regions

It is well established that moisture limits crop production in dry regions more than any other factor; but it is equally true that effective fertilizer use enhances the water utilization efficiency of crops. A delicate balance exists in the dry regions between the soil fertility level and the moisture available. Too much fertility with limited moisture can be responsible for crop overstimulation, but too little nutrient will leave the crop without sufficient root vigour to seek out and utilize moisture stored in deeper soil horizons. Thus, nitrogen and phosphorus needs for transforming low-value grasslands of semi-arid regions into productive range for animal production are being recognized increasingly around the world.

It has been a commonly employed practice to keep land in weed-free fallow for several months and up to two years in semi-arid regions prior to planting the next crop. During this fallow period moisture is stored to some depth in soil where it is safe from evaporation loss, while mineralization of nutrients, especially nitrogen, is occurring simultaneously. The following crop then benefits from both stored soil moisture and additional nutrients. Unfortunately, fallowing by traditional methods has not been very efficient in storing moisture because of rapid evaporation losses, 20% storage of rainfall received being a fairly representative figure. Recent experience with minimum tillage (conservation tillage, ecofallow, etc.) has indicated substantially greater moisture intake and storage when the previous crop's residues are left on the surface with minimal soil disturbance, as well as increased erosion control. This enhanced moisture storage gives an opportunity for expanding the land area available for continuous crop production without fallow or, at least, with less time in fallow providing the additional required fertilizer nutrients are provided.

5.V.3.5.5 Transformations Accompanying Correction of Soil Acidity and Alkalinity

One of the first recognized improvements of land for agricultural pursuit in modern times came with liming of inherently acid soils. This amendment transformed the acid native forest lands of much of northern Europe and eastern USA into productive cropland following removal of the trees. It was soon learned, however, that 'lime and lime alone makes for a rich father and poor son', meaning that the improved soil environment for biological processes enhanced the rapid mineralization of soil nutrients for the near term while hastening exhaustion of these native reserves. At the other extreme, the use of sulphur products along with appropriate fertilization has made possible the reclamation of sodic (alkali) soils. Sodium on the soil exchange complex is replaced by calcium, with eventual improvement in soil structure and internal drainage making possible the economic production of crops.

5.V.3.6 Long-term Effects of Fertilizers on Soil Properties and the Environment

5.V.3.6.1 Beneficial Results

Judicious use of fertilizers and lime has a number of beneficial effects on soil properties and the surrounding environment beyond the obvious enhanced food and fibre production. By improving growth there is more crop canopy and more root growth below the surface to protect the soil against the erosive action of wind and water. The end result is more fresh organic material left in the soil as roots and as top residue, with all the benefits of such residue on macro- and micro-biological activity and ultimate soil structure synthesis.

Further, increased yields from fertilizers applied to better soils decreases the need for cropping erosive soils that should be left in permanent cover.

A foremost benefit claimed of organic farming systems over farming with chemical inputs is an increase in soil aggregation from enhanced biological activity. Many experiments have confirmed this effect, albeit with organic amendment rates that would allow treatment of but a small fraction of the earth's arable lands.

The fertilized crop will use slightly more water than where no fertilizer is applied as yield is increased, but the water is used notably more efficiently in terms of kilograms produced per unit of water consumed. Judicious fertilizer use is thereby a significant factor in the conservation of water resources committed to agricultural pursuit.

Long-term plots of the Rothamsted Experimental Station in England, the Askov Field in Denmark, the Morrow plots in Illinois, the Sanborn Field in Missouri and many others have clearly shown that, far from damaging soils as some have claimed, the proper use of fertilizers and lime will actually improve them over time.

5.V.3.6.2 Detrimental Effects

Although claims have been made that certain fertilizers are responsible for 'hardening' soils and making them less permeable to water, the effect, where measured, usually has been the consequence of poor management by the farmer, such as too much traffic over the soil surface or pulling anhydrous ammonia shanks through the soil when too wet. Extended use of nitrogen and phosphate fertilizers, especially the former, can, however, cause an acidification problem. As a general axiom, each kilogram of fertilizer nitrogen (other than sodium and potassium nitrates) requires about 1.8 kg of lime for its neutralization, ammonium sulphate three times that amount. The correction of such acidification in surface soils thus entails periodic liming of soils that are not inherently calcareous, which is easily carried out at nominal cost in most locations. In ignoring the acidity induction, however, farmers will find that subsoils eventually become acid as well. If this is allowed to progress until the subsoil is strongly acid with associated manganese and aluminum toxicities, crop root development is inhibited and productivity declines. Moreover, correction of subsoil acidity is not readily accomplished since calcium from applied lime is very immobile and does not leach readily from the zone of placement. It is now a generally recognized rule that ammonium sulphate should not be used as a nitrogen source for tropical region soils of low buffer capacity because of the acidification problem.

Long-continued, traditional use of the major nutrients nitrogen, phosphorus and potassium can bring about nutrient imbalances in soil with effects other than toxicity that are deleterious to crop growth. Overloading the soil with phosphorus, especially in neutral to alkaline soils, can induce zince deficiencies. When this problem is corrected by building up soil zinc levels with fertilization, an iron shortage may develop owing to the high phosphorus and zinc levels. A problem is thus created for which economic soil treatment measures do not presently exist. Similarly, loading soils excessively with potassium and nitrogen has been found to cause low magnesium levels in forages to the particular detriment of animals consuming the forage, i.e. one of the causes for hypomagnesemia or 'grass tetany'.

Great concern has been expressed about the effects that fertilizers may have on the nutrient pollution of surface and ground waters (FAO, 1972; OEEC, 1983). Certainly a prime causative factor in the eutrophication of surface waters is the concentration of nitrogen and phosphorus in the water, although other factors like depth, temperature and clarity of the water and presence or absence of a carbon source may be equally involved. The more fertile the adjacent watershed from which water drains into streams and lakes, the more nitrogen and phosphorus likely to be found there. Concentrations of these elements needed to trigger the eutrophication process, other things being favourable, are small indeed (of the order of 0.3 ppm nitrogen and 0.01 ppm phosphorus). Other elements are involved as well, such that precise definition of the role of a single element is virtually impossible. Nonetheless, as nutrients go, it is most commonly the concentration of phosphorus that is rate-limiting in fresh water, since even rainfall generally contains more than that amount of nitrogen.

People do not appreciate the interference with their water sports imposed by aquatic plant growth. Potentially more serious, however, is the contamination by nutrients of groundwaters that serve as potable water sources. Problem areas have been recognized in several countries of the world where nitrate nitrogen has accumulated excessively in the groundwater. When this concentration exceeds 10 or 11 ppm public health authorities become especially concerned because of its possible contribution to methemoglobinemia in infants. Overuse of fertilizer nitrogen with the excess moving downward to the groundwater has been documented as partially responsible for recognized high groundwater nitrate in a number of areas (OECD, 1983). At concentrations an order of magnitude higher, adults may be affected adversely as well, and animals at even higher levels. There are no recorded cases of other fertilizer nutrients leaching excessively to groundwater and contaminating it for human use, in part because none other of those commonly applied has the mobility to move with water through soil material as nitrate does.

Concern has been expressed in recent years about the environmental consequences of overloading the nitrogen cycle. Whereas biologists have estimated the total natural fixation of nitrogen in the earth's system (symbiotic, non-symbiotic, associative, lightning) to be of the order of 100×10^6

tonnes annually, it will not be long before fertilizer nitrogen fixing industrial plants equal that figure. Beyond water pollution, which can be controlled in large degree by judicious fertilizer use, no other hazard with certain consequences is apparent. Question has been raised concerning the impact of N_2O coming from denitrification of fertilizer nitrogen in soil may have on stratospheric ozone (Crutzen, 1974). This gas can diffuse to the stratosphere, where it breaks down into elemental N_2 along with small amounts of NO and NO_2, which then react with ozone (O_3) converting it to O_2. Depletion of ozone in the stratosphere would allow greater penetration of ultraviolet light from solar radiation to reach the earth surface, with probable adverse effects on both climate and living organisms. Estimates have been made from 'slight' up to 11.5% as a likely reduction in ozone from the added production of 100×10^6 tonnes fixed nitrogen per year (CAST, 1976); but substantial monitoring remains to be done in documenting the actual effects of N_2O on the earth's ecosystems.

Otherwise, most of the NH_3, NO and NO_2 released by plant or soil that escapes to the atmosphere is returned to the soil in rain, if not reabsorbed directly by plant or soil. Increased fallout of nitrogen compounds during recent years in Europe and North America for which the internal combustion engine is largely responsible is well documented. The compounds of nitrogen are of relatively small magnitude in the 'acid rain' problem compared with those of sulphur in most regions of concern. On a global scale, moreover, the anthropogenic nitrogen contribution is considered to be small compared with amounts produced by natural phenomena such as forest fires and electrical storms.

5.V.4 MEANS OF SOIL FERTILITY EVALUATION

In transforming any given soil to a higher fertility state that permits a higher level of crop production, some means of soil fertility evaluation is essential to indicate the kind and amount of nutrients needed. Not only should the system be capable of designating treatments that achieve the most economic yield; it must also prevent excessive accumulation of nutrients that can become environmental pollutants. Accordingly, from the standpoints of economy, energy and resource conservation and environmental quality, it is essential that fertilizers be used in the most efficient manner possible. A large background of research and experience is required for effective nutrient prescription.

5.V.4.1 Soil Classification Information

Detailed soil maps prepared under modern systems of soil classification can be very helpful to the farmer in projecting at least some of his fertilizer needs.

Thus, soils of the Ustoll group, Mollisol order, will generally be near neutral in reaction and high in the basic elements of plant nutrition, including potassium, calcium and magnesium, and the farmer need not concern himself with them in his fertilizer purchase. Variations will exist at the soil type level as found on detailed maps which designate such factors as depth, texture and organic matter content of the soil. These can assist in making a closer projection of needs that will generally involve little more than nitrogen and phosphorus. On the other hand, soils classified as Udults of the Ultisol order are almost certain to be rather strongly acid in reaction, with problems of aluminum and manganese toxicity in crop production. They are always low in the basic elements and low in organic matter content, with attendant low buffer capacity. The usual requirements of the fertility programme in this case is neutralization of the exchangeable aluminum by lime and phosphate, with further likely need of nitrogen, potassium and one or more of the micro-nutrients. The major difficulty from the classification standpoint with this latter group of soils in tropical regions is that little detailed soil surveying has been done and information on the specific soil type, so useful in temperate regions, is not likely to exist at the field or farm level in such regions.

5.V.4.2 Field Trials with Fertilizers

The surest expression of fertilizer needed to transform a low-fertility soil to one that affords the most economic yield comes with field fertilizer trials as long as the trials include sufficient coverage of rates and kinds of nutrients likely to limit yield. Such trials indicate the nutrient needs of similar soils in the locality that have had similar past management. The FAO Fertilizer Programme (FAO, 1981b) has conducted thousands of such trials in the past 20 years that indicate nutrient needs by regions. These can be very helpful to farmers as they initiate fertilizer treatments for soil improvement. Not only have the trials been useful for delineating likely nutrient shortages, but they have also served the added extension function of displaying results to farmers in the respective localities. The technically advanced farmer himself could conduct such trials on the different soils of his land, although few would have the time, facilities and competence to properly carry through the operational details and interpret the results.

Field trials are expensive and time-consuming, but are the absolutely essential foundation to any soil fertility evaluation programme. They cannot, however, be depended on indefinitely to express the kind and amount of nutrients that should be used by farmers of a given region. As fertilizer use becomes commonplace and varied amounts are used on different fields, varied residual effects begin to appear, even on soils of a given classified unit. Further, the varied fertility status at the soil type level from different pedogenesis makes region-wide projection tenuous. It is important, then, that thorough site characterization be made with each fertilizer trial so that transferability will exist for other sites where quick testing methods have revealed similar soil characteristics.

5.V.4.3 Soil and Plant Analysis

Ideal as the field trial method is for the specific soil and field in delineating treatments needed for soil fertility transformation, it is obvious that time and cost limitations preclude its use for every soil situation. The best substitute then becomes soil and/or plant analyses that have as their base a sound background of field experiments. Quick testing procedures on a representative soil sample are capable of diagnosing nutrient shortages in the soil before a crop is planted, thereby permitting correction in that year by proper fertilization.

Soil testing, to be meaningful, must have its methods correlated to the soil conditions that exist in the region of the service and must be calibrated with field yield response data for the major crops of that region. Therewith comes the required background already cited of extensive field trials measuring crop response to the different fertilizer nutrients. Without this correlation and calibration foundation, and unless soil samples are collected so as to truly represent the site, the measurements made will have little meaning and the associated recommendations are likely to be totally erroneous. Different crops having varied yield potentials, total nutrient needs, and nutrient extraction capabilities will have different calibrations; thus the crop to be grown is a prime consideration in fertilizer recommendations.

Soil testing does, of course, serve several useful functions beyond prescribing kinds and amounts of fertilizers to be applied. In dry, irrigated regions it will delineate salinity and sodic problems that limit crop production and project the treatments required for correction. The service will also recognize the likelihood of Al and Mn toxicity potentials in acid soil regions and the quantity of lime needed to alleviate them. Without correction of these problems, any recommended fertilizer treatments are not likely to provide the response expected. A foremost requisite of soil testing is that the fertilizer recommendations therefrom be provided by totally impartial individuals, such as appropriate governmental or farmer-owned organizations who have nothing to sell (OEEC–EPA, 1956).

Plant analysis can serve as a useful adjunct to soil testing services in resolving some soil fertility problems. The service, to be meaningful, must have the capability of measuring concentrations of most of the essential plant nutrients because of likely physiological interactions within the plant, where one nutrient can accumulate in excess, so depressing the uptake and utilization of others. Diagnosing nutrient needs from plant analysis is complicated by varying values with stage of growth, plant part, crop variety or hybrid, and attendant climatic factors. Moreover, the results obtained can seldom be used for fully correcting the problem in the existing crop.

Soil and plant analyses involve complicated chemical procedures. The quality of the recommendations for fertilizer treatment coming from them can be no better than the quality of the correlation and calibration work from which they are derived. Accordingly, only the best trained individuals in a developing country should be assigned the task of setting up a soil and plant analysis service. The laboratory will require considerable expensive equipment and several qualified technicians to run the routine analyses. Total costs can be substantial, but they are quite insignificant in comparison with the returns that can come from the efficient operation of a laboratory in transforming nutrient-deficient soils to highly productive soils.

5.V.5 CONCLUSIONS

The introduction of fertilizers has transformed many regions of inherently low productivity into agriculturally effective regions. Examples are: bush fallow to continuous cropping in tropical rainforest; low-quality range on Cerrada soils of South America to effective cropland by correction of aluminum toxicity/ phosphorus deficiency; and conversion of the desert to intensive agriculture as irrigation is also introduced.

Judicious fertilizer use will improve the quality of good arable soils over time by the greater organic return from crop residues and resulting stimulated micro-biological activity. Furthermore, the increased productivity of the better arable lands achieved by fertilizers reduces the need for cropping of lower-quality erosive areas, which are better left in permanent grass or forest cover.

Organic systems of soil-fertility maintenance are invaluable for transforming low-productivity lands to agricultural use and should be used to the extent that is possible. Limitations are: inadequate quantity of organic wastes for treating more than a fraction of all arable land; salts or toxic components in some wastes; and inadequate water for legumes in dry farming regions.

The combination of reduced tillage and added fertilizer can extend the potential area of cropped land in dry regions by the moisture conservation and nutrients afforded when compared with traditional systems of land management.

Excessive fertilizer use can result in nutrient pollution of the environment, as with eutrophication of surface waters and nitrate contamination of groundwater. Such transformations in associated waters are reversible in a relatively short period with application of judicious fertilizer management practices.

Soil acidification as induced especially by nitrogen fertilization can usually be compensated at modest cost by liming, except in certain regions where no surface exposure of limestone bedrock exists. Where acidification is allowed to proceed to the state that the subsoil as well becomes strongly acid, the resulting acidity and aluminum toxicity problem in crop production is virtually irreversible.

Non-judicious fertilizer use can result in soil nutrient imbalances that may or may not be readily reversible. Examples are: copper toxicity from over-use of copper fertilizers and Bordeaux mixture; iron deficiency in neutral to alkaline soils induced by excessive phosphorus and zinc fertilizer application.

A system for soil and plant analysis needs to be established for prognostic purposes in any given agricultural region in the interests of economy and of resource and energy conservation, and ultimately the prevention of environmental pollution by fertilizer.

5.V.6 REFERENCES

Council for Agricultural Science and Technology (1976) *Effect of Increased Nitrogen Fixation on Stratospheric Ozone*. CAST Tech. Report 53, Iowa State Univ., Ames, IA.

Crutzen, P. J. (1974) Estimates of possible variations in total ozone due to natural causes and human activities. *Ambio*, **3**, 201–210.

Food and Agriculture Organization (1972) Effects of intensive fertilizer use on the human environment. *Soils Bull.*, **16**, SIDA and FAO, Rome.

Food and Agriculture Organization (1981a) *Production Yearbook*, Vol. 35, and *Fertilizer Yearbook*, Vol. 35. United Nations, Rome.

Food and Agriculture Organization (1981b) *FAO Fertilizer Programme: 20 Years Increasing Crop Yields*. FAO, Rome.

National Research Council (1972) *Soils of the Humid Tropics*. Agric. Board, NRC, Nat. Acad. Sci., Washington.

Nelson, L. B. (1975) Fertilizers for all-out food production. In: Martin, W. P. (Ed.), *All-Out Food Production: Strategy and Resource Implications*. ASA Spec. Pub. No. 23, 15–28. American Society of Agronomy, Madison, WI.

Organization for European Economic Cooperation (1956) *The Organization and Rationalisation of Soil Analysis*, 11–17. OEEC–EPA Project 156, Paris.

Organization for Economic Cooperation and Development (1983) *Background Document on Agriculture and the Environment*. ENV/SE/83.7. OECD, Paris.

Pimental, D., Hurd, L. E., Bellotti, A. C., Forster, M. J., Oka, I. N., Shales, O. D., and Whitman, R. J. (1972) Food production and the energy crisis. *Science*, **182**, 443–449.

US Department of Agriculture (1980) *Report and Recommendations on Organic Farming*. US Govt. Printing Office, Washington, DC. (Restricted).

Land Transformation in Agriculture
Edited by M. G. Wolman and F. G. A. Fournier
© 1987 SCOPE. Published by John Wiley & Sons Ltd

CHAPTER 5.VI
Plant Protection and Land Transformation

L. BRADER

5.VI.1 INTRODUCTION

Plant protection plays a significant role in optimizing the productivity of a given crop. Agricultural pests, which include all organisms such as fungi, bacteria, viruses, insects, mites, nematodes, weeds, rodents and grain-eating birds, that live on and/or compete with plants determine, to a varying degree, if crops can be grown economically in certain situations. Usually agricultural pests inflict considerable damage to crops and represent a significant production constraint. Effective plant protection thus becomes essential to minimize the losses caused and to ensure that full benefit is drawn from other production inputs.

Indirectly, agricultural pests and plant protection are therefore of importance to land transformation. They determine partially the productivity of the transformed land. Especially when effective chemical pest control measures were lacking, crop rotations and other agricultural practices—and thus the ways in which the land was used—were, to a large extent, aimed at reducing pest incidence. After the widespread introduction of pesticides in some parts of the world, such practices could be changed and a tremendous intensification of agricultural production could take place. Systems developed, such as mono- and multi-cropping, that would have been impossible without the new control measures. Certain crops could be grown in places where this would have been impossible without effective crop protection.

However, it should be stressed that the intensification process has also rendered most crops more susceptible to pest attack, and in fact crop losses continue to be very high. A new crop environment has been created. In addition, intensive use of pesticides in itself considerably affects the land and its biotic environment. Identification of the negative aspects of this type of transformation of the agro-ecosystem will in the future lead to a gradual change in plant production and protection practices. Within the framework of so-called integrated pest management systems renewed attention is given to the significance of certain agricultural practices and crop varieties grown to reduce pest incidence.

5.VI.2 CHANGES IN CROP PROTECTION

Phytophagous organisms were present on this planet long before man appeared and started to use certain plants for his well-being, be it for food, shelter or clothing. The notion of plant pests only became applicable from the time that it was recognized that these organisms competed for the same materials that man considered as useful. However, pests are rarely very apparent in natural ecosystems, and the growing of crops as such has created much more favourable conditions for the development of pest organisms.

From recent studies on the perception of pests by traditional farmers, it may be concluded that a certain amount of damage caused by the so-called sedentary pests was, and often still is, considered an integral and unavoidable part of the traditional agricultural production process. These farmers seem to have a notion that the crop ecosystem should be rather diversified. For example, in certain African situations a small amount of food may be left on the field after harvest for the rodents, the latter being considered part of a 'healthy' ecosystem.

However, epidemics and plagues were certainly recognized in early times, and their threat must have had a strong impact on crop production. One of the oldest quoted examples is the locust plague which appeared at frequent intervals (about every 5–7 years) in large parts of Africa and Asia. Such plagues could cover enormous areas and, depending on the weather conditions, could last for a couple of years.

Thus, the struggle with plant pests has certainly taken place from the time that crops were grown deliberately, i.e. from about 8000 BC onwards. Only during a relatively short period of time has man felt totally confident that he could fully master certain pest organisms. This occurred during the period from roughly 1950 to 1970 when confidence in the newly available insecticides led some to believe that a permanent solution to certain insect problems had been found. Matters turned out to be more complex than foreseen, and only gradually are systems being developed to use these new powerful tools optimally, both from an economic and an ecological point of view.

Uncertainty about the effectiveness of pest control efforts has continuously led man to seek for improvements in plant protection. Three periods may be described. For each of these some aspects of particular relevance to land transformation will be discussed, although it should be stressed that these aspects are not necessarily the major characteristic of, or limited only to, the period concerned.

Until the 1940s certain agricultural practices were to a considerable extent aimed at the reduction of the impact of crop pests. These included matters such as crop rotations and field clean-up after harvest. Most pests could not be directly controlled, or only inadequately so. In some cases, existing or newly introduced pests made it uneconomical to grow certain crops or crop varieties. This period clearly demonstrated the importance of the way in which the land is used (cultural practices) in reducing crop losses in a preventive manner.

The development and widespread use of synthetic organic pesticides drastically changed this approach. The idea gradually arose that plant and animal pests were of secondary importance. The effectiveness of these new control means made it possible to change land-use considerably. However, this period also revealed a succession of changes which had considerable ecological consequences. With respect to land transformation, it should be noted how the intensive use of pest control measures affected the agro-ecosystem.

During the last two decades the idea of integrated pest management has emerged. It aims at the combined use of pesticides and other control factors. Elements of this approach are gradually being introduced into current agricultural research and development activities. Emphasis is again placed on preventive measures, such as cultural practices, varietal resistance and maintenance of natural enemies. The importance of the crop condition to withstand pest outbreaks is highlighted again and the idea of plant health is gaining recognition. Thus a more stable agro-ecosystem is created, reducing the need for curative control measures.

However, before discussing certain aspects of these three periods in more detail, a general discussion of crop losses seems useful. This will place plant protection in proper perspective in relation to agricultural productivity.

5.VI.3 CROP LOSSES

Spending any effort to reduce the impact of crop pests certainly requires a proper definition of the magnitude of the problem. Crop loss assessment serves this purpose and aids in evaluating progress resulting from control activities. But experience has clearly shown the difficulties in collecting data that are really useful. Recent efforts to systematically collect comprehensive crop-loss data on a global basis have produced few results, and there are no clear indications that much progress can, or will, be made in the near future. Crop-loss information must therefore continue to be collected from data that become available from various crop improvement experiments. In this respect, special attention can be drawn to the data generated by the International Agricultural Research Institutes, such as IRRI in the Philippines, CIAT in Columbia and IITA in Nigeria. An analysis of some of these data on damage caused by insect pests to major food crops clearly confirmed that estimates published by Cramer in 1967 still provide good order-of-magnitude estimates (Cramer, 1967), although for developing agriculture, Cramer's data are rather on the conservative side (Brader, 1982).

None the less, these estimates providè good insight into the overall magnitude of plant protection problems, and, as shown in the following, the loss figures are already such that there seems to be little need for additional efforts to prove the urgency of continued implementation of proper crop protection measures. The estimates of Cramer have been applied to the 1981 agricultural production figures given in the 1981 FAO Production Yearbook (FAO, 1982). Prices given in the same FAO publication have been used (Table 5.VI.1) to translate the losses into money values.

It can be seen in Table 5.VI.1 that the total annual cereal losses are of the order of 838 million tonnes, or the equivalent of about US$147 × 10^9. Total annual food crop losses would be almost 1400 million tonnes, worth more than US$300 × 10^9. Adding to this the non-food crops, the value of the total losses would be over US$348 × 10^9. It must be noted that post-harvest crop

Table 5.VI.1 Amount of agricultural produce lost to pests in 1981 and its approximate money value

		Calculated losses			
	Actual production (kt)	(% of potential production)	Amount (kt)	Approximate price per t (US$)	Approximate value (US$ billion)
Wheat	458 195	23.9	143 873	180	25.9
Rice	413 785	46.4	358 338	200	71.7
Maize	451 704	34.8	241 210	150	36.2
Barley	158 488	20.2	40 097	140	5.6
Rest cereals	181 656	23.1[a]	54 497	140	7.6
Total cereals	1 663 828		838 015		147.0
Potatoes	256 978	32.3	122 578	120	14.7
Other root crops	304 589	23.1[a]	91 377	100	9.1
Pulses	42 403	23.1[a]	12 721	500	6.4
Vegetables	351 961	27.7	134 801	400	53.9
Fruits	282 526	23.1[a]	84 758	400	33.9
Oil crops (oil equiv.)	52 984	32.5	25,485	500	12.7
Sugar (raw)	92 046	45.3	76 214	370	28.2
Total food crops	3 047 315		1 389 008		305.9
Cocoa	1 670	45.9	1 416	2 000	2.8
Coffee	5 846	44.4	4 665	2 500	11.7
Tea	1 845	32.2	874	1 500	1.3
Cotton lint	15 301	33.9	7 834	1 950	15.3
Other fibres	6 482	24.0	2 042	600	1.2
Tobacco	5 334	30.8	2 374	3 700	8.8
Rubber	3 807	25.0	1 268	1 100	1.4
Total non-food crops	40 285		20 473		42.5
Overall totals	3 087 600		1 409 481		348.5

[a]Estimates.

losses have not been included in the figures. They are certainly of significant importance, but reliable data to calculate these losses are not available. In addition, attention should be drawn to the significant losses that can be caused by various pest organisms in grasslands and forests.

It is quite evident that the magnitude of crop losses is enormous. The conclusion must be drawn that this extremely high level of wasted agricultural produce should be unacceptable to society as a whole, especially in view of recurring food shortages. Examples are given below of the impact of these losses, and efforts to reduce them, on land use. In fact, these examples also show that for virtually all categories of agricultural pest organisms, particularly in recent years, man has still not been able to reduce these losses significantly. Part of this failure is certainly due to the fact that inadequate recognition is given to the importance of cultural practices and the need to improve the overall plant health situation.

5.VI.4 PLANT PROTECTION UNTIL THE 1940s

Until the 1940s, as noted earlier, efforts to reduce crop losses were mainly based on the use of natural control practices in the broadest sense. From the middle of the nineteenth century these efforts were gradually complemented by pesticides of botanical and, in particular, inorganic origin. But plant protection maintained as its basis a complex of efforts aimed at the augmentation of the overall resistance of the agro-ecosystem to the development of pest organisms. Pest control strictly speaking (i.e. the deliberate control of specific pest organisms through pesticide applications) was of rather limited importance and mainly supplementary. With respect to land transformation, two matters are of particular relevance:

(1) large-scale outbreaks of pests which in certain cases resulted in the impossibility, at least temporarily, of growing specific crops or crop varieties and made agriculture in the region concerned permanently more difficult;
(2) the adoption of crop rotations to reduce pest carry-over from one crop to the other.

Both elements had, of course, significant impacts on agricultural practices in general. The spread of pests is still a major concern in current plant protection practices, but modern control measures generally allow the reduction of the overall impact, be it sometimes at very high cost. Pests continue to lead in a number of cases to restrictions on the growing of crops in certain regions.

5.VI.4.1 Spread and Outbreaks of Pest Organisms

The sudden spread of new pests is well known for most groups of pest organisms. Fungal diseases that can be easily spread by wind and water, and for which the causal organisms can in certain cases survive for long periods in the absence of the host plant, are very well known in this respect. But the introduction of new weed species, as well as the spread of insects and nematodes, with or without their host plant, has also frequently occurred. The subsequent outbreak of pest organisms occur, in particular, in situations where agricultural crops have been introduced rather recently, such as in the USA and Australia.

Ainsworth (1981) has given an overview of the major disease epidemics and their impact. The epidemic of the potato blight in Europe caused by *Phytophtora infestans*, which appeared in Europe for the first time in the 1840s, is certainly one of the most notable cases. It caused massive famine in Ireland and resulted in the migration of a large portion of the population to the USA. Coffee rust (*Hemileia vastatrix*) was accidentally intoduced around 1869 in Sri Lanka and wiped out coffee production in that country. The crop was replaced by tea. The same disease organism spread accidentally to Brazil in the 1960s and is now considered the major threat to the coffee industry of Latin America.

The introduction of vine mildew (*Plasmopara viticola*) from North America into France in the 1870s threatened wine growing in that country for many years. The losses caused were overcome through the breeding of resistant varieties by hybridizing European stocks with American. Dutch elm disease was introduced into North America from Europe in 1931 and has since then been destroying American elms. The reintroduction of a more virulent strain of the same disease into England in 1970 is still devastating elm trees in that country. Chestnut blight (*Endothia parasitica*) was accidentally brought into the eastern USA in 1904 with nursery stocks from Asia. Within a few years it virtually eliminated the American chestnut.

Certain endemic diseases could reach epidemic proportions after, for example, periods of favourable weather conditions. Epidemics of ergot of rye (*Claviceps purpureum*), for example, had caused many human deaths until the nineteenth century. In recent years some disease epidemics have been created by the release of new varieties with augmented disease susceptibility, and considerable losses have sometimes occurred (Lupton, 1976). An example of this is leaf blight (*Bipolaris (Helminthosporium) maydis*) in hybrid maize production. This susceptibility came from the male sterile line and had been introduced to such an extent in commercial varieties that losses estimated at one billion dollars were suffered in the USA during the years 1971–73 before the cause was discovered and action could be taken. The same phenomenon occurred with the release of hybrid varieties of pearl millet in India that proved very susceptible to downy mildew (*Sclerospora graminicola*). As a result the national pearl millet yield in India was reduced by more than 50% in 1973.

The still ongoing efforts to contain the spread of pear fireblight in West

Europe caused by the bacteria *Erwinia amylovora*, including the destruction of various wild hosts, show the continued impact of disease epidemics in modern agricultural practice.

In discussing introductions of insect pests Simmonds and Greathead (1977) note that certain characteristics favour spread. Small insects, for example, can readily be transported on plant material or passively by air currents, and polyphagous species have more chances to survive in their new surroundings. But they also note that in many instances pests were distributed by man when he transported them together with their host plants into new surroundings.

The grape Phylloxera (*Viteus vitifoliae*) was introduced in the middle of the nineteenth century from North America into Europe and nearly destroyed the wine industry in France. This aphid attacks the roots and was estimated to have destroyed by 1884 the vines over an area of 1.2 million hectares. The total monetary loss at that time amounted to US$2 billion. In fact this has also become the classic example of the successful control of an insect by breading resistant varieties. In this case resistant American rootstocks were used.

Until the end of the nineteenth century cotton had been grown in North America for about 100 years without suffering from significant insect damage. However, in 1892 the boll weevil (*Anthonomus grandis*) entered Texas from Mexico and has since rapidly spread over a large part of the American cotton belt. This has led to the disappearance of the slow-growing sea-island cotton varieties in the southern part of the USA. In addition the boll weevil became a permanent cotton pest that has dominated production practices until today. It is now considered the most costly insect in the USA. Moreover, control measures taken have led to a series of secondary effects, which have increased further the costs of cotton production.

The cottony cushion scale *Icerya purchasi* was introduced around 1870 into California and became such an important pest that in the 1880s it threatened the citrus industry in the western United States. The introduction of an effective predator of the scale, the ladybird beetle (*Rodolia cardinalis*) re-established a more balanced agro-ecosystem and reduced the pest to below economically significant levels. However, this pest has continued to be the determining factor in the citrus production practices of this area.

In certain instances locally occurring species have become pests of introduced crops. For example, the European corn borer *Ostrinia nubilalis* fed originally on *Artemisia* spp. before it became a pest on maize. It was later introduced into North America. The colorado beetle (*Leptinotarsa decemlineata*) provides a similar example. It became a pest of potatoes when the latter was introduced from South America into North America. Later the beetle was introduced into Europe where it is still considered a pest requiring specific plant quarantine measures.

Weeds constitute a major constraint to agriculture and they have been widely introduced into new areas with inherent economic damage. A study of

500 weeds of North America showed the following origin: native 39%, Europe 35%, Eurasia 13%, tropical America 3%, Asia 2%, undetermined 8% (Burnside, 1979). An excellent example of an introduced weed from Europe to the USA is St. John's wort (*Hypericum perforatum*). In Europe this weed is not of economic significance, but in its new environment it became one of the worst grassland weeds. It was finally brought under control by the introduction of Chrysomelid beetles (*Chrysolina* spp.) that became very effective 'pests' of the weed.

Weeds were sometimes introduced as ornamentals or as fodder crops. As such, prickly pear (*Opuntia* spp.) became a major problem in Australia. By 1925 *Opuntia* had invaded about 25 million hectares, rendering them unsuitable for agriculture. It has also become one of the most outstanding examples of effective biological control of weeds. During the 1920s the deliberately introduced lepidoptera *Cactoblastis cactorum*, and some other biocontrol insects, succeeded in gradually reducing prickly pear to an insignificant weed.

The water hyacinth *Eichhornia crassipes*, which originates from Central and South America, has very nice flowers; this is one of the main reasons why it is now a major waterweed in North America and parts of Africa and it is still spreading further. It may completely block waterways and has caused enormous economic losses. In the course of the 1960s the weed *Parthenium hysterophorus* was probably introduced into India with food aid shipments. Since then it has spread continuously. Many other cases of pest introduction have been reported in the literature and many have certainly occurred unnoticed. The spread of various plant parasitic nematodes belongs to the latter category.

The spread of pest organisms continues to be of major concern even in today's modern agriculture. It has not only been greatly facilitated by increased communication, but also by agriculture itself. Agricultural practices in general offer very favourable conditions for newly introduced pest organisms and augment their chances of reaching pest status. However, the major reason that newly introduced organisms often become more important pests than in their country of origin is the absence of effective natural control mechanisms. The result may be a much faster population build-up, unless biotic mortality factors are also introduced. The examples given of the prickly pear and St. John's wort show that the agro-ecosystem can indeed be made more resistant to pest development. Similar examples exist for the biocontrol of introduced insects (e.g. the cottony cushion scale in citrus growing in the western USA mentioned above).

5.VI.4.2 Crop Rotations

In the older literature crop rotations are often highlighted as useful plant protection measures, especially for the control of soil living organisms.

Wardle and Buckle (1923) note, while discussing crop rotations with respect to insect control, that traditionally crops have been and are grown in a particular order. They indicate that maintaining soil fertility is generally the major reason, but that there might also be some benefits for insect control; however, these authors do not give further details. Crop rotation is very briefly mentioned by Smith (1948), and he concludes that

> Crop rotation in practice probably has little influence upon the degree of insect infestation over the whole area, chiefly because of the variation in crop rotation on different farms and because insect pests are rarely specific to one plant host and also have strong migratory powers.

This is probably correct with the exception of certain soil insects. For example, rotating corn with soy-beans is considered an effective measure for the control of corn rootworms.

Crop rotations are often mentioned in connection with the control of noxious weeds. Rotations using crops with different life histories will help to reduce weeds with life cycles that are not adapted to the cultural practice of the cultivated crops. Winter annual weeds, such as weed bromegrasses, will be of little importance under cultural practices normally applied to summer annual row crops, such as corn. Many summer annual weeds will not thrive under the cultural practices of autumn planted cereals (Koch, 1970). For example, wild oats (*Avena fatua*), a summer weed, is abundantly present in a continuous succession of summer cereals, while its population does not increase in a cycle where the summer cereal is grown only once every four years.

The significance of crop rotations has been studied in particular for soil-living disease organisms, including nematodes. A detailed review has been published by Curl (1963), from which a number of examples are drawn here. He notes that crop rotation as a control measure has been directed chiefly, and with the greatest success, towards control of root diseases, many of which cannot yet be economically reduced by fungicidal treatment. It would be applicable, in particular, to the so-called root-inhabiting fungi that are regarded as specialized parasites, and normally do not survive long in an active state in the absence of living host plants.

Studies on the control of different foot and root rots of wheat in the USA led to the conclusion that the impact of these diseases may be reduced through crop rotation. But in that case wheat should be preceded for 3–6 years by non-susceptible crops. Maize root rots, caused primarily by species of *Pythium*, *Helminthosporium* and *Fusarium*, have been reduced by crop rotation in many places in the USA. Similarly crop rotations have been used with success to obtain profitable yields with peas in spite of fungi that caused pea root rot.

Various rotation experiments have been carried out to control verticillium wilt (*Verticillium albo-atrum*) of potatoes. Based on numbers of infested tubers, it was demonstrated that 3- and 4-year rotations gave practically complete control. The reduction of *Plasmodiophora brassicae*, the cabbage club root fungus, requires a host-free period exceeding four years. Such a long period is needed as resting spores survive for quite a long time in the soil.

Of the various measures recommended to control nematodes, crop rotation has been and often still is one of the most satisfactory. The best known examples in this respect are the sugar beet nematode (*Globodera schachtii*) and the potato root eelworm (*Globodera rostochiensis*). In Great Britain the sugar-beet eelworm Order of 1943 allowed the enforcement of crop rotation on infested and non-infested land; and although it was quite difficult to apply the same rules with good results to various situations, this legislative procedure served to draw attention to the value of rotation in reducing the rate of spread of the parasite and subsequently preventing serious crop losses.

Comparable rules have been established for potato-growing in Western Europe for the control of the root eelworm. For example, in the Netherlands the following regulations are currently enforced. Susceptible potatoes may only be grown in those fields where (i) for the last three years no potatoes have been grown; or (ii) for the last time, three years ago, susceptible potatoes have been grown and a soil disinfection has been carried out and notification given to the plant protection service; (iii) for the last time, three years ago, resistant potatoes have been grown; or (iv) for the last time, two years ago, a resistant potato variety has been grown and the field has been disinfected before or after the growing season and has been notified to the plant protection service. Resistant potatoes may be grown on fields where (*a*) for the last two years no potatoes have been grown, or (*b*) for the last time, two years ago, potatoes have been grown and a soil disinfection has been carried out and the plant protection service notified.

Crop rotation is seldom included in recommendations for controlling diseases caused by viruses, particularly where the virus has a wide host range. The best evidence of virus control by rotation is found in studies with tobacco mosaic and wheat mosaic viruses. It was found in field experiments in Illinois that cropping with non-susceptible crops may reduce damage from wheat mosaic, but rotation was not considered entirely adequate as a control measure. The incidence of virus diseases of Brassica crops in Great Britain was reduced by omitting one crop for a short period.

In a recent publication (Palti, 1981) the various reasons for applying crop rotations were listed in the following way:

(1) General agronomy: better use of nutrients and desirable effects on soil texture, with deep-rooted crops alternating with crops with shallow roots; water economy, in particular conservation of water in years of fallow.
(2) Weed control: row crops in which in-growth tillage for weed control

could be practised alternated with crops sown by broadcasting, crops likely to stifle by their rapid growth and dense foliage (beets, potatoes) alternated with crops making slow and sparser growth (onions).

(3) Soil-born pathogens were held to be amenable to 'starving out' by depriving them of their hosts for several years in succession.

The same author also cites the example that:

> The Incas passed a law that potatoes must not be grown on the same land more than once in seven years. It so happens that this is the period required for nematode populations to diminish to a level which makes it possible to produce another potato crop. The Spanish conquerors of Peru, contemptuous of a law that to them seemed senseless, repealed it with disastrous results.

However, the economic need to further augment agricultural productivity has in the last 40 years strongly decreased the reliance on crop rotations as a way to reduce pest impact. The availability of relatively cheap and effective synthetic organic pesticides has favoured this development considerably.

5.VI.5 WIDESPREAD USE OF SYNTHETIC ORGANIC PESTICIDES

The rapid development of the synthetic organic pesticides since the 1940s runs parallel with the strong increase in agricultural productivity. It would be incorrect to conclude that the latter was determined by the former, but the availability of these new protection measures certainly made the very rapid changes in agriculture easier. Various cultural practices could be changed without significant increases in pest problems. The increased percentage of mono-cropping of large areas is, of course, the best example where the use of pesticides has relegated pest problems to a secondary position in planning decisions. Economic requirements and technological developments were the major forces behind these changes.

In view of man's continued struggle to become more independent of agricultural and other pests, the confidence placed in these new tools was well justified, although it can now be easily argued that mistakes were made. But, with the knowledge originally available it would have been impossible to advise how plant protection matters could have been better solved. Consequently, it was not surprising that within plant protection the emphasis shifted from the use of a number of crop improvement measures, to the control of individual pest species through the intensive use of pesticides. This occurred particularly for those groups of pest organisms for which effective pesticides became available (i.e. insects and weeds). Thus, pesticides have contributed enormously to the way in which land use has changed in modern agriculture.

This contribution can probably best be appreciated by briefly reviewing the developments in pesticide usage. The statistics for this are generally expressed on the basis of the money value of the annual purchases. The best information

Table 5.VI.2 Pesticide usage on major crops in 1982 in US$ million

Crop	Herbicides	Insecticides	Fungicides	Totals
Fruit and Vegetables (including vines)	425	1160	1300	2885
Maize	1140	440	40	1620
Rice	490	645	380	1515
Cotton	325	1020	40	1385
Soy-beans	980	130	50	1160
Wheat	650	100	345	1095
Sugar-beet	315	105	45	465

is available from the US market, and the USA also best illustrates shifts of usages in modern agriculture. Some global figures however, provide a useful introduction (Woodburn, 1983).

The 1982 global market in crop protection chemicals amounted to an end-user value of US$13.3 billion. Of this, herbicides represented 39.5%, insecticides 32.7%, fungicides 21.9% and others 5.9%. Over 70% of the herbicides are used in North America and Western Europe. Insecticides are used in particular on crops in tropical and sub-tropical regions, while fungicides are mainly used in high-value crops. An overview of the usage on some major crops is given in Table 5.VI.2.

Table 5.VI.2 shows the significant use of pesticides on fruit and vegetables, including vines. The latter crop is, of course, the main reason for the fact that this group covers over 40% of the world fungicide market. Maize, soy-beans and wheat are together responsible for over 50% of the herbicide market, while fruit and vegetables and cotton each consume about one-quarter of the insecticides sold. Over the recent years the market share of herbicides has been growing considerably, in particular at the expense of insecticides.

The average annual real increase in the world pesticide market since 1970 has been around 4.5%. This growth rate was substantially higher at the beginning of the period; since 1978 the annual growth rate has been about 4%. In particular, the use of non-selective herbicides has increased in recent years and is directly related to the extension of reduced tillage practices in agriculture. A continued growth of pesticide usage is expected, at about 5% per annum. Towards 1990 the total agricultural pesticide market is expected to amount to US$20 billion (in 1982 $ terms). The geographical spread of pesticide consumption will not change significantly from today's figures, as shown in Table 5.VI.3.

The use of pesticides in US agriculture has been reviewed by Eichers (1982). The US farmers' expenditure on pesticides increased from less than US$200 million in 1951 to nearly US$2 billion in 1976. Use on a constant

Table 5.VI.3 Percentage share of total agricultural pesticide market

	1982	1990
USA	33	31
Western Europe	24	25
Japan	11.5	10
Eastern Europe	9	10
Rest of world	22.5	24

dollar basis increased about 5.5 times over these 25 years. By 1976 the farmers used annually an average of about 2.75 kg of pesticide for each hectare of agricultural land treated. In 1982 herbicides represented 57% of total agricultural pesticide use in the USA, insecticides 30% and fungicides 13%. It must be noted that pesticides are practically not used on pasture or range land.

Over the period 1966 to 1976 insecticide use increased by about 17%, fungicide use was up by about 43%, while herbicide use almost quadrupled. The latter figure demonstrates the gradual change in use and treatment of arable land—for example, the extension of no-tillage practices. Such changes would not have been possible without the availability of non-selective herbicides. From 1966 to 1976 the area treated with herbicides in the USA doubled from 40 to 80 million hectares, while the quantity of active ingredient used per hectare rose from 3 kg to 5.4 kg. Leading products in 1976 were atrazine and alachlor, which together accounted for about one-half of the market, while the third-ranking 2,4-D had dropped from its leading place in 1966 when it represented 35% of all herbicides used.

The total area treated with insecticides increased between 1966 and 1976 from 17 to 30 million hectares, but the amount of active ingredient applied per hectare decreased from 9 kg to 6.2 kg. They were applied to 76% of the tobacco, 60% of the cotton, 55% of the peanuts and 38% of the maize grown in the USA in 1976. Because of problems with insecticide resistance and restrictions on use due to persistence in soil and water, the share of organochlorine insecticides diminished from 60% in 1966 to 25% in 1976. Organophosphates increased from 27% to 50%, and the other major group, the carbamates, increased from 9% to 19% of total insecticide use.

The area treated with fungicides increased between 1966 and 1976 from 1.9 to 4.3 million hectares, while the amount of active material used per hectare decreased from 20 kg to 11 kg. Over that period there have been relatively few changes in the types of fungicides used.

Eichers emphasizes the significant increase in agricultural productivity achieved through pesticide use, and he concludes:

The reasons for farmers' increased use of pesticides are rather clear: to reduce the labour required for production, to increase the amount of food available while using less land, to reduce waste and maintain or create a quality of food preferred by consumers, and to reduce the risk of crop failure.

It is estimated that in 1982 about 22% of all pesticides were used in developing countries; the Latin-American region showed the fastest growth rate. Pesticides are mostly used on so-called 'cash crops' in developing countries. For example, the high figures given above for maize mostly relate to the production in the USA. Maize produced for food by the small farmer receives very little, or no, crop-protection inputs, notwithstanding the potentially high losses caused by various pests.

While considering the above figures on pesticide use, it should be born in mind that currently the major tasks in tropical agriculture are the production and distribution of enough food for the rapidly expanding world population, plus the improvement of small farmers' welfare through a gradual increase in their earnings. The most important tropical food crops—such as rice, maize, sorgum, millet, pulses, roots and tubers—are almost exclusively grown by small farmers; thus the requisite increases in food production have to be realized by this group.

However, during recent decades the development of pest control technology has been determined mainly by the needs of highly capital-intensive agriculture in the Western world. Since the 1940s the change from an emphasis on the use of cultural practices and resistant varieties to reduce pest incidence, to chemical pest control, was economically justified in this system and was generally adopted. This technology was then widely introduced into tropical agriculture to protect cash crops.

The tendency now is to apply the same approach to small farmers' food crops in an attempt to achieve the required food production increases. Before doing so, however, adequate attention should be paid to various specific aspects of current agricultural production, to ensure that permanent beneficial results are achieved. The criterion to be used to evaluate the results should be real progress in the plant health situation. In other words, an agricultural production system must be established where the minimum of plant protection inputs and efforts are needed to achieve the healthiest product. This goal can be achieved, but only if both the useful characteristics as well as the limitations of pesticides are well understood.

In this respect, we must draw on the experience gained so far and strive for well-balanced changes in land use in the developing countries. However, it should at the same time be stressed that these countries do not yet take adequate advantage of the benefits that could be obtained through the proper use of these useful tools. For example, limited and well-adapted use of herbicides would not only reduce the current enormous losses caused by weeds, but would also help to overcome the shortage of hand-labour in critical growing periods, as well as part of the drudgery of hand-weeding.

Thus, over a period of 40 years synthetic organic pesticides have become an integral part of modern agriculture. They have directly contributed to a more productive use of farmland and have allowed the maximum benefit to be derived from the investments made in other production factors. However, their intensive use has also revealed a number of shortcomings which led to negative changes in the agro-ecosystem. As a consequence, the use of certain products has decreased, and the use pattern of pesticides has and will gradually be further changed in the future. More details of these developments are discussed in the following section.

5.VI.6 INTEGRATED PEST MANAGEMENT

Regular use of pesticides may result in a number of unforeseen side-effects, which reduce the benefits to be gained. The study of these side-effects has indicated that pesticides can be used most effectively in reducing pest populations, when proper consideration is given to other crop production and protection factors. In fact, we should return to the concept of the period before 1940 that the total plant protection situation needs to be considered in order to optimize control results. If a specific pest is controlled, then it should be known how these activities affect the overall system. The integration of chemical control with other control measures is needed, and integrated pest management has emerged as a new phase in crop protection.

The side-effects referred to above concern negative changes in the agro-ecosystem that reduce its capacity for plant production. In the context of plant protection and land transformation they require further discussion. Side-effects include:

(1) the emergence of pesticide-resistant strains of pest organisms;
(2) effects on the balance of organisms in the agro-ecosystem and the subsequent emergence of new pest problems;
(3) the pollution of the environment by pesticides.

Pesticide resistance occurs in all groups of pest organisms (CAST, 1983). Surveys have revealed that among arthropods (insects, mites and ticks) at least 428 species had developed resistance to one or more insecticides that were once effective against them. In particular in cotton growing, where insecticides are used very extensively, this has led to situations where the major pest could not be adequately controlled during certain periods—for example, in Peru and Mexico. In other situations, such as in Central America and Sudan, the cost of cotton production has increased enormously. Similar problems have occurred in other crop situations where pesticides were frequently used. In Central America, Turkey and Sudan, an additional effect was found as malaria-transmitting mosquitoes also became resistant to the products commonly used in treating cotton. This in turn entailed considerable costs and difficulties for malaria control. Of the above reported 428 species of arthropods, about 60% are of agricultural importance. Resistance has been reported in more than 90 species of plant pathogens, 36 species of weeds and 2 species of nematodes.

Pesticide-resisting strains occur as a result of the selection of resistant individuals within the existing population. The speed of development of resistance depends on many factors, the major ones being the selection pressure exerted by the pesticide and the development rate of the pest organism concerned. This rate is generally higher under tropical conditions, and widespread application of certain pesticides can therefore lead to the appearance of pesticide-resistant populations within a period of only 2–3 years. The situation is further complicated by the phenomenon of cross resistance, i.e. resistance to a specific pesticide entails simultaneous resistance to related pesticides. The speed of occurrence of pesticide-resistant strains can only be reduced by more selective and reduced use of the pesticides concerned. The lesson to be derived from this is that excessive use of pesticides should be avoided, and that they should only be applied when the pest situation really warrants it.

Pesticides are chemical compounds deliberately released into an ecoystem (usually an agro-ecosystem) for the purpose of eliminating or significantly reducing specific organisms (pests) that adversely affect crop productivity, or animal and human health. These pest organisms are part of an intricate ecological system and are often a food source for, or provide shelter to, a multitude of insects, fungi, bacteria etc. It is useful to note, however, that of the insects encountered, only a very small percentage of the species feed on higher plants. By far the majority live on other insects or detritus.

Experience has taught us that the application of a pesticide, in addition to the effect on the pest to be controlled, will almost always simultaneously affect many of the so-called non-target organisms. This in turn will have a destabilizing effect on the ecosystem, and as a consequence may negate part of the objective of the pesticide application (i.e. an improvement of the plant health situation).

Cotton growing in Sudan provides a good example of the potential ecotoxic effect of pesticides. Currently the cotton whitefly *Bemisia tabaci* is the most important cotton pest in the Gezira in the Sudan, causing every year severe damage associated with very heavy pest populations in late season. This contrasts strongly with the situation before the large-scale application of the pesticides. Then, the whitefly was only an occasional pest and infestations were limited to early and mid-season. A combination of factors is certainly responsible for this change; but there is now clear evidence that the reduction of the natural biological control agents has been the major cause for the increased pest incidence. The development of pesticide-resistant pest populations has further worsened the situation. Following introduction of the boll weevil into the USA, mentioned earlier, application of extensive control measures has led to the emergence of the bollworm (*Heliothis virescens*) as a new major pest, due to the elimination of natural control agents.

Spider mites are the most typical man-made pests in apple orchards; the best-known species in Western Europe is the fruit-tree red spider mite, *Panonychus ulmi*. From a number of experiments it could be concluded that the absence of predatory mites was the major cause for the abundance of the spider mites in modern apple orchards. However, only when it became evident that certain fungicides used for the control of diseases were toxic to these predators could this knowledge be used for practical purposes.

With respect to disease control the situation seems to be different to the extent that only a limited number of comparable examples have been studied. Bollen (1982), in discussing the microbial balance, notes that in the natural environment suppression of sensitive species will soon result in an increase of resistant species competing for the same substrates. He also notes that pathogenic fungi are often parasitized by other fungi and that fungicides may effect this parasitism. Reference is made to the high level of white mould damage caused by *Sclerotium rolfsii* in peanuts sprayed with benomyl. This has probably resulted from the inhibition of the effective parasite *Trichoderma viride*.

The use of herbicides definitely leads to shifts in weed problems. Weed species less susceptible to the herbicide most commonly used will gradually dominate the specific agro-ecosystem. Parker (1977) notes that 'interspecific selection by herbicides is now by far the most important factor resulting in changes of weed problems'. He cites the example of herbicide use in small grain cereals in the United Kingdom, where first *Stellaria media* and *Galium aparine* emerged as new weed problems. After a change in the herbicide used *Polygonium* spp. emerged, and this was followed by the dominance of mayweeds (*Matricaria* and *Tripleurospermum* spp.). Meanwhile there was a continuous increase of wild oats (*Avena fatua*), requiring the use of specific herbicides, while for the grassweed *Alopecurus myosuroides* other new control measures were needed. In maize in the USA there have been corresponding changes resulting in increases again in annual grass weeds, particularly *Digitaria sanguinalis*, *Setaria* spp. and *Panicum dichotomiflorum*.

The examples cited clearly demonstrate that the use of pesticides, although primarily aimed at solving one or more specific pest problems, in many cases will have additional effects. These effects vary considerably depending on the type of pesticide used and the composition and functioning of the agro-ecosystem concerned. With insecticides and fungicides the impact on parasites and predators has to be considered in control operations, in weed control the relative effectiveness of the herbicides *vis-à-vis* the various plant species determines shifts in dominance.

The overall effect is that the agro-ecosystem tends to become more resistant towards the chemical control measures commonly applied. In other words, the system has been made somewhat less suitable for current agricultural practices and, in that sense, the land use possibilities have been changed, however slightly.

The impact of pesticides on the environment has been discussed in many fora and has been the strongest factor leading to gradual changes in plant-protection practices. Although numerous studies have been published, the exact cause–effect relationships, short and long term, have rarely been demonstrated.

It has become clear that the environmental impact of a pesticide is determined by, in addition to its toxicity, the persistence of the compound in the environment, and its ability to accumulate in living organisms. Many organochlorine insecticides show a high degree of persistence and are lipid soluble, and thus were detected in soil, water and various organisms at unexpectedly high levels. Various observations have shown that these types of compounds can persist in the soil for many years.

Accumulation in living organisms can be exacerbated in the terrestrial environment by the so-called food-chain effect: lower organisms being eaten by higher organisms, and the latter thus accumulating increased dosages of the compounds concerned. For example, various studies have been carried out on the impairment of reproduction of raptorial birds as a result of eggshell-thinning. It has now been accepted that metabolites of DDT, and in particular DDE, are the most likely cause of shell-thinning. As the current most widely used pesticides do break down quite rapidly, such chronic effects are unlikely to occur. However, the possibility of a direct impact on the environment cannot be excluded, and this aspect is, and must be given careful consideration in the development of new compounds.

The total impact of the continuous efforts to reduce the effect of pest organisms in agriculture can best be evaluated by studying crop loss data. Such data are available from the USA and are summarized in Table 5.VI.4. The figures indicate that the relative importance of crop losses has not decreased since the 1940s. On the contrary, an increase is noted in insect and disease damage, while only in the case of weeds has a significant reduction occurred. It is rather difficult to explain these trends in detail. They certainly confirm that synthetic organic pesticides alone, in the way they have been

Table 5.VI.4 US agricultural crop losses (percentage of potential production) caused by pests (Pimentel, 1978)

	1942–51	1951–60	1974
Insects	7.1	12.9	13.0
Diseases	10.5	12.2	12.0
Weeds	13.8	8.5	8.0
Total losses	31.4	33.6	33.0

used so far, have not enabled us to reduce pests to insignificant levels. This is, in fact, rather surprising as it can be readily shown in well-planned experiments that most organisms that are amenable to chemical control can generally be very well reduced to very low levels.

However, this is only a short-term effect and, in most cases, inadequate attention has been paid to the long-term effects. The build-up of pest resistance is one adverse phenomenon, but probably more important is the elimination of natural control organisms in the case of insects, mites and fungi. To a certain extent the opposite of what is aimed at is achieved: pests are controlled in such a way that simultaneously their reproduction possibilities are increased by reducing the effect of natural control factors (Brader, 1980). It should also be realized that important groups of organisms such as viruses and bacteria cannot be controlled by chemical means, while effective control of various soil organisms is always difficult to achieve. Moreover, as mentioned in the introduction, intensification of various agricultural practices has made crops more conducive to pest development.

The newly developed integrated pest management (IPM) programmes have been designed to avoid the above-mentioned shortcomings. Within IPM two types of control measures are applied:

(1) crop-protection measures that have a long-term effect and render the agro-ecosystem more resistant to pest development;
(2) short-term measures that are applied when the pest reaches the so-called economic damage threshold (i.e. the level at which the crop losses caused equal the cost of the control measure).

The long-term measures concern modified cultural practices and, in particular, the introduction of pest-resistant crop varieties. The cultural practices may include matters such as changed sowing dates, modified fertilizer use, clean-up of crop left-overs and appropriate rotations. IPM requires regular monitoring of pest populations to ensure that control measures are only applied when really needed. In addition, when control measures are applied maximum care is taken that the activity of the so-called natural enemies (parasites and predators that usually destroy a large portion of the pest organisms) is reduced as little as possible. This can be partly achieved by the choice of the pesticides used and the timing of their application. In addition, increased progress is made on the production of selective control means (for example, mass-reared parasites and predators).

IPM programmes are increasingly applied in a multitude of crops, including cotton, rice, soy-beans and various orchard and glasshouse crops. Their introduction into practice has shown that pesticide usage can usually be reduced by at least 50%, while maintaining or even improving production results. Indeed these programmes clearly provide a model for making the best use of pesticides.

5.VI.7 SUMMARY AND CONCLUSIONS

In traditional agriculture production practices have evolved that rendered the crops rather resistant to pest attack. This, along with other factors, led to the implementation of well-adapted cropping systems and rotations. Under such conditions certain losses occurred, but these were considered unavoidable. The occurrence of pest epidemics as a result of weather patterns or the introduction of new pest organisms has from time to time upset this system and caused major catastrophies. But, in general, these systems were rather stable. Gradually pesticides have been developed that allow better control and have, in particular, reduced the significance of agricultural pest epidemics. Originally pesticides were extracted mainly from plants or were of inorganic chemical origin. Since the 1940s the synthetic organic pesticides have become the major control measures. The introduction of these new tools has permitted considerable change in agricultural practice.

However, the widespread use of pesticides has also revealed a number of shortcomings. These include the emergence of new pest problems, the development of pesticide-resistant pest populations, and pollution of the environment. Combined with the intensification of agriculture, which has generally rendered the crops more susceptible to pest attack, these effects have resulted in a situation where crop losses continue to be high. In some cases, where the same crop varieties are grown on a large scale, for example the so called High Yielding Varieties, the results, temporarily, may be real pest epidemics.

Recognition and analysis of the potential shortcomings of pesticide use has also led to a much better understanding of the functioning of agroecosystems. The emergence of new pest problems was due to the elimination of so-called natural enemies, parasites and predators that to a large extent proved to control the development of the population of pest species. It has thus become clear that every effort should be made to safeguard these useful organisms. Pesticide resistance results from the selection of resistant specimens within the existing pest populations under the continuous pressure of large-scale and intensive use of the same type of pesticides. Reduction of use and eventually changing the types of pesticides may considerably alleviate the development of resistant strains. The environmental impact of certain pesticides is mainly due to the combined fact that the compounds degrade very slowly and can accumulate in body tissues. Consequently, products with different physico-chemical properties are now widely used.

In addition to these factors directly related to pesticides, more attention is being given to the resistance of the crop as a whole to pest development. The important criterion to be considered is the need to maintain the economic viability of crop production. For example, increased fertilizer and water use augments the susceptibility to pest attack. But, as these factors are so essential for crop production, they cannot be changed for the sake of plant protection alone. The same holds for current cropping practices which allow for the optimal use of farm machinery. On the other hand, matters such as the genetic susceptibility of the varieties grown offer excellent opportunities to reduce pest incidence. Breeding of pest-resistant varieties has thus become of major importance in recent years.

In traditional agriculture a combination of practices were willingly or unwillingly used to render the crop less prone to pest attack. Modern pest-control measures were initially so effective that the major attention could be shifted to the acute pest control problems and the overall susceptibility of the production system to pest attack did not require special concern. Other production factors largely determined the way in which agriculture has been developed in recent decennia. However, new experience and knowledge have shown the existence of limitations, and it has become clear that a permanently healthy production system requires an integration of various plant-protection practices. Thus a healthier and more stable agro-ecosystem will be established in which optimum use can also be made of modern pest-control measures through which maximum benefit can be drawn from land transformation efforts and other production inputs.

5.VI.8 REFERENCES

Ainsworth, G. C. (1981) *Introduction to the History of Plant Pathology*. Cambridge University Press, Cambridge.

Bollen, G. J. (1982) Fungicide resistance and microbial balance. In: Dekker, J., and Georgopoulos, S. G. (eds), *Fungicide Resistance in Crop Protection*. Pudoc, Wageningen.

Brader, L. (1980) Advances in applied entomology. *Ann. Appl. Biol.*, **94**, 349–365.

Brader, L. (1982) Recent trends of insect control in the tropics. *Ent. Exp. Appl.*, **31**, 111–120.

Burnside, O. C. (1979) In: Ennis, W. B. (ed.), *Introduction to Crop Protection*. American Society of Agronomy, Madison, Wisconsin.

CAST (Council for Agricultural Science and Technology) (1983) *The Resistance of Agricultural Pests to Control Measures*. CAST, Ames, Iowa. Report 97.

Cramer, H. H. (1967) Planzenschutz und Welternte. *Pflanzenschutz Nachrichten, Bayer, Leverkusen*, **20**, 1–523.

Curl, E. A. (1963) Control of plant diseases by crop rotation. *Bot. Review*, **29**, 413–479.

Eichers, T. R. (1981) Use of pesticides by farmers. In: Pimentel, D. (Ed.), *CRC Handbook of Pest Management in Agriculture*, Vol. II. CRC Press, Boca Raton, Florida.

Koch, W. (1970) *Unkrautbekämpfung*. Verlag Eugen Ulmer, Stuttgart.

Lupton, F. G. H. (1977) The plant breeders contribution to the origin and solution of pest and disease problems. In: Cherrett, J. M., and Sagar, G. R. (Eds.), *Origins of Pest, Parasite, Disease and Weed Problems*. Blackwell Scientific Publ., Oxford.

Palti, J. (1981) *Cultural Practices and Infectious Crop Diseases*. Springer-Verlag, Berlin.

Parker, C. (1977) Prediction of new weed problems, especially in the developing world. In: Cherrett, J. M., and Sagar, G. R. (Eds.), *Origins of Pest, Parasite, Disease and Weed Problems*. Blackwell Scientific Publ., Oxford.

Pimentel, D. (1978) Socio-economic and legal aspects of pest control. In: Smith, E. H., and Pimentel, D. (Eds.), *Pest Control Strategies*. Academic Press, New York.

Simmonds, F. J., and Greathead, D. J. (1977) Introductions and pest and weed problems. In: Cherrett, J. M., and Sagar, G. R. (Eds.), *Origins of Pest, Parasite, Disease and Weed Problems*. Blackwell Scientific Publ., Oxford.

Smith, K. M. (1948) *A Textbook of Agricultural Entomology*. Cambridge University Press.

Wardle, R. A., and Buckle, P. (1923) *The Principles of Insect Control*. Manchester University Press.

Woodburn, A. (1983) The pesticide world market, current trends and development of new products. *Gifap Bull.*, **9**, (9), 1–6.

Land Transformation in Agriculture
Edited by M. G. Wolman and F. G. A. Fournier

CHAPTER 6
Criteria for Observing and Measuring Changes Associated with Land Transformations

C. CHEVERRY, F. FOURNIER AND S. HÉNIN

6.1 INTRODUCTION

Previous chapters have examined the factors and conditions present when land transformation processes were engendered by agricultural development. These land transformations affect the environment through changes in a number of parameters, which naturally constitute criteria for appraising the transformations themselves if it is possible to observe, measure and interpret variations in the parameters.

Parameters are to be found at various levels in the terrestrial environment. They may concern the edaphic environment itself, relating either to its constituent elements (chemical elements, ions, molecules) or to the internal organization of these constituents. But the consequence of a land transformation may also affect the environment through the behaviour of fluids (including gases) that circulate above or in the soil, or through the behaviour of plant and animal populations supported or nourished by the area in question. This change of behaviour then constitutes a criterion for judgement.

Land transformations may also affect the environment in a more comprehensive manner, notably in terms of climate or the form of the landscape. These, too, are criteria for assessing and evaluating the transformations.

One aim of this chapter is to illustrate criteria that are useful for evaluating and monitoring land transformations, with the aid of a number of specific examples taken from chapters of this report and case studies prepared by SCOPE national committees (especially the French committee) as part of the SCOPE project on land transformation. It should be noted at the outset that these criteria do not all possess the same value. Some draw attention to a transformation while it is actually taking place or to the dangers it will eventually lead to: these are 'warning signal' criteria. Others provide immediate and better information about the mechanisms involved (they have a diagnostic value) and thereby possess a superior operational value (they have a prognostic value).

The second aim of the chapter is to provide a list of available criteria; the list cannot be exhaustive but it will have a definite indicative value. Stress will be laid on the methodological problems involved in using these criteria: analytical problems, problems of sampling in time and space, for example.

Finally, an attempt will be made to determine to what extent the proposed criteria may be used to forecast possible controls over transformations, and consequently over environmental change.

6.2 CRITERIA RELATED TO THE CONSTITUENTS OF SOILS AND WATER

Four examples will illustrate the use and advantages of this type of criterion. A listing will then be given of the main constituents, the variations of which in space and time may serve as a reference. Points will also be made regarding the precautions to be taken when using this type of criterion.

6.2.1 Examples

6.2.1.1 Rapid Increases in Copper and Zinc Content of Soils (Coppenet and Golven, 1984)

In the last 20 years or so there has been an intensification of pig farming in Brittany, the corollary of which is that large amounts of liquid manure are spread on the land. One consequence is a marked increase in the 'assimilable' copper and zinc content (extraction through EDTA) of the upper soil horizons. Table 6.1 shows the results of monitoring (between 1973 and 1982) of 64 farms in Finistère on which large amounts (higher than 100 m^3/ha per year) of liquid pig manure were spread. The results refer to the ploughed horizon (0–30 cm).

There is a manifest increase in copper and zinc content. If this rate of

Table 6.1 Recorded Cu and Zn concentrations (mg per kg of earth) in Finistère

	1973	1979	1982
Assimilable Cu	3.8	5.1	5.9
Assimilable Zn	3.2	6.2	7.4

increase is maintained, it is possible that, within a century or two, certain soils in the region may become toxic for plants because they contain excessive amounts of these elements.

The same phenomenon can be observed in several regions of Europe.

This change in the contents of the two elements may exist, from the chemical point of view, either in an 'exchangeable' state on the clays, or in association with organic products (chelation), or in the form of saline combinations of varying degrees of solubility. The change is linked to a profound transformation in Breton agriculture, the main characteristics of which are as follows.

In the past Breton farmers practised a diversified and not very intensive form of normal husbandry. The bulk of the animal fodder came from the farm. Since 1960 animal husbandry has become more intensive and specialized, notably owing to the construction of units housing hundreds of pigs. Some of the fodder is now brought from outside the region and includes mineral additives enriched with copper and zinc, for these elements have a favourable effect on the physiology of the animals. But the pigs only retain a small proportion (5%) of the copper and zinc they ingest; the rest is eliminated in their urine. The urine is collected with the faeces and stocked in the form of liquid manure, i.e. not mixed with straw as it was in the past. The manure produced in this way is spread on the soil but often on surfaces that are too restricted, thus causing a gradual concentration of copper and zinc in the upper part of the soils where these elements are fixed.

In this example, the 'change in the copper and zinc content of the soil' criterion is both a symptom of the effects of the transformation (a 'warning signal' criterion), and a factor in diagnosis, since the source of the dangerous elements and the mechanism whereby they are accumulated are in this case fairly well known. This criterion is a signpost to prognosis ('the accumulation of copper and zinc in the soil may become toxic over a century') and to forms of treatment: among the latter, it will be advisable to reduce concentrations of the incriminated chemical elements in the animal feed, and to improve the organization of spreading (choice of dates, dosage, land).

6.2.1.2 Increases in Nitrate Content in the Highest Level of Groundwater

Analyses of water from wells in the Platte River Valley, Nebraska (USA) have revealed a recent rapid increase in nitrate content. The average concentration of $N(NO_3^-)$ in surface water rose from 3 mg/l in 1950 to 18 mg/l in 1980 (Schepers *et al.*, 1984). This increase is a criterion to indicate the consequences of a growing use of irrigation and mineral fertilization in order to intensify corn (maize) farming. The soils in this valley are in many cases shallow, light in texture, sandy, or sandy and loamy. The highest groundwater level is near to the surface.

Inadequately controlled irrigation of corn (maize) feeds this groundwater and boosts the lixiviation of the dissolvable elements near the base of the profiles, beyond the zone of root implantation.n. The level of mineral fertilization has risen considerably (see Chapter 5.V). During 1981, for example, an average of 189 kg/ha of mineral nitrogen was added annually on most of the 17 000 ha in the area concerned. Some of this nitrogen reaches the groundwater in the form of nitrate ions, in addition to the organic nitrogen in the soil which has been mineralized during the same period.

The increase in the $N(NO_3^-)$ content of the water thus constitutes a criterion that is symptomatic of the consequences of a certain form of intensification of agriculture, focusing on:

(1) the rate of increase (from 3 to 18 mg/l of $N(NO_3^-)$ in 30 years);
(2) a comparison of the concentrations that have been attained with a reference value—in this case 10 mg of N per litre—beyond which the water is judged to be unsuitable for human consumption.

This criterion is also useful for making a prognosis since it makes possible the observation that 'water in a given area, at the present rate of increase, is already, or in *x* years will be, unsuitable for consumption'. But this $N(NO_3^-)$ criterion is not an explanatory criterion, since in this case it does not allow either the origin of the incriminated nitrogen or the reasons why it is leached to be identified. Several sources are possible: nitrogen from mineral fertilizers, of course, but also nitrogen from the mineralization of organic matter, as well as nitrogen from rainwater, or nitrogen re-injected by the irrigation water itself.

A research of the respective importance of these different sources can only be effected using high-precision chemical methods, and especially a study of the stable isotope ^{15}N. These methods of isotopic geochemistry require not only expensive apparatus, but also the ability to interpret their results; interpretation is a highly sensitive process because of the isotopic fractioning that occurs during the nitrogen cycle (Hauck, 1982; Mariotti, 1982).

An excellent 'warning signal' criterion, 'change in $N(NO_3^-)$ content' is thus not really an operational criterion since it is not sufficiently explanatory. To reduce the risks involved in this land transformation, the Nebraska authorities (Hall County Water Quality Special Project) used a series of additional observations and measurements. These measurements, carried out with the help of farmers in the region and a technical support team, ensured regular monitoring of soil water content, of soil nitrate content over a sufficient depth, and of the rooting depth of the corn. Coupled to the use of soil cartography and a simulation model, these measurements made it possible to effectively improve irrigation and fertilization management and gave grounds for hope that the concentrations of $N(NO_3^-)$ in the water could be stabilized.

This and the previous example concerned elements in the chemical sense of the term; other criteria involve combinations of elements or ions.

6.2.1.3 Increase in the Electrical Conductivity and SAR of Extracts of Water Taken from the Soil and of the Highest Level of Groundwater

This third illustration concerns what happens, unfortunately, in many arid or semi-arid regions when irrigation is practised, even using water that is only slightly salty. In this particular instance the land transformation is the result of a desire to improve agricultural production in regions where there is insufficient precipitation. Poorly controlled drainage often causes gradual salinization of soils and water. This transformation has often left its mark on the landscape after the irrigation network has been operating for a few years, but changes in the electrical conductivity of the soil or the water constitute a better criterion for making a more accurate judgement of the land transformation in terms of salinization.

This conductivity is measured using either water extracted from soil samples (including the well-known 'saturated paste extract') or the groundwater itself. The ions in the water reduce the resistance of the water when an electric current is passed through it, and hence change its electrical conductivity. This is a criterion relating to a combination of elements.

The criterion is often used as a 'warning signal'. In Pakistan, for instance, the electrical conductivity of soils and water increased considerably in the late 1950s as a result of the intensification of irrigation in the Indus valley and the poor quality of the drainage associated with it. A government project, revealed the consequences of the appearance of this symptom (see Chapter 5.I).

This symptom may also appear a long way downstream of the transformed system (the off-site effect). The Colorado river in the United States provides an example of this. The electrical conductivity of the water in the river was increasing and appeared likely to continue to do so if appropriate measures had not been taken (observations made at Imperial Dam). This was a consequence of an apparently successful transformation upstream; the drainage system established in the cultivated and irrigated areas successfully protects the soil against becoming too saline for plants. But in the course of time the drainage water has salted the waters of the Colorado river itself. The negative consequences of the land transformation have simply been transferred downstream. Because of its simplicity and flexibility, the 'change in the electrical conductivity of water' criterion here makes it possible not only to draw attention to the phenomenon but also to follow spatially, with the water vector, the downstream advance of the system's negative consequences.

Another criterion of the same type is often associated with the above. This has to do with the sodifying capacity of water, or the SAR (sodium absorption ratio). An example from France, a non-arid region, illustrates the usefulness of this criterion in diagnosing, and even providing a prognosis of, the consequences in the course of a land transformation (Conesa and Schenck, 1982).

In the Alsatian part of the Rhineland plain, there has been an increased use of irrigation, using groundwater from the plain. The irrigated soils are starting to show signs of structural degradation: blocking of porosity and a reduction in permeability.

A survey was carried out of water from 43 wells stretched along a 100 km SSW–NNE axis, a disposition which corresponds to the direction of the groundwater flow in this part of the Rhine plain. From measurements of concentrations of Na^+, Ca^{++} and Mg^{++}, it was possible to calculate the SAR level of the water in each well. Four conclusions emerged from this survey:

(1) The SAR levels of the 43 samples were relatively high (3.6 on average, with a maximum of 7.4) in comparison with data found in the international literature.
(2) Soils irrigated with water with a high SAR showed signs of structural degradation.
(3) The application of Hénin tests confirmed a reduction in structural stability and permeability (Hénin and Monnier, 1960).
(4) The SAR values fell quite regularly from SSW to NNE. A large-scale sylvinite (Na, K, Cl) mining area is active upstream along the flow axis. The KCl is separated from the NaCl, then sold as fertilizer. The NaCl subsists in a residual form, capable of causing sodic pollution.

The 'change in time and space of water SAR' criterion is thus a 'warning signal' in this case. But it also has an explanatory value, making it possible to trace the origins of sodic pollution spatially: this pollution originates upstream of the region under consideration and is propagated longitudinally by the groundwater. It is thus possible to explain the phenomenon itself: the sodium from the irrigation water encourages the dispersal of the clay colloids in the soil, and thus their structural degradation.

6.2.1.4 Reduction of the Proportion of Organic Matter in Soils

Molecules constitute another, structural, mode of combination of chemical elements. Organic molecules of the soils, which result both from plant and animal activity (humic acids) and from specifically human activity (pesticides) play an all-important role in the various physical, chemical and biological properties of the soil, both as regards water dynamics, resistance to erosion, and the dynamics of metallic elements (chelation phenomena), or of elements necessary for plant nutrition. Changes in the soil's organic molecule content will thus in many cases provide a criterion for assessing the consequences of land transformation under the influence of agriculture. One such case, relating to the physical properties of soils, will illustrate this point.

In many world regions, à reduction in organic-matter content is observed following intensification of agriculture: clearing of forests, ploughing of former permanent meadow, less frequent use of organic manures, etc. In some cases the result is a degradation of the physical properties of the soils. Especially in the case of soils with a loamy texture, this degradation seems to take the form of local manifestations of compaction either through pressure exercised by animals (trampling) or by farm machinery (ruts). The reduction in a soil's organic-matter content is a noteworthy criterion for assessing the negative consequences of a land transformation; but in practice it is an extremely sensitive criterion to use, as is shown by the example of a long-term test carried out in eastern France (Boiffin *et al*., 1982).

The test was carried out on undrained loamy soils in a climate of intensive and fairly abundant rain (900 mm/year). The soils were previously covered by meadow. Three crop rotations were established and compared, with care being taken to maintain a constant ploughing depth (23 cm):

Rotation A: corn (maize)—wheat—rape seed—wheat (all straw removed)
Rotation B: corn (maize)—wheat—rape seed—wheat (straw ploughed in)
Rotation C: corn (maize)—wheat—rape seed—wheat (straw removed)—followed by four years of temporary meadow (fescue, clover) at two levels of fertilization (level 1:110 units of nitrogen per hectare per year; level 2:165 units).

At the beginning of the test, the organic matter content in the plough horizon was 10.5, expressed as C‰. Table 6.2 shows the values recorded in the ploughed horizon after 14 years.

The change in the percentage of organic matter is pronounced and, after 14 years, varies significantly depending on the treatment:

(1) It proved to be impossible to maintain the initial organic level, whatever the treatment.

Table 6.2 Results of test described in section 6.2.1.4

Level of fertilization	Rotations A	B	C
1	8.3	8.8	9.5
2	8.6	9.1	9.5

(2) Periodic cycles of temporary meadow (4 years) nevertheless make it possible to maintain an organic level only slightly lower than the original level.
(3) Ploughing in straw slows down the impoverishment in humus.

During this test the influence of treatments on the structural stability of the ploughed horizon was also monitored. These soils are highly sensitive to beating, which compromises the success of sprouting from seed. Table 6.3 shows the evolution of structural stability registered by Hénin tests: the structural instability index (I_s) increases and the percolation index (K) diminishes when the percentage of organic matter diminishes in the order of treatments C—B—A. Statistically the values are significant, but the differences are nevertheless not great.

The reduction in organic matter content is thus in this case the criterion of the consequences of an agriculture in which circumstances have led to the development of specialized holdings and to the disappearance of certain traditional sources of organic restitutions (manure, long-term meadows). But the differences in values, which are not pronounced even after 14 years, are only significant as a result of the special precautions taken during the test.

A similar illustration of the use of the 'evolution of proportion of organic matter' criterion is provided by the test carried out since 1976 in Illinois (USA) on 'mollisols' ('Morrow Plots', cf. Daniels, 1982).

The percentage of organic matter has fallen everywhere since 1976, but at rates which differ according to the crop rotation chosen. This drop is accompanied by greater sensitivity to erosion and a tendency to produce lower yields. But many interactions must be considered. In certain conditions (addition of $CaCO_3$, N, P_2O_5, K) the organic-matter content on the Morrow Plots has even risen recently.

More generally, the role of organic matter seems difficult to assess without an in-depth analysis of the climatic and technical context within which the variations of the properties of land are revealed, induced by variations in the amounts of these forms of organic matter which they contain (Boiffin *et al.*, 1982). This type of criterion is thus a 'warning signal' criterion of an 'explanatory' kind, but one which can only be used after considerable

Table 6.3 The evolution of structural stability (see section 6.2.1.4)

Level of fertilization	$\log_{10} I_s$			$\log_{10} K$		
	A	B	C	A	B	C
1	1.90	1.82	1.72	1.55	1.73	1.81
2	1.87	1.80	1.73	1.63	1.75	1.92

precautions have been taken; and to take account of the transformations described here (structural degradation) it will sometimes be advisable to use criteria relating to the morphological characteristics of the soils themselves ('organizational' criteria—see later).

6.2.2 Other Criteria Related to the Constituent Elements of Soils

The aim here is not to draw up an exhaustive list of all the chemical elements, ions and molecules present in soils and soil water, whose variations in space and time could be taken as criteria. We wish simply to draw attention to a number of criteria that have not yet been described—especially pH, heavy metals, phosphorus, and the exchangeable elements in the soil.

6.2.2.1 pH-Value

Variations in pH may be the criterion of a poorly controlled land transformation. The drop in a soil's pH-value following an intensification of nitrogenous and phosphate manuring, without the simultaneous addition of lime amendments, is an example of this (see Chapter 5.V). Most mineral nitrogenous fertilizers are acidifying, especially ammonium sulphate. If there is no correction by liming, acidity will gradually reach deep into the soil and will eventually create risks of aluminic or manganic toxicity. The lowering of pH here appears as a 'warning signal' criterion.

6.2.2.2 Heavy Metals

Copper and zinc have already been used to illustrate the role of monitoring heavy metals as a criterion for evaluating the consequences of land transformation. Other metals may also be usefully taken into account:

(1) those indispensable for plant growth and development, but which soon become toxic when their concentrations in the soil rise: Cu and Zn (already discussed), Co, Mn, Mo;
(2) those not indispensable for plant production and to varying degrees toxic to living organisms: Cd, Hg, Pb, Ni, Se, Cr, Sn.

The total content of these elements is used to evaluate the evolution of land, for their increase often reveals the consequences of intensified agriculture, involving a greater use of fertilizer and other products, which leads to the dissemination of heavy metals in the environment.

6.2.2.3 Phosphorus

The soil survey in Finistère cited in section 6.2.1.1 shows a pronounced enrichment over a 10-year period of the soil's 'assimilable' phosphorus content in the surface horizons. The 'assimilable' P^2O^5‰ content in the 0–30 cm horizon (Dyer method) rose (average of 64 holdings) from 0.44 in 1973 to 0.69 in 1982.

This enrichment is linked to the intensification of animal husbandry and to the increase in restitution to the soil through spreading. A content of 0.35‰ is already considered optimal in the loamy soils of this region. Enrichment beyond this level thus constitutes the criterion of a transformation increasing the risks of a zinc deficiency (if the soil has a non-acid pH; see Chapter 5.V) or of the transfer downstream, through erosion, of earthy particles particularly rich in phosphorus (eutrophication of the water).

6.2.2.4 The Soil's Exchangeable Elements

Cationic exchange capacity (CEC) is a good indicator of a soil's ability to fix cations, some of which play a favourable role in plant nutrition (K^+, NH_4^+) or in the physical behaviour of soil (the flocculent role of Ca^{++}, Mg^{++}).

The diminution of a soil's CEC expressed in me/100 g of soil may thus be considered as a criterion of the lowering of fertility; it is often linked to a land transformation involving a drop in the percentage of organic matter.

For the sake of completeness, the four cations that influence the saturation rate may also be used as criteria in so far as a variation in their content reflects the results of farming operations (especially calcium, potassium and magnesium) or affects the physical properties of the soil (sodium). Proof of this may be found in Chapter 5.V or elsewhere in this chapter. It is significant that at Grimari Station in the Central African Republic the levels of calcium and magnesium fell, in the first 15 cm of soil, 40% below that observed under the pre-existing savannah, when the latter was cleared and turned to agricultural use. This is a 'warning signal' criterion; the transformation of the organic matter and its reduction constitute the explanatory criterion.

6.2.3 Precautions to be Taken When Criteria Related to the Constituent Elements of Soils are Used

The determination by analysis of the chemical element and molecule content of soil or soil water does not, *a priori*, seem to present insuperable difficulties. Further consideration shows, however, that the application of this type of criterion is less simple than it appears.

6.2.3.1 Sampling Problems

Whatever the end in view, the results of analyses of soil samples carried out in laboratory conditions are only fully significant if the samples are representative. A simple example will illustrate the inaccuracies that may occur as a result of carelessness. A comparative analysis of areas of ploughed land may reveal a variation in the amount of a given element contained in it. This positive or negative variation may depend on an enrichment or an impoverishment of the milieu in the element concerned, but it may also be due to a variation in the depth to which the areas of land were ploughed. If additional determinations made in the layers beneath the arable land also show changes in amounts of the same element, it is no longer possible to establish a balance by taking account of the surface layers alone.

It is thus necessary to pay attention to the points where samples are taken and to relate the analytical data to what is known of the volumes of earth in question in order to form a better judgement of the overall evolution of the environment.

6.2.3.2 Problems of Analytical Determination

Monitoring concentrations of chemical elements may concern either the 'total reserve' or the 'mobile' (or assimilable') reserve, i.e. the quantity of elements capable of being absorbed or put in solution, even by agents with a low dissolvent capacity (cf. the case of Cu and Zn described earlier). It is interesting to establish whether the state in which the chemical elements are found makes them immobile or, to a greater or lesser degree mobile. The problem here is to understand the mechanisms whereby an element passes from one state to another, since the different elements which may serve as criteria change state in different ways.

6.2.3.3 Problems Arising from Concentrations and Flows

Contents or concentrations are static data which reveal deficiencies or excesses of a given element. Flows of these substances through a given volume of soil are, on the other hand, dynamic data which involve quantities in relation to time. Flows and concentrations thus make it possible to identify the elements in the 'entrance–exit' balance and to define the stationary or non-stationary nature of the system concerned (permanent or variable regime). *A priori* the problem appears simple, but here too we are faced with the question of the immobility or mobility of the elements and of the evaluation of assimilable elements, discussed above.

6.2.3.4 Problems with the Determination of Organic-matter Content

Particular attention must be paid to this problem, firstly because it is not the organic-matter content but the carbon content that is being measured directly, the former being derived from the latter using a coefficient; and secondly because a distinction must be made between several forms of organic matter. Agronomists usually distinguish two forms: those consisting of the residues of living creatures, easily separated from the mineral fraction by a rapid analysis method such as density or mechanical sorting, and molecules resulting from microbial activity and stemming from the more or less direct transformation of so-called free organic matter.

Methods of dosing do exist. They are used to a greater or lesser degree, but agronomists are able to calculate balances from them and have been able to formulate equations representing the speed and order of magnitude of the change and the transformation of these two types of substance.

Many more problems could be discussed here. Thus in spite of the apparent simplicity of their application, the criteria relating to elements, to ions, and *a fortiori* to molecules, should be used with care both in reasoning and in calculations.

6.3 CRITERIA RELATED TO THE ARRANGEMENT OF THE SOIL'S CONSTITUENTS

The structure and the porosity of a soil are examples of parameters that depend on the organization of the soil's constituents (mineral or organic). The evolution of these parameters is often used as a criterion, as G. Monnier points out:

> The evolution of the conditions of land use in most agricultural regions in the last few decades reveals a number of dominant characteristics. They include land transfers in connexion with reallocation procedures, simplification of farming methods, increasing use of heavier and more powerful farming equipment, and an increase in drained surfaces. Either directly or indirectly, these developments are liable to lead to transformations of the physical properties of cultivated land, some of which correspond to degradation: beating, compacting.

Three specific examples will be taken to illustrate the criteria relating to the organization of the constituents of soils, and then a list will be made of the criteria of this type that are available for application to changes linked to land transformation. The diversity and wealth of possibilities for using the criteria relating to the arrangement of the constituents of the soil will be shown.

6.3.1 Examples

6.3.1.1 Formation of Ruts and Reduction in Porosity (Monnier and Fies, 1982)

Several hundred kilometres of electricity transmission lines are installed each year in France. The cost of this equipment is such that those responsible systematically take the shortest route, without attaching to much importance to the effects on the land crossed. The case of the implantation of large pylons for high-tension lines illustrates this approach.

Heavy equipment must be brought to each site on many occasions throughout the year, whatever the hydric state of the soil. The passage of these machines causes the soil to become compacted, revealed by two symptoms:

(1) ruts, the shape of which is recorded by asperimetric measurements:
(2) a diminution in the porosity of the soil under the ruts, recorded by measurements of bulk density.

This compaction caused by rolling initially lowers the 'structural' porosity linked to the existence of fissures and galleries of biological origin (Table 6.4). However, it may also reduce the 'textural' porosity which corresponds to the gaps left by the arrangement of the granulometric constituents. The extent of these compactions varies with the textural type of the soil: loamy soils are particularly sensitive to it. It also varies with the hydric conditions of the soil when rolling takes place: for a given material there is a 'range' of water content within which the soil shows itself to be particularly sensitive to compacting. Each time a machine passes such effects may be created.

The criteria for assessing these transformations are the appearance of ruts and the reduction in the soil's porosity. These are 'warning signal' criteria, but they are only partially explanatory. A more precise diagnosis must be based on a better understanding of the specific role of the texture, humidity, organic-matter content and other parameters. This diagnosis requires the use of additional experimentation, especially dynamic (Proctor) or static (oedometry) compaction tests. Here, only more thorough diagnosis of this kind can make it possible to attempt a prognosis, and above all to advise on effective preventive or curative treatments.

Table 6.4 Comparison of the aeration porosity ('structural' porosity) beneath the ruts and between two ruts (Monnier and Fies, 1982)

	Under one rut	Between two ruts[a]
Not rolled	20%	21%
One passage of the roller[b]	11.8%	15.2%
Four passages	4.2%	5.4%

[a]The two ruts are 1.80 m apart. The measurement is thus made at 90 cm from the axis of each rut.
[b]Pressure on the soil is 5.2 kg/cm^2.

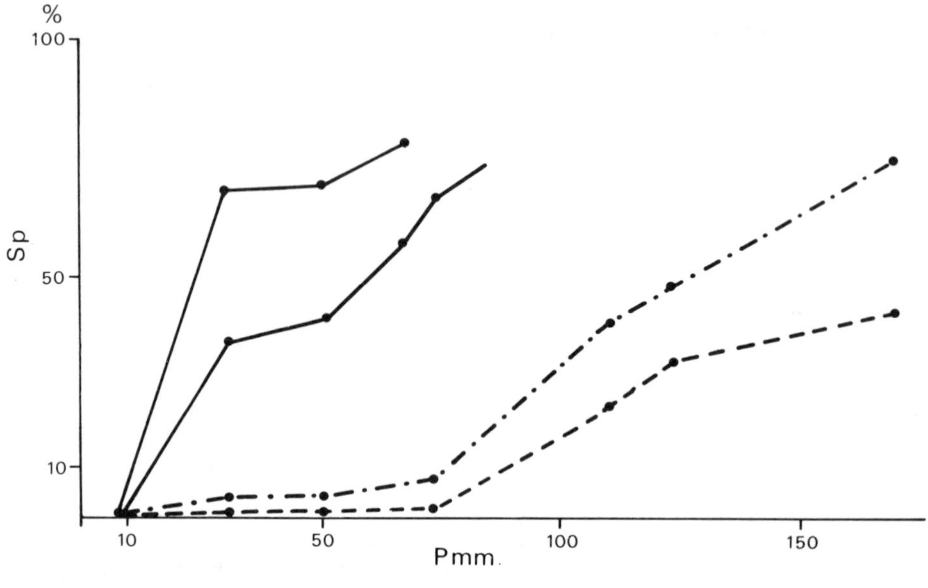

Figure 6.1 Evolution of the percentage of soil covered by continuous crusts (first indicator of beating), S_p, as a function of the cumulative depth of rainfall from the moment of the soil's exposure to the rain, *P*
——— Evolution of the unintercepted rate of the energy of drops
·—·—· Evolution when the energy is intercepted to 30%
— — — Evolution when the energy is intercepted to 70%

6.3.1.2 The Appearance of a Beaten Crust (Boiffin, 1984)

A beaten crust is often observed on the surface of loamy soils. This structural degradation is linked to the action of rain (Figure 6.1).

This criterion reveals an increased sensitivity in the milieu to rainfall action as a cause of disaggregation of soil structure, notably by bursting processes. This is a warning signal that draws attention to a phenomenon which may have serious consequences: increased runoff, return to a massive state from the fragmentary states created by working the soil, and bad crop implantation.

6.3.1.3 Structural Degradation (Centre Technique Forestier Tropical, 1969)

When mechanized groundnut farming began at Sefa in Casamance, Senegal, around the 1950s, the structural stability and permeability of the tropical ferruginous soils under dry forest cover were satisfactory. In the soil's upper horizon and beneath the forest, the Hénin instability index ($\log_{10} I_s$)[1] was much lower than 1 (between 0.43 and 0.57 for five samples), and the permeability (evaluated by the *K* coefficient of Darcy's law) varied between 30.7 and 47.0 mm/hour.

Deforestation, then mechanized groundnut farming, increased the value of $\log_{10} I_s$ to 1.21 after two years' farming, then to between 1.42 and 1.77 (the effect of six years' farming). Parallel to this, permeability suddenly became slight—20–30 mm/hour at the end of the first year's farming—and subsequently did not return to its former level.

Further examination of the 'clay and loam' fraction revealed that beneath forest cover, only from 7–9% of this fraction was dispersed, whereas six years later, after cultivation, from 13.5–20% had become dispersed. As granulometric analysis indicated that elements smaller than 20 μm made up 15–20% of the soil's surface horizon, this meant that almost all the fine elements were dispersed.

Thus we have here a criterion which is both a 'warning signal' since it changed with the land transformation (deforestation and bringing into cultivation), and 'explanatory' when combined with other data from the physical analysis of the soil.

6.3.2 Criteria Related to the Arrangement of the Soil's Constituents

Criteria related to the arrangement of constituents usually concern changes in this form of arrangement: the formation of a rut, the appearance of a beaten crust. But they may also concern the properties of the soil particles themselves, determining the behaviour of a given soil in response to actions tending to cause change. In this case the criterion is the sensitivity of a soil to a given action, a sensitivity which will be assessed using a scale of reference. We shall, for example, speak of sensitivity to compaction or of structural stability. The criterion here will be a soil's change of position on the reference grid. A soil will, for example, have become more sensitive to compaction after transformations that have brought about a marked diminution of the percentage of organic matter. Consequently a distinction will be made here between these two cases.

6.3.2.1 Changes in Forms of Arrangement of the Soil's Constituents

These transformations can be attributed to three main categories of phenomena:[2]

(1) *Disaggregation phenomena under the action of water, inducing structural changes*. Their criteria may be (i) the appearance on the surface of the soil, following a series of downpours, of sheeted structures which harden when they dry (these are the beaten layers analysed above); (ii) the return to a massive state of fragmentary layers through the action of tools used when the soil is too wet, and the appearance of a continuous structure; or (iii) the appearance of continuous structures in the subsoil, due, for example, to inadequate drainage of these layers.

(2) *Compaction phenomena*. Here the criterion may be the appearance of ruts (an example cited above) or the appearance of deformations as a result of trampling by animals' hooves. These are morphological criteria, but they may also be evaluated by physical analysis, especially of an increase in bulk density.
(3) *Fragmentation phenomena.* The appearance of a broad network of shrinkage cracks in clayey ground may be the consequence of sodification of the milieu, since the exchangeable sodium encourages the clays to disperse in the presence of water, and then to fragment in this way when they dry out.

For each of these categories the list could be made longer. Among the criteria linked to disaggregation through the action of water, the appearance of rills or even of new arrangements might be mentioned, as in a case cited by Daniels (1982) in the USA. Ill-adapted farming practices caused considerable erosion which entirely removed the ploughed horizon. A new surface horizon was reconstituted, the organization of which (size of aggregates, porosity, etc.) was new and thus revealed the unfolding consequences of the land transformation.

6.3.2.2 Changes of Behaviour of Soils and their Sensitivity to an External Action

Monnier distinguishes three main groups of properties:

(1) *Structural stability*. This denotes the resistance of aggregates to disaggregation through the action of water. The problem is how to standardize evaluation methods. The French school (Hénin and Monnier, 1960) attaches considerable importance to the intrinsic properties of the soil (its mineral and organic constitution), particularly considering that a change in the organic status of land plays a major role in the evolution of its structural stability.
(2) *The behaviour of soils in relation to mechanical actions*. This involves the assessment of consistency limits, notably by Atterberg tests; and behaviour under compacting, assessed either by dynamic compacting tests (the Proctor test) or by static compacting tests (oedometry).
(3) *The tendency of soils to fissure*. The tests used reveal first of all the role of texture, and notably that of clay. But the organic matter and the cations attached to the clays (through the flocculating or dispersing properties of the cations) also play a role which makes it possible to use these tests as criteria in certain cases of land transformation.

The expressions of these properties may be shown on texture diagrams for given soils.

6.3.3 Conditions Under Which it is Possible to Make Better Use of Criteria Related to the Arrangement of the Soil's Constituents

6.3.3.1 The Diversity of this Type of Criterion

The number of illustrations and the list of criteria concerning the arrangement of constituents could have been considerably longer. Here one last example, revealing the diversity of these criteria, will be presented.

A team of research workers from south-western France chose as a criterion the notation of clods of soil after ploughing (Hutter *et al*., 1982), according to a pre-established code. These authors were particularly interested in the smoothing and crumbling phenomena manifested in the clods, smoothing being an indicator of land wastage. The processing of these notations through factor analysis of the correspondences enabled the team to point out the influence of irrigation and previous cultivation on the stability of ploughing. Thus it appears that in irrigated conditions, maize–sorghum as previous crops cause considerable smoothing, unlike cereals. This difference disappears if there is no irrigation.

Here the notation of an organizational feature (clods produced by ploughing) and processing through calculation of the marks given constitute the criterion of consequences of a land transformation (introduction of irrigation, modification of rotations).

6.3.3.2 Use of these Criteria Through a Spatio-temporal Process

Boiffin (1984) adopted this approach in his study of the formation of beaten crusts in loamy French soils. After making a careful typology of the various morphological stages of these crusts, Boiffin carried out large-scale mapping (several square metres) of their spatial distribution and then monitored them over a period of several months, punctuated by rainy periods. The association of these three processes—typology, cartography and monitoring—over a period of time enabled him to understand more clearly how beaten crusts are formed, and thus to use the 'appearance of a beaten crust' not only as a symptomatic criterion, but also as a criterion that could be used in making a diagnosis.

6.3.3.3 A Synthetic Criterion: Examination of the Cultural Profile

The concept of cultural profile, defined by Hénin and Monnier in 1956, meets the need to pay attention to *the overall physical state of the soil* when judging the effects of a change in farming practices, techniques or systems. Within the soil profile attention is focused on the entity formed by the superposition of differentiated layers of earth through the action of farm implements, of root systems, and that of factors reacting to the two previous actions. This examination is an extremely reliable synthetic criterion for an experienced observer.

Finally it should be pointed out that in the case of land transformations involving physical modifications to the soil, criteria related to the arrangement of the soil's constituents are, because of their diversity and wide range of applications, often more operational than criteria related to a single constituent element of the soil, especially organic matter.

6.4 CRITERIA RELATED TO THE WATER AND GAS COMPONENTS OF THE ECOSYSTEM

The behaviour of water and gas in soils and on land may constitute a highly reliable criterion for assessing a land transformation and its consequences. The four examples chosen here relate either to *regime* characteristics or to *balance* characteristics and are significant in this respect. The very nature of the criterion means that it is unnecessary to draw up a list of parameters or to indulge in methodological considerations.

6.4.1 Examples

6.4.1.1 Increase in the Water Runoff Coefficient

In Madagascar, during storms or cyclones, the annual runoff coefficients may exceed 60% and the instantaneous flows reach 1350 l/s/ha (during a decennial flood) under secondary vegetation after burning and in undulating areas. Under natural or artificial forest (after reforestation with eucalyptus), on the other hand, the same coefficients and flows are far lower: 20–30% and 500 l/s/ha) (Bailly, 1982).

Results of the same order have been obtained in Guyana in similar relief conditions in a study area of ten catchment basins, where the total measured flow varies from 11% to 50% of annual rainfall depending on the use to which the basins have been put (reforestation, natural new growth, pasture, orchard).

This criterion related to the behaviour of water reveals the consequences of an extremely widespread form of land transformation: the destruction of dense tropical forests. 140 million hectares will have been affected between 1975 and the years 2000, in addition to which it is foreseen that 70 million hectares of open forest will also have been destroyed. Reforestation during the same period will only amount to an estimated 15 million hectares. Unfortunately the use of this criterion is thus highly topical.

6.4.1.2 Rise in the Level of Superficial Groundwater (Shannan, Chapter 5.I)

A rise in the water level of the root area signals an unfavourable environment for the roots of most cultivated plants, notably because of lack of oxygen. In semi-arid or arid regions, it also points to a secondary salinization of the land.[3]

The example of Pakistan is highly revealing from this point of view. Over a large part of 15 million irrigated hectares, the originally deeper-lying groundwater has risen rapidly—by 30 cm annually. As a consequence, in 1978, over 54% of the total irrigated surface, the groundwater was at a depth of less than 3 m.

This depth has been considered as the threshold above which risks of waterlogging and salinity become very acute. This 'rise in groundwater level' criterion has led the Pakistan government to launch a large-scale drainage programme (the SCARP programme) without which 50% of the irrigated surfaces would now be unfit for cultivation.

6.4.1.3 Modification of the Flood Régime

In the west of France, rural land was until 1960 structured by a network of hedges consisting of wooded banks skirted by ditches: the *bocage*. Socio-economic considerations induced the farmers in this region to regroup their land-holdings, extend plots, construct access roads, etc. Consequently there has been a massive levelling of the wooded banks. A change in the flood and low-water régime in the elementary catchment basins (several dozen hectares) is one of the criteria which reveals the consequences of this profound transformation of the landscape.

A comparative study of two elementary basins, one subject to reallocation, the other not (Mérot, 1978; Carnet, 1978), has made it possible to describe these modifications. Flood hydrograms (Figure 6.2) show that at this spatial scale at least the destruction of the *bocage* leads to a greater rate of discharge during flooding and smaller flow rates at low water.

6.4.1.4 Reduction of the Water Drainage Coefficient

The same type of criterion was used by Gachon *et al.* (1982) in medium-altitude mountain areas of the Massif Central (France), during experiments carried out on lysimetric plots. The water balance, and notably the drainage coefficient, revealed the consequences of changes in agricultural systems and changes in the conception and intensity of mineral manuring. Gachon and his co-workers have thus shown that in cultivated meadowland subject to crop rotation, drainage fell by between 5% and 10% when the level of nitrogenous manuring was raised. Under permanent meadow, on the other hand, drainage is almost constant and independent of the level of nitrogenous mineral manuring. A high compensatory production of white clover compensates for the absence of this fertilization, and helps to maintain the level of water consumption.

Changes in the behaviour of water in soils may thus constitute a criterion, as indicated by the four cases chosen as examples. The same applies to another

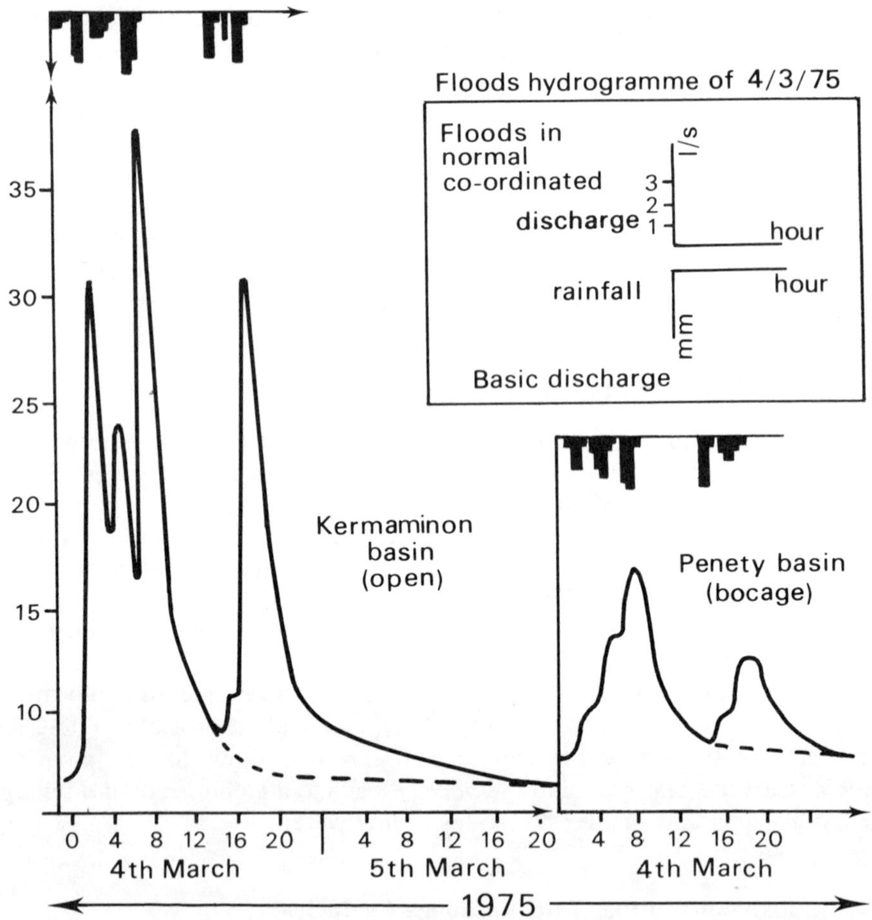

Figure 6.2 Comparison of the behaviours of a *bocage* catchment basin and an open catchment basin. The two basins have the same area (32 ha) and their geological and geomorphological characteristics are closely comparable. They are less than 4 km apart and are subject to virtually the same precipitation (Mérot, 1977)

environmental fluid, gas. The appearance of a specific gas such as hydrogen sulphide may reveal a milieu that has become extremely reducing subsequent to a transformation which has increased waterlogging.

However, the criteria related to water and to gases are usually only of symptomatic value, and possess little explanatory value. In the last example, the Massif Central in France, the drainage coefficient criterion nevertheless makes it possible to analyse the favourable role of nitrogenous fertilization using the water consumption of meadow plants. As a rule, however, the criteria relating to fluids are not very useful for making diagnoses, nor consequently for making prognoses, unless they are compared with climatic data recorded at frequent intervals.

6.5 CRITERIA RELATED TO THE BEHAVIOUR OF PLANT AND ANIMAL SPECIES

The behaviour of plant and animal species often provides information about the consequences of a land transformation of agricultural origin. The criteria may be:

(1) the disappearance or appearance of species;
(2) the modification of the respective share of various species in a given ecosystem (impoverishment, diversification);
(3) the modification, within a given species, of certain behaviour characteristics: the localization of roots in the case of a plant species, the sensitivity of a species to environmental changes.

Three examples have been selected to illustrate the application of these criteria.

6.5.1 Examples

6.5.1.1 Impoverishment of Forest Flora and Fauna (Bonneau, 1982)

In many forests, economic imperatives have made it necessary to attach great importance to productivity. The result has been a trend towards uniformity and increasing density: forests are created by plantation; there are no clearings; they consist of a single type of tree; all the saplings are regularly spaced and of approximately the same height.

Under dense plantations of resinous trees, the natural flora is impoverished. The litter, consisting of plant matter of a single origin, may decompose less well: the range of species of microflora (bacteria, fungi) and microfauna (microarthropods, earthworms, insect larvae) which cooperated to fragment, alter and mix the fallen leaves in the soil, is reduced in size and variety and becomes less efficient. Animal life in the forest diminishes—notably the bird population, which prefers a mixed forest of irregular structure. The forest is more exposed to attacks from parasites. The lack of diversity in the herbaceous and ligneous flora and in the plant structures reduces the number and variety of potential pest-destroying predators.

Thus the impoverishment both of the natural flora beneath the forest canopy and of the ground-level fauna are here the criteria of the consequences of a profound change in forest management.

6.5.1.2 Diversification of Flora and an Increase in the Value of Plant Species as Animal Food (Loiseau, 1982)

This criterion is closely related to the previous one, but in a sense converse to it. In the pastureland of the Monts Dore in central France, a diversification of the flora has been observed on parts of the common land used for pasturing ewes, with the appearance of red fescue and white clover which constitute nutritious forage. This corresponds to the organic manuring of the commons by local sheep farmers. In places where the farmers have not applied this new strategy, i.e. have continued to spread organic manure outside the commons, the vegetation remains relatively undiversified, with spreading of spikenard, followed by heather; this has led to the abandonment of this land and a consequent regression in the quality of the pasture.

An improvement in the potential of pastureland is here the criterion of the consequences of a change in the organization of pasturing.

6.5.1.3 Disappearance or Appearance of Plant and Animal Species

The flora of the Netherlands has declined sharply since 1950, as shown in Table 6.5. The forms and mechanisms of this change in the number of species have been identified in two regions of the Netherlands which have undergone profound transformations:

(1) In the region of the dunes, the exploitation of fresh water from the groundwater has led to a marked drop in the water level. The number of species of phreatophytes and birds has fallen considerably. Some are unable to survive the desiccation.
(2) In the region of the polders, intensive land use has led to transformations of the same type, notably the decline of certain species of birds as shown in Table 6.6.

Many more illustrations could be given of the ways in which plant or animal

Table 6.5 Decline of the Dutch flora (de Molenaar, 1980)

	Number of species	%
Total number of species	1502	100
Number of species that have diminished	824	55
Diminution:		
50%	108	7
51–70%	423	28
71–90%	182	12
91–100%	111	7

Table 6.6 Decline of a bird species in the polder region (Eyerman, 1975)

Species	Extensive land use	Intensive land use	Decline (%)
Oystercatcher	33	0	100
Lapwing	133	4	97
Black-tailed Godwit	120	2	98
Redshank	53	4	92

species can be used as a criterion. Nor are such criteria invariably based on the number (appearance, disappearance, diversification, impoverishment) of species or on their 'quality' (as fodder, for example). They may also be based on the behaviour of a given part of a plant. The 'foliar diagnosis' approach is a well-known illustration of this. But the localization and state of roots are also reliable criteria.

Thus in the marshes of western France, examination of cultural profiles carried out in meadowland showed the existence of a mat 7 cm thick consisting essentially of root debris (from fodder plants) and a little earth. This mat represents a system almost independent of the subsoil, from which it can be separated as if it were placed on a substratum. This criterion enabled Hénin and Fournier (1982) to make a diagnosis of the consequences of developing these clayey lands won from the sea by a network of embankments and drained by ditches. Improvement of this land was carried out by eliminating obstacles to deep root development (asphyxia, excessive chloride content). The first step in the treatment is to lower the water table. Thus the 'appearance of a rooty mat' criterion is in this case a symptom but, since it has some explanatory value, it also points towards diagnosis, prognosis and forms of treatment.

6.6 CRITERIA RELATED TO GENERAL CHANGES IN THE ENVIRONMENT

The parameters that can be used as criteria for assessing and evaluating land transformation are largely related to components of the terrestrial environment: soil, water, plant and animal populations, etc. In some circumstances, however, a land transformation has consequences that may make a more comprehensive impact on the environment; in this situation some of the general characteristics of the environment may be affected, such as the form of the landscape or climatic behaviour. These changes may be included among the criteria for assessing and evaluating land transformations, as the following examples reveal.

6.6.1 Examples Related to the Shape of the Landscape

6.6.1.1 The Appearance of Sinkholes (Unesco, Land Subsidence, to be published)

Since the 1950s, large numbers of sinkholes have appeared in Alabama as a result of land subsidence due to the lowering of the groundwater level, a phenomenon which is itself the result of excessive pumping of underground water for various reasons (agriculture, human consumption, etc.). Over 4000 sinkholes have appeared in the insufficiently consolidated clayey deposits surmounting carbonated rocks. Most notable of all are 1700 holes spread over 36 km^2, an average of almost 50 holes per km^2. Their size varies greatly: from 1 m to 90 m in diameter and 0.30 m to 30 m in depth.

Because of their shape and density, these sinkholes create a landscape that is both a 'warning signal' criterion and a 'diagnostic' criterion, for it is easy to connect the pumping of the water and the appearance of fields of sinkholes.

6.6.1.2 The Appearance of Salt 'Lunettes' (Unesco, Land Subsidence, to be published)

Small hillocks of earth formed by the wind are sometimes found on the edges of depressions in North Africa. Analysis reveals that they are composed of pseudo-sand formed by the aggregation of particles of clay and salt crystals. Clumps of grassy vegetation support the first stages of accumulation of these pseudo-sands. The growth and multiplication of the 'lunettes' creates a distinctive landscape.

This 'landscape' criterion reveals the very high degree of salinity of the soils in the region. The salts bring distinctive structural properties to the soil's surface crust, and pseudo-sands are created. The movement and accumulation of the latter takes the form of a landscape of salt 'lunettes'.

6.6.2 Examples Related to the Climate

6.6.2.1 Evapo-transpiration (Choisnel and Payen, Document prepared for the SCOPE Project to be published by the French Ministry of Environment)

Evapotranspiration may act as a criterion for judging the effects of transformations in agriculture. The oasis effect (an increase in evapo-transpiration) is an illustration of this. In Mediterranean areas, when irrigation is introduced on to small and scattered plots of land, evapo-transpiration becomes abnormally strong. The mechanism of this phenomenon is well-known: in the non-irrigated zones, the surface temperature of the soil remains higher than the temperature of the air for most of the day. The temperature of an irrigated plot, on the other hand, is always lower than that of the air temperature: the convection flow is then directed towards the soil and evapo-transpiration increases.

Focusing on this oasis effect by monitoring evapo-transpiration constitutes a 'warning signal' criterion since it draws attention to a certain waste of water. It also makes it possible to define an irrigation strategy: in the dry Mediterranean region the irrigation of these small plots should aim to bring water to the soil in sufficient quantities to maintain a level of evapo-transpiration at most equal to the net radiation. The oasis effect is thus also an explanatory criterion which leads to a proposal for a solution.

More generally, a modification of the surface energy balance may reveal the consequences of transformations linked to agriculture which bring about changes in the hydric state of the surface of the soil, in the phenological state of the vegetation, in the albedo of the soil–vegetation system, or specific modifications of the wind régime.

The destruction of the *bocage* in western France is relevant here (see section 6.4.1.3). Land reallocation led to the disappearance of the hedges which surrounded the fields; climatic measurements were carried out simultaneously on areas that remained as *bocage* and on the reallocated areas, using batteries of apparatus posted at different heights. The comparison of climatic balances (Missonnier *et al*., 1982) shows that the *bocage* acted as an energy trap and increased the continentality of the climate; this had diminished in the redistributed areas. Furthermore, the potential evapo-transpiration of the open (reallocated) plots has not changed in relation to that of the closed plots of the *bocage*.

Here too the usefulness of certain climatic parameters is apparent as criteria for evaluating the consequences of land transformation in agriculture.

6.7 CONCLUSIONS

This chapter does not claim to be an exhaustive presentation of all the environmental parameters that may be used as criteria for judging the evolution of land. It seeks above all to highlight the types of parameter and criteria which may be useful in environmental monitoring.

For this reason, a number of concluding remarks are called for, the first of which is that parameters relating to the soil are more widely taken into account than other environmental parameters. This is because the consequences of land transformation in agriculture primarily affect soils, by causing their properties to vary.

It appears that monitoring is not simply a question of recording changes in the properties of soils. It may be carried out not only by focusing on modifications in the cycles of chemical elements and the structural properties of soils, but also by taking account of modifications in elements external to soils, such as water, or supported by soils, such as animals and plants. Finally, such general characteristics of the environment as climate and landscape may also be affected.

Today there is a tendency to wish to monitor the evolution of the earth's surface on a global scale. Teledetection, which is used for this purpose, takes account of the two general criteria, climate and landscape, among others. But if teledetection now makes it possible to draw up periodic land-occupation balances and to record surfaces seen from satellites with a certain precision, it cannot provide a complete explanation of environmental mechanisms, nor of the cause-and-effect relationships between the phenomena that exist in the environment. However, analyses are essential for decision-makers, and it is to further this work of analysis that a series of parameters have been discussed which emerge from the preceding chapters (especially those that evaluate knowledge of agricultural practices which cause changes in land use).

6.8 REFERENCES

Bailly, C. (1982) Quelques données sur les effets de la transformation ou de l'aménagement des forêts et des prairies dans le monde tropical. In: *La Transformation des Terres*. Ministère de l'Environnement, numéro spécial janvier 1982, Paris, 55–73.

Boiffin, J. (1984) *La Dégradation Structurale des Couches Superficielles du Sol Sous l'Action des Pluies*. Thèse, Institut National de Paris-Grignon, Paris.

Boiffin, J. *et al.* (1982) Systèmes de culture et fertilité des sols limoneaux: effets de l'enfouissement des pailles et de la prairie temporaire en champ d'essai de longue durée. In: *La Transformation des Terres*. Ministère de l'Environnement, numéro spécial janvier 1982, Paris, 74–106.

Bonneau, M. (1982) Evolution du sol sous forêt après enrésinement. In: *La Transformation des Terres*. Ministère de l'Environnement, numéro spécial janvier 1982, Paris, 107–123.

Carnet, C. (1978) *Etude des Sols et de leur Régime Hydrique en Région Granitique de Bretagne: une Approche du Rôle du Bocage*. Thèse, Université de Rennes.

Centre Technique Forestier Tropical (1969) *Techniques Rurales en Afrique*, **12**, 15–16. Secrétariat d'Etat aux Affaires Etrangères, Paris.

Choisnel, E., and Payen, D. *Caractérisation du Climat par l'Etude du Bilan Hydrique du Sol et du Bilan Energétique de Surface—Aspects Macroclimatiques et Microclimatiques*. Document préparé pour le projet SCOPE Transformation des Terres.

Conesa, A., and Schenck, C. (1982) Effet de l'irrigation à l'eau salée sur l'évolution des sols dans les périmètres irrigués en Alsace. In: *La Transformation des Terres*. Ministère de l'Environnement, numéro spécial janvier 1982, Paris, 149–161.

Coppenet, M., and Golven, J. (1984) *Etude Lisière–Sol–Plante. Bilan de Dix Années de Suivi dans une Soixantaine d'Exploitations Intensives du Finistère*. INRA, Station de Quimper.

Daniels, R. B. (1982) US Contribution, SCOPE Land Transformation Project. SCOPE Report.

Eyerman, C. (1975) de problematiek in onze weidevogelgebieden. *Het Vogeljaar*, **23**, (3), 105–107.

Gachon, L., *et al.* (1982) Bilan de l'eau et des éléments minéraux N et K en moyenne montagne granitique du Massif Central. Variations en fonction du système cultural.

In: *La Transformation des Terres*. Ministère de l'Environnement, numéro spécial janvier 1982, Paris, 162–173.

Hauck, R. D. (1982) Isotope ratio analysis in investigations using stable nitrogen tracers. In: *Methods of Soil Analysis*, Part 2 (rev. edn) (Ed. Al L. Page *et al.*). Am. Soc. Agron., Madison, Wisconsin, USA.

Hénin, S., and Fournier, F. (1982) Définition et méthode d'étude du problème. In: *La Transformation des Terres*. Ministère de l'Environnement, numéro spécial janvier 1982, Paris, 48–52.

Hénin, S., Monnier, G., *et al.* (1960) *Le Profil Cultural: Principes de Physique du Sol.* SEIA, Paris.

Hutter, W., *et al.* (1982) Etude du système de notations au labour sur un dispositif avec rotations. In: *La Transformation des Terres*. Ministère de l'Environnement, numéro spécial janvier 1982, Paris, 124–141.

Loiseau, P. (1982) Parcours et terres incultes de montagne: évolution de la valeur pastorale et processus d'évolution. In: *La Transformation des Terres*. Ministère de l'Environnement, numéro spécial janvier 1982, Paris, 201–211.

Mariotti, A. (1982) *Apports de la Géochimie Isotopique à la Connaissance du Cycle de l'Azote*. Thèse, Université de Paris VI.

Mérot, P. (1977) Le bocage et l'eau. *Penn ar bed, B rest*, **11**, 154–159.

Mérot, P. (1978) *Le Bocage en Région Granitique: une Approche de la Circulation des Eaux*. Thèse, Université de Rennes.

Missonier, J., *et al.* (1982) Bocage et aménagement rural: conséquences de l'arasement des talus boisés dans les régions bocagères de Bretagne. In: *La Transformation des Terres*. Ministère de l'Environnement, numéro spécial janvier 1982, Paris, 224–246.

Molenaar, J. G., de (1980) *Bemesting, Waterhuishouding en Intensivering in de Landbouw en het Natuurlijk Milieu*. RIN rapport 80/6, Leersum.

Monnier, G., et Fies, J. C. (1982) Dégats occasionnés aux terrains agricoles lors de l'implantation de pylones porteurs de lignes électriques de transport. In: *La Transformation des Terres*. Ministère de l'Environnement, numéro spécial janvier 1982, Paris, 249–266.

Schepers, J. S., *et al.* (1984) Influence of irrigation and nitrogen fertilization on groundwater quality. *Proc. Int. Assoc. Hydrological Sciences*, Symposium HS2, Hamburg and Koblenz, FRG, 1983.

Unesco. *Land Subsidence*. IHP Report to be published.

6.9 NOTES IN THE TEXT

(1) The factors in the Hénin stability index (I_s) are: (i) the average percentage of stable aggregates (this percentage is the arithmetical average of three percentages of aggregates stable in water: without pre-treatment, after pre-treatment with alcohol, after pre-treatment with benzene); (ii) the dispersed 'clay and loam' fraction; and (iii) the 'rough sand' fraction. Then

$$I_s = \frac{\text{(ii)}}{\text{(i)} - 0.9 \times \text{(iii)}}$$

Stability is considered good when $\log_{10}(I_s)$ is less than 1.

(2) After Monnier: Document prepared for the SCOPE Project and to be published by the French Ministry of the Environment.

(3) Shanan uses the expression 'the twin menaces: waterlogging and salinity'—see Chapter 5.I.

Land Transformation in Agriculture
Edited by M. G. Wolman and F. G. A. Fournier

CHAPTER 7
Conclusions and Recommendations

M. G. WOLMAN AND F. G. A. FOURNIER

7.1 AN APPRAISAL

On a planet which is 70% ocean, but with a human population of 3600 million growing at 1.7% per annum land must be regarded as a limited resource. Although finite in quantity, it must provide the foundation for human nutrition and virtually all the space for other human activities. Mankind, although nominally part of the 'natural' environment, has nevertheless been able to transform both the surface cover of the land and the top few metres of the earth's crust itself. In part such changes reflect deliberate attempts to garner resources and in part are accidental impacts which are by-products of those processes. The rates of metamorphosis, the depths to which it penetrates, and the degrees of irreversibility induced are a result of the growing capacity of human societies, mediated by access to energy and technology, to bring about changes in their environment.

The ability to effect major alterations can be seen, in the long perspectives of history, to have had several turning points: the control of fire, the discovery of agriculture and development of the use of fossil fuels mark steps between episodes rather like the climbing of a staircase with wide treads—each plateau phase, though, marking higher intensity of land transformation than the one before. In this respect, the technological developments of the nineteenth century almost certainly outrank all those that went before, for steam transportation opened up the world to form a global system in which the growing of crops for export permitted a degree of specialization based on soils and climate which had been unknown in the earlier days of low volumes of international trade and of high degrees of local self-sufficiency. Perhaps the most significant transformation of the last century was the conversion of the natural grasslands of the mid-latitudes to a granary which nourished the burgeoning urban–industrial populations of the core areas of factory-based production of goods in Europe and North America.

In the nineteenth century and earlier, much land transformation was extensive; that is, agriculture often expanded into new or previously uncultivated lands. At the same time, where populations grew rapidly, or good agricultural land was limited, intensive transformations occurred. In contrast, this century has seen an intensification of agriculture in the developed world through mechanization and the use of fertilizers, crop breeding, and biocides based upon the development and application of new knowledge. In the developing world, similar intensification of agriculture is under way.

A growing world population, expanding urban and commercial use of land, and recognition of potential long-term and ecological changes associated with the intensification of agriculture, have all given rise to a concern that land transformations in agriculture in the modern world may markedly affect the capacity of agriculture to continue to sustain and enhance human society on the globe. It is these concerns that prompted the SCOPE effort reported in this volume.

History provides the backdrop against which to measure change, and to evaluate present trends. From the standpoint of human society, the historical review reminds us that land transformations in agriculture have been essential to human social and cultural development. Agriculture provides the basis for urbanization and civilization. Until this century, our use of the land was extensive. Farmers used hand tools, traditional varieties of crops, manure and not chemical fertilizers, and allowed the land to lie fallow periodically. Over time, new crop varieties and fertilizers were introduced, fallow periods shortened, and draught animals replaced manual labour. Even at these stages, it is important to note that forests were replaced by farmland throughout much of the world and, in humid temperate regions such as Europe, vast wet areas and swamps were transformed into cultivated land. In arid and semi-arid regions, irrigation transformed the landscape from desert to agriculture, creating systems maintained for centuries.

Until the dawn of the twentieth century, the only way to increase production was to increase the area of land under cultivation. Only in this century has technological progress allowed an increase in the speed of opening new lands to agriculture, and an increase in the production of surface units. These advances in technology coincided with an increased demand for agricultural products from a rapidly growing population.

The improvement of agricultural practices, the implementation of modern techniques, and mechanization, induced the opening of new arable lands to agriculture. Thus, projections in Chapter 2 suggest that between 1975 and 2000 the area of land under cultivation will increase by 20% (300 Mha), while the area of land under forest will decrease by almost 15% (from 4100 to 3500 Mha), as the area of land devoted to non-agricultural activities increases from 400 Mha to 600 Mha. Generalizations at the global scale mask vital differences between regions and countries. In some developing countries abundant reserves of land permit expansion of cultivation to increase production. In the industrialized countries and in many developing countries, all arable land is already occupied, and an increase in production can be obtained only by an intensification of agriculture on land that is already cultivated.

Because of the significance of intensification in agriculture, Chapter 5 and its sub-divisions deal specifically with agricultural practices intended to increase production per unit area in an effort to judge their impacts on land transformation. These practices involve changes in properties of the land and the soil. The effects are beneficial when potential productivity is maintained, or improved with limited negative impacts on the environment. However, some practices may result in harmful changes in the land, leading to deterioration of the soil and to losses in productivity. All deterioration is not necessarily permanent. However, over time some changes may become exceedingly difficult to reverse technologically, and may actually be irreversible owing to the high costs involved. An attempt is made here to summarize briefly some of the significant impacts of major agricultural practices and, in some cases, the prospects for reversibility. These conclusions draw upon the text, the case studies and some additional sources.

7.1.1 Irrigation

Over the last decades, perhaps the most important form of land transformation has been the expansion of irrigated land. In 1965, irrigated areas covered 149 million hectares. By 1975, irrigation was practised on 223 million hectares, and this figure is expected to rise to 273 million by 1990. It is estimated that irrigated lands, which presently occupy 15% of the world's total cultivated area, produce up to 30% of the global agricultural production. Over large areas of the world, irrigation has assured production by overcoming uncertainties due to unreliable or insufficient rainfall.

Though the rate of increase in irrigated land has been the highest during the last 50 years, irrigation is as old as recorded history. It should be stressed that irrigation has given rise to very stable production systems which endured for thousands of years, such as those in Mesopotamia, Egypt, China and Sri Lanka, and for comparatively long periods in Syria, Iran, and Indonesia. While some of these areas have been plagued with problems of drainage and salinization, under proper maintenance, constructive transformation of the land by irrigation has taken place over vast tracts of land for long periods.

In contrast, irrigation has sometimes resulted in transformations which, in the long run, have caused a deterioration of the environment. One such major change has been modification and disturbance of the local or regional salt and water balance, resulting in waterlogging and salinization. Prior to irrigation, a stable water balance is developed between annual rainfall, vegetation, surface water and groundwater. Irrigation may produce a threefold to tenfold increase in water input compared with natural conditions. With poor management, this large change in the water balance may lead to deleterious land transformations such as waterlogging, saline and alkaline soil development, and land subsidence. These effects may be disastrous. For example, it is estimated that more than 5 million of the 15 million hectares of irrigated land in the Indus Basin in Pakistan—that is, a third of the surface—are affected by secondary salinization of the soil and waterlogging. In Iraq, the proportion amounts to 50%.

Over most of these lands, changes in the salt and water balance are reversible. The main remedies are effective water control through optimum irrigation scheduling, reduction of water losses, drainage and effective groundwater management. Lessons learned from serious land transformation problems faced by earlier civilizations and progressive modern technology have made it possible to analyse the problems, predict land and water behaviour and design remedial measures. Such measures have been implemented in Pakistan and elsewhere. The task, however, is great and requires large investments. While no precise figures are available for the world as a whole, it is estimated that of the total irrigated area (223 million hectares in 1975), 45 million hectares are in need of rehabilitation and 78 million need to be drained in order to maintain a satisfactory salt and water balance.

A principal lesson to be learned from the history of irrigation is that governments must recognize these potential hazards early in the project planning stages and establish competent operational management structures prepared to heed the early warnings that signal deteriorating conditions. A number of well-established 'early warning' observations are described in Chapter 6. There are no perfect solutions and governments will always have to choose among policy options that trade off losses in production and farmer's income in order to achieve long-term stability and permanence. Good planning is essential to ensure survival of an irrigation project and the society dependent on it. In the past, emphasis in planning has been placed on the need to attain maximum yields per unit of land. Greater weight must be given to maximizing economic returns on water and minimizing long-term hazards. In carrying out the economic analysis of a project, the investments required to prevent long-term project failure must be included in the stream of costs, or the stream of benefits must be adjusted to take into account a probable reduction in yields as the hazards advance.

7.1.2 Mechanization

A controversy has spread between those who argue that only hand-tool and animal-draught technologies are appropriate levels of technology for developing countries, and those in favour of mechanical power. There is really no contest here. Farming based on hand-tool technology seldom exceeds a subsistence level, and the area that can be cultivated by a single family is limited to 2 ha. The lack of draught animals, for instance, in the lowlands of Africa is a major constraint to the spatial expansion of agricultural land. Feeding the world's population, including areas now at the subsistence level, will require an expanded technology. This need not imply a disregard for proven ways or appropriate methods adapted to farming at small scales. In places, hasty changes in mechanization and soil handling in recent years have placed increasing pressure on the environment, resulting in deleterious changes in soil structure which may prove difficult or costly to reverse. Nevertheless, mechanization is an integral part of the inputs needed to increase agricultural production in developing countries throughout the world. It accounts for about one-third of the commercial energy used in agriculture, second only to the energy used in making fertilizer. Although agricultural production itself uses only about 4.5% of total commercial energy in the developing countries, while the percentage remains constant, a fourfold increase will be needed to meet food demands in many of them.

Rural poverty is often the result of the low productivity of manual labour in agriculture. The objective then is to increase productivity and, in turn, to determine the role that mechanization can play in solving rural development problems. If farmers are to feed themselves and others, subsistence agriculture must gradually be transformed into commercial agriculture. Agricultural income must provide access to means of production and generate enough surplus to create demand for non-agricultural goods and services.

Though it is generally agreed that mechanization is an indispensable input to rural development, there is a lack of clearly defined strategies for agricultural mechanization. A balance is required between increasing production and the environmental hazards of non-adapted tillage practices. Conservation, management and appropriate rotations should be the basis of mechanized farming. A major priority for research is the definition of soil environmental requirements needed in diverse regions of the world to enhance and sustain the productivity of the soil, and the design of 'preventive' husbandry measures to achieve these objectives.

7.1.3 Soil Conservation

The need to expand land in agriculture and to intensify its cultivation has, in a number of instances, led to degradation of the soil. Agricultural land use need not in itself cause degradation of the land. Through experience and research, agricultural practices have been established which maintain stable systems of production. At the same time, mismanagement of the land, whether driven by necessity or exploitation, leads readily to deterioration of the resource base.

Although water erosion dominates the process of degradation, other types of degradation—such as wind erosion, salinization, alkalinization, acidification, compaction and soil sealing, and decrease of biological activity—may be important. All may be important, and all may lead to a temporary or permanent decrease in soil productivity.

Because data on erosion and its relation to productivity are severely limited in most countries, this SCOPE study did not attempt to assess the rate of degradation or loss of productivity for specific areas around the world. Similarly we cannot say which deterioration of the land area degraded each year is reversible: estimates vary widely from 500 to 6 million hectares. A recent study noted in Chapter 2 suggests that, in the developing world, unchecked soil erosion could reduce the area of potential rain-fed crop land by 18%, and sustainable potential rain-fed crop productivity by as much as 29% by the year 2000. The larger figure reflects the fact that some of the crop land is projected to be degraded from very productive to much less productive land. Over the 20-year period from 1980 to 2000, this would represent a potential loss of roughly 18 million hectares each year. It is implied that if proper action is taken, degradation by erosion is preventable, yet if land is degraded, the result will not be reversible if excessive erosion is allowed to proceed. In this and other analyses, however, two major problems are unresolved. The first is the relation of soil loss to erosion on agricultural fields, and the second, and most important, the relation of soil loss to decline in potential productivity.

As the case study of soil erosion in Iowa, USA, in this volume indicates (see Chapter 13), the pattern of erosion and deposition on agricultural lands, as well as the relationship between erosion and loss in productivity, are complex and not well established. Wide variability is to be expected. Regional and local data on erosion and productivity throughout the world are very limited and insufficient to establish a firm basis for estimating future productivity losses, or the reversibility of erosional effects, particularly on the more productive soils. Thus, this volume does not include a country-by-country assessment of the impact of land erosion as the data do not permit accurate assessment at that level. Only through such assessments in the future will it be possible to plan remedial measures and to estimate the investments required to maintain and restore productivity levels.

It is clear that erosion poses severe hazards in a large number of areas. Moreover, on steep marginal lands the effects are likely to be nearly irreversible. Perhaps most distressing is the fact that the last several decades have seen, at best, modest adoption of soil conservation techniques in many parts of the developing world where they are particularly needed. Population pressure, costs to farmers, and the uncertainty of benefits, along with the absence of political will, have limited adoption of conservation measures despite the availability of proven technologies in most instances. As Chapter 5 and its sub-divisions indicate, new approaches to conservation of soil and water based on watershed units have shown promise in Africa and elsewhere.

It is too early to judge whether this promise will be fulfilled on a scale commensurate with the problem. Experience suggests that far stronger inducements are needed than those employed thus far. Farm management must be based on techniques designed to improve productivity beyond low levels currently attained in many parts of the developing world.

Soil conservation must be an integral component of intensified agriculture—the choice of adequate conservation measures will need to be based on an assessment of the form and intensity of the degradation process, and the choice of management practices adapted to the environmental conditions, economic feasibility, and the social acceptability of the proposed control techniques. It is vital that land resources be protected. Yet it must be realized that farmers will do so only when they are given the motivation and the means to do so.

7.1.4 Fertilizers and Crop Protection

The use of inorganic fertilizers, particularly nitrogen, is the largest single factor responsible for increasing the realized and potential productivity of agriculture throughout the world. The introduction of fertilizers in agriculture has not only enhanced the productivity of many good soils, but has transformed many regions of inherently low productivity into agriculturally effective regions. Examples of this transformation include shifting from bush fallow to continuous cropping in tropical rain-forest, shifting from low-quality range on Cerrado soils of South America to effective crop land by correction of aluminum toxicity/phosphorus deficiency, and conversion of the desert to intensive agriculture as irrigation is introduced.

The application of fertilizer influences land transformation in two ways. First, judicious use of fertilizer improves the quality of good arable soils over time by increasing the organic return from crop residues, thus stimulating micro-biological activity. Second, increased productivity achieved with fertilizers on good arable lands reduces the need to crop lower-quality areas better left in permanent grass or forest cover.

If the world's population is to be fed in the future, increased productivity will be required on the better arable lands. Until recently, much of the world production of inorganic fertilizers has been used to increase agricultural production in the developed world. In the absence of opportunities for expanding crop land, much more extensive use of fertilizers is essential to increase agricultural productivity throughout the world. Organic systems of soil fertility maintenance are invaluable for transforming low-productivity lands to agricultural use and should be utilized to the extent possible. The symbiotic system involving silk, pigs and fish ponds in China described in Chapter 5.IV illustrates superbly the realization of such an opportunity. There are, however, severe limitations in exclusive reliance on organic systems. Most important, the quantity of organic wastes is sufficient to treat only a small fraction of all arable land. In addition, some wastes contain excessive salts or toxic components, and in dry farming regions, water for legumes to be grown as green manure is inadequate. For the latter, the combination of reduced tillage and added fertilizer can extend the potential area of cropped land in dry regions through moisture conservation and the use of inorganic nutrients beyond what is possible under traditional systems of land management.

Poor management and inappropriate use of fertilizers can have harmful effects on both land and water. For example, excessive fertilizer use can result in the pollution of surface waters and nitrate contamination of groundwater. Some of these transformations—for example, groundwater contamination below permeable soils—may be reversible in a relatively short period by careful management of fertilizer application. In contrast, eutrophication of surface waters may be difficult to reverse and require significant changes in land use practices.

Injudicious fertilizer use can result in soil nutrient imbalances that may or may not be readily reversible. Soil acidification induced especially by nitrogen fertilization, for example, can be compensated by liming, at modest cost, except where lime is unavailable. However, where acidification is allowed to proceed to the point where the subsoil becomes strongly acid, the resulting acidity and associated aluminum toxicity to crop production are virtually irreversible. Other examples where reversibility is difficult include: copper toxicity from over-use of copper fertilizers and Bordeaux mixture, and iron deficiency in neutral to alkaline soils induced by excessive application of phosphorus and zinc.

Technically, increasing use of fertilizers throughout the world is essential and can be done without producing significant harmful effects. Maintaining the balance between productivity increases and environmental quality again requires appropriate information about soils and plants, and careful management.

Like fertilizers, pesticides and other techniques for crop protection enhance productivity and lessen the need to extend crop land for agriculture. Over time, traditional agricultural practices (including cropping systems and rotations) evolved that rendered crops rather resistant to pest attack. The occurrence of pest epidemics as a result of particular weather patterns, or the introduction of new pest organisms, from time to time, upset the system and caused catastrophies, but in general these systems were reasonably stable. Gradually, pesticides were developed that allowed better control and have, in particular, reduced the significance of agricultural pest epidemics. Originally, pesticides were mainly extracted from plants or were of inorganic chemical origin, but since the 1940s synthetic organic pesticides have become the major control measures. These new tools have permitted considerable

changes in agricultural practices. But crops more susceptib
under intensified agriculture and crop losses continue to be
losses from pests do continue to take a heavy toll of proc
billion tonnes each year—so their reduction can make a ma
world food resources.

Despite the successful application of pesticides, their widespread use has also revealed a number of problems, including the emergence of new pest problems, the development of pesticide-resistance, and pollution. Recognition of these shortcomings is leading to the development of new products, and new approaches to pest management. Reduction of use and changing the pesticide may alleviate the development of resistant strains, and new compounds that degrade more rapidly may lessen potential environmental impacts. Moreover, breeding of pest-resistant varieties of plants provides opportunities to reduce losses from pests without sacrificing the productivity gains associated with increasing use of fertilizers and irrigation water, both of which also increase susceptibility to pests. It is now recognized that maintaining long-term productivity requires integrating a variety of techniques to manage pests. This involves manipulation of crops and management of agricultural operations along with appropriate use of pesticides.

Plant protection and herbicide use considered together represent techniques for enhancing productivity. Because of their chemical composition, both pose potential problems for land and water pollution. Through accumulation in the soil, and alteration of soil biota, both may result in transformation of the land. Because herbicides have provided a means of weed control without cultivation, they have also permitted development of conservation tillage (sometimes called 'no-till' agriculture) over large areas in some regions. Such tillage constitutes a significant transformation of land use and significantly reduces runoff and erosion, thereby reducing pollution of surface waters. Concern has been expressed that increased herbicide use and reduced runoff would lead to pollution of groundwater, but this has not yet been the case.

Additions to the soil essential to modern agriculture all pose real, or theoretical, prospects of creating deleterious transformations of the land, potentially leading to irreversible changes inhibiting the ability to sustain agricultural production into the future. Such changes are documented in this volume, and a few enumerated here. Nevertheless, both history and current practice indicate that the majority of such potential negative impacts can be controlled at the outset by proper management, or reversed with reasonable effort after careful analysis, often using known techniques.

7.2 RECOMMENDATIONS

Land transformation is the name given here to a dynamic process in which human beings alter the surface and sub-surface of the landscape, creating positive and negative impacts, some reversible, others irreversible. The process involves physical, chemical and biological changes affecting land, water and air over time. It has, of course, involved major changes in social organization as well. For millions of years, human beings have so altered the land both intentionally and inadvertently. The drive to do so is inherent in human cultural and economic activities.

Land transformation in the modern world is characterized by the rapidity and scale of change. Rapid change and large scale require planning and the careful use of knowledge to mitigate the undesirable effects of positive actions, and to reduce the negative influence which some transformations may pose on options for future uses. Planning, in turn, requires an assessment of the potential of various lands for different uses with different levels of inputs. The concept of 'suitability' describes such evaluations, provided it is recognized that suitability may be a function of current cultural and technological opportunities. As these change, so suitability may change. Moreover, the biological and physical consequences of different agricultural activities are often subject to prediction and processes may be manipulated to mitigate unwanted effects.

Four criteria are particularly relevant to an evaluation of the prospects of land transformation in agriculture. Initially, productivity, or potential productivity, may be the primary focus. Over a long time scale, three additional criteria must be considered: sustainability, reversibility and fragility. None of these is easy to define without qualifications, but each has been touched upon in the text and case studies and emphasized in this concluding chapter. Virtually all transformations in agriculture have affected the sustainability, reversibility and fragility of some lands in some places. Irreversible effects have been associated with major irrigation projects throughout human history and from the exploitation of lands subject to rapid erosion from heavy rainfall. Fragility simply expresses the degree to which a system is buffered by natural processes which absorb man-made impacts. Under traditional agricultural practices, alterations of the natural system by human beings were limited. Through centuries cropping patterns and rotations developed sustainable agriculture adapted to local conditions of soil and water in many temperate and humid regions. The scale and rate of modern change raise the spector of exceeding the buffering capacity of the land and soil, reducing the sustainability of agricultural enterprise, and degrading land and soil to irreversible (or nearly irreversible) states.

The review here suggests that while irreversible results from the intensification of agriculture throughout the world are clearly possible, the increase in agricultural productivity needed to supply food and to improve the living standards of the world's population can be met by the creation of a sustainable agricultural base without irreversible damage to the land. Thousands of hectares of irrigated land are subject to irreversible degradation through waterlogging and salinization, and permanent loss of productivity from erosion is conceivable. Nevertheless, proper land evaluation and management—coupled with the knowledge needed to reverse some current trends—can provide more than enough produce from a sustainable world agriculture. Moving from world, to region, to country, to local scale, the prospects become increasingly grim. Some lands are naturally poor for agriculture, and others have been degraded so as to make the process of reversing the damage expensive and/or beyond the capacity of those who dwell on them. Thus locally and at the country level, such land cannot sustain present, let alone growing, populations. There is, of course, no reason why each country or region must be self-supporting in agriculture. Since many lack the resources to buy food, realistic appraisal is grim for those who must, for some time, make do with the environment in which they find themselves.

On a country-by-country basis, insufficient information exists with which to quantify the degree to which erosion has produced irreversible effects that permanently reduce potential productivity. Critical studies are needed to evaluate the relationship of land erosion to potential productivity in a variety of different environments throughout the world. However, the evidence indicates that change for the worse is occurring in the very areas where the need for improvement is the greatest.

Irreversible effects are not solely confined to irrigation and erosion. Transformation of range lands to thorny shrub lands is often irreversible. In more humid areas, reversible transformations from crop land or grassland to shrub are currently under way in a number of developed regions where agricultural villages have now become bedroom communities and the landscape has reverted to shrub or forest. The degree of reversibility of spontaneous or induced transformations of the land is a function of cost, technology, and in some instances social organization.

A central role in both planning and developing land transformations in agriculture must be given to observation and monitoring of change. For the surface itself, remote sensing provides a new and powerful tool for measuring—and in some cases evaluating—changes not only in land use but also in land change and the environment. As Chapter 6 illustrated, specific criteria involving properties and processes on the land and soil have been developed which can be used to monitor direct and indirect effects of changes in land use. These range from measures of chemical constituents and organic material in the soil, to more complex measures of soil structure and hydrological behaviour. A hierarchy of observations and analysis is needed: an objective set of data, temporal or spatial correlations of cause and effect in processes, and ultimately a thorough knowledge of the specific physical, chemical and biological processes relating agricultural practices and resulting effects. Given the complexity of climatic and pedologic relationships around the world and the varieties of agriculture, much remains to be known about how specific processes operate under varying natural and agricultural conditions. In many humid tropical regions, little basic data is available and few observations extend over significant periods of time.

Actions will be taken and change will occur with inadequate knowledge and before predictions about possible consequences can be reliably drawn. What is needed is not a presumption of infallible planning, but, in the absence of research, pragmatic mechanisms that permit the process of trial and error to be observed and recorded, with the results fed back to adjust operations. The management of pesticides provides an excellent illustration of a complex process that demands practice and experience, with continual modification designed in response to new knowledge.

Throughout this volume the emphasis has been upon a set of physical, chemical and biological phenomena. However, not only are the pressures which lead to land transformation generated by society, but the solutions to present and potential dilemmas are dominated by social and behavioural phenomena and not primarily by science and technology. The major social pressures leading to intensification of agriculture and to transformations of the land stem from the need to feed existing and new populations and the desire to raise the living standards of both. Traditional agriculture cannot fulfil this need. Whether or not the resultant transformations do lead to sustainable agriculture is a matter of choice—degradation is not fore-ordained.

The litany of phrases, 'the potential exists to sustain a high level of agriculture while maintaining environmental quality', 'degrading processes on the land are in many instances reversible', 'appropriate evaluation of land capability, and proper management, can conserve land and increase productivity', are sober, accurate and reasonable statements; yet they are difficult to achieve.

While farmers make up 30–40% of the world's population, and perhaps 75% of the population in the developing world, like any other sector of society they alone do not control their own destiny. Farmers, urban dwellers, governments, private individuals, commerce and industry, and innumerable international agencies in pursuit of innumerable objectives, create the environment to which, and within which, farmers respond. Thus the course of land transformation in agriculture is not solely the responsibility of farmers, and often not within their control. At the same time, it is farmers who decide what to plant, how to practice their trade, and what to do with their products. Thus, despite the complexity of the cultural, economic and social environment, for land to be 'appropriately managed' and degrading practices controlled, farmers must be given incentives to enlist in the effort. This task has proven to be far more difficult than the acquisition of the scientific information and know-how which contribute to an optimistic view of the results of the rapid land transformations in agriculture currently under way.

Section II
Case Studies

Land Transformation in Agriculture
Edited by M. G. Wolman and F. G. A. Fournier

CHAPTER 8
Land Transformation in Israel

D. H. K. AMIRAN

8.1 INTRODUCTION

The countries around the Mediterranean have experienced far-reaching changes in land quality and land use. The centres of very early, highly developed civilizations of the western world were here. Together with their great achievements they brought about intensive use of the land—sometimes excessive use.

The political changes that affected the area during a millennia of its history involved similar changes, for better or for worse, in husbanding of the land. Extensive areas of olive groves—famous since antiquity—fruit orchards and vineyards, many of them grown on carefully terraced slopes, indicate but some of the prominent achievements of mediterranean agriculture. By contrast, periods of decline, being sometimes periods of depopulation, brought about very adverse effects, in particular abandonment of cultivated land and disrepair of the terraces. In consequence there was erosion of the soil, and downwash into the valleys of the lowland plains; there was excessive sedimentation in these valleys, choking the river and wadi beds, and sometimes followed by formation of swamps owing to lack of drainage; and not infrequently malaria affected these swampy areas. This whole syndrome of land deterioration is illustrated in many mediterranean countries.

A classic case is Israel of the nineteenth century. Israel showed some significant achievements in agriculture and land development in antiquity, and the deterioration is easy to follow in detail. Two factors bring about particularly significant effects of processes of land transformation. One factor is the country's dense population and its high rate of urbanization. The population density, exclusive of the Beersheva district, the largest area of which comprises the Negev Desert, was 517 per square kilometre by the end of 1980, and 86.7% were classified as urban (90.4% of the Jewish population). Demands for transformation of non-urban land to urban land uses are therefore particularly pressing.

The second factor is the great geographical diversity of Israel. Its elements are plains and mountains, the latter a veritable mosaic of geomorphic units, upon which are superimposed two different climatic areas—the mediterranean climate in the north and the arid climate of the Negev Desert in the south. The Judean Desert on the east flank of the mountains and the Rift Valley of the Jordan owe their aridity to their topographical location in the rain-shadow of the mountains. This diversity intensifiies the effects of land transformation, especially if the changes occur near the border of aridity, being the transition from mediterranean to desert climate, an area particularly sensitive to change.

This diversity of landscape, the sensitivity to changes near the border of aridity, and especially the long and varied history of the land of Israel, acted upon by a variety of civilizations, and its extensive documentation, make Israel a particularly apt topic for a study in land transformation.

8.2 LAND TRANSFORMATION IN ANCIENT TIMES

Although evidence for land transformations in ancient times is often of a general nature only, it is sufficient to illustrate some major trends. There is evidence to show that the natural vegetation of the uplands of Israel, from Galilee in the north to beyond Hebron in the Judean hills in the south, were forests in the natural state (Gradmann, 1934, especially map 1; Zohary, 1955; 1970). They were covered by a mediterranean mixed forest. One of the major land transformations of Israel in antiquity was the removal of this upland forest, presumably to obtain building materials and fuel, possibly in part by setting fire to it for purposes of hunting, or even to clear the ground for settlement. One less 'legitimate' reason for deforestation was the need for materials for military purposes, especially siege machinery. Historical sources have it that the Roman command, when besieging Jerusalem in 70 AD, had to send troops to provide logs and wood for their siege works as far as 90 stadia (18 km) from the city, as all the area nearer to Jerusalem had already been deforested long ago.[1]

It is assumed (Reifenberg, 1955) that this deforestation of the uplands resulted in a standard series of consequences: erosion of much of the topsoil, which in most areas was of the terra rossa type, elsewhere rendzina; excessive sedimentation in the adjacent coastal plain, resulting in waterlogging in the valleys; local creation of swamps, which eventually became foci of malaria; which, becoming endemic, weakened the population and its capacity for work, and sometimes led to desertification of villages, and even to abandonment of cultivation of land. There were at certain times secondary factors fostering abandonment. The coastal plain of Israel was an inter-regional highway of traffic between Syria in the north and Egypt in the south-west. It was used by, amongst others, many armies and marauding bands which plundered and extracted tribute or forced labour. Nomadic tribes from the deserts around Israel contributed their share to this development. They sent occasional or regular raiding parties (*ghazzus*) to obtain livestock and protection money—a kind of forerunner of modern mafias.[2]

As a result of all this the Romans had to apply drainage works to swamp areas of the intrinsically fertile coastal plain, at least on the southern Carmel coast (Nir, 1959) and at the Poleg valley south of Netanya. These drainage works fell subsequently into disrepair and considerable areas turned into swamps infested by malaria. This situation persisted into the nineteenth century.

However, long before this—apparently as early as the Second Temple period (the second half of the first millennium BC)—a vast project of improvement in land transformation was undertaken. This was a project rather improbable from the point of view of the twentieth century. Instead of reclaiming the plains land, which would have required drainage technology and conditions of security that apparently were not available to the population of those times, the rocky upland hillsides were transformed into agricultural land. Agriculture, therefore, centred on the uplands—much of it karstic, where soil was very shallow, if soil indeed had remained at all—instead of in the lowland plains where there were relatively deep soils. This land transformation involved a tremendous investment of manual labour to create agricultural terraces, which in certain areas covered more than 50% of the total land surface (Ron, 1966). The construction of terraces did benefit from a roughly horizontal stratification of the prevalent limestone rock. They were relatively narrow and accumulated a cover of terra rossa soil, on which traditionally olives and some fruit trees were grown, with cereals (mainly wheat) below them. The size and topography of the terraces excluded any mechanized agriculture, and do so even today in the limited areas where ancient terraces are still cultivated. A mule or donkey is the only means used for ploughing beyond manual labour.[3]

Obviously, to terrace the uplands of Israel, especially the Judean Mountains, would be impossible and unreasonable under present socio-economic conditions: the standard of living, and therefore the cost of labour, would prevent both their installation and their cultivation. The philosophy behind these large areas of agricultural terraces, in Israel, in countries around the Mediterranean at large, and in many countries of the Far East (Sorre and Sion, 1934, pp. 41–42 and 46–48; Spencer and Hale, 1961) is based on minimal cost and a nearly unlimited availability of manual labour—about the opposite of conditions obtaining in the late twentieth century, at least in developed countries. That in many countries terraces were the prevailing form of land use illustrates not only the basic socio-economic difference between the past and the present, but no less the well-known fact that before the industrial revolution there were but few alternatives to employment in agriculture.

The active period of this upland transformation into terraces and their full cultivation seems to have continued up to about the Arab conquest (seventh century AD). Throughout these times and until the nineteenth century the uplands were the important region in Israel's agriculture, and therefore its rural settlement; the lowlands and their plains with their rich natural resources for agriculture ranked a poor second, or were barely cultivated. In the second millennium AD most of the terraces were gradually abandoned and deteriorated by disrepair. Today, only a few of these terraces are still cultivated, mainly by Arab villages marginal to the mainstream of modernization.

Coincident in time, but not in space, with the creative stage of land transformation in the uplands there occurred an extraordinary development in desert agriculture in the central Negev in the south. It was initiated apparently by the Nabateans in the first century BC and might have persisted up to the Arab conquest in the seventh century. Considering both the area to which it was applied and the agricultural technology employed, it was exceptional. Both 'Nabatean agriculture' and mediterranean terrace cultivation had in their days a wide regional distribution; but unlike upland terracing, Nabatean-type agriculture had a special intellectual quality, based on a detailed understanding of the mechanism of desert nature (Evenari *et al*., 1971; Kedar, 1967; Amiran and Kedar, 1959; Yair, 1986).

The Nabatean introduction of agriculture into this fully arid environment was apparently subsidiary to their main activity, which was trading along caravan routes. On some of the slopes fruit trees were grown, apparently, and possibly vineyards. But more important, in order to utilize the rare and highly irregular rainfall, they installed their fields in the wadi beds, with proper channelling of the rainfall from the slopes to each individual field. The slopes were specially treated to channel the runoff, and apparently to facilitate washdown of soft soil elements as well, which would accumulate in the wadi bed as substratum for cultivation.

In view of the rarity and sparseness of rainfall, there was an extraordinarily large ratio between the drainage area on the slope and the individual field on the valley bottom to which it provided runoff water—between 17:1 and 30:1, with an average of about 20:1. Owing to the strong floods which sweep the major valleys, Nabatean agriculture was restricted mainly to secondary wadis. The Nabateans, furthermore, devised a special type of cistern, adopted to the conditions of the prevailing types of rock, which passed the incoming runoff water through a strainer compartment, reduced evaporation, and was fitted with stairs to draw the water at successively lower levels without stirring up the bottom deposit.

By all these means the Nabateans managed to live in a desert which today has an average annual rainfall of 80–90 mm.[4] With the cessation of their caravan trade, their agriculture and their towns[5] were abandoned and never revived (except today at an experimental farm at Avdat).

8.3 RECLAMATION AND REASSESSMENT OF THE LOWLANDS

The type of extensive agriculture characterized by the slope terraces of the uplands persisted with little improvement until the late nineteenth century. It was only then that the next decisive step in land transformation and improvement of land use was taken.

Not bound by tradition nor by resignation to the run-down state of the natural resource base of agriculture, the leaders of the Zionist settlers who returned to their ancestral homeland made an unbiased and modern reappraisal of the conditions of the land. They appreciated the agricultural potential of the soils of the coastal plain, provided they were properly drained, which in turn would permit the eradication of malaria.[6] They understood the need for irrigation under the climatic conditions of the inhabited northern part of Israel.[7] Its mediterranean-type climate had too unreliable a rainfall on which to base a stable agricultural society. Furthermore, the potential advantages of the warm summer in a climate that has exclusive winter rains could be utilized only under irrigation.

Consequently, projects of drainage and reclamation of the plains were initiated—a lengthy, difficult and costly process, as was gradual development of groundwater resources for irrigation. The latter development, which began in the 1880s, culminated in the 1950s and 1960s with the inauguration and eventually full operation of the National Water Carrier.

All this resulted in one of the most far-reaching revolutions in land transformation in Israel. The lowlands—fertile and productive regions in antiquity—which had more and more deteriorated over time, and which by the nineteenth century were a rundown and partly abandoned area, became again in the twentieth century the prime agricultural region in which the economic core areas of Israel developed. Drainage and land works reclaimed the coastal plain, including the Haifa Bay between Haifa and Acco (Acre), which had been swampy and highly malarious owing to the obstructed drainage of the Na'aman River. Also reclaimed were the inland basins—the large Yizre'el Basin (Plain of Esdraelon) and the adjacent basins of Harod and Beth Shan to the east, sections of the Middle Jordan Valley north of here towards Lake Kinneret (Lake Tiberias or the Ṣea of Galilee), and finally the large and potentially very fertile Hule Basin in the northern part of the Jordan Rift Valley.

The inland basins had a variety of soils with a considerable percentage of heavy alluvial soils, in part the remnants of the swamps. Cultivation here, therefore, was mixed, including field crops such as cereals, fodder crops and vegetables, and later (since the 1960s) cotton. In contrast the coastal plain has a narrow stretch of heavy alluvial soils in the east at the foot of the mountains, but most of it consisted of light sandy soils, partly the weathering product of former stretches of coastal dunes.

Table 8.1 Citrus production and exports

Planted areas (ha)	
1922	29 000
1930	106 500
1948/49	125 000
1961/62	360 000
1979/80	396 000
Production (t)	
1948/49	272 700
1961/62	532 500
1979/80	1 542 700
Exports ($ million)	
1949	18.0 (63%)
1960	48.2 (22.25%)
1970	94.1 (12.1%)
1980	233.6 (4.2%)

Sources: Statistical Abstract of Palestine 1937–38, and Statistical Abstract of Israel for relevant years.

These light sandy soils, which are rather permeable and poor in mineral content, had been considered a poor agricultural resource in the past. They were but barely used for agriculture, and much of them carried a secondary forest of spontaneous growth in which the oak (mainly *Quercus ithaburensis*) prevailed. The last sizeable remnants of this forest were cut down to fire Turkish railway engines in the First World War. The availability of irrigation put a premium on these light soils. They permitted a vast extension of citrus plantations, which by the turn of the century gave Israel's major export crop and which retain their primacy among agricultural exports to the present (Table 8.1). Irrigation was at first from local groundwater wells, and is today from the National Water Carrier. The modern development of water resources permitted the southward extension of citrus acreage and of other crops. Today, the Gaza area 52 km south of Tel Aviv is one secondary centre of citrus growing (see Figure 8.1).

Citrus is but one element in Israel's agriculture. As agricultural economics,

---

Figure 8.1 The development of areas of citrus orchards and built-up areas in Israel between 1922 and 1967. Information was obtained from the *Atlas of Israel* and survey maps, and the map was kindly drawn by Mr N. Z. Baer of the Department of Geography of the Hebrew University of Jerusalem

(1) Area of citrus orchards in 1922
(2) Area of citrus orchards in 1937
(3) Area of citrus orchards in 1967
(4) Urban built-up area in 1922
(5) Urban built-up area in 1937
(6) Urban built-up area in 1967

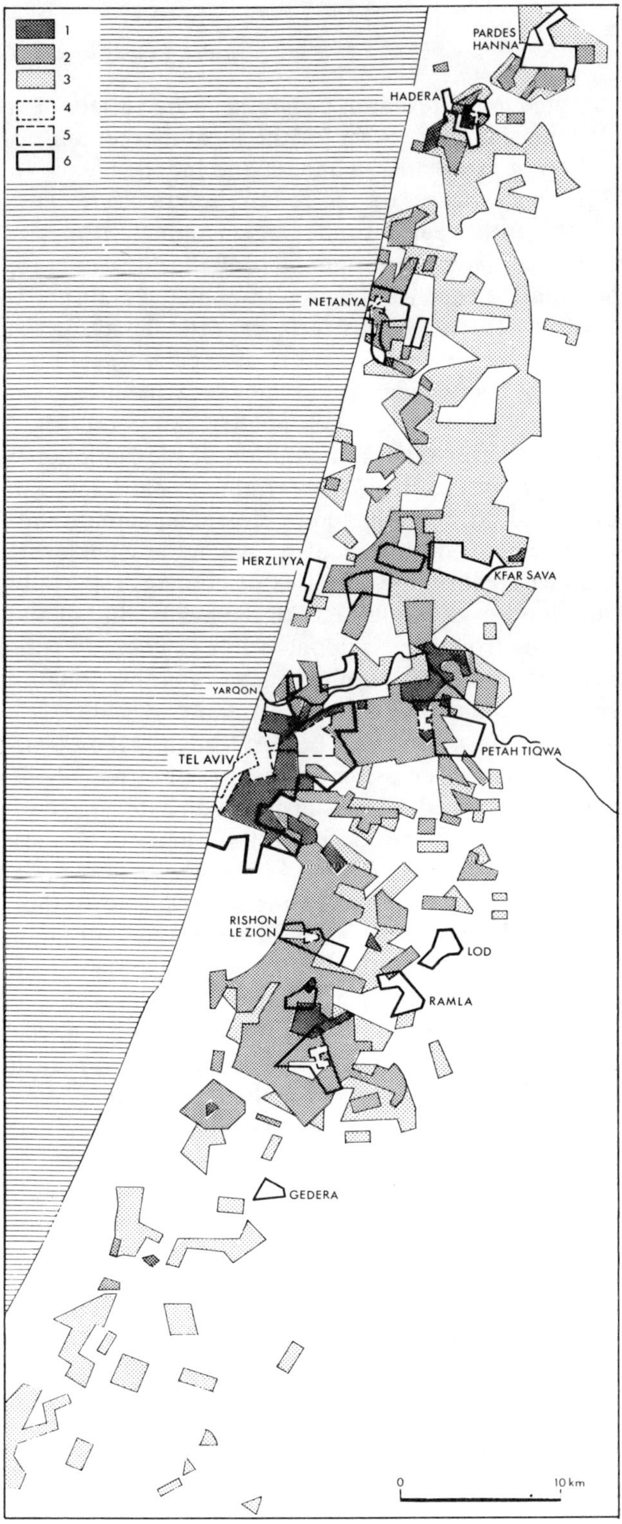
1
2
3
4
5
6
PARDES HANNA
HADERA
NETANYA
HERZLIYYA
KFAR SAVA
YARQON
TEL AVIV
PETAH TIQWA
RISHON LE ZION
LOD
RAMLA
GEDERA
0
10 km

genetics and agrotechnical research progressed, the variety of crops grown in Israel expanded considerably, both for the local market and for export.

The termination of the British mandate and independence for the State of Israel in 1948 put high stresses on the country's agriculture. The great wave of immigration in the early years of independence made it more than hard for the Israeli farmer to satisfy the demands of the local market. But by the mid-1950s the situation eased and agricultural planners turned their attention again on 'industrial' crops and on farm exports. Therefore, in addition to cereals for the local market, and a variety of vegetables and fruit, 'industrial' crops such as cotton, peanuts and sometimes sugar beets were now grown, as well as cattle (mainly dairy) and much poultry and pond fish.[8]

In the early years, before the inauguration of the National Water Carrier, farm practice had progressed from dry farming to partial irrigation with what was known as 'mixed farming', including livestock, cereals and fodder crops, vegetables and possibly fruit as well, all grown in the same farm unit. This was a kind of insurance against the vagaries of climate, especially the seasonal amount and distribution of rainfall.[9] With the availability of an expanded and reliable water supply from the National Water Carrier, agriculture started to specialize and focus on a rather limited range of crops, benefitting both from the advantage of specialization and the economy of scale.[10]

Much later, when after 1967 the Lower Jordan Valley south of the Beth Shan Basin again became accessible to the Israeli farmer, here too a radical land transformation occurred. This is an area with a distinctly arid climate, the result of its location in the rain-shadow of the uplands of Samaria and Judea a thousand metres and more below them.[11] For generations prior to 1967, this hot and sticky area was unavailable as agricultural land to the fellaheen and their simple agricultural technology (there were some very minor exceptions based on mediocre springs at the outlet of Wadi Far'a). Here the same regional advantages for out-of-season agriculture prevailed as the Israeli farmer had learned to use in the Negev; water was provided by systematic drilling on the fault borders of the Rift Valley, and an intensive agriculture has developed in the Jordan Valley since the late 1960s. It uses as a soil base the alluvial cover spread by the large and small valleys coming down from the uplands over the saline marls and underlying rocks of the rift bottom. In this respect the Jordan Valley has a definite advantage over the Negev, where sterile *hamadas* are the standard cover of the area and arable soil occurs in strictly limited locations. Incidentally, under essentially similar conditions, the Ghor project has been developed across the river in Jordan.

It is important to note that present-day Israeli farmers were not the first to engage in agriculture in this potentially fertile region. In antiquity, farmers used a sophisticated technology widely used throughout the Middle East; they installed kanats—underground tunnels for conducting water—in order to cultivate areas in the Phasaël region at the outlet of the Far'a Valley. These fell into disrepair, and cultivation ceased here apparently more than a millennium ago (Cressey, 1958; Evenari *et al*., 1971, pp. 172–178; Porat, 1970; Ilan, 1973; Troll, 1963).

The Lower Jordan Valley is thus another striking example illustrating that a certain region with a distinct agricultural potential was utilized in antiquity by farmers to whom kanat technology was available. It reverted to anecumene for well over a thousand years when backward farmers employing simple, 'traditional' methods of farming had neither the means nor the initiative to cultivate it. In recent years this same area—apparently even larger in size than in antiquity—forms a flourishing agricultural district thanks to research that has developed a technically advanced agriculture.

There is one specialized aspect of land transformation that deserves special mention—*fish ponds*. Apart from the insignificant amounts of sea fish landed by Arab fishermen at Jaffa and Acco, there was no supply of fresh fish in Israel. In the mid-1930s a few fish ponds were installed in the Kurdaneh swamps in the Haifa Bay using the saline water of the nearby springs. At the time carp of near-marketable size were imported from ponds in Yugoslavia and sold in Israel when they reached proper size. This was a profitable business, as carp for 'gefillte fish' commands a good market for the Jewish shabbat and holy-day table. The Kurdaneh enterprise, however, was an ephemeral project.

In the mid-1950s fish-ponding was revived, this time at first in the Beth Shan Basin. When the water resources of the basin were being developed, some large saline springs had to be reckoned with. Their water, as was that of the Kurdaneh springs mentioned above, was too saline to be used in agriculture. Part of it was mixed with good-quality water to an admissible level of salinity and was used for irrigation. The more saline water was used for fish ponds. In this way two sub-marginal resources were utilized. A few years later fish ponds were integrated into the agricultural layout of the Huleh Basin on completion of its drainage: surplus water was available here and used in the ponds.

Pond-fisheries soon appeared to be one of the most remunerative branches of Israeli agriculture, owing to the low input of labour. Fish ponds, therefore, appeared in most of the plain areas of northern Israel, as far south as the Hadera region of the coastal plain.[12] In the coastal plain, ponds were even installed on dune sand quite near to the beach. The sand was made impervious after a short time by lining the pond with chicken manure. Poultry manure, in fact, was a major element of feeding. Poultry farming and fish breeding, therefore, were often connected at the village (or at least at the regional) level. Fish breeding remained a preferred branch of agriculture for many years and receded only recently (Table 8.2). Unlike the early venture at Kurdaneh and its successor in the Beth Shan Basin, most fish ponds now use regular water from the National Water Carrier, augmented by rainwater. As

Table 8.2 Fish ponds

	Area (ha)	Volume of production (t)
1948/49	1 500	2 509
1969/70	5 400	12 000
1978/79	4 100	13 900

Source: *Statistical Abstract of Israel* for relevant years.

this involves high water losses by evaporation, the Water Commissioner is not in favour of expanding pond acreage, and actually attempts to restrict it.

The striking progress of lowland agriculture in various regions of Israel as treated above had negative repercussions on land use and the agricultural economy of the uplands. It has been mentioned that most of the upland terraces were gradually abandoned and deteriorated by disrepair throughout the second millennium AD. It is a telling expression of the state of deterioration of Israel prior to Zionist reclamation that, irrespective of this deterioration of upland areas, the uplands were still the most important agricultural region and the (economically speaking) 'leading' part of the country, if this term is applicable in the present context; and the region was definitely the most densely populated area until very late in the nineteenth century. All this is a striking measure of the decline and deterioration of the area over many centuries, which came to an end less than a hundred years ago.

The striking advances of lowland agriculture, and the standard of living they allow, today make upland farming on terraces a very poor second best. This applies not only to the resulting standard of living for the farmer, but also to the cost of the product, and therefore its ability to compete in the market. When labour was still relatively inexpensive, terrace agriculture was maintained in the Arab sector, and even slightly expanded beyond the nineteenth-century acreage. Similarly, a number of Jewish villages cultivated upland areas, mainly on terraced land, some of it improved and modernized. Today, when labour is nearly the most expensive element in agriculture, there is further regression of upland agriculture, the progressive element of which concentrates on specialized crops for which there is a regional advantage in the uplands, and to a large extent in the development of recreation as a regional resource, for which the uplands have a definite climatic advantage. These include rest houses, and even hotels, restaurants, picnic areas, recreational forests, swimming pools, sports grounds, and the supporting agricultural production. All this is in addition to generally small—but often highly specialized—industrial plants, which in many upland villages are today the most important branch in the village economy (particularly in kibbutsim).

To sum up, by the second quarter of the twentieth century the century-old pattern of land use in Israel had been radically transformed. The plains were again the country's principal agricultural area, in a modern agricultural technology and economy. The uplands, which for many centuries had been the important agricultural region and the important region of settlement, reverted to second rank in regional status, with the resort industry and an increasing residential land use by a commuter population assuming gradually more importance as the major regional resource base (see section 8.6).

8.4 PRIMARY RESOURCE BASES FOR AGRICULTURE (OTHER THAN SOIL)

The *mise-en-valeur* of land for agriculture in Israel until the 1950s had proceeded on a conventional pattern, expanding by a series of agro-technical improvements. The completion of the National Water Carrier in the 1960s, which led to both a much greater rentability and reliability of farming,[13] and therefore created considerable capital resources, permitted the natural resource base to be used in a new and unconventional way. This involved a number of phases, each one with special regional emphasis.

A sober appraisal of the possibilities created by the difference between the winter climates of the northern Negev (and better still the Arava section of the Rift Valley further to the south) and that of the European market, and to a lesser degree that of the populated north of Israel, brought about the *mise-en-valeur* of these areas which, until about 1950, had been considered entirely unsuited to agricultural use. Parts of the desert south of Israel with a fully arid climate were transformed into intensively cultivated agricultural land. The key to this development was the inter-regional differences in winter climates (Table 8.3). Even in mid-winter, the average daily maximum temperatures are high enough to permit the harvesting of vegetables and fruits, as well as flowers which were for a time part of the region's exports. The desert climate, which by its very nature has minimal and very infrequent rainfall,[14] therefore has minimal cloudiness as well. This even gives it an inter-regional advantage over the north of Israel, with its mediterranean climate and winter rains.

For this resource base for 'out-of-season' agriculture, suitable soils and water had to be identified, capital for basic investment provided and the marketing structure organized. By definition, neither water nor soil for agriculture are usually available in a desert; but here the special geographical conditions of the Arava as a section of the Rift Valley become a positive asset. Groundwater could be located by drilling at certain places on the fault borders of the rift; it could then be piped to the area to be cultivated. The soil component is available at rather special locations, different from the norm in Israel. Most of the Negev desert is upland plateaux, at 400–1000 m elevation,

Table 8.3 Average daily temperatures (°C)

Average daily temperature						Average daily maximum temperature						Average daily minimum temperature					
Oct	Nov	Dec	Jan	Feb	Mar	Oct	Nov	Dec	Jan	Feb	Mar	Oct	Nov	Dec	Jan	Feb	Mar
Hazeva, Arava, Israel (30°48′ north at −140 m) 1967																	
25.9	20.1	16.4	13.7	14.5	16.6	31.4	25.7	22.3	19.5	19.8	22.6	20.4	14.7	10.5	7.8	9.2	10.6
														7.4[a]	2.0[a]	5.5[a]	
Tel Aviv, Ben Gurion International Airport, Israel (40 m) 1951–60																	
21.9	17.9	14.6	12.7	13.0	14.3	29.2	24.6	20.4	18.5	19.1	21.0	14.6	11.2	8.7	6.9	6.9	7.6
Köln, FR Germany (45 m) 1931–44 and 1951–52																	
10.4	6.4	2.6	1.9	2.7	6.0	14.1	8.9	4.8	4.4	5.6	10.1	6.7	3.8	0.3	−0.7	−0.3	1.8
St Albans, Hertfordshire, England (83 m) 1941–70																	
10.5	6.3	3.9	2.9	3.3	5.5	14.5	9.4	6.9	6.8	6.4	9.6	6.5	3.1	1.0	0.0	1.2	1.4

[a] Absolute minimum figures.

Sources: Hazeva and Tel Aviv—Israel Meteorological Service. Köln—*Tables of Temperature, Relative Humidity, Precipitation and Sunshine for the World. Part III—Europe and the Azores*. London, Meteorological Office, HMSO, 1972, p. 46. London—*Averages of Temperature for the United Kingdom 1941–70*. London, Meteorological Office, HMSO, 1976, p. 16.

covered with *hamada* surfaces. They drain to the Rift Valley in the east, where the Arava is at between 100 and 300 m below sea level.[15] On the rare occasions when the wadis are in flood (these floods often being rather violent), the steep drop down to the Rift Valley followed by a sharp easing of gradient causes heavy sedimentation in the valley beds in the Arava. The sandy elements of these sediments form the soil base for the modern agriculture of the region. Unfortunately the areas so cultivated are located in the wadi beds and thereby subject to the same floods which deposited the soil. Arava agriculture is therefore high-risk farming, exposed to flood risk and occasional killing frost, in addition to the normal range of risks, natural and economic, under which all farmers operate.

The north-western Negev, less arid than the Arava, was developed slightly later, when water for irrigation became available from the National Water Carrier. As this was available to the individual farmer in strictly controlled amounts,[10] and in view of the high rates of evaporation in the region, greenhouse farming in a strictly controlled environment became prevalent, thereby reducing the amount of water applied; the water was further re-used by recirculation throughout the agricultural season. As a result, an average family greenhouse farm required 2000–3000 m^3 of water each year, compared with the 10–15 thousand required by a 'normal' farm with open-field agriculture. The north-western Negev, like the Arava, produces vegetables and fruit for the European market and for the major inhabited northern half of Israel.

Greenhouse agriculture detaches itself from soil as a basic natural resource. The greenhouse not only farms a minimal area (under the conditions of the north-western Negev, one-tenth to one-fifteenth of the area of an open-field farm); it does not even farm the local soil, but a sand layer mulched with peat, which is normally imported (mostly from Ireland). Here Israeli agriculture has come full-circle: from traditional dry-farming for which soil was the decisive resource base, to greenhouse farming which transports soil and water in minimal amounts to areas with the commercially most advantageous climate as the decisive resource base.[16]

Israeli farmers did have ample experience with greenhouse farming prior to its widespread introduction into the north-western Negev. Again, agro-climatic considerations had been decisive. The crop grown in greenhouses on a considerable scale were flowers for the export market. It was due to individual initiative that intensive flower farming, mainly of roses, on a commercial scale was introduced at Shachar Village in the Lachish region in the earlier 1960s. The location of the area is not far from the border of aridity,[17] but is still well within the area of mediterranean climate. Greenhouse crops benefitted from the moderate cloudiness and associated high temperatures of the area, while the cooler temperatures at night and during stretches of cool weather were countered by heating the greenhouses whenever necessary. This permitted—exceptional years excluded—the roses to ripen for the Christmas market in Europe. Farmers in the Lachish region, and those of other areas in Israel who followed suit, use the local soil for greenhouse farming, with some added mulch, not sand trucked to the greenhouse.

8.5 IMPLICATIONS OF INTENSIVE APPLICATION OF CHEMICALS ON THE RESOURCE BASE OF AGRICULTURE

Greenhouse farming, north-western Negev style, with sand as soil had the additional advantage of being very suited to agriculture with an intensive input of chemicals. One should remember that until the early twentieth century agriculture in Israel was not only *dry*-farming but farming almost without fertilizers. Not much livestock was kept in the farm unit, and the way it was kept did not bring about the accumulation of manure. In the days prior to the introduction of trucks and of mechanized transportation in general, camels were still kept in great numbers, roaming over meagre pastures and dispersing their droppings.[18] Following the near-total deforestation of the country in antiquity, charcoal and dwarf bushes (especially *poterium spinosum*[19]), as well as dried manure, were used for heating and cooking. To this day, the Beduin use dried camel droppings in open fires in the preparation of coffee and ceremonial meals.

After the First World War modern agriculture, with Jewish farmers in the lead, applied proper fertilization with manure, both of cattle and poultry, as well as of sheep where kept. Livestock rearing was thus an essential component of 'mixed farming'.

The application of farm manure is, however, rather labour consuming, so by the 1960s the application of chemical fertilizers became prevalent. They had the additional advantages of being more pleasant to handle, they could be mechanically applied, and in their selection and combination could be adapted to the specific requirements of soil and crop. The intensive use of chemical fertilizers and soil conditioners supplied the soil with elements to replace those withdrawn by the crops. Chemical fertilizers therefore permitted the intensification of land use by speeding crop rotation, or even dispensing with it. The farmer could cultivate the same field with the same crop, season after season, without intervening fallow. An outstanding example is cotton, which was known as being rather demanding on the soil. For many years Israeli farmers have grown successive crops of cotton, year after year, without any apparent damage to the soil, owing to adequate fertilization.[20] At the same time Israeli agriculture adopted pesticides; these were even more convenient for mechanized application, often being sprayed from small aircraft.[21]

Nevertheless, the intensive use of chemical fertilizers and pesticides places a premium on light and permeable soils. The heavier and more impervious a soil is, the more chemical residues it will retain and accumulate over time. The ideal soil for this type of 'chemical agriculture' is a highly permeable one, light and chemically 'neutral'. Pure sand fits these specifications best. 'Chemical agriculture' therefore places a premium on sandy soils, especially pure dune sands of the coastal zone and even more of inland areas.[22] It was for this reason that the large areas of dune sand of the north-eastern Sinai[23] and the adjacent areas of the north-western Negev assumed prime rank for the most modern agriculture, exceeded only by greenhouses operating on allochthonous pure sand.

Chemical agriculture, however, involves another environmental risk, which became evident in Israel in recent years. Irrigation and the rains in winter wash remnants of chemical fertilizers and pesticides into the ground and bring about their enrichment in the groundwater. As groundwater is used for supplying the National Water Carrier, over the years this has become a matter of concern. In particular, nitrate concentrations in the groundwater of certain areas of the coastal plain did reach near-critical values (Shuval, 1980), especially for those sectors of the population with limited tolerance (mainly babies).

A move away from chemical pesticides and fertilizers towards biological methods of pest control and organic fertilization seems due. Should this materialize, it would have implications for land use and land transformation. Whereas recent, most intensive, types of agriculture, with their preference for specific types of light soils, rotated certain lands out of use, and thereby tended to bring about a slight decrease in the total area cultivated, biological pest control and organic fertilization should bring about a renewed expansion of cultivated land.

The need for a very careful and comprehensive appraisal, and for periodic reappraisals, of the environmental implications of land transformation and of development projects is illustrated by the drainage of the Huleh swamps (Karmon, 1960). The Huleh Basin is the northern-most part of the Rift Valley in Israel between the Kfar Gil'adi area in the north and former Lake Huleh in the south. This small and shallow lake (maximum depth barely more than 5 m) was dammed behind a basalt barrier of Neogene–Quaternary age.[24] Upstream, i.e. to the north of the lake, there was a transition to a larger swamp area. This apparently formed only in relatively recent times as a result of the obstruction of the drainage of the Jordan river created by the inadequate opening of a bridge built *ca* 1260. The swamp was certainly not in existence as late as the first century BC when a flourishing settlement existed at Tell Anafa, recently excavated and located in the midst of what later were the swamps.[25]

The drainage of the Huleh swamps had been an objective of Zionist settlement policy since very early this century. This was the area with the richest surface water resources in Israel. It was obvious that a large amount of water, a precious resource in Israel, was wasted by evaporation from the swamps. Underneath the swamp was deep alluvial soil, rich in organic content, and under this rather thick deposits of peat. The swamp made the Huleh not only unusable for agriculture, but over the centuries it became infested with malaria; and after the drainage of the Kabbara swamps near Hadera in the coastal plain in the 1880s, the Huleh remained the most malarious area of Israel.

It was only in the 1950s that the drainage of the Huleh could be realized, and both the swamps and the shallow Lake Huleh to the south of them were drained and disappeared from the map. (It is ironical that one of the reasons for the drainage project had disappeared in the meantime, when spraying with DDT by the British Army during the later years of the Second World War had eradicated malaria.) On completion of the drainage the Huleh Basin came under agriculture, including some areas of fish ponds (discussed in section 8.3). The water draining off its rich organic soils enriched Lake Kinneret (Lake Tiberias or the Sea of Galilee), 280 m lower and 20 km to the south, with large amounts of phosphorus and nitrate minerals to which the effluents of pesticides and other chemicals used by the farmers were added. This changed the limnologic, trophic balance of the Kinneret, the major intake of the National Water Carrier. For a while, eutrophication of Lake Kinneret seemed threatening and some people had second thoughts on the wisdom of having drained Huleh and eliminated the swamp, which had acted as a natural strainer, preventing too high a mineralization of the Jordan and Lake Kinneret. Some extremists even considered re-flooding the Huleh Basin.[26]

Today a more balanced and sober view prevails and the limnology of the Kinneret is under close control. But the drainage of the Huleh certainly should teach planners a lesson about the need for a most careful and comprehensive appreciation of the environmental consequences of development projects, even if it appears obvious that the project is justified.

8.6 URBANIZATION OF AGRICULTURAL LAND

Land transformations dealt with so far relate only to different types of agricultural land use, to *mise-en-valeur* of land not formerly cultivated, or to abandonment of cultivated land. However, in recent decades more than 90% of the Jewish population of Israel (86.7% of the total population) are urbanized (Table 8.4).

It is obvious that an urban population of this size places considerable demands on land, and by necessity on agricultural land, either land actually cultivated or land potentially suited for cultivation, or rather both. Most cities, and particularly large and fast-growing ones, developed originally as central places of prosperous agricultural areas. The very reason which made the cities

Table 8.4 Percentage of urban population in Israel

	Total	Jewish
1951	–	76.0
1955	70.8	76.4
1961[a]	77.9	84.6
1972[a]	85.3	90.4
1980	86.7	90.4

[a]Census figures.
Source: *Statistical Abstract of Israel* for relevant years.

grow and expand leads, therefore, to them demanding increasing amounts of the same land on which early growth depended. The larger the city, the more far-reaching tend to be the urban transformations of agricultural land. For obvious reasons, in competition for land between the city and agriculture the city prevails.

These general principles apply in Israel as they apply elsewhere. Transformation to urban land use is, therefore, the most extensive and the most serious in the Tel Aviv metropolitan area, inhabited by close to $1\frac{1}{2}$ million people, roughly one-third of the total population of Israel. A map of 1878 shows Jaffa, the ancient nucleus of Tel Aviv, in its setting of citrus orchards and vineyards (Sandel, 1980). Practically every hectare of them is today within the built-up area of the conurbation. The same applies to Haifa and to the other cities of Israel, although to a lesser degree.

As stated before, the state of the Tel Aviv metropolitan area refers not only to the largest item of transformation of agricultural land to urban use, but is the most serious as well. The reason is the transformation to built-up land of prime citrus orchards, producing for a hundred years Israel's first-ranking agricultural export crop. The same, although to a lesser degree, applies to other cities of the coastal plain. For the same reason, transformation to urban use is less serious from the point of view of the national economy in the case of the upland towns, such as Jerusalem, the Carmel sections of Haifa, Safed, Bethlehem, Nablus, or some of the New Towns like Karmiël or Ma'alot. Jerusalem today covers 50 times the area of the Old City, within the walls of which all Jerusalem was confined until 1860; but with one insignificant exception all of the New City is built on what in mandatory terminology was classified 'rocky ground'—some of it never cultivated before, the rest unused for agriculture for many centuries and unsuited to agricultural cultivation under present socio-economic conditions.

In recent years a new element of land use emerged in the uplands, viz. a sizeable number of non-agricultural settlements, most of them classified *rural*, a few of them towns. The majority of their gainfully employed population are commuters, working in the metropolitan areas or medium-size towns. The proximity and easy access of the Tel Aviv metropolitan area from places in Samaria, that of Haifa from Galilee, and Jerusalem from the Judean Highlands, make commuting convenient. By the end of 1981 there were some sixty such non-agricultural settlements, established in the preceding 10 years, including six towns.[27] Most of these 'rural settlements' were classified as *community settlements* (yishuv kehilati) or *industrial villages* (kafat, the abbreviation of Kfar ta'asiyati), adding to the variety of standard village types in Israel: *moshav*, the cooperative village; kibbuts, the communal village, all these on state-owned land;[28] and the *moshava* as well as the Arab *village* which are regular villages with private ownership of land and no in-built cooperation.[29]

Several reasons combined for this significant expansion of non-agricultural settlement into the upland districts of Israel. The high ratio of urbanization makes urban build-up a necessity. At the same time the adverse effects of crowded urban environments and of deterioration in many an older section of cities are a strong impetus to move to a better and cleaner physical and social environment. With rising standards of living they induce many a family to migrate from urban to rural areas. In this respect the uplands have several advantages over the lowland plains. Their summer climate is more agreeable than that of the warm and often humid plains. Considering rational land use, the rocky uplands do not compete with agricultural land. They are therefore cheaper and more easily obtainable for residential housing than the agricultural land of the plains. Finally, for a certain sector of population there are political reasons for moving to Samaria and Judea.

Not all the working population of the non-agricultural villages commute to places of work elsewhere. Apart from the *industrial villages* a number of the other rural places, even some of those that engage in agriculture, have one or more industrial plants providing some local employment. In most cases they are of small size, often quite specialized and sophisticated: electronics, fine mechanics and science-based industry are amongst branches represented.

To sum up, we find a whole range of motives for the spread of the population engaged in urban occupations into non-agricultural rural places and towns in the uplands, most of them commuting to work. These motives include: easy commuting accessibility; rational use of rocky land unsuited for agricultural use, which is therefore relatively cheap; an advantageous climate; a rural environment conducive to a better quality of life; and finally, at least in Samaria and Judea, motives of political ideology. This last motive is of particular practical importance. It applies not only to part of the population who moved into Samaria and Judea,[30] but to the present government of Israel (in power since 1977). In order to advance Jewish settlement in these areas, people who move there enjoy considerable advantages in mortgages, loans and often subsidized costs of construction and infrastructure. These economic advantages result in the migration into these upland places of many families whose motives are pragmatic rather than ideological.

The problematic aspects of transformation of agricultural land to urban land use, therefore, centre in the lowland plains, the agricultural region of relevance in the Israel of today. Here it is a matter of concern, and should be a matter of even graver concern to the national planner the responsibility of whom is planning for future development.

But the effects of expansive urbanization concern not only the urban areas proper. Cities and their urban population have a socio-economic spillover effect, the geographical expression of which is urban sprawl. Urbanites move as residents into villages, creating conflicts in land use with the activity of farmers (cf. Chatham and Griffin, 1958, especially p. 202). In addition, urbanites, especially the inhabitants of large cities, require considerable areas of the rural realm as recreational areas—for hiking, picnic areas, sports grounds and a variety of activities—all of which tend to clash with agricultural activity.

An additional essentially urban contender for rural land is industry, which for reasons of its own[31] either relocates from cities into the countryside or is established at rural places, sometimes by or for the rural population, in an age of diminishing employment in agriculture resulting from increasing mechanization. While agriculture mainly requires soil of good productive quality, towns require a certain area, in which good drainage and a relatively moderate topography are of advantage but not a *sine qua non*. Soil of productive quality is of no concern. Logic should, therefore, allot agriculturally productive soils to agricultural use and locate cities off such soils. For a great number of reasons, real-life conditions are different, and sometimes the exact opposite. As good agricultural land is limited in extent, this must be a matter of concern.

8.7 THE *SEVEN EARTHS* OF ISRAEL

The heading of this section borrows a phrase from the title of a classic paper by Edward Higbee (1952). Israel is an outstanding example illustrating that the same earth—the same natural resources—are put to different uses by different people, employing different technologies. Israel illustrates that land of great value to one population was entirely useless to another one, and was even considered anecumene, i.e. outside the area potentially cultivable. Different people employing different technologies have been using the land of Israel in seven different ways.

The *First Earth* apparently was the coastal plain and the inland basins, including locations in the Jordan[32] and Beth Shan Basins where very early agricultural settlements have been studied by archaeologists. Agriculture in these areas, apparently, was the regular type of cultivation—mainly dry-farming with occasional irrigation using the water of local springs. At least in the coastal plain, settlement and cultivation required occasional drainage to remove swamps. Drainage works were installed by Roman engineers in the first centuries AD.

In part concurrent with this stage, but agro-technically much more complicated and as far as the size of the enterprise is concerned much more ambitious, was the *mise-en-valeur* of the *Second Earth*. This was the transformation of the deforested and denuded upland slopes into agricultural terraces (discussed in section 8.2). This was part of a general mediterranean culture trait, as were the crops cultivated on the terraces: olive trees, other fruit trees such as figs or apricot, grapes, and cereals, mainly wheat. Installation of these terraces required an extraordinary input of manual labour; their maintenance required a well-functioning social structure to ensure proper maintenance of terrace walls and of drainage spillways. It is assumed that this terrace culture dates to the first millennium BC and to most of the first millennium AD. Evidence of ancient settlement proves that both the First and the Second Earth supported a rather well-to-do population.

Concurrent in part with these two cultures, but in an entirely different region of Israel, we find the *Third Earth*, the Nabatean desert agriculture (also discussed in section 8.2). In many ways this is intellectually the most fascinating 'Earth', highly unconventional and innovative. Whereas the First and Second Earths cultivated the normal cultivable land of Israel—its standard ecumene being the northern half of Israel with mediterranean climate where the semi-humid climate created natural soil and adequate winter rains permitted cultivation—the Third Earth was in the Negev desert. Except for oases, where water is provided locally by nature, the Negev, like all deserts, is considered anecumene, i.e. land not only uncultivated, but unsuited for agriculture owing to the absence of both soil and water, a consequence of the arid climate. With the exception of some occupation in prehistoric times—in the third millennium BC, which did not in any way approach the sophistication of Nabatean agriculture[33]—the central Negev was definitely anecumene after the demise of this culture until the 1950s, and concerning the core area of the Third Earth remains so to this day. The Nabatean settler was an admirable observer of nature and had an acute understanding of its mechanism. He chanelled water, and with it soil-forming elements of the ground, on to wadi beds, a location rightly avoided by farmers in other regions of Israel. He developed a technique by which the rare and extremely irregular rains of the desert soaked into the field and thus permitted crops to grow. Nabatean agriculture was not discontinued owing to a deterioration of nature nor of its resources, but to changes of a socio-economic nature which shifted regional preferences. This sophisticated agricultural management of the arid Negev is a part of a wider culture trait. Evidence of similar types of ancient agriculture is found in various areas of the Middle East, from Yemen to Cyrenaica in present-day Libya.

Whereas the First and especially the Second and Third Earths of Israel represent periods of prosperity and development, the *Fourth Earth* is the period of regression and of deterioration of the land resource discussed earlier. It began late in the first millennium AD and lasted to the late nineteenth century. For more than a thousand years the Third Earth reverted to desert. Farmers of the Fourth Earth made but poor and reduced use of the First and the Second Earths, both of which deteriorated sadly and to an increasing degree. It is highly significant that, at least during the later stages, farmers of the Fourth Earth had considerably reduced, or even abandoned, cultivation of the fertile First Earth, the plainlands, most of them having withdrawn to cultivating upland terraces. Even here, cultivation fell considerably short of that of the Second Earth, both by area cultivated and by deterioration of the terraces and their maintenance. In the absence of irrigation, heavy (mostly alluvial) soils or basaltic soils in the north-east were the preferred substratum for agriculture. These were the centuries of decay when the land of Israel turned into an impoverished, malaria-stricken country, suffering both from maladministration and from a backward population. At this stage the total cultivated area reached its minimum and the population declined. The under-use of the plains with their naturally fertile resource base is illustrated by the distribution of urban centres. In the nineteenth century six of the ten first-ranking towns, including the country's largest one, were located in the uplands. By the end of the British mandate (1948), four only of the ten first-ranking towns, including the second ranking one, were still upland towns.[34]

The pioneer settlers of the outgoing nineteenth century, and to an increasing degree their successors during the first half of the twentieth century, brought about decisive improvements, creating with enterprise and modern agricultural techniques the prosperous *Fifth Earth*. Large-scale projects reclaimed the land which had deteriorated throughout centuries. Irrigation was introduced at an ever-increasing scale, and for the first time modern mechanized farm implements became standard equipment. Consequently, and in contrast to the preceding stage, emphasis in cultivation reverted from the uplands to the plains. As irrigation expanded so did the extent of cultivated land, both in area—mainly into the southern parts of the coastal plain—and in type of soil, with the *mise-en-valeur* of the light sandy soils. Throughout this period farming diversified, expanded and consolidated.

The next stage, the *Sixth Earth*, dates to the second half of the twentieth century. In common with the Second Earth of terrace construction and cultivation in the uplands, but somewhat less than the Third Earth of the Nabatean farmers of the Negev, the Sixth Earth represents a highly unconventional and novel approach to land use and to agriculture (discussed in section 8.4). Its technique is highly mechanized and involves, therefore, a pronounced reduction in manpower employed. In contrast to all previous approaches, land now ranks as a secondary resource base. By now subsistence agriculture is a matter of the past. The market directs agricultural planning. The possibility of introducing out-of-season agriculture in the Negev in order to produce during the winter fresh products for the European market makes the climate—especially high temperatures coupled with clear skies—the first-ranking resource, and shifts regional preferences once more into the Negev, although into areas different from those of Nabatean agriculture. Soil is cultivated irrespective of the inherent risk of flooding, and water is piped to the optimal place of use rather than made use of where it occurs in nature.

Elsewhere in Israel, the National Water Carrier ranks water higher than soil as a resource in farming. It is used where each cubic metre gives the highest return. In extreme cases, soil too is used as an *artificial* resource, not *in loco*, but at the climatically optimal location in greenhouse farming. This again tends to give regional preference to the north-western Negev with its clear skies. In addition to advanced genetic research improving strains of crops, chemicals have largely taken over as fertilizers and pesticides, both for convenience of handling and in particular to intensify cultivation. This obviates, if so desired, the need for crop rotation. The intensive use of chemicals has placed a premium on light, permeable and chemically neutral soils which retain a minimum of the chemicals applied. Pure sandy soils, including dune areas formerly considered of no agricultural value, are now particularly in demand. Therefore, the coastal dunes came into the orbit of agriculture, as did the inland dune areas of the northern Negev. The increase of intensity of cultivation, more discriminating as regards the properties of the soil, restricts cultivation to the more desirable soils and reduces to a certain degree the total cultivated area. Yields and size of crop are no longer determined mainly by the size of area cultivated, but rather by modern agricultural technology, its chemicals and genetic improvements.

After the end of the First World War a very different type of land use and land transformation emerged, making gradually increasing demands for often potentially fertile and even cultivated land. This is the urbanized *Seventh Earth*. As about 90% of the population require this type of land use today, the tendency for and pressure of its expansion are considerable. To quote Higbee (1952): 'What is of value to urban man is tended to satisfy his wants and suit his tastes; the rest is neglected.' In view of the fact that much of the land involved in urban transformation is of prime agricultural value, this is a matter of considerable national concern.

8.8 REVERSIBLE AND IRREVERSIBLE CHANGES OF LAND USE

An attempt was made in section 8.7 to show how different populations employing different agro-technologies utilize the same land. We have seen times of improvement and of positive changes in land use alternating with times of deterioration, and even of desertification. The example of Israel might be important in showing that in a region, without any physical changes involved, deterioration in land use with negative effects on the resource base might follow periods of significant improvement and be followed in turn by another positive change, provided that the period of deterioration did not destroy the resource base, and thereby bring about irreversible changes.

In surveying the range of changes in land use in Israel, one can note many important but reversible changes in land use, and irreversible ones as well. Swamps in the lowlands and valley beds which were clogged have been drained, malaria was eradicated, upland slopes which were deforested and denuded were transformed into agricultural terraces; many of them were recently reforested and serve now as recreation areas. But examples are not missing on the debit side. Upland deforestation brought about soil erosion which can be but partly redressed by re-afforestation. In the conditions of Israel, soil of natural quality and depth does not reform, at least during the period of human planning. Overuse of the coastal plain aquifer has, in three areas,[35] brought about penetration of seawater into the groundwater—a change if not completely irreversible, at least irreversible for a prolonged period of time. The same would apply to the excessive enrichment of the coastal aquifer by nitrates, at least locally, should the suspicions of some experts materialize. This is a typical case of environmental overload, leading to severe economic consequences concerning acceptance by the market of the products grown under these conditions.

No negative environmental change bringing about worsening conditions has been *planned* intentionally. They are the unintentional results of incompetently or incompletely planned actions by man. As they result from intentional actions they are examples of *planning by default*.

The most recent, areally considerable and qualitatively serious, case of planning by default concerns urban expansion. No one intentionally brings about the expansion of the built-up area of cities, that of the urban fringe, nor of urban sprawl into the rural countryside on to valuable agricultural land. In a majority of cases the built-up area even succeeds land actually under agricultural cultivation. In the case of urban expansion, planning of the city and agricultural planning are the task of and are in the competence of different agencies, which not only base their activities on very different premises, but also operate at vastly different speeds, and generally at very different economic levels, and under very different socio-economic pressures. For obvious reasons, in a conflict of interest between city and agriculture, the city prevails, representing much more powerful interests, both economical and political. The result is therefore a special case of planning by default, created by an inbuilt mechanism of compartmentalized planning which lacks coordination to balance two very unequal contenders in order to safeguard the optimal long-term national interests.

The transformation of agricultural land to urban uses, though being of minor importance in terms of the areas involved, is today in many countries the most active process of land transformation. This certainly is the case in Israel. Of all the processes of land transformation this is by realistic appraisal the most irreversible one. In view of the global nature of present-day urbanization, all this applies *grosso modo* in most advanced or strongly developing countries anywhere.

8.9 REFERENCES

Amiran, D. H. K. (1960) Two types of border aridity in Palestine. *Compes rendues, 18th Int. Geograph. Congr.*, Vol. II, Rio de Janeiro, 461–465.

Amiran, D. H. K., and Kedar, Y. (1959) Techniques of ancient agriculture in the Negev of Israel. *Comptes rendues, 18th Int. Geograph. Congr. 1956*, Vol. I, Rio de Janeiro, 206–217.

Amiran, D. H. K., and Shahar, A. (1961) The towns of Israel: the principles of their urban geography. *Geograph. Rev.*, **51**, 348–369.

Atlas of Israel (1970). Elsevier, Jerusalem and Amsterdam.

Bar-Yosef, O. (1973) Early man in the Jordan Valley: the excavations at 'Ubeidiya. *Archaeology*, **28**, 30–37.

Bitan, A. (1974) Climatological aspects in locating settlements in arid regions. *Geoforum*, **20**, 39–48.

Bitan, A. (1982) The Jordan Valley project: a case study in climate and regional planning. *Energy & Buildings*, **4**, (1), 1–9.

Chatham, R. L., and Griffin, P. F. (1958) Urban impact on agriculture in Santa Clara Country, California. *Annals of the Ass. of Amer. Geographers*, **48**, 195–208.

Cressey, G. B. (1958) Qanats, Karez and Foggaras. *Geog. Rev.*, **48**, 27–44.

Eliav, M. (1977) The Rafah Approaches in the history of Jewish settlement. *Cathedra*, **3**, 117–208 (in Hebrew).

Evenari, M., Shanan, L., and Tadmor, N. (1971) *The Negev*. Harvard University Press, Camb., Mass.

Gradmann, R. (1934) *Die Steppen des Morgenlandes*. Engelhorn, Stuttgart (Geogr. Abh., 3R, H6).

Herbert, S. (1978) Tel Anafa. *Israel Exploration Jl*, **28**, 271–274.

Herbert, S. (1982) Tel Anafa—a further report. *Israel Exploration Jl*, **32**, 59–61.

Higbee, E. C. (1952) The Three Earths of New England. *Geog. Rev.*, **42**, 425–438.

Ilan, Z. (1973) *The Jordan Valley and the Samarian Desert*. Am Oved, Tel Aviv.

Karmon, Y. (1960) The drainage of the Huleh Swamps. *Geog. Rev.*, **50**, 169–193.

Kedar, Y. (1967) *The Ancient Agriculture in the Negev Mountains*. Bialik Intitute, Jerusalem (in Hebrew).

Nir, D. (1959) Artificial outlets of the Mount Carmel Valleys through the Coastal 'Kurkar' Ridge. *Israel Exploration Jl*, **9**, 46–54.

Porat, Y. (1970) Chain wells in the Jordan Valley. *Teva ve-Aretz*, **12**, 128–132 (in Hebrew).

Reifenberg, A. (1955) *The Struggle Between the Desert and the Sown*. Bialik Institute, Jerusalem.

Ron, Z. (1966) Agricultural terraces in the Judean Mountains. *Israel Exploration Jl*, **16**, 33–49 and 111–122.

Sandel, T. (1980) Karte der Umgebung von Jaffa, 1:31 000. *Zeit d. Deutsch. Palästina Vereins*, **2**, Tafel III. Reproduced in *Atlas of Israel*, op. cit., sheet XI/4, map B.

Shuval, H. I. (1980) *Water Quality Management Under Conditions of Scarcity: Israel as a Case Study*. Academic Press, New York (especially chap. 13.III).

Sorre, M., and Sion, J. (1934) *Géographie Universelle*, VII, Pt. I: *Méditerranée*. Armand Colin, Paris.

Spencer, J. E., and Hale, G. A. (1961) The origin, nature and distribution of agricultural terracing. *Pacific Viewpoint*, **2**, 1–40.

Statistical Abstract of Israel (1981) Vol. 32, tables XIII/2 (pp. 382–383) and XIII/4 (p. 385).

Troll, C. (1963) Qanat-Bewässerung in der Alten und Neuen Welt. *Mitteil., Oesterr. Geograph. Ges. Wien*, **105**, 313–330.

Yair, A. (1983) Hillslope hydrology water harvesting and areal distribution of some agricultural systems in the northern Negev Desert. *J. of Arid Environments*, **6**, 283–301.

Zohary, M. (1955) *Geobotany*. Merhavia, Pt. IV-b (in Hebrew).

Zohary, M. (1970) Vegetation of Israel and adjacent areas. In: *Atlas of Israel*, op. cit., sheet VI/1, map A.

8.10 NOTES IN THE TEXT

(1) Josephus Flavius: *de Bello Judaico*, V, paras. 522 and 523. I am indebted to D. Bahat of the Department of Antiquities, Israel, for this reference.

(2) Jerusalem, a walled city, was subjected to a Bediun raid as late as 1820 and the Turkish governor held for ransom. By the middle of the nineteenth century the large inland plains of Yezre'el (Esdraelon of the English Bible) and Beth Shan were frequently raided by Bediun from Trans-Jordan.

(3) A recent attempt to create broader terraces, connected by ramps to allow the movement of mechanical equipment from terrace to terrace, and to cultivate these by irrigation, was not overly successful.

(4) There is no indication that in Nabatean times rainfall was more ample than today.

(5) Most of Nabatean settlements in the Central Negev were towns.

(6) Lands of the coastal plain were relatively readily available as the great majority of the population lived in the uplands, in distinct contrast to present-day conditions. They had abandoned the lowlands for being swampy and malarious, as well as for suffering from poor security. They, and no less the rich absentee landlords who owned much of the land, were only too glad to sell it to whomever made a worthwhile bid for it.

(7) There was little thought of settling the arid south at the time. The major exception was a project for settling the el-Arish area in Sinai; this was investigated by a mission of experts in 1903, who eventually submitted an adverse report cf. Eliav, 1977).

(8) Sugar cane was grown successfully in the Huleh Basin, but was discontinued as being uneconomic for its high water consumption. Poultry—chicken and turkey—are today the main source of animal protein in the Israeli diet.

(9) *Atlas of Israel*, op. cit., no. 1, sheet IV/2, for maps of inter-annual variation of rainfall; and sheet 13 in the forthcoming third edition.

(10) The amount of water available to every village, and beyond that to every agricultural unit, is determined annually in very strict allotments by the Water Commissioner, a senior official of the Ministry of Agriculture.

(11) These topo-climatic conditions make it for man physiologically harder than the climate of the Negev. See Amiran (1960) and Bitan (1974).

(12) See *Atlas of Israel*, op. cit., no. 1, sheet VIII/2, maps B and D.
(13) The latter is of critical advantage under the climatic conditions of Israel.
(14) 18–87 mm per year during a 10-year period at Hazeva in the Arava.
(15) The base level of erosion is the Dead Sea, the level of which has an elevation of approximately −400 m.
(16) A secondary advantage of the unconventional agriculture of the Negev is that it cultivates 'empty' areas not in use, as these lands could not be cultivated by agricultural technologies employed previously. On a different level, pest control is much easier in a controlled environment. Finally, the proximity of the farmhouse and the greenhouse—as it were, the *field*—enables the farmer's wife and children to work in the greenhouse for short periods whenever they can afford to do so.
(17) The average annual rainfall is slightly less than 400 mm.
(18) Until the First World War the camel was the main means of transport in Israel and throughout the Middle East. Vast numbers of camels were kept. This was the essential reason for the persistence and for the social prestige of the nomadic Beduin before the war. They not only provided camels for long-distance transportation; they also organized, serviced, guided and protected the caravans—and by mutual tacit agreement robbed them, or at least extorted protection money, if they did not hold travellers for ransom. The advent of motorized transportation—the bus and in particular the truck—made the camel obsolete, and with it the traditional way of life of the Beduin. After the First World War bediun society underwent a severe economic crisis and decrease in social status. Herds of camels decreased, both with the Bediun and the villagers, the fellaheen (Table 8.5). The Bediun soon adopted the modern way of life and integrated into its economy.

Table 8.5 Camels in Israel

1922	19 225
1932	23 317
1942/43	29 736
1958	11 000
1964	10 500
1971	10 000

Sources: *Statistical Abstract of Palestine*, 1937–38, and *Statistical Abstract of Israel* for relevant years.

(19) *Poterium spinosum* represents the last stage in deterioration of the climax vegetation of a mixed mediterranean forest. Until the 1940s it was collected in large loads, transported by women on their heads, to be used for domestic heating and especially to fire lime kilns, which demanded large amounts of the shrub.
(20) In 1979/80, 24% of the field crop acreage of Israel was in cotton. The total cotton acreage exceeded that of citrus by 57% (*Statistical Abstract of Israel*, 1981).
(21) A number of commercial firms operate a fleet of small aircraft for agricultural spraying operations.

(22) These are beyond the reach of salt spray which by aerial transportation reaches the dunes nearest to the coast.
(23) Here intensive farming went on for over a dozen years in the 'Yamit Region'. It ceased when the region was returned to Egypt under the peace settlement in April 1982.
(24) See *Atlas of Israel*, op. cit., no. 1, sheet III/1, Geological Map, by L. Picard.
(25) Herbert (1978) reports: 'Luxury products . . . abound in the Hellenistic levels . . . (and) reinforce this impression of wealth and conspicuous consumption in the Hellenistic period.' This is certainly a very different picture of the malarious swamp existing there for many centuries until 1950. Cf. Herbert's further report (1982). According to Karmon (1960, p. 173), settlement in the Huleh Basin continued uninterrupted until the thirteenth century.
(26) In view of the experience gained one should be more than careful when considering proposals to mine the apparently thick peat deposits of the Huleh Basin. This has been suggested as a possible energy source to fuel a power plant. Presumably, such mining operations would result in surface subsidence which, even if slight, would have serious consequences for the drainage of the basin.
(27) Two additional towns had been established in Galilee in the 1950s and 1960s. By 1980, Karmiël, the larger of the two, had a population of over 13 000.
(28) The few collective moshavim, *moshav shitufi*, adhere to the moshav principle in community structure and to that of the kibbuts in the operation of their economy.
(29) Some Arab villages retain to this day *mush'a* ownership, whereby village land is owned by all land-owning families in common. The individual lots are redistributed for cultivation to all property-right owners every 1–3 years to enable every family in its turn to benefit from cultivating fields more favourable for soil or location.
(30) It must be noted again that the same process of establishing non-agricultural, as well as agricultural, settlements progresses in recent years in the uplands of Galilee, including mountainous Upper Galilee, where no political motives are involved.
(31) The factors might be fiscal, such as taxation and cheap prices for land, or a more pleasant environment, availability of labour and sometimes their wage levels, accessibility, and other factors.
(32) The oldest settlement traced in Israel is the Ubeidiya site slightly south of Lake Kinneret in the Rift Valley, where man lived about one million years ago.
(33) In the present context, this term refers to successive periods as well, until Byzantine times, i.e. the time span from the first century BC to the later part of the first millennium AD.
(34) Ranking data are for 1875 and 1944 respectively. Cf. Amiran and Shahar (1961, especially p. 358).
(35) The Tel Aviv area, that of Gaza, and Emek Hefer near Hadera.

Land Transformation in Agriculture
Edited by M. G. Wolman and F. G. A. Fournier

CHAPTER 9
Influence of Large-scale Farming Methods on Soil Exploitation in Czechoslovakia

M. BARTOŠ

9.1 INTRODUCTION

Agriculture takes up the largest share of land in Czechoslovakia: it covers approximately 54% of the entire land area (together with forestry, it affects 91% of the country's territory). Hence, as the major landscape-forming factor, it bears a great responsibility for the utilization and protection of the environment for future generations.

The essence of agriculture is its biological nature, i.e. the utilization of the productive ability of animate organisms (plants and animals), which represent the basic means of production. The entire production process in crop cultivation and livestock farming must proceed in harmony with the laws of biology, which to date are still insufficiently known, and man still cannot always use them correctly to meet his needs. To an important extent the production of plants and animals is affected by further factors, particularly technology and chemicals, which have a supplementary function and cannot yet supersede the actual biological production process, the essence of which is the unity of the productive organism and the productive environment.

The negative effects of wholesale technologies in crop and livestock farming can be summed up as follows. At the present stage of maximum application of the scientific and technical revolution in all production spheres, agriculture, too, is making a marked effort to introduce, wherever possible, industrial forms of large-scale production, in line with the principles of integration, concentration and specialization of production in order to attain maximum economic advantage. In that sense, highly efficient machinery and the massive use of chemicals dominate in crop cultivation and livestock farming, including melioration and land reclamation. This, together with an intelligent and comprehensive harmonizing with the biological principles of production, has brought great success in all branches of farming, particularly initially. However, the rate of mechanization, use of chemicals and other factors in many ways outstripped the results of research and the possibility of a more thorough testing of findings in practical application. This is now leading to very serious worries over the disruption of the components of the natural and human environment, and in consequence also of the ecological balance of the entire biosphere. This is further aggravated by the impact of wholesale technologies used in industry, building and energy production, for example.

In some regions, the negative effects of mass-production techniques in agriculture and other production spheres are already reaching such disastrous proportions that they represent an imminent danger to the basic conditions of life and the independent existence of plants and animals, and thus, finally, also man and the entire biosphere. These damaging effects become manifest in the deteriorating quality of the basic natural components: water, air, nutrients, food and soil.

9.2 BRIEF HISTORICAL SURVEY OF FARMING

The Slav period of the country dates back roughly to the fifth century AD, when there was already a developed agriculture and a clan system. The fertile soil favoured population growth. Life concentrated in the clan villages, primarily in the plains cleared of forest in the river basins of the Elbe, Ohře, the lower Vltava, the lower reaches of the Berounka, Svratka and Dyje.

Towards the end of the tenth century, new villages began to form on the basis of neighbourly communities. At that time, the state of production forces in agriculture favoured individual farming: the primitive collective cooperation in the clan communities became a hindrance to economic initiative, and individual families left them to make use of new farming techniques. In the new communities, arable land and gardens were owned privately; woods, meadows and pasture land were owned jointly by the community.

In the ninth and tenth centuries, the annex-farming system was superseded by a fallowing system. Crop rotation was introduced: one field was laid fallow, the second was planted with winter crops, the third with spring crops. On the land belonging to a village, each farmer owned and tilled his own fields—the work was done concurrently by all—and the approach to one's field involved crossing the neighbour's land. The land was divided into as many lots as the village had farmers.

Woods were left standing only in localities where farming was difficult or impossible or where the soil was not fertile enough—on stony, dry, shallow or steep ground, in the hills or in the border mountains. Also, farmers avoided forests on shifting sands (the woods in the Elbe Basin), wetlands regularly flooded by the rivers, and the inaccessible swamps and swamp forests, such as existed, for example, in the South Bohemian Basin.

As late as the end of the twelfth century, the territory stretching from the Bohemian Forest down to Southern Moravia was covered by deep, swampy, virgin forest, flooded repeatedly each year by the waters of the river Lužnice. The greater part of the territory was covered with wetlands, moors, swamps and deep forests. Extensive land reclamation in these areas facilitated their use for farming and cultivation, and in the lowlands also for human settlements. The main role in this was played by the construction of a system of fish ponds which began at the beginning of the thirteenth century.

Under the feudal system, the settling of land formerly not farmed occurred in two ways: either by the growth of the rural population, or by a deliberate and controlled colonization, which started in the thirteenth century when foreign colonists came to the country. The colonists (primarily Germans) received better legal status than the Czech subjects. The Emphyteutic Law facilitated hereditary land ownership. The reverting of land to the Crown upon the owner's death was abolished.

The swift growth of towns, the development of trade, crafts and mining in the sixteenth century led to a rising demand for farm products. Agriculture could meet this only by extending the area under cultivation, primarily by further deforestation of the territory. At the time, roughly 45% of Bohemia was still covered by forests.

At the beginning of the seventeenth century, productivity and general farming standards were comparatively high and were not surpassed for another 200 years. For instance, the wheat yield in the most fertile regions amounted to 12 metric hundredweights per hectare.

Fruit and vine growing reached high standards and was extensive in climatically suitable regions—around Prague, at Mělník, Litoměřice and in South Moravia. Hops, too, were grown on almost all feudal estates, which also had their own breweries; but even in those times the typical hop-growing regions of today clearly emerged: Žatec, Louny, Rakovník, Ústí, etc.

The improved crop-rotation and crop-alternation systems, with constant cultivation and harvesting each year, exhausted the soil more than the old fallowing system had done. Fertilizing intensity was low. As a rule, only one-sixth of the fields were fertilized, i.e. manure was applied to a field only once in six years. In the first half of the nineteenth century, it was still usual to fertilize a field only once in six years, with 250–280 metric hundredweights of manure per hectare, i.e. equivalent to 42–47 hundredweights per hectare annually. The cultivation of arable land by the large estates of the day is shown on Table 9.1, which shows that the crop ratios remained virtually unchanged until the middle of the nineteenth century.

The yields of individual cereals in the period 1700–1850 are shown on Table 9.2: representative data are taken from the Třeboň Estate (South Bohemia) and the Lovosice Estate (North Bohemia–Elbe Basin). Taking into account the level of fertilization (if one can call it that) and of mechanization,

Table 9.1 Crop ratios on large estates (percentage of arable land)

Year	Cereals	Other; husks	Feed grain	Root crops	Fallow lying
1776–1800	63.3	2.3	4.6	2.0	27.8
1801–1820	64.8	2.7	6.0	2.0	23.5
1821–1840	65.8	2.8	7.6	2.9	20.9
1841–1850	60.2	4.1	13.2	3.1	19.4

these yields are remarkably good, which indicates the enormous amount of human labour which went into cultivation. On changing the corvée system and abolishing seerdom under Joseph II at the end of the eighteenth century, the feudal estates lost a substantial part of their labour and were forced to lease the land and entire farms to tenant farmers.

The crop rotation system of farming ceased to meet economic needs by the beginning of the nineteenth century and was substituted by a more intensive crop-alternation system. The period beginning with the middle of the nineteenth century is described as the 'agricultural technical revolution', because new technology (such as the steam engine, the threshing machine, the crusher, etc.) facilitated the introduction of new production methods. Until that time, unregulated water conditions in the soil led to waterlogging or sub-surface erosion of the land and prevented intensive farming. Therefore, land reclamation schemes were introduced, swamps and wetlands meliorated, rivers and streams regulated, and flood-dams were built along river banks.

These measures, carried out mainly in the beet-growing lowlands, led to rising crop production. This is evident from the increasing areas of root crops and feed-grains, which in Bohemia, for example, had amounted to only 8% in 1848, rose to 29.9% by 1875 and to 37.2% in 1937. The most marked change in the intensity of cultivation appears in the decrease of fallow fields from 21.6% to 4.8% in 1875, while by 1937 fallow-lying fields had disappeared entirely.

Table 9.2 Grain yields on large estates (tonnes per hectare)

Year	Třeboň Estate				Lovosice Estate			
	Wheat	Rye	Barley	Oats	Wheat	Rye	Barley	Oats
1701–1750	1.20	0.87	1.26	0.96	1.36	0.97	1.17	1.17
1751–1765	1.28	1.08	1.26	0.85	1.63	1.08	1.33	1.32
1776–1800	1.54	1.39	1.26	1.13	1.93	1.36	1.43	1.59
1801–1820	1.54	1.39	1.20	1.30	1.87	1.97	2.10	1.82
1821–1840	1.66	1.58	1.38	1.59	2.70	2.40	2.08	2.11
1841–1850	1.74	1.75	1.46	1.80	2.95	2.37	3.35	3.56

Table 9.3 Crop cultivation in Bohemia and Moravia (thousands of hectares)

	1790	1820	1841	1871	1883	1896	1910
Arable land	3 568	3 516	3 823	3 862	4 098	4 092	4 093
Gardens	78	including meadows			104	103	103
Perm. meadows	641	711	807	857	707	706	703
Pastures	627	619	658	715	423	418	412
Vineyards	32	32	17	24	13	13	12
Total farmland	4 846	4 938	5 305	5 458	5 345	5 332	5 323
Total woodland	1 974	2 286	2 282	2 254	2 291	2 294	2 307
Total non-farm land	1 111	707	344	219	295	305	301
Percentages							
Arable land	71.6	71.2	72.1	70.8	76.8	76.8	76.9
Gardens	1.6	–	–	–	1.9	1.9	1.9
Meadows	13.2	15.6	15.2	15.7	13.2	13.2	13.2
Pastures	12.9	12.5	12.4	13.1	7.9	7.9	7.8
Vineyards	0.7	0.7	0.3	0.4	0.2	0.2	0.2

The proportion of cereals fell from 66% in 1848 to 57% in 1937; the main decline took place in rye, oats and partly also in barley, whereas the wheat-growing area increased by 50%.

The intensity of cultivation can be seen best from the growing yields per hectare, which in the period 1800–1937 more than trebled.

An idea of the changes in cultivation areas that occurred in the farming landscape during the eighteenth and nineteenth centuries can be gained from the figures in Table 9.3. A more detailed survey of the exploitation of arable land is given on Table 9.4, which shows crop structures in Bohemia in the period 1848–1914.

Of course, changes in the crop structure affected the look of the open-farming landscape, but had a far less negative impact on it than the later eras at the end of the nineteenth century, and particularly in the first half of the

Table 9.4 Crop structure in Bohemia (percentages)

	1848	1871–75	1876–80	1881–85	1885–90	1891–95	1896–1900	1901–05	1906–10	1911–14
Cereals	66.0	61.2	61.2	61.5	61.7	61.0	61.2	61.3	59.8	59.9
Root crops	4.3	18.7	19.1	19.3	18.8	20.3	20.4	20.1	18.5	18.6
Feed-grain	3.7	11.2	11.5	11.6	12.0	12.2	12.4	12.8	15.7	15.9
Other	4.4	4.0	4.2	4.5	4.9	4.8	4.7	4.6	4.9	4.7
Fallow	21.6	4.9	4.0	3.1	2.6	1.7	1.3	1.2	1.1	0.9

Table 9.5 Crop yields in Bohemia (tonnes per hectare)

	1848	1870	1871–80	1881–90	1891–1900	1901–10	1911–14
Wheat	1.08		1.20	1.36	1.40	1.65	1.85
Rye	0.95		1.06	1.14	1.15	1.45	1.67
Barley	0.96		1.07	1.37	1.39	1.64	2.00
Oats	0.81		0.90	1.08	1.05	1.28	1.71
Cereals (total)	0.95		1.06	1.22	1.24	1.51	1.81
Potatoes	6.0–7.0		7.00	7.30	7.80	9.90	9.50
Sugar-beet	12.00	16.00	17.00	20.80	22.40	27.00	26.20
Peas		0.68	0.81	0.72	0.76	1.04	1.20
Lentils		0.61			0.76	0.79	0.85
Rape		0.83	1.07	1.02	1.08	1.24	1.34
Poppy		0.35			0.59	0.76	0.83

twentieth century, when unilateral agricultural exploitation made itself fully felt.

All measures carried out in farming had a single goal—to raise yields—which is evident from Table 9.5, showing the trend of yields per hectare in Bohemia in the period 1848–1914. The largest increase was for sugar-beet. As sugar-beet had proved the economically most viable crop, the greatest attention was devoted to it and it was cultivated on the best land. Also, the cultivation area increased substantially.

After the freeing of the peasants in 1848, land holdings were carved up, farms were parcelled, individual fields sold to other holders or shared out among inheritors. For instance, when farm land was being shared among several brothers and sisters, so as not to put anybody at a disadvantage concerning soil fertility, locality or distance, the original lots were sub-divided longitudinally. Long and narrow strips resulted, which were economically most unsuitable (South Moravia). Other factors, such as the building of railways, river regulation, industrial construction, the growth of towns and villages, contributed to the splitting up of land-holdings.

The outcome of these changes, taking place over a number of years, was many split and scattered lots, with a large number of small parcels of land in different, often widely scattered, locations belonging to one farm. Further, it resulted in odd-shaped fields.

Considerable changes in land-holding occurred in Czechoslovakia in the period of 1918–38, on the basis of the Land Reform Act of April 1919. In accordance with that Act, the land of the gentry and large estate-owners was confiscated, i.e. all arable land exceeding holdings of 150 ha and other land over and above 250 ha, totalling 4 081 000 ha.

Under the Allocation Act of 1920, only about 1 800 000 ha were shared out, as follows: 400 000 ha were taken by the state which set up state-farms; 600 000 ha went to the provinces and communities in the form of provincial estates; and 790 000 ha were allocated to small-holders and middling peasants.

The scattered land-holding determined the character of the farming landscape. Millions of parcels and strips of various shapes, which persisted until 1948, formed a colourful mosaic. This image of the typical Czech landscape, however, could not continue: it had to be transformed to comply with the new, collective type of farming.

9.3 COLLECTIVIZATION OF AGRICULTURE FROM 1945

After the Second World War, Czechoslovak agriculture underwent marked changes. These changes occurred during a historically short period—not only with regard to the introduction of new production methods, but primarily with regard to the introduction of a new organization of production relations. The socialization of agriculture and rural life was completed within a very few years, unequalled in other countries. The transformation of the Czechoslovak village into cooperative-based farming communities cannot be compared with any similar schemes in the world.

The National Assembly passed an Act concerning the setting-up of Unified Agricultural Cooperatives (JZD). The government concentrated all its attention on the socialist transformation of agriculture, on the transition from individual small-holding farming to cooperative farming on a large scale. That historic document was preceded by efforts made since the end of the Second World War to implement the Košice Government Program and new land reform, designed to eliminate all large estates and to allocate their land to farm-workers and small-holders.

The land allocation was carried out in cooperation with the peasants themselves, according to the proposals of the Farmers' Committees set up under the auspices of the local People's Committees. The 'Hradec Program'—calling for the carving up of estates of more than 50 ha—was directed against the village bourgeoisie and could be implemented fully only after 1948.

Through the land reform, small and middle size peasant farms acquired more than 1 600 000 ha of farm land and, concurrently, a state socialist sector emerged, which in 1948 farmed roughly 6% of all the land. The class composition in the villages changed: the number of farms with more than 20 ha fell to 26.3%, whereas in 1930 it had amounted to 35.1%. The originally founded Farming Cooperatives (JZD) were comparatively small units with areas of 200–300 ha of farm land, as is evident from Table 9.6.

The state farms, which immediately after the Second World War farmed only a small proportion of farm land, quickly expanded after the year 1948

Table 9.6 Development of farming cooperatives

Year	Number of JZD	Average area (ha)	Proportion of total farm land (%)
1949	28		4.3
1951	3 138	355	14.6
1952	5 848	353	29.7
1954	6 502	276	26.5
1956	8 061	263	30.8
1958	10 816	420	66.5
1962	7 912	550	59.6
1964	7 135	592	58.6
1966	6 328	616	55.6
1968	6 328	625	55.5
1970	6 270	638	56.0
1972	5 318	766	57.6
1974	3 099	1 343	59.0
1976	1 959	2 183	61.1
1978	1 813	2 340	60.8
1980	1 747	2 474	62.4

(Table 9.7). The state took over breed improvement and plant improvement farms, church estates, large farms and vacant farm land in the border regions. During the period of collectivization, they served as exemplars for the organization of large-scale agricultural production.

Farm land in less productive regions, primarily in the border hills, was given

Table 9.7 Development of state farms

Year	Number of SF	Average area (ha)	Proportion of total farm land (%)
1952	192	3.201	8.3
1954	188	3.857	10.0
1956	182	4.949	12.1
1958	164	6.090	13.5
1960	465	3.105	15.5
1962	359	3.096	16.4
1964	314	4.426	19.3
1966	346	4.265	20.6
1968	343	4.273	20.5
1970	336	4.264	20.0
1972	305	4.720	20.4
1974	290	4.952	20.3
1976	214	6.514	19.9
1978	191	6.864	20.1
1980	203	6.864	20.1

Table 9.8 Workers permanently occupied in agriculture

Year	Number of people	Index 1936 (%)
1939	3 277 000	100
1946	2 565 000	78
1950	1 922 000	59
1955	1 839 000	56
1960	1 386 000	42
1965	1 195 000	37
1970	1 183 000	36
1975	990 000	30
1980	894 000	27

Table 9.9 Hectares per worker in agriculture

Year	1936	1948	1950	1955	1960	1965	1970	1975	1980
Hectares	2.35	3.34	3.89	3.97	5.35	5.96	6.00	7.11	10.65

to the state farms. The great labour shortage in those regions resulted in their large-scale production being of a rather low-intensity character. The substantial and constant dwindling of labour in farming is shown in Table 9.8 (its causes are outside the framework of this study). Agriculture today employs only one-third of its pre-war manpower. Thus, the land area per worker has greatly increased, as shown on Tablc 9.9.

From the above brief survey it is obvious that increasing demands are placed on the farmland of Czechoslovakia. With a declining number of farm workers and an increasing population and rising nutritional standards, it is required to produce an ever-growing volume of staple foods.

9.4 TRANSITION TO LARGE-SCALE PRODUCTION IN CROP CULTIVATION

9.4.1 Dynamics of Land Exploitation (Table 9.10)

At 1 January 1980, the total land area of Czechoslovakia, which amounts to 12 788 094 ha, was divided up as follows:

Farm land	6 923 549 ha,	(54.14%)
Woodland	4 534 617 ha,	(35.46%)
Other	1 329 928 ha,	(10.40%)

The present condition of, and the future outlook for, the use of land is not too encouraging. Since 1945, agriculture has lost approximately 500 000 ha of farm land, 490 000 ha of which was arable. The problem of resettling the

Table 9.10 General dynamics of land exploitation (hectares)

	Farm land	Woodland	Other
1950	7 505 846	4 350 525	930 723
1955	7 417 000	4 371 094	1 002 640
1960	7 362 031	4 371 799	1 054 264
1965	7 187 866	4 445 160	1 228 429
1970	7 103 080	4 445 563	1 239 451
1975	7 042 142	4 477 505	1 268 447
1980	6 923 549	4 534 617	1 329 928

border regions after 1945 caused difficulties with ensuring the farming of land in those parts. On long, uncultivated land in the border regions, mostly at higher altitudes, meadows, pastures and even arable land turned, by natural succession, into woodland, which until recently caused considerable losses of farm land in those parts.

Table 9.11 Annual absolute losses of farm land and arable land (hectares)

Year	Farm land	Loss	Arable land	Loss
1948	7 748 215	–	5 218 000	–
1950	7 505 846	–	5 096 000	–
1953	7 215 403	–	5 007 000	–
1955	7 417 000	–	5 156 000	–
1959	7 388 577	26 546	5 153 231	9 959
1960	7 362 031	35 237	5 143 272	11 813
1961	7 326 794	30 881	5 131 459	11 021
1962	7 295 913	58 541	5 120 438	11 337
1963	7 237 462	24 764	5 109 101	2 441
1964	7 212 698	24 832	5 106 606	17 362
1965	7 187 866	27 676	5 089 298	17 671
1966	7 160 190	15 812	5 071 627	18 127
1967	7 144 378	12 384	5 053 500	14 745
1968	7 131 994	15 235	5 038 755	13 362
1969	7 116 759	13 679	5 025 393	14 838
1970	7 103 080	10 193	5 010 555	12 177
1971	7 092 887	16 289	4 998 378	8 946
1972	7 070 598	6 081	4 989 432	9 094
1973	7 070 517	10 186	4 980 338	13 815
1974	7 060 345	18 203	4 966 528	23 260
1975	7 042 142	38 169	4 943 268	35 272
1976	7 003 973	14 260	4 907 996	1 445
1977	6 989 712	13 681	4 906 552	8
1978	6 976 031	24 446	4 906 554	15 321
1979	6 950 585	27 986	4 891 223	33 923
1980	6 923 549			

Particularly ominous is the substantial loss of farm land since 1937, which by 1979 represented a total of 1.32 million hectares of farm land and 1 million hectares of arable land. A survey of farm and arable land losses in absolute figures per year is given on Table 9.11. From this table and Figure 9.1 it can be seen that, despite legislation to protect the agricultural land, its losses are still too high. Although an amended law on the protection of agricultural land came into force in 1976, in recent years losses have risen sharply, mainly by the conversion of fields into woodland and land serving other purposes, accounting for 74% of the losses. However, 'these losses of farm land reflect also the introduction of order into the exploitation of farm land, implemented by a process of physical stock-taking of farm land areas by all socialist farming organizations.

The transfer of agricultural and arable land to other purposes since 1948 is shown on Table 9.12. In the majority of cases, high-quality, flat, arable land is taken from agriculture, because building on it requires relatively small investment.

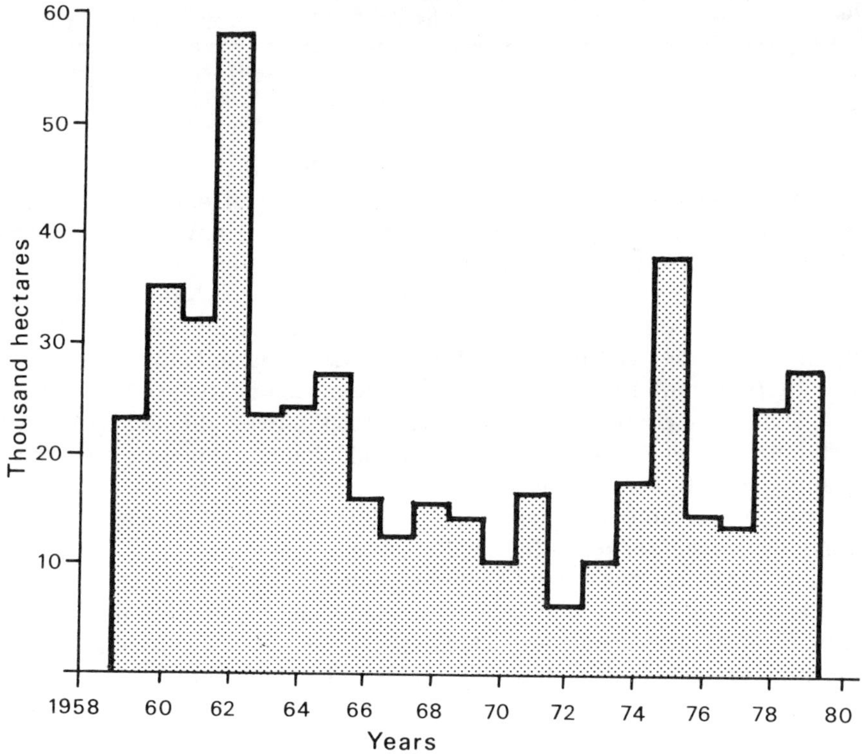

Figure 9.1 Trends in absolute losses of farm land in Czechoslovakia

Table 9.12 Transfer of farm land and arable land to non-farm use (hectares)

	Construction		Mining		Other purposes	
Year	Farm land	Arable	Farm land	Arable	Farm land	Arable
1963	8 137	4 664	1 764	1 082	28 611	5 973
64	10 248	6 570	1 894	1 407	30 879	7 283
65	9 299	5 390	1 357	692	24 590	7 075
66	9 191	5 458	1 175	899	16 394	3 267
67	8 557	4 870	1 049	555	13 773	2 815
68	7 045	3 902	617	311	13 525	2 454
69	8 816	4 681	456	332	11 797	2 582
70	6 549	3 713	383	280	9 443	2 118
71	6 596	3 783	451	315	14 574	1 721
72	8 099	4 676	376	258	6 504	1 148
73	5 729	3 624	295	137	8 511	2 038
74	7 399	4 722	91	56	16 546	2 330
75	9 382	5 390	305	179	36 754	4 247
76	7 051	4 045	177	70	15 737	2 650
77	5 957	3 769	532	311	15 183	2 938
78	8 925	5 173	722	498	23 865	4 344
79	6 292	3 524	787	505	27 900	4 350

It should not be forgotten that a large part of permanent land losses is caused by the farmers themselves. The effective use of machinery in crop cultivation requires fields of 50–80 ha. Their formation, however, is affected by the network of ways and roads and the structure of the ground. Hence on small fields which cannot be combined with others to provide larger areas, cultivation is uneconomical and these areas are often left unused or only partly used. Another factor contributing to losses of farm land is the lack of care that farming enterprises pay to the maintenance of cartways and paths; if they are badly passable, vehicles drive directly over the fields, causing tremendous and unregistered losses of arable land, amounting to several thousands hectares nationwide. Moreover, the land deteriorates under the heavy machinery. Untended roads also result in many fields not being limed, as heavy lorries cannot reach the fields.

Two further categoreis of land should be mentioned which, though belonging to agriculture, are not farmed at present: these are sites temporarily exempt from farming, which on 1 January 1980 totalled 16 432 ha, and temporarily uncultivated land, which totalled 85 822 ha, representing 1.24% of the total farm land.

Sites temporarily exempt from farming are used as handling areas for building projects—in particular the construction of roads, telecommunication

Table 9.13 Farm and arable land areas per inhabitant (acres)

Year	Farm land	Arable land
1936	56.20	40.80
1950	64.80	46.70
1955	56.65	39.40
1960	53.60	37.50
1965	50.40	35.70
1970	48.93	34.47
1975	47.76	33.53
1980	45.10	31.79

lines, barrage construction, housing estates, industrial and produce-processing plant building. Temporarily uncultivated land includes land held by socialist farming organizations which, for various reasons, cannot be farmed at the moment (waterlogged, overgrown with shrubs, steeply sloped, thin topsoil, etc.), but which can be reclaimed and again serve agriculture. In some parts of Czechoslovakia, the area of temporarily unfarmed land is increasing. As a rule, it is made up of small enclaves of land in hilly, wood-covered regions, meadows scattered among the forests, and lots on steep slopes. With the contemporary heavy machinery, they cannot be cultivated or even reached, or are too far away from the farms to make their cultivation viable.

However, farm land and arable land registers not only losses, but in some years also increments. Non-farm land, by deforestation or reclamation, becomes farm land.

What, then, is the outlook for future farm-land trends? It may be assumed that, after 1981, annual losses will amount to roughly 5500 ha of farm land. However, this will be possible only provided all land-protection measures are adhered to and no more land of high fertility, often in vegetable-growing areas, is seized for the construction of housing, power stations, factories, warehouses, railway stations, etc.

Permanent losses of farm land result in a diminished farmed area per inhabitant, as shown on Table 9.13.

9.4.2 Reallotments (Tables 9.14–9.16)

Reallotment changes not only the property or usage rights, but above all the economic use of land. Land reallotment in general is to provide conditions more favourable for cultivation.

Table 9.14 Trends in farm land structure (hectares)

	1937	1948	1960	1972	1980	Difference 1937–80
Arable land	5 894 722	5 870 125	5 134 114	4 989 432	4 857 300	−1 037 422
Hop fields	–[a]	–[a]	8 880	9 591	12 219	+ 12 219
Vineyards	17 015	18 639	24 466	35 720	45 343	+ 28 328
Gardens	185 356	209 172	213 908	215 718	221 706	+ 36 350
Orchards	–[a]	–[a]	48 888	78 403	77 376	+ 77 376
Meadows	1 135 310	1 153 455	1 082 098	961 202	864 974	− 270 336
Pastures	875 406	848 629	820 780	786 532	844 681	− 30 725
Total farm land	8 107 809	8 100 020	7 333 134	7 706 598	6 923 599	−1 184 210

[a]Hop and orchard areas for 1937 and 1948 are included under other categories.

Table 9.15 Percentage representation of main categories of farm exploitation, taking farm land = 100

	Arable land	Meadows	Pastures
1937	72.0	14.0	10.8
1948	72.5	14.2	10.5
1980	70.2	12.5	12.2

Until 1948, farm land, mostly privately owned, was split into small strips of fields in scattered locations, often quite distant from the farm buildings, sometimes even belonging to the cadastre of another village. Typical, too, were the impractical shapes of fields, particularly strips, wedges, etc., and the bad accessibility for lack of roads.

The farm land was divided into 33 million lots averaging 7 ares in Bohemia and Moravia and 4 ares in Slovakia. There were 1 507 000 individual farms.

These land-holdings badly hampered farming. In particular, they either impeded or prevented the mechanization of work in the fields, the use of sophisticated cultivation techniques, and the effective application of pesticides. In addition, they caused losses of farm land (2–3% on the average on field boundaries) which represented 240 000 ha and made melioration difficult.

Therefore, attempts had been made in the past to overcome these disadvantages by rearranging private land-holdings by various means. Thus, for example, the so-called 'rounding-off' consisted in a voluntary exchange of lots between neighbours in order to form larger and continual holdings. 'Consolidation' aimed mostly at improving the shape and accessibility of fields, 'separation' merged lots of individual farms into larger wholes.

The rationalization of land-holdings met the needs of private smallholders well, but became unusable after 1948, when, under the socialist system, farmers began to merge their fields to form collectively farmed larger areas. This transition to socialist large-scale farming called for a new kind of reallotment, which would facilitate, primarily, the joint organization and operation of agricultural production on the basis of cooperation among small and middling farmers, as well as the introduction of progressive, large-scale farming methods, providing room for the application of research findings and modern technology to ensure high productivity.

In Czechoslovakia, reallotment can be divided roughly into the following stages: 1948–52, 1952–60, 1960–71, and 1971 up to the present day.

The first stage after the passing of the Land Unification Act in 1948 mainly consisted in the ploughing-under of boundaries dividing individual fields, as one of the first steps of collectivization. Though this first unification of the

Table 9.16 Structure of cultivated areas (thousands of hectares)

Year	Cereals	Wheat	Barley	Corn	Husks	Sugar-beet	Potatoes	Bulk feed-grain	Energy-providing feed-grain	Oil crops	Fibre crops
1936	3 363	893	633	102	133	154	714	807	104	19	22
48	3 020	869	858	142	81	182	551	711	355	53	31
49	2 842	797	572	129	75	194	569	946	215	55	34
50	2 755	755	614	129	90	221	660	826	225	60	33
51	2 638	740	634	139	97	244	666	826	214	68	60
52	2 639	766	624	139	109	233	669	862	178	74	65
53	2 634	748	649	149	119	232	636	734	309	81	67
54	2 526	712	629	147	119	215	630	769	289	82	56
55	2 568	720	642	160	108	216	621	841	313	64	63
56	2 633	722	668	183	101	222	630	876	305	60	63
57	2 640	742	670	169	96	227	629	809	353	57	62
58	2 596	738	669	180	93	234	607	871	373	59	64
59	2 567	720	672	187	92	242	585	857	447	70	65
60	2 494	652	707	195	104	242	569	831	567	76	62
61	2 472	643	696	201	104	252	515	952	527	82	54
62	2 497	673	694	237	121	260	508	806	619	72	54
63	2 466	720	692	213	137	259	503	959	483	72	58
64	2 506	831	686	186	153	259	491	959	457	72	59
65	2 448	826	667	161	142	230	444	950	489	69	48
66	2 521	892	690	151	138	232	437	1 041	443	64	43
67	2 550	929	712	150	111	206	408	975	515	67	43
68	2 602	999	712	138	93	197	372	967	525	68	45
69	2 641	1 054	780	127	70	183	325	1 064	492	50	39
70	2 613	1 081	803	128	58	183	338	1 083	494	48	33
71	2 676	1 103	851	142	48	189	332	1 055	478	64	28
72	2 756	1 197	854	148	37	192	321	883	587	65	29
73	2 782	1 235	873	169	34	199	305	918	528	72	33
74	2 756	1 276	867	167	–	205	280		1 450	45	31
75	–	1 183	980	158	–	218	251		1 410	45	30
76	2 724	1 278	857	204	–	221	240		1 394	63	30
77	2 733	1 287	856	203	–	218	237		1 386	72	33

land was of fundamental economic and political importance to collective farming, some indicators even then signalled the negative effects this would have on the environment. The ploughing-under of boundaries and obsolete grassy paths and the elimination of other territorial and vegetational obstacles often entailed the pollution of surface waters and streams by topsoil.

The second stage represented a gradual transition to a higher form of reallotment, described as 'technical economic reallotment'. This facilitated the new organization of the farm land by unifying the fields of each farm to form large areas and shapes suitable for mechanized cultivation, but keeping within the framework of existing field paths, roads and water systems, such as streams, drains, ditches, etc. Obstacles in the way of field unification—such as obsolete paths, hollow ways, high boundaries with no erosion-preventing function, shrubs, isolated trees—were eliminated on the basis of simple earth-work projects.

The only road to economic and successful farming was seen in the complete clearing of the land of all biological obstacles. The reallotted territories were not considered from the water-management angle at all, and earlier reclamation was disregarded. For instance, reallotment damaged existing drainage systems, disrupting the water management of extensive areas, which led to waterlogging or water oozing to the surface. Damage was done by abolishing high-banked field paths, which sometimes stood in the way of the smooth cultivation of unified fields, but at the same time served the natural draining of surface water and flood waters, or into which opened forgotten drain outlets, for which no substitutes were built. By deep ploughing and, above all, by the new method of deepening the topsoil, old drains were either crushed or interrupted, which resulted in extensive waterlogging because the drainage systems worked in reverse and collected water into spots from where there was no way for it to escape. The waterlogged fields could either not be farmed at all and had to be left fallow, or they were farmed after long waiting periods.

Very often, land reallotment disregarded the erosion-preventing function of high boundaries reinforced by turf and shrubs. Soil, deprived of turf cover, was washed away into ditches, streams and rivers.

Gradually but inexorably, reallotment eliminated the scattered vegetation from the landscape. In particular, in the rich farm regions, which were poor in woodland, this had an extremely negative effect on micro-climatic, hydrological and soil conditions, and, above all, on the living, recreational and aesthetic value of the landscape.

Major changes occurred in the structure of cultivation in the post-war years, owing to the increasing proportion of technical crops (mainly sugar-beet) and feed-grains and the falling proportion of cereals (particularly rye and oats) and potatoes. The shift from cereals to technical crops and feed-grains on arable land resulted—despite increased yields per hectare—in a lower cereals production than in pre-war years.

9.4.3 Some Positive and Negative Phenomena Caused by Large-scale Farming

All measures carried out in post-war years (and which are still being carried out in the sphere of cultivation) had one aim—to increase crops. Despite the substantial loss of farm land, crops increased, mostly owing to the changed structure, and to cultivating crops with higher yields per hectare, largely through higher doses of fertilizer. In recent years there has been a gradual transition to the use of combined fertilizers, which are more effective under unfavourable climatic conditions. Until recently, all measures taken to increase yields (fertilizing, new, higher-yield varieties, pesticides, new cultivating methods, better machines) were considered from one angle only—that of increasing yields. However, all the above measures brought secondary consequences, which were frequently under-estimated. Yields were forced up at any cost, regardless of the risks involved. It therefore is important to reassess the resulting situation and to reconsider the numerous risks incurred by constantly increasing the demands on agricultural production. (Large-scale farming is often regarded as the chief polluter of the environment, a view with which one cannot agree.)

Table 9.17 shows the trend of yields per hectare of the major field crops. The per-hectare yields of cereals were particularly favourably affected by the new varieties cultivated, among which an increasing proportion is made up of highly productive Soviet-bred varieties, especially wheat.

At present, crop cultivation is oriented towards greater concentration and specialization. In crop cultivation, concentration concerns mainly cereals, because they have the highest unit-yield from the area under cultivation, the products are easily stored and cultivation and harvesting is highly mechanized. Many farming enterprises therefore cultivate cereals three or four times in succession, although the fertility of the soil is not up to it. Figure 9.2 shows how cereals cultivation has varied over the years: in recent years the area has been about 55% of the total. This should mean that cereals may be planted twice in succession at the most. However, the small quantity of interim crops causes cereals to be grown more often.

Farmers mostly concentrate on growing wheat, because this is economically the most attractive. However, wheat is also the most demanding as to preparatory crops and individual varieties, and individual cultivation methods. Therefore, wheat is sometimes grown in entirely unsuitable climatic regions. Harvesting then is often delayed and crops lose their full value.

Thus, with increasing integration and specialization, the principles of correct crop rotation are not always respected, with a negative impact not only on crops, but also on the environment. This includes mainly the following:

(1) On raising the quotient of cereals on arable land, the proportion of crops improving soil fertility declines, gradually impairing soil quality.

Table 9.17 Crop yields per hectare (tonnes per hectare)

Year	Cereals	Wheat	Barley	Corn	Husks	Sugar-beet	Potatoes	Bulk feed-grain	Energy-providing feed-grain	Oil crops
1936	1.58	1.63	1.61	2.31	1.49	30.86	14.35	–	–	1.09
48	1.58	1.64	1.54	2.09	1.44	23.58	11.01	–	–	0.70
49	1.83	1.95	1.79	1.87	1.45	22.04	10.15	–	–	0.87
50	1.71	1.89	1.68	1.70	1.23	28.47	12.35	–	–	0.77
51	1.99	1.80	2.18	1.39	21.39	21.35	11.05	–	–	1.04
52	1.81	2.05	1.82	1.53	0.98	20.55	11.84	–	–	0.76
53	1.92	2.08	1.92	2.88	1.30	24.11	15.26	–	–	0.76
54	1.70	1.56	1.77	2.59	1.26	26.11	13.19	–	–	0.73
55	1.99	2.04	2.01	2.44	1.45	28.51	12.73	–	–	1.05
56	2.07	2.13	2.11	2.18	1.48	20.67	15.29	5.11	18.06	1.13
57	1.97	2.06	2.04	2.65	1.16	29.83	13.94	4.46	18.36	0.97
58	1.87	1.83	1.79	2.73	1.22	29.91	10.91	4.95	–	1.07
59	2.16	2.29	2.19	2.80	1.23	20.56	10.87	4.60	–	1.28
60	2.32	2.33	2.48	3.05	1.43	34.63	9.02	4.50	21.40	1.10
61	2.31	2.60	2.28	2.47	1.39	28.03	10.40	4.84	14.83	1.28
62	2.32	2.45	2.53	2.35	1.42	22.84	9.89	3.91	14.66	1.02
63	2.31	2.46	2.35	2.93	1.41	31.42	12.96	4.64	18.80	0.90
64	2.12	2.22	2.10	2.65	1.29	29.31	15.60	4.37	18.70	0.83
65	2.18	2.42	2.14	2.79	1.45	26.10	8.46	5.55	17.34	1.26
66	2.34	2.53	2.34	3.27	1.28	34.10	13.40	5.65	21.31	1.38
67	2.58	2.71	2.73	3.03	1.70	37.61	14.83	5.32	18.30	1.46
68	2.84	3.16	2.98	3.38	1.47	41.78	17.56	5.26	21.52	1.25
69	3.00	3.10	3.23	3.99	1.79	32.16	15.90	5.28	21.18	1.20
70	2.77	2.95	2.84	4.09	1.68	36.96	14.21	5.79	24.00	1.54
71	3.30	3.53	3.36	3.98	1.89	31.27	13.94	5.53	20.15	1.76
72	3.17	3.37	3.12	4.44	1.70	36.11	15.81	5.23	27.95	1.81
73	3.51	3.78	3.40	4.03	1.69	31.09	16.75	5.92	23.58	1.84
74	3.83	3.99	3.92	4.01	1.30	39.73	16.32	5.68	–	2.08
75	3.43	3.57	3.20	5.49	1.37	35.62	14.24	6.03	–	2.06
76	3.49	3.77	3.40	3.60	1.25	24.61	17.55	5.70	21.00	1.80
77	3.85	4.07	3.77	4.70	1.68	38.40	16.04	7.88	29.25	

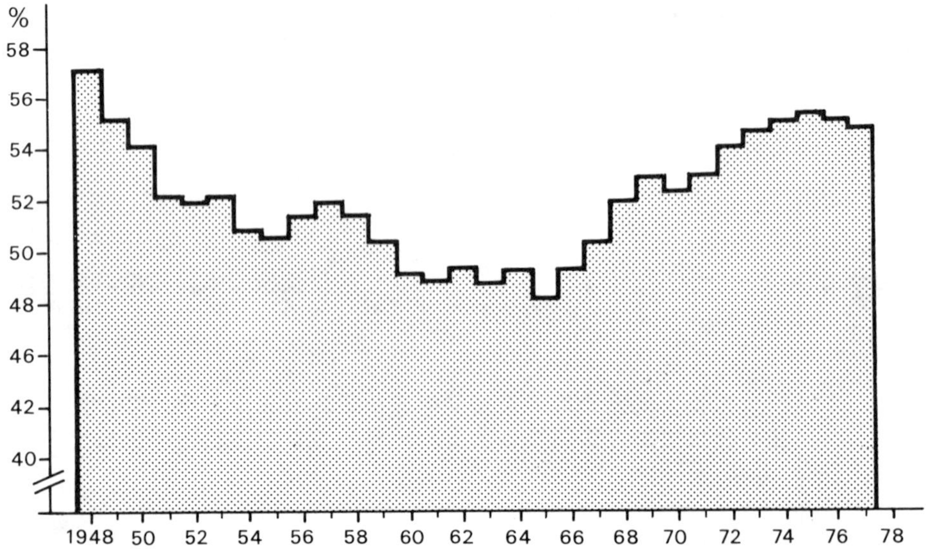

Figure 9.2 Cereals cultivation (by area) as a percentage of total cultivation

(2) On growing cereals in too long a succession, there is a risk of so-called soil fatigue, multiplication of damaging factors (e.g. pests and plant disease), and increasing seasonal weeds following winter crops.
(3) By repeatedly growing crops with similar biology and cultivation, vegetation factors in the soil are exploited unevenly and unilaterally, which reduces yields.
(4) Repeated cultivation of similar spring crops leaves the soil unprotected by vegetation for a longer period, resulting in undesirable moisture evaporation, in a risk of water and wind erosion of the soil, and the runoff of unused nitrates and residual pesticides.
(5) In connection with a repetition of similar or identical crops, the effect of unsuitable machinery on soil fertility makes itself felt, in particular the excessive use of heavy transport, cultivating and harvesting machinery.

9.4.3.1 Chemicals in Cultivation

The application of synthetic fertilizers has increased very quickly throughout the entire period since the Second World War. Among the socialist countries, Czechoslovakia ranks high on the list of users of synthetic fertilizers. The trend is shown in Table 9.18 and Figure 9.3.

However, the constantly rising doses are not matched by the per-hectare yields, which are only increasing slowly, as demonstrated in Figure 9.4. For

Table 9.18 Application of synthetic fertilizers (kg/ha)

Year	Total	Nitrates	Phosphates	Potassium
1936	13.1	3.1	6.6	3.4
48	18.4	4.9	7.9	5.6
49	21.3	5.6	7.3	8.4
50	28.2	7.7	9.2	11.3
51	27.4	7.2	8.6	11.6
52	30.5	6.7	8.9	14.9
53	36.8	8.1	10.7	18.0
54	44.6	10.7	12.2	21.1
55	46.7	13.4	13.2	20.1
56	53.5	12.7	15.6	25.2
57	57.2	12.6	15.3	29.3
58	70.2	17.6	18.0	34.6
59	67.0	18.3	20.2	29.5
60	68.3	20.1	21.9	26.3
61	76.6	19.7	24.8	32.1
62	80.1	22.1	26.1	31.9
63	93.4	24.5	30.8	38.1
64	115.2	32.0	35.6	47.6
65	126.1	34.7	35.8	55.6
66	130.9	39.2	36.1	55.6
67	133.4	40.4	35.1	57.9
68	158.5	50.3	41.9	66.3
69	168.6	56.8	44.3	67.5
70	182.4	59.5	49.7	73.2
71	195.5	62.0	52.1	81.4
72	202.4	66.9	51.9	83.6
73	200.9	66.5	54.0	80.4
74	222.0	72.6	59.4	90.0
75	242.8	76.7	65.8	100.3
76	230.6	84.7	61.2	84.7
77	245.6	90.2	68.2	87.2
78	253.7	88.7	69.7	95.3

the season 1980/81, farms were to be supplied with fertilizers corresponding to 255 kg of pure nutrients per hectare, which, apart from the GDR, is the highest rate in the CMEA countries.

Compared with pre-war farming, the quantities of fertilizers used have increased 25 times, but yields have only risen by roughly 2.5 times. Although still higher doses of synthetic fertilizer are being considered for future application, salination is appearing on as much as 30 000 ha. Because in recent years the function of organic fertilizer has been under-estimated, the soils are unable to absorb the large quantities of nutrients supplied. Even the vegetation cover is unable to utilize all the available nutrients, and thus a

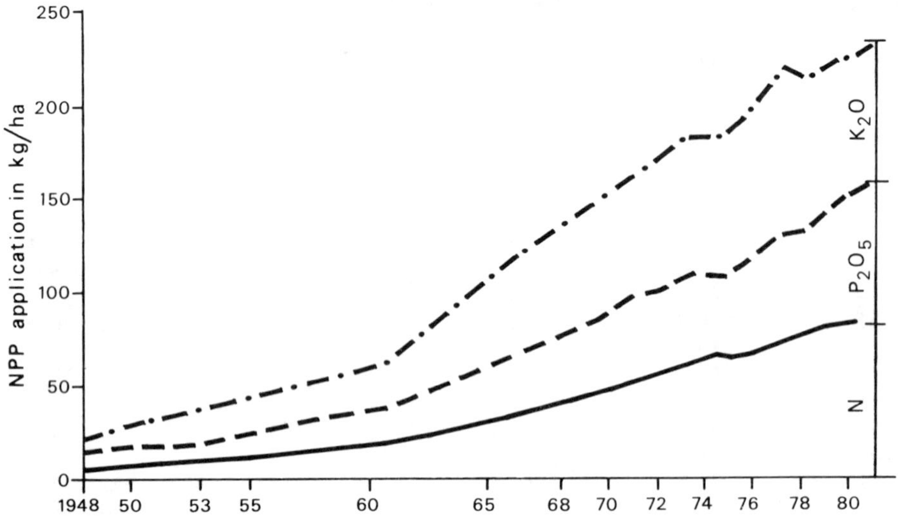

Figure 9.3 Trends in the application of artificial fertilizers in Czechoslovakia

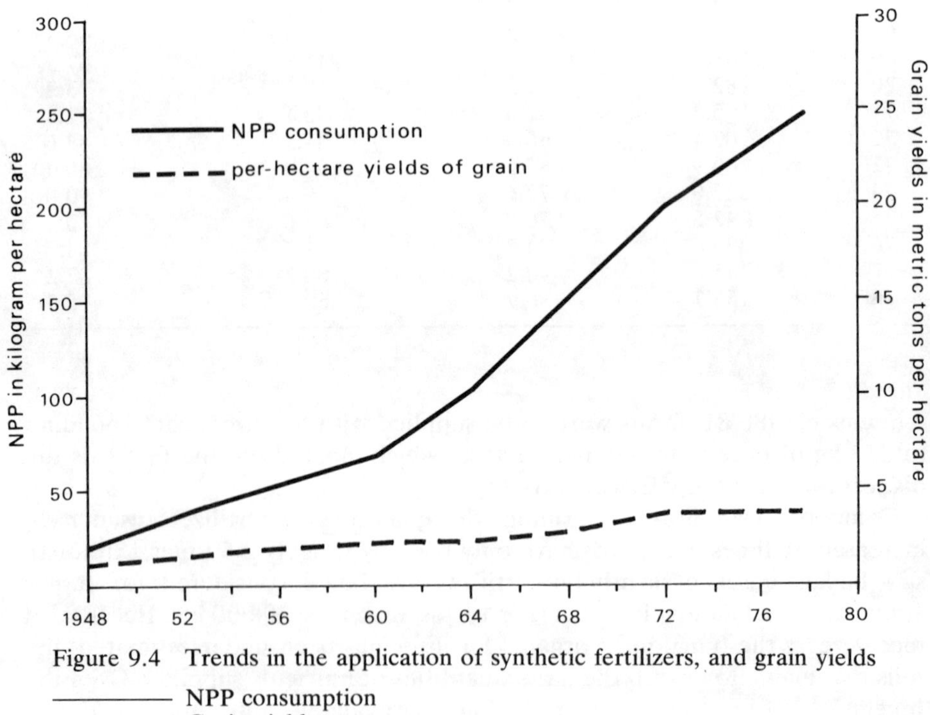

Figure 9.4 Trends in the application of synthetic fertilizers, and grain yields
——————— NPP consumption
—·—·—·—Grain yields

Table 9.19 Application of chemical pesticides

	1969–71	1975	1976	1977
Total consumption of pesticides (t)	6 078	13 302	14 239	14 487
Application of pesticides (kg/ha)	0.856	1.889	2.033	2.073

substantial part of these substances reaches surface waters and groundwater. This threatens the quality of water in all reservoirs and rivers with eutrophy caused by phosphates and unabsorbed fertilizer; the quality of drinking water is often spoiled by nitrates, which reach surface water, and groundwater by runoff from fields.

The use of plant-protecting chemicals is also increasing, as shown in Table 9.19.

These tremendous quantities of chemicals applied over large areas of land lead to the pollution of water and the environment as a whole.

9.4.3.2 Adding to the Size of Fields

The steadily increasing areas of individual fields also raise the risk of damaging the environment. Within a relatively short space of time, the mosaic of tiny strips, squares and wedges changed into fields stretching beyond the horizon, often exceeding 200 ha.

The heavy machinery used on large fields has a negative effect on soil fertility and general crop yields, owing to the soil compaction. It also facilitates erosion and unproductive water evaporation.

As a consequence of the removal of boundaries, paths, ditches, copses and other obstacles impeding the surface water from running off, the risk of erosion on large fields is growing. Today, more than 720 000 ha of soil are threatened by water-caused erosion. Large, uninterrupted areas also encourage wind erosion and dust.

On too-large fields, irregularities in soil conditions are difficult to even out, in particular on slopes and uneven ground (e.g. saturation with nutrients, destruction of foci of persistent and resistant weeds, pests and diseases, differentiated soil treatment, etc.). This leads to uneven conditions of growth and development in crops, irregular vegetation and yields, and a general irregularity of soil fertility, an irregular influence on evaporation, soil erosion and weed infestation. Shortcomings in cultivation have a far more marked and negative effect on large areas than on smaller fields, distributed over the land in a mosaic.

From large crop-covered areas, footpaths through the fields, grass-covered banks, shrubbery and trees have disappeared, so small, highly valuable living areas in the environment have become inaccessible. Together with isolated trees or clumps of trees, local names of individual localities disappear and the landscape becomes anonymous.

The biological protection of vegetation on extensive fields is greatly limited, as they no longer provide any refuge for small wildlife—birds, insects and rodents—which ensure a certain biological control of the environment. Small game slowly disappears from the land farmed with wholesale methods.

Huge fields, chemicals and highly efficient machinery, with its noise and width of implements, play their parts in the contemporary decline of game (hares, for example). Unending mono-cultures offer a single-plant diet, which ends abruptly when the entire field is harvested, and the animals cannot adapt quickly enough to other available food resources.

It is obvious that the mere increase in the size of individual fields has a strong impact on the ecology of a landscape.

9.4.4 Measures Needed to Combat Undesired Interference with the Environment

Guidelines for the Preparation and Endorsement of Projects of Reallotment with Regard to its Effect on the Environment was drafted in 1975. These guidelines, approved by the Ministry for Construction and Technology, the Ministry of Agriculture and Nutrition, the Ministry of Forestry and Water Management, the Ministry of Culture, the Home Office, the Ministry of Health and the Czech Geological Office, serve the staff on People's Committees at all levels, the staff concerned with nature conservation and the protection of the environment, as well as planners. They are guidelines for the preparation and planning, assessment and endorsement, implementation and supervision of all land reallotments, not only in order to protect the human environment, but also to ensure its purposeful formation and to increase desirable favourable effects.

The guidelines have to be applied wherever reallotments are undertaken, i.e. wherever there is a change in the exploitation of land of any type and for whatever reason. They unify and supplement the principles controlling environmental protection. They will be amended and extended in accordance with new findings of science, research and practical experience, and to cover further types of activities affecting the exploitation of territories and the environment.

In future, therefore, reallotment will be planned in accordance with the methodology issued by the Ministry of Agriculture and Nutrition and with the *Guidelines*. These determine procedures to be followed with regard to the following problems:

(1) wind erosion;
(2) water erosion;
(3) the watershed;
(4) the biology of the environment concerned;
(5) the social function of the environment.

Complying with the guidelines should help to avoid wholesale negative effects on the landscape such as those which occurred in the past. Thus, the open-farming landscape should fulfil not only its productive function, but also a function for the whole of society, for present as well as future generations.

9.5 TRANSITION TO WHOLESALE LIVESTOCK FARMING

Livestock farming, like crop cultivation, underwent a gradual change of its internal structure during the long-term process of collectivization. The basis of livestock farming was provided by domestic animals coming from private breeders and sheds to the large cooperative or state farms.

The trend of livestock production during the past 30 years can be divided into three stages. In the years 1945–49, small-scale husbandry still prevailed, whilst the basis was being laid for socialization of that branch of the economy. During the years 1949–60, socialist production methods recorded their first successes in livestock farming. Since 1960, a modern wholesale production has developed.

In the post-war period, a substantial increase was recorded, mainly in the breeding of pigs and poultry. In the period from 1948 to 1975, animal production increased two and a half times. The ratio in the number of animals bred changed considerably. Since the end of the 1950s, qualitatively new wholesale methods began to develop, accompanied by large investments and the development of socialist agriculture.

9.5.1 Trends in Farm Animal Density

The population density of farm animals is an indication of intensity of livestock farming. In the post-war period 1949–63, the number of farm animals rose quickly. Figures for livestock per 100 hectares had exceeded the pre-war level as early as 1953 in Bohemia and Moravia in 1950, in Slovakia in 1952. Since 1954, the number of farm animals per 100 hectares of farm land remained on roughly the same level until 1965, and has been showing a mild upward trend ever since (see Table 9.20).

This trend is connected with an accelerated turnover of herds, but also with substantial changes in the structure of livestock herds. These consisted mainly in a rising number of pigs and poultry and a decline in the number of horses.

Table 9.20 Number of livestock per 100 ha of farm

Year	Livestock per 100 ha	Year	Livestock per 100 ha
1920	85	1953	105
25	93	54	97
		55	98
30	91	56	101
31	91	57	102
32	91	58	101
33	95	59	100
34	93		
36	99	60	103
37	105	61	104
		62	105
47	78	63	105
48	74	64	104
49	83	65	105
		66	101
50	96	67	102
51	95	68	103
52	98	69	103

9.5.1.1 Horses

The marked decline in the number of horses bred (see Table 9.21) in connection with the spreading use of tractors in the past 30 years, also affected—apart from all the undeniable advantages which this brought about—the exploitation of the land.

The use of increasingly heavy machinery on the land leads to soil deterioration, as previously described. When horses were used as tractive force, these problems did not arise to any marked degree. Also, it is worth mentioning that even today there remain small areas of land that are not used for agriculture, as the contemporary machines cannot work small and irregularly shaped lots, nor too steeply inclined or overly wet ones, where formerly all work had been done by horse-drawn implements. Frequently, there is a missing link between contemporary strong machinery and human strength.

The curtailment of horse-breeding was not properly utilized to extend cattle-raising, though more feed had become available. In Czechoslovakia, horses are not bred for meat.

9.5.1.2 Pigs

The main characteristic of animal breeding in the post-war years was the fast increase in the number of pigs and poultry (Tables 9.22 and 9.23) and the slow

Table 9.21 Numbers of horses kept

Year	Horses (thousands)	Year	Horses (thousands)
1920	567	1957	542
25	702	58	517
		59	456
30	666		
31	646	60	389
32	666	61	330
33	659	62	292
34	659	63	254
35	654	64	227
36	663	65	204
37	645	66	188
38	645	67	177
		68	166
47	653	69	
48	630		
49	628	70	144
		71	
50	629	72	
51	605	73	100
52	572	74	84
53	559	75	
54	544	76	62
55	543	77	57
56	543	78	53

increase in cattle. The great demand for meat caused the fast development of slaughter pigs. Slaughtering weight was relatively high, and only in recent years, responding to consumers' wishes, has the slaughtering weight been reduced. Between 1952 and 1970, the number of piglets raised from one sow has risen from 9.1 to 17.5. New wholesale breeding techniques have evolved and through their adoption the number of pigs has almost doubled. Pig breeding is done in large concentrations, relatively independent of the land. Compared with pre-war days, the number of pigs per 100 hectares of arable land has doubled, in Slovakia even trebled.

Currently, pigs are concentrated in large-capacity breeding and finishing houses, comprising from 5000 to 20 000 head. Of course, this increased number of pigs and their strong concentration has its effect on farm and arable land. Apart from the sties taking up much land (which, owing to their noxious smell, have to be well apart from human habitations), the main problem is the disposal of manure. Particularly during the vegetative period, liquid manure, though kept in huge storage tanks, is difficult to dispose of and

Table 9.22 Numbers of pigs kept

Year	Pigs (thousands)	Year	Pigs (thousands)
1920	2 926	1957	5 369
25	2 470	58	5 435
		59	5 283
30	2 696		
31	2 509	60	5 687
32	2 570	61	5 962
33	3 381	62	5 895
34	2 962	63	5 897
35	2 678	64	5 845
36	3 160	65	6 139
37	3 538	66	5 544
38	3 712	67	5 305
		68	5 601
47	2 944	69	
48	2 566		
49	3 242	70	5 036
		71	5 935
50	4 218	72	6 093
51	3 802	73	6 086
52	4 234	74	6 256
53	4 918	75	6 630
54	4 174	76	6 686
55	4 771	77	6 820
56	5 285	78	7 510

is often carted away in tank-trucks to distances of over 50–60 km. In addition, the land in the vicinity of the pig-breeding stations is often over-manured and there is a risk of soil contamination by parasites and germs, causing infectious diseases.

It is important, therefore, to arrange cultivation around pig stations in a manner facilitating the optimum application of liquid manure over the largest possible area.

Table 9.23 Percentages of pigs and poultry against total livestock

Year	1936	1953	1955	1960	1966	1968	1974
Pigs and poultry	25.8	26.5	28.0	32.1	34.9	33.7	38.6

Table 9.24 Numbers of poultry kept

Year	Birds (thousands)	Year	Birds (thousands)
1920	22 772	1957	23 876
25	29 009	58	24 250
		59	25 364
30	29 900		
31	31 124	60	27 569
32	32 124	61	28 157
33	34 124	62	28 805
34	36 124	63	28 032
35	37 730	64	30 093
36	38 902	65	28 840
37	41 748	66	27 752
38	42 254	67	29 446
		68	31 208
47	12 321	69	
48	13 478		
49	16 393	70	34 870
		71	38 238
50	17 794	72	39 170
51	18 206	73	39 157
52	18 690	74	41 232
53	20 787	75	
54	21 094	76	43 065
55	22 540	77	44 142
56	23 367	78	44 774

9.5.1.3 Poultry

During the last 30 years, the number of poultry kept has doubled (Table 9.24). Particularly in recent years, the production of broilers and eggs increased substantially. Wholesale methods for breeding broiler chickens were introduced. These, in contrast to pig-raising, did not in any essential way affect farm land exploitation. So far, there does not appear any need for special cultivation methods for disposing of manure.

Also, there has been a marked increase in the breeding of ducks. Wholesale geese-breeding has not proved as successful, owing to certain problems encountered.

9.5.1.4 Cattle

Throughout all post-war stages, the number of cattle increased slowest of all (Tables 9.25 and 9.26), though owing to the falling number of horses some

Table 9.25 Head of cattle per 100 ha of farm land

Year	Head
1934–38	55.4
1949–53	57.1
1954–58	56.8
1959–63	60.3
1964–68	62.5
1969–73	61.4
1974–78	66.1

Table 9.26 Trends in head of cattle (absolute numbers)

Year	Cattle total	Year	Cattle total
1910	4 377 149	1956	4 106 755
		57	4 133 866
20	4 139 010	58	4 090 845
		59	4 183 467
25	4 487 222		
		60	4 302 513
30	4 256 225	61	4 387 350
31	4 247 686	62	4 518 291
33	4 141 744	63	4 507 175
34	4 213 380	64	4 479 806
35	4 104 272	65	4 436 166
36	4 079 039	66	4 388 984
37	4 376 045	67	4 462 073
38	4 705 809	68	4 436 691
		69	4 249 376
47	3 974 851		
48	3 274 967	70	4 222 576
49	3 663 009	72	4 348 785
		72	4 466 340
50	4 212 756	73	4 466 557
51	4 303 262	74	4 555 889
52	4 376 382	75	4 654 236
53	4 445 371	76	4 553 377
54	4 082 428	77	4 653 956
55	4 040 645	78	4 758 301

feeding resources had become available. The pre-war level in the number of cattle was reached again only in 1971, when the pre-war dairy production was also surpassed. Greater meat production was attained by increasing the number of cattle, as well as by raising the slaughtering weight.

The contemporary trend in cattle-farming is oriented towards the largest possible productivity by setting up high-capacity cattle stations. The aim is to achieve a labour productivity of 60–80 dairy cows, or 350 calves, 350–400 heifers or 400–800 steers per worker. However, the concentration of farm animals over and above a certain limit produces a series of new problems and ecological phenomena. In connection with the housing of farm animals, ecological problems are at present acquiring ever-greater significance, as entirely new trends in construction strongly affect the biotop of stable and sty.

High livestock concentrations in one locality (e.g. Žehušice, 1741 dairy cows; Polerady, 5000 steers; Milotice, 3430 calves; etc.) call for gigantic supplies of bulk feed. Such quantities, despite the introduction of special cultivation methods, cannot be grown in the vicinity, and have to be transported over considerable distances, sometimes over tens of kilometres. This is all the more important wherever large cattle stations are located in the foothills of mountain regions, as transport costs over uneven ground sharply increase. Also, such stations are often located in regions where formerly cattle freely grazed on ranges (e.g. in the border mountains), whereas today it is necessary to concentrate fodder in one spot, sometimes in quite distant places.

The composition of livestock feed in super-large cattle stations has been changed, feed mixes replacing bulk feed, which is more difficult to handle. Very often, as much as 75% of the grain produced is fed to livestock and poultry, which leads to an incessant increase in the grain cultivation area, with all accompanying phenomena described in the preceding section.

9.5.1.5 Sheep

The least favourable post-war trend appears in the number of sheep farmed (Table 9.27). Although the pre-war density per 100 ha had been reached as early as 1952 (10.9 head) and the increase continued until 1954 (14.4 head), after 1955 their numbers began to fall. In recent years the number of sheep is rising again, owing to increased consumer interest in mutton and lamb, sheep cheese and other dairy produce.

Over the territory of Czechoslovakia, sheep farming is distributed irregularly. It is concentrated mainly in the hills and mountains of Slovakia. Sheep are used, too, to graze off the grass on airfields, in water-protection zones, etc. In recent years, the number of small sheep-farmers has been increasing again, and they are using a number of small areas, gardens and fallow fields as pasture for their animals.

Table 9.27 Numbers of sheep kept

Year	Sheep (thousands)	Head per 100 ha	Year	Sheep (thousands)	Head per 100 ha
1920	878	10.8	57	956	13.2
25	744	9.2	58	889	12.3
			59	817	11 1
30	506	6.5			
31	438	5.7	60	727	10.0
33	381	4.9	61	646	8.9
34	395	5.1	62	603	8.3
35	424	5.5	63	527	7.3
36	453	5.8	64	527	7.4
37	486	6.3	65	568	8.0
38	534	6.9	66	614	8.6
			67	670	9.5
47	491	6.6	68	770	10.9
48	386	5.2	69	906	12.8
49	459	6.2			
			70	977	13.9
50	531	7.2	71	–	–
51	596	8.1	72	–	–
52	800	10.9	73	–	–
53	982	13.4	74	842	11.6
54	1 017	14.4	75	805	11.4
55	1 017	14.2	76	797	11.4
56	1 000	13.7	77	841	12.0

9.5.1.6 Goats

Until the Second World War the keeping of goats was fairly common, but with the progressing specialization of farming it lost its importance. The number of goats fell sharply and no attempt was made to farm them on a larger scale. Today, about 82 000 goats are kept in the whole of Czechoslovakia.

9.5.1.7 Domestic Animals

The falling number of 'domestic' animals kept by individual rural inhabitants (cattle, sheep, geese, goats) led to the common pastures, owned by almost every village, falling into disuse. These pastures, the 'village commons', usually covered less fertile grazing ground near the village. After the transition to wholesale cattle farming, these village commons lost their function, natural dissemination covered them with brush or otherwise changed their plant-life, and frequently they became a source of weed infestation. Often, these unused pastures are ploughed up and used as arable land, though their quality is low.

9.5.2 The Trend Towards Concentration in Livestock Production

At the very begining of the collectivization of Czech agriculture, it became obvious that it was absolutely essential, for economic and operational reasons, to transfer the livestock from their small, inferior sheds scattered throughout the whole village into communal sheds and stables, which would have at least some mechanization.

In that first stage of collectivization, larger farms and estate stables for 20–30 head of cattle were used for the purpose. Where there were no such facilities available, barns were converted to house the animals in the villages and, if there was enough space, larger cowsheds for 50–70 animals were added to them. The newly built or converted stables in the villages, with too many animals crowded in and insufficient technology, though serving their economic purpose, did not bring any marked improvement in working conditions. Moreover, they had a negative effect on the village environment.

During the next stage of development, the institute 'Agroprojekt' in Prague and its branch offices drafted, in 1954, so-called 'development plans' for all Unified Agricultural Cooperatives (i.e. for every village). These had an agricultural part, which included a calculation of the numbers of cattle and other livestock that could be supported by the available land. The technical part contained the selection of a site and planned the necessary sheds for the envisaged number of animals.

The adapted sheds in the villages were abandoned, including the adjacent grounds which had served as paddocks. Those buildings, yards and gardens, left to their fate, soon became derelict and an eyesore in the villages. The new cattlesheds, made to house 96–105 head of cattle, mostly without any storage rooms, were built on farm land, in order to comply with the extremely strict hygienic, veterinary and fire-prevention regulations.

The plan contained the distribution of the various kinds of animals into production centres. At the time, larger concentrations were already proposed:

500–600 dairy cows	or
400–600 calves	or
600–700 steers	or
200–400 sows	or
1500–3000 pigs	or
10 000 hens.	

As an alternative, the concentration of pig-finishing into three locations with 12 500 head each and two poultry breeding stations with 100 000 birds each was proposed in some districts.

Based on the results of these studies, principles and methods for all districts were to be derived. Compiled by design offices, university departments and research institutes, these studies were of key importance for territorial projections of agricultural production and its relationship to comprehensive district planning.

Though these territorial district studies undoubtedly brought some positive results, they could not be considered a final solution, in view of changing concepts and unclarified long-term problems related to economics. For this reason, work on them was stopped. However, the studies did show that it was just as unrealistic to approach problems of human settlement only from the angle of agricultural production, as to under-estimate or disregard farming when planning settlements.

In 1963—the next evolutionary stage—the Farm Production Authorities compiled so-called General Regional and District Plans as long-term outlines of agricultural production up to the year 1970. However, these were not comprehensive and did not approach farming investments and rural construction from the district-territorial angle. They were only a sectional back-up document.

The period of agricultural construction after 1960 was already characterized by attempts to introduce specialization and concentration of crop cultivation and cattle-farming, permitting wholesale production methods. Objects built in those days were already equipped with more up-to-date machinery, animals were mostly standing on grids without beds of straw. This raised labour productivity, e.g. one attendant tended 30 dairy cows. The sheds were at a distance from the villages and, to make construction cheap, were built on flat sites, often on the best arable land. This procedure in turn led to the construction of housing for the farm workers nearby, as well as of further farm buildings and installations, resulting in many cases in a shift of the village to the new site, outside the original settlement, which became deserted and neglected.

This state of affairs showed the need for approaching the development of villages by country planning methods. In 1958, the State Committee for Construction of that time (SVV) issued Regulation 22 on building documentation, including JZD buildings, which stipulated that all construction work had to be preceded by so-called technical–economic analysis (TER). This was a simplified method of country planning to be employed in individual villages.

Similarly, construction work at state farms had to be based on so-called technical–economic projects (TEP). These methods of planning and designing buildings could be applied only from 1960, when, concurrently with the territorial reorganization of Czechoslovakia's administration, the first mergers of Farming Cooperatives (JZD) to form larger economic units were taking place. As a result, the TER method, which was applicable only to one JZD within one village, no longer fully corresponded to the problems at hand, not meeting the need of merged Unified Farming Cooperatives.

Therefore, in this period of transition it became necessary to approach problems from the aspect of broader production relations. To this were added further, unclarified views on eliminating the difference between industry and farming and between town and rural areas.

On 22 November 1960, the Central Committee of the Communist Party of Czechoslovakia decided that all these matters should be treated in a pilot project covering ten selected districts, in the form of Territorial District Plans, part of which should be a territorial pilot plan for the economic territory covered by a single farming organization—i.e. the merged JZD. The Territorial District Plans included:

(1) a project for the territorial organization of the district and its settlement structure;
(2) a project for the territorial organization of farming production and construction.

Farming enterprises was proposed in a way that would facilitate maximum application of wholesale farming methods and technology.

The livestock farms, however, turned out to be mere shadows of the original concepts and had numerous operational, technical, architectural and constructional faults and shortcomings—such as un-reinforced access roads, connecting ways and handling areas, neglected (often devastated) immediate surroundings of the farms without any greenery, uninspired building designs, which were often left unfinished. Drains and lighting were bad, dung-heaps and dungwater-pits too small, hygienic facilities for the workers insufficient and frequently not functioning.

These shortcomings had several causes, the most important of which were (i) a noncomprehensive approach to design and construction, (ii) non-compliance with designs and plans, and (iii) limited funds which resulted in economizing measures. Architectural design was underestimated or suppressed for fear of increasing costs.

9.5.3 Brief Operational–Economic Evaluation of Large Cattlesheds

To date, no strict limits have been set as to how big a concentration of domestic animals is supportable for individual regions. There is no doubt that—despite the difficulties encountered—wholesale cattle farming will increase. A brief economic assessment is given below.

9.5.3.1 Calf-raising Stations (over 500 animals)

In 1977, in the Czech Socialist Republic alone, there were 69 calf stations with a capacity exceeding 500 animals. The daily weight increment during the milk-feeding stage was 616 grams, during the plant-feeding stage 706 grams. Production costs per kilogram amounted to 16.26 crowns. A serious problem was the high number of animals lost by forced slaughter and perishing, which averaged 9.07%.

9.5.3.2 **Raising Heifers** (capacity over 300 animals)

In 1977, there were 61 heifer stations with capacities exceeding 300 animals. The average capacity was 620 animals. The daily weight increment was 558 grams and production costs per kilogram averaged 19.48 crowns. Forced slaughter accounted for 1.89%, perishing 0.47%.

9.5.3.3 **Large-capacity Dairy Farms** (exceeding 300 animals)

In 1978, large-capacity dairy farms contained a total of 113 358 animals, corresponding to 8.6% of the total in the country. In one-quarter of them, annual milk yields were very low (less than 2600 litres per cow), while another 25% attained good milk yields (more than 3400 litres per cow). 30–40% of the dairy cows were bracked out annually, so that they did not even reach their maximum lactation age.

9.5.3.4 **Cattle-fattening Stations** (over 500 animals)

In 1977, there were 78 large cattle-fattening stations, averaging 754 animals. The daily increment in weight amounted to 767 grams and the costs of producing 1 kg of meat averaged 16.98 crowns. Out of the average number of steers, 9.84% had to be slaughtered prematurely and 1.21% perished. The highest losses were among animals stabled on grids, the lowest among those stabled on deep straw.

9.5.4 Guidelines for Assessing Wholesale Farming Buildings

In 1972, the *Guidelines for Assessing Buildings Serving Wholesale Farming Production with Regard to Environmental Protection*, which were compiled jointly by four ministries, were issued. After testing, these guidelines were to be transformed into generally binding legal regulations.

Plans for farming construction have to be considered very carefully. Decisions on location and erection of new large-scale farming operations must be preceded by a comprehensive analysis of all their possible effects on and consequences for the environment. The main responsibility for this activity will rest with the People's Committees at all levels. It will be primarily in their hands to minimize the negative impacts on the environment.

9.5 CONCLUSIONS

The goal of the socialist system, which is interested in permanently high production and not short-term profit, must be to strive for the optimum, not the maximum, in managing its natural resources. Therefore, the urgent need to eliminate possible conflicts between economic and biological reproduction is moving into the foreground.

Efforts must increase to optimize economic development while keeping in view the ecological aspects of the natural environment and the healthy progress of human society. This involves a constant battle to suppress one-sided and biased approaches inclined to give priority to group or local interests.

9.7 BIBLIOGRAPHY

Houška, V., *et al*. (1971) *Vývoj zemědělství a výživy v ČSSR*. SEVT, Praha.
Jůva, K., *et al*. (1978) *Pozemkové úpravy*. SZN, Praha.
Jůva, K., Klečka, A., and Zachar, D. (1975) *Půdní fond ČSSR*. Academia, Veda.
Kabrhel, J., *et al*. (1975) *30 let vývoje československého zemědělství a jeho další perspektivy*. Horizont, Praha.
Kudrna, K. (1979) *Zemědělské soustavy*. SZN, Praha.
Lom, F. (1957) *Vývoj osevních ploch obilovin a sklizní od 16. století v Čechách*. Sborník ČAZ, Praha.
Lom, F. (1960) *Soustavy hospodaření v zemědělství*. Sborník VŠZ, Praha.
Lom, F. (1972) *Přehled dějin zemědělské výroby v Čechách*. VŠZ, Praha.
Mezera, A., *et al*. (1979) *Tvorba a ochrana krajiny*. SZN, Praha.
Šimon, B. (1951) *Technicko-hospodářské úpravy půdy*. Brázda, Praha.
Zachar, D. (1970) *Erozia pôdy*. Vydavatelstvo SAV, Bratislava.
Žák, L. (1947) *Obytná krajina*. Mánes—Svoboda.

Čísla pro každého (1968–78) FSÚ, ČSÚ, SSÚ, SNTL/Alfa.
25 let socialistického zemědělství (1974) FSÚ, Praha.
Soupis vyhodnocení velkokapacitních stájí pro skot (1974) MZVž ČSR, Praha.
Statistická ročenka Československé socialistické republiky. (1975, 76, 77, 78) SNTL/Alfa.
Statistická ročenka (1963 to 1977) MZVž, ČSR, SSR.
Statistická ročenka o půdním fondu v ČSSR (1965 to 1980) ČÚGK, SÚGK.

Land Transformation in Agriculture
Edited by M. G. Wolman and F. G. A. Fournier

CHAPTER 10

Effects of Intensification of Agriculture on Nature and Landscape in the Netherlands

W. B. Harms, A. H. F. Stortelder and W. Vos

10.1 INTRODUCTION

In the Netherlands, agriculture largely determines the quality of nature and landscape. During this century, as agriculture has changed, so have nature and landscape. Now we are faced with the special situation of a small and crowded country (Table 10.1) with an extremely productive agriculture (Table 10.2) and an environment that is stressed by it. The per-hectare yields of products such as winter wheat, potatoes and sugar-beet are the highest in the world and milk production is still rising (Table 10.2). The Netherlands also has one of the highest percentages of agricultural land in Europe (Table 10.3), and potentially an exceptional diversity of natural ecosystems, because of its location and physical conditions.

The change in Dutch agriculture is an intensification. This will be discussed further in section 10.2. The discussion will be limited to changes that have occurred during the twentieth century.

The direct or indirect effects of this intensification on nature and landscape are manifested in three ways:

(1) on the biotic components of ecosystems;
(2) on the anthropogenic elements and structures in landscape that have a specific cultural and/or historical value;
(3) in the visual aspects of landscape: the scenery.

These three categories are discussed in sections 10.3–10.5. The intensity of the effects varies regionally: in various types of landscapes we are confronted with various effects.

The effects and the causes of the intensification are placed in an international context in section 10.6, while to obtain a preview of future developments, developments in rural planning policy are discussed in section 10.7.

Table 10.1 Population and civilian labour force within the EEC (*Selected Agri-figures*, 1980)

	NL	B	L	D	F	I	UK	IR	DK	EEC	USA
Total population (millions)	13.9	9.8	0.36	61.3	53.3	56.7	55.9	3.2	5.1	259.7	218.5
Density of population (per km^2)	377	323	138	247	97	188	228	46	119	170	23
Total active civilian labour force (millions)	5.0	4.0	0.15	25.7	22.1	21.5	26.0	1.1	2.6	108.2	100.4
Share, ditto, as a percentage of total population	36.2	40.5	42.2	41.9	41.4	37.9	46.6	35.2	50.7	41.7	46.0
Workers in agriculture, fishery and forestry (millions)	284	118	8.4	1.608	1.907	3.090	654	229	216	8.114	3.937
Share, ditto, as a percentage of total labour force	5.6	3.0	5.6	6.3	8.6	14.4	2.5	20.2	8.3	7.5	3.9
Number of agricultural workers per 100 ha of agricultural land	14	8	6	14[a]	7[a]	23[a]	5[a]	4	7	10[a]	2
Annual decrease of the number of agricultural workers, 1968–78 (%)	2.1	5.2	4.6	4.4	4.7	3.5	2.6	3.0	2.9	1.9	1.9

[a]Low productive grassland areas have been subtracted from total agricultural land.

Table 10.2 Average yields of some agricultural products within the EEC (*Selected Agri-figures*, 1980)

		NL	B/L	D	F	I	UK	IR	DK	EEC	USA
Farm produce (100 kg/ha)											
Wheat	1966–68	44.2	38.6	38.7	21.6	22.8	38.6	39.8	45.8	32.7	18.1
	1976–78	57.5	46.5	45.5	43.5	24.8	46.4	47.1	50.6	38.8	20.8
Rye	1966–68	30.1	31.2	30.6	19.3	17.9	26.7	18.5	31.7	30.8	14.1
	1976–78	35.1	34.4	35.2	28.5	22.1	32.1	24.7	34.5	33.9	15.0
Barley	1966–68	37.6	36.2	34.5	32.0	15.0	35.8	37.4	38.4	35.5	22.0
	1976–78	45.4	46.0	41.1	35.5	25.8	40.3	42.9	37.6	38.3	24.6
Oats	1966–68	39.8	35.5	33.0	35.8	13.6	32.2	30.5	37.9	10.7	17.7
	1976–78	47.3	32.8	35.6	29.5	17.6	37.6	37.3	31.8	31.8	18.1
Potatoes	1966–68	337	313	283	204	118	247	264	239	241	236
	1976–78	340	293	272	229	165	281	265	230	258	293
Sugar-beet	1966–68	467	481	473	426	381	375	372	415	421	393
	1976–78	472	472	451	427	457	324	408	386	430	454
Dairy produce											
Milk (kg per cow)	1966–68	4.260	3.733	3.709	2.977	2.697	3.729	2.404	3.948	3.393	4.008
	1976–78	4.920	3.730	4.196	3.326	3.242	4.602	2.996	4.700	3.872	5.035
Fat content of milk (%)	1966–68	3,83	3,33	3,74	3,70	3,60	3,77	3,58	4,24	3,74	3,68
	1976–78	3,97	3,42	3,83	3,76	3,55	3,80	3,52	4,25	3,79	3,66

Table 10.3 Area and distribution of the agricultural area in use; number and size of holdings within the EEC (*Selected Agri-figures*, 1980)

		NL	B	L	D	F	I	UK	IR	DK	EEC	USA
Area in 1978 (millions of hectares)		3.70	3.05	0.26	24.86	54.91	30.13	24.41	7.03	4.31	152.65	936.31
Agricultural land		2.17	1.45	0.13	13.18	32.20	17.55	18.46	5.72	2.93	93.77	446.38
of which arable land (including temporary grasslands and fallow)		0.86	0.72	0.06	7.44	17.15	10.02	6.82	0.99	2.63	46.69	203.56
of which horticulture		0.12	0.04	0	0.58	2.05	2.38	0.25	0.01	0.02	5.47	2.82
of which permanent grasslands		1.19	0.68	0.07	5.16[a]	12.99	5.15[d]	11.38[e]	4.71	0.28	41.61	240.00[f]
Forests		0.31	0.61	0.09	7.22	14.56	6.04	2.06	0.31	0.50	31.70	288.00
Total number of holdings (>1 ha) (×1000 holdings)	1970	164	130	6.9	1083	1421	2173	312	267	143	5701	5701
	1977	137	99	5.2	859	1148	2007	262	257	124	4899	2409[a]
	1978	134	97	5.0	844	1126	1985	260	254	122	4827	2370[a]
Average area per holding (ha)[g]	1970	13.0	11.6	19.4	11.7	21.0	7.7	56.8	17.7	20.7	15.5	169
	1977	15.0	14.5	25.4	14.3	25.4	7.9	65.6	21.7	23.5	17.7	177
	1978	15.2	14.8	25.9	14.6	25.9	7.9	66.0	22.5	23.9	17.9	180

[a]Farms having sales of $1000 or more.
[b]Includes 1.4 million ha 'Viehweiden'.
[c]Includes 3.2 million ha 'pacages et parcours'.
[d]Includes 4.0 million ha 'pascoli'.
[e]Includes 6.4 million ha 'rough grazings'.
[f]Includes 220 million ha 'non-forest pasture and range'.
[g]Excludes holdings less than 1 ha.

From this discussion it is clear that the conflict between agriculture and nature and landscape conservation is far from over, while it is precisely this polarity that harms the interests of an important part of agriculture (mostly the less productive part) as well as the quality of nature and landscape.

10.2 INTENSIFICATION OF AGRICULTURE

At the beginning of the twentieth century, agricultural incomes rose because of increasing production brought about by land reclamation and improved biological techniques (e.g. in plant and livestock breeding), coupled with improved health.

After the Second World War the emphasis in agricultural developments shifted from an increase in production to an increase in productivity, especially in output per person and per hectare, resulting in higher farm incomes even though food prices remained low and production costs rose.

This increase in productivity has mainly been achieved by the following processes:

(1) increasing use of chemical fertilizers and pesticides;
(2) increasing use of fodder concentrates;
(3) replacement of labour by machines;
(4) improved farm management;
(5) improved external conditions of production.

These processes will be discussed briefly.

Fertilizer application has been raised to maxima of 400–500 kg of N per hectare per year for clay and sandy soils, and 250 kg of N per hectare per year for peat soil. Compared with other countries within the European Community, the use of nitrogenous fertilizers is very high in the Netherlands (Table 10.4).

The increased use of fodder concentrates and roughage, together with a higher density of livestock, has led to excessive production of manure, which is difficult to dispose of and is often dumped on maize fields and pastures, to the detriment of the environment.

Mechanization has made great strides (Table 10.4). In dairy farming, for instance, the introduction of the milking machine has had a great impact on labour productivity. More and more farmers are forced to milk indoors, and this has consequences for the distribution of land: it becomes desirable for at least two-thirds of the land to be located close to the farm buildings. But mechanization has led to more investment and often to a higher burden of debt for small farmers. Recently, the introduction of the milk tank for the convenience of the dairy factories has forced many small farmers out of business because they were unable to invest more in machines. The same is true for other types of farm machinery, such as harvester-threshers (com-

Table 10.4 Farm machinery and use of fertilizers within the EEC (*Selected Agri-figures*, 1980)

	NL	B L	D	E	I	UK	IR	DK	EEC	USA
Farm machinery, 1977										
Agricultural tractors (×1000)	164	110	1.453	1.382	910	497	125	186	4.827	4.370
Milking machines (×1000)	76	51	459	383	115	113	44	69	1.309	11
Ditto, per 100 milch cows	35	49	85	37	32	29	63	46	1	
Harvester threshers (combines) (×1000)	6.5	10	171	154	30	58	4.5	41	474	535
Ditto, per 1000 ha of grainland	28	22	32	16	7.2	16	12	23	18	7.4
Fertilizers (kg per ha of agricultural land										
Nitrogenous fertilizers	217	114	100	55	46	63	48	127		
Phosphate fertilizers	42	71	66	57	36	32	36	48		
Potash fertilizers	54	99	89	47	17	22	42	60		

bines). The new agricultural technology is visible in the landscape in the form of landmarks such as new farm buildings and tower silos.

The improvement in the external conditions of production has been fostered by the national government by means of the Land Consolidation Act (1924, subsequently modified in 1938, 1954 and 1975). This government intervention was necessary because the Dutch farmer was tied to an inherited traditional agriculture, and therefore there were difficulties in introducing rationalization.

Land consolidation is the most important policy instrument to improve the structure of Dutch agriculture, because in many cases plots are small and scattered for historical reasons, and there are problems of water control and the size of the farms. After land consolidation, modern technology can be properly applied. Land consolidation is mainly concerned with the following:

(1) Enlargment of parcels where the individual plots are small. Irregular fields surrounded by hedges make farming time-consuming.
(2) Accessibility. Agricultural roads are improved by being surfaced and the road network is enhanced.
(3) Water control. Nearly half of the agricultural land is subject to waterlogging in winter. By canalizing streams and draining the land, the carrying capacity of the land for cattle and machines is improved. The growth of plants also benefits, thus increasing production.
(4) Enlargement of holdings. Many holdings are too small and therefore uneconomic. In 1947 the economically desirable size was 10 ha; today the norm is 30 ha for arable farming and 25 ha for dairy farming. In a land-consolidation project, land is bought on the free market and used for the necessary enlargement of the remaining holdings. As in other European countries, one of the results has been that the number of holdings of less than 20 ha has decreased rapidly, while the number of larger holdings (mainly those larger than 50 ha) has increased (Table 10.5).

Now more than one-half of the agricultural land-consolidation projects have been completed or are in progress (Table 10.6).

What kinds of intensification can be expected in the future in arable farming and dairy farming?

In arable farming, intensification is apparent from the increase in potatoes and beet in the crop rotation scheme (25% in 1965 and 43% in 1977). This development is made at the expense of grains, especially rye and oats. Maize has also become important in relation to dairy farming (0.4% in 1965 and 16% in 1977 in crop rotation schemes). More uniformity in crop rotation can be expected in favour of potatoes and beet. Other farms, already specialized, will aim at enlarging their size and increasing their crop yields. These developments require a high standard of quality and uniformity of the

Table 10.5 Changes in the size of farmholdings within the EEC between 1970 and 1978 (*Selected Agri-figures*, 1980)

		NL	B	L	D	F	I[a]	UK	IR	DK
Classification of holdings according to size of holding (number)										
1970	1– 5 ha	42 497	44 172	1 481	405 768	325 671	1 487 043	57 586	54 604	16 579
	5–10 ha	39 155	33 059	994	232 713	250 466	387 213	39 282	56 435	30 077
	10–20 ha	52 079	33 480	1 547	267 785	354 826	182 188	49 788	81 544	43 971
	20–50 ha	27 881	17 089	2 619	157 557	369 610	80 174	80 545	59 439	144 084
	>50 ha	2 507	2 597	298	19 295	120 351	36 845	84 277	14 912	8 689
	Totals	164 119	130 397	6 939	1 083 118	1 420 924	2 173 463	311 478	266 453	143 400
1978	1– 5 ha	32 478	28 200	962	281 223	221 000	1 352 354	36 148	39 091	13 910
	5–10 ha	27 970	19 895	581	161 229	169 000	345 028	32 779	42 482	22 495
	10–20 ha	39 887	25 901	795	194 070	243 000	165 624	40 718	76 343	33 453
	20–50 ha	30 443	19 107	1 975	177 986	347 000	85 768	69 119	73 241	41 413
	>50 ha	3 599	3 687	689	29 134	146 000	36 293	81 438	22 558	10 993
	Totals	134 377	96 790	5 002	843 642	1 126 000	1 985 067	260 202	253 715	122 264
Ditto (%)										
1970	1– 5 ha	26	34	22	37	23	68	18	20	11
	5–10 ha	24	25	14	21	18	18	33	21	21
	10–20 ha	32	26	22	25	25	8	16	31	31
	20–50 ha	17	13	38	15	26	4	26	22	31
	>50 ha	1	2	4	2	8	2	27	1	6
	Totals	100	100	100	100	100	100	100	100	100
1978	1– 5 ha	24	29	19	33	20	68	14	15	11
	5–10 ha	21	20	12	19	15	18	13	17	19
	10–20 ha	30	27	16	23	21	8	16	30	27
	20–50 ha	23	20	39	21	31	4	26	29	34
	>50 ha	2	4	14	4	13	2	31	9	9
	Totals	100	100	100	100	100	100	100	100	100
Average annual increase or decrease of holdings between 1970 and 1978 according to size of holding (%)										
	1– 5 ha	−3.3	−5.5	−5.3	−4.5	−4.7	−1.2	−5.7	−4.5	−2.2
	5–10 ha	−4.1	−6.2	−6.4	−4.5	−4.8	−1.4	−2.3	−3.5	−3.6
	10–20 ha	−4.1	−3.2	−8.0	−4.0	−4.6	−1.2	−2.4	−0.2	−3.4
	20–50 ha	+1.1	+1.4	−3.5	+1.5	−0.8	+0.8	−1.9	+2.6	−0.8
	>50 ha	+4.6	+4.5	+11.0	+5.3	+2.4	−0.2	−0.6	+5.3	+3.0
	Average all groups	−2.5	−3.7	−4.0	−3.1	−2.9	−1.1	−2.3	−0.5	−2.0

Table 10.6 Completed land consolidation projects since 1924 (in 100 ha) (Algemene Milieustatistiek, 1977/78)

	1960	1970	1975	1976	1977	1978
Groningen	30	156	244	370	380	370
Friesland	60	308	674	780	818	872
Drenthe	92	336	665	735	785	785
Overijissel	252	613	824	873	902	939
Geiderland	212	519	847	891	1 040	1 070
Utrecht	70	143	143	143	143	143
Noord-Holland	49	236	303	313	321	321
Zuid-Holland	–	35	115	208	208	208
Zeeland	215	687	814	814	814	814
Noord-Brabant	218	396	1 051	1 189	1 238	1 248
Limburg	145	256	451	451	451	451
Nederland	1 341	3 684	6 130	6 766	7 098	7 220

products and therefore homogeneity in soil conditions (physical as well as chemical). Because of the economies and capacity of machines and their requirements regarding the size of land parcels, further enlargement of plots can be expected.

In dairy farming, intensification will involve stepping up the use of fertilizers and the stocking rate per hectare of grassland. Developments in biological techniques will lead to an increase in production, a more even production of grass over the seasons, improved land use and fewer crop failures. This can be achieved by increased use of fertilizers, irrigation and re-sowing poor grassland. The consolidation of buildings and holdings will be improved by reallotment, enlargement of parcels and water management. These labour-saving developments will lead to the expulsion of the workforce or, if the workforce is retained, then farm management must intensify still more.

10.3 EFFECTS ON NATURE

No form of land use has had more influence on nature than agriculture, and therefore ecosystems have changed considerably as a result of agricultural intensification. Large-scale changes in agriculture have only occurred since the Second World War, and thus it is only recently that agriculture has grown from controller of nature and landscape into their greatest threat.

We can distinguish two important groups of causes: (i) interventions in the sphere of land development; and (ii) changes in the intensity of land use.

For the benefit of nature conservancy, efforts are being made to preserve the diversity of species of plants and animals by maintaining spatial variation.

In contrast, land development techniques aim to equalize conditions as much as possible.

In former times the possibilities for man to manipulate his environment were restricted; a farmer was therefore forced to adapt his farming system to the environmental conditions. In an area with a great variety of dry and wet habitats and with different soil types, the resulting great spatial variety was expressed in several kinds of agricultural land use; each land use type was intrinsically stable. These two features of land use—spatial variety and stability—accentuated the diversity in the occurrence of plant and animal species, i.e. their communities and habitats. An important feature of agriculture in the past that supported this differential land use was the limited availability of extra minerals from outside the systems. There was a more or less closed system, in which manure was the most restricting element.

10.3.1 Interventions in the Sphere of Land Development

In order to increase food production and productivity, there has been large-scale interference in the system in recent decades, which has had far-reaching consequences for nature. The most important will be dealt with here.

10.3.1.1 Interference in Water Management, to Regularize the Water Table

Generally this has meant levelling off the differences between dry and wet areas and decreasing the spring floods by canalizing streams and lowering the water table in areas where it used to be high. It goes without saying that the ecosystems that are typical for these moist areas have deteriorated greatly. The list below illustrates the threat to the nearly 100 plant communities (associations) in the Netherlands that are dependent on moist or wet conditions (after Westhoff, in: de Molenaar, 1980):

facing extinction	1%
gravely threatened	46%
threatened	34%
hardly threatened	19%

10.3.1.2 Soil Disturbance

As a result of measures like deep ploughing (disturbance of B horizons) of sandy soils, levelling the relief, etc., existing and potential spatial differences in the soil are greatly decreased. The disturbance of B horizons can destroy a perched water table. Levelling the relief often wipes out micro-differentiation in ecosystems. These processes have often occurred and still do occur in Pleistocene areas in the eastern part of the Netherlands.

10.3.1.3 Enlargement and Alteration of the Parcels

Although these processes are also relevant for cultural and historical aspects of the landscape, they play an important part for nature conservancy. The shape and size of the parcels are determined and delimited over most of the Netherlands by hedges, wooded banks, lines of trees and ditches. These landscape elements are themselves important for many species of organisms, and generally speaking their removal leads to a decline in several species of plants and animals. The enlargement of parcels often coincides with the disappearance of the line elements (i.e. hedges, lines of trees). Larger land units mean that the animals have less chance to survive. Hedges and wooded banks are also important for the regulation of micro-climate. However, their real function for agriculture is questionable.

10.3.2 Changing the Type and Intensity of Land Use

The decline in the flora and fauna must also be attributed to fundamental changes in land use. Formerly the rule was for each area to be farmed differently, according to its physical constraints, but the pattern remained stable. Modern agriculture has upset this stability: nowadays farming is the same throughout the country, but is characterized by its mutability. This has been fatal for wildlife.

A closely related development is the increasing use of fertilizer and the growing input of minerals in the form of concentrates. These minerals finally end up on the land. The eutrophication of the soil and of ground and surface water is stimulated by this process, and so not only agricultural areas but also natural sites are influenced. The farmer has changed from a collector of manure into a manure spreader (Figures 10.1 and 10.2). The habitats of organisms that are adapted to infertile environments are vanishing. The various types of grassland consists of plant communities that are characterized by a low species diversity (only a few highly productive species).

Mainly as a result of the processes in agriculture mentioned above, more than half of the Dutch flora is threatened with extinction. Between 1950 and 1970, 36 species are presumed to have become extinct, while 260 species were under threat (in 1950 the number of threatened species was 174). This decline refers to the number of find-spots expressed in standard inventory grid squares of 5 × 5 km (Table 10.7). The decline is seen in individual species as well as in the vegetation, which has clearly been impoverished. An extreme example is the decline in the Cirsio-Molinietum (hay-fields, very rich in species, with a high water table). The decline of these hayfields, mainly caused by intensification of agriculture since 1900, is as much as 99.9%.

Another aspect of agricultural intensification that has a negative influence on nature is the substitution of machinery for manpower. The consequence is

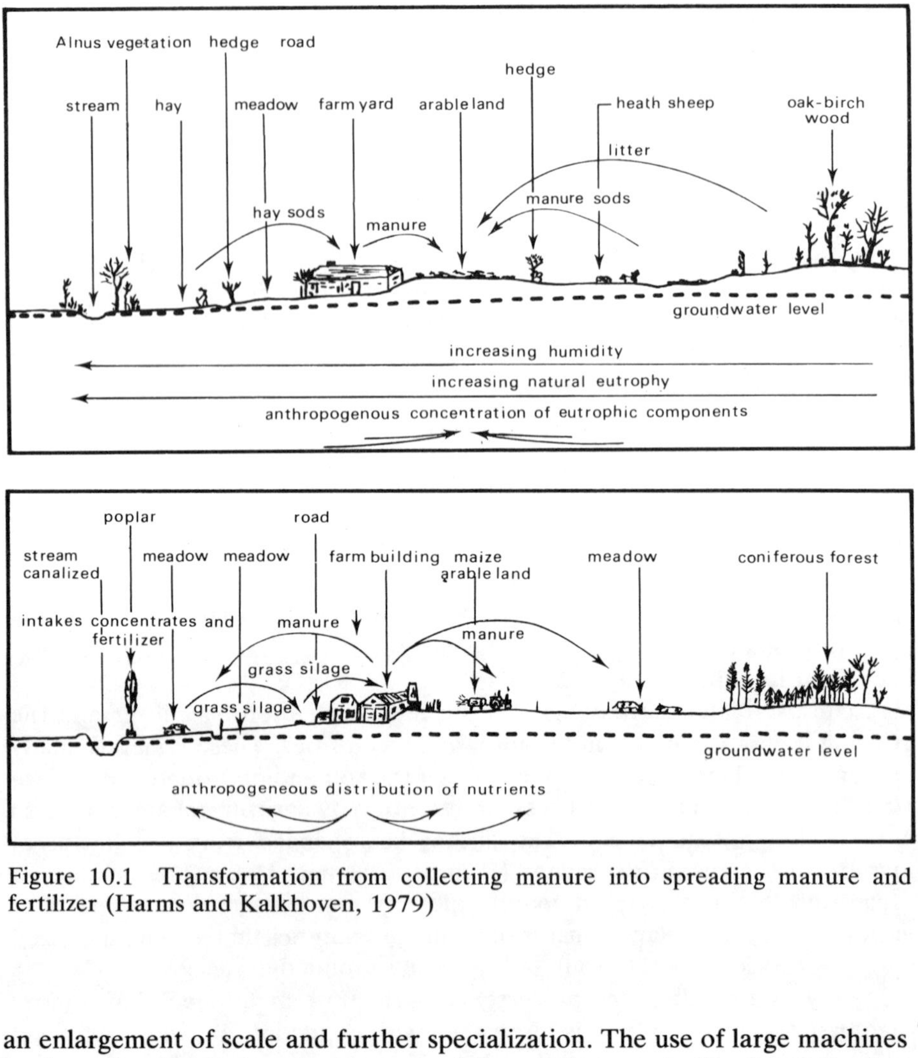

Figure 10.1 Transformation from collecting manure into spreading manure and fertilizer (Harms and Kalkhoven, 1979)

an enlargement of scale and further specialization. The use of large machines leads to local soil compaction, especially on the peat soils. Specialization diminishes spatial differences (a narrowing to one or two crops).

The withdrawal of manpower also means that the farmer has little or no time to manage certain landscape elements, i.e. clearing ditches mechanically, cutting hedges, mowing verges, maintaining agricultural roads by the use of sods, etc. These landscape elements are no longer maintained, or else are managed with the help of pesticides or machines. This has had negative consequences for wildlife.

Finally, the consequences of the accelerated grassland management in spring need to be mentioned. The result of early spring mowing is that the

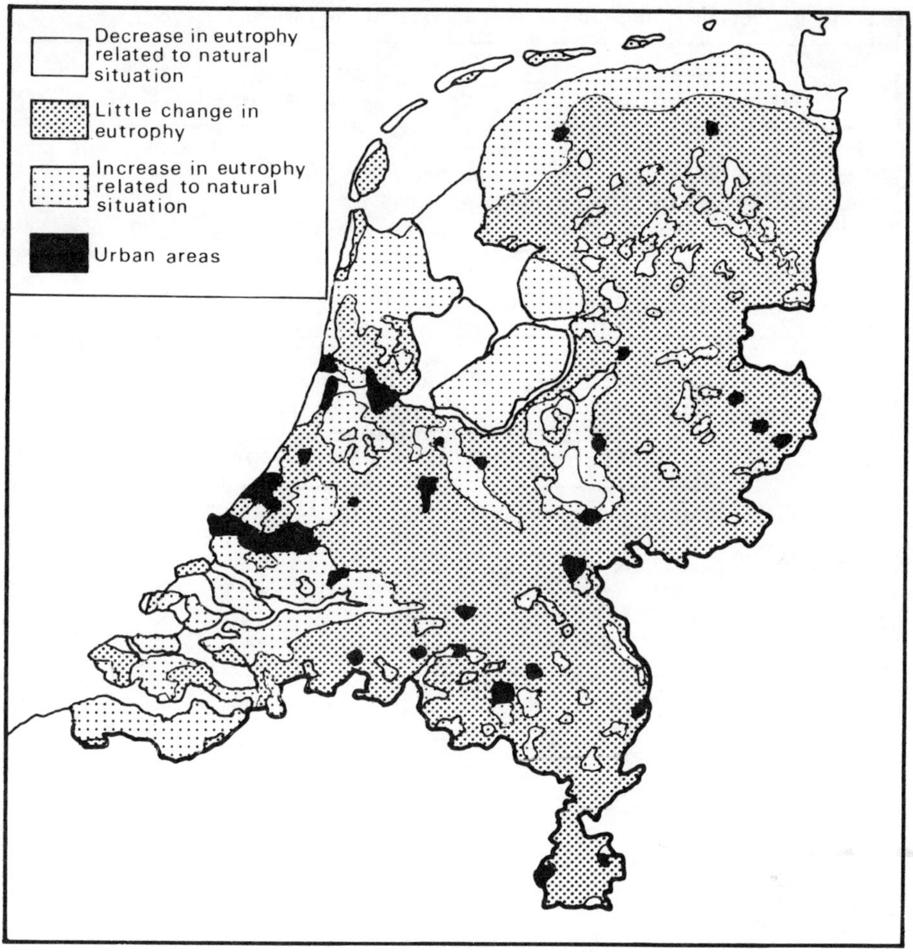

Figure 10.2 Change in eutrophy anthropogenic interference (situation before and after 1975) (van der Maarel and Dauvellier, 1978)

plant species are not given the chance to flower and to propagate. Moreover, the intensification of grassland management, together with the lowering of the groundwater level, have severely threatened the meadow birds. The Netherlands is an important area for grassland birds: see Table 10.8 and Fig. 10.3.

Table 10.9 shows the difference in density of breeding pairs on intensively and extensively managed agricultural areas. In the last 30 years, the number of breeding pairs of some species of this group has decreased markedly (in some areas by up to 90%).

Table 10.7 Decline in the Dutch flora since 1950 (de Molenaar, 1980)

	Number of species	%	%
Flora:			
Total number of species	1502	100	
Number of species that have diminished	824	55	100
diminished by 50%	108	7	13
51–70%	423	28	51
71–90%	182	12	22
91–100%	111	7	13

Table 10.8 The population (breeding birds) of some characteristic bird species of grassland areas in central and western Europe as percentages of the populations (de Molenaar, 1980)

Black-tailed godwit	90%
Lapwing	50%
Redshank	50%
Ruff	75%
Snipe	40%

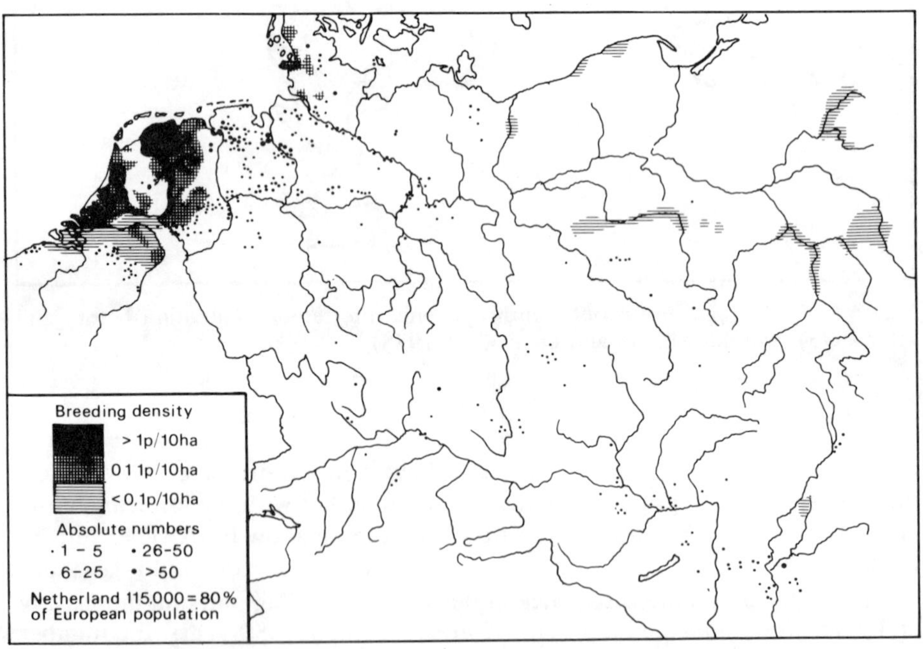

Figure 10.3 Distribution of the black-tailed godwit as a breeding bird in Central Europe, 1966–75 (Glutz von Blotzheim *et al*., 1977)

Table 10.9 Density of breeding sites of some bird species of grassland areas, in pairs/100 ha, in the polder 'de Schermer' (Eyerman, 1975)

Species	Extensive management	Intensive management	Decline (%)
Oyster-catcher	33	0	100
Lapwing	133	4	97
Black-tailed godwit	120	2	98
Redshank	53	4	92

10.4 EFFECTS ON CULTURAL AND HISTORICAL ELEMENTS AND STRUCTURES

The landscape itself is the expression of both natural and cultural processes that are operating today or have operated in the past. Therefore the landscape is like a palimpsest, on which every generation has put its own marks, sometimes partly or entirely erasing the signs made by the preceding generation.

The landscape can be compared to a history book: the elements and patterns in a landscape may tell its history. Examples of elements are farms, mills, bridges and archaeological sites (e.g. burial mounds, stone circles). Examples of patterns are settlement patterns and road patterns. To trace changes in the landscape pattern, old topographical maps of the Netherlands have been compared with recent maps (dating from about 1970). Attention was paid to land ownership, accessibility, cultivation and urban sprawl. The changes are shown in Figure 10.4.

Because agriculture has developed along only a few lines in the Netherlands (crop-growing or large-scale livestock farming), regional variations are being ironed out. This levelling process is also seen in the following cultural and historical aspects of the landscape:

(1) Parcels. Parcel size has increased: the parcels have become more rectangular to facilitate cultivation with heavy machinery.
(2) Patterns of ditches, hedges and wooded banks. These are intimately associated with parcels and parcel patterns, and have therefore declined markedly.
(3) Road patterns. Old roads have often been paved and widened. Almost inaccessible areas have been opened up. Transport by boat—formerly the norm for large parts of west Netherlands—has become exceptional.
(4) Settlement patterns and architecture. Because of research on the most efficient ways of building, and improved communications, the enormous regional variety in farm architecture has given way to a few types of farms. The traditional way of building is often no longer practicable.

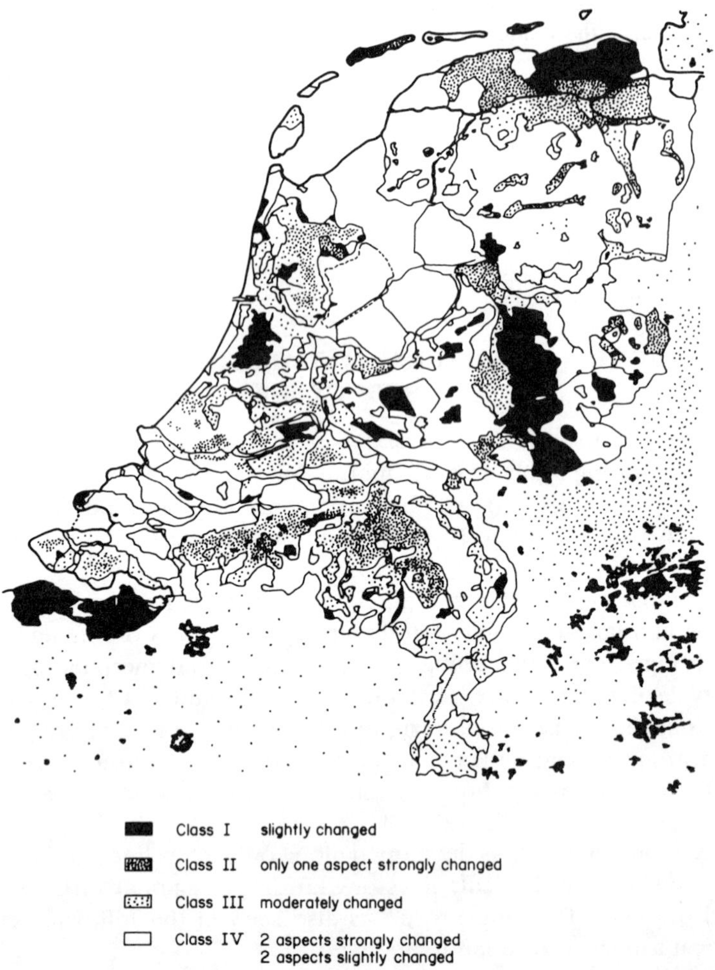

Figure 10.4 Degree of change of landscape pattern (Jaarverslag Rijksplanologische Dienst, 1974)

Often the settlement patterns have also changed, especially in land consolidation areas. Patterns in which farms were clustered have changed enormously because of the re-siting of farms (especially in the eastern and central parts of the Netherlands).

(5) Land use. The diversity of land use types has also diminished strongly because of the disappearance of mixed cultivation. The contrast between wet and dry areas on a farm, or near and far parcels, has vanished, thanks to improved drainage and land consolidation, respectively. This, and the loss of the farming system's self-sufficiency, can be mentioned as causes

of the disappearance of several land use types. Some traditional activities (e.g. harvesting thatch) are no longer economic and have almost disappeared. Where old arable land has been converted into grassland, these 'essen' (interesting from the archaeological point of view) have become difficult to distinguish in the landscape and archaeological research has become more difficult.

The developments mentioned above pose a threat to the regional identity of the rural area. Research should be done on regional rural architecture, and areas should be managed more in accordance with their regional features. This would enable the 'living' landscape to continue to evolve and maintain its diversity.

10.5 EFFECTS ON THE SCENERY

Landscape is a source of sensual perception. The information in the scenery is picked up and interpreted by people. The interpretation is influenced by prior experience, knowledge, memory and other factors.

There are three main criteria by which the perceptions of scenery can be classified:

(1) cognitive aspects (e.g. legibility, orientation and identity);
(2) emotional aspects (e.g. beauty, naturalness, history, security and concept of place, symbolic significance, stress and hazards);
(3) behavioural aspects (e.g. accessibility, flexibility and congruence between form and function).

The changes in the scenery resulting from the intensification of agriculture have to do with convergent developments in the rural areas. As explained earlier in this chapter, the process is one of levelling: characteristic features become blurred.

Typical extensive, open landscapes have become more closed (e.g. because farms have been re-sited); closed landscapes have been made more open by enlargements of parcels and removal of hedges and small woods (Figure 10.5).

Since the 1970s, research has been done on the visual analysis and perception of different landscapes (Coeterier, 1979; Landschapsbeeld, 1976; de Veer and Burrough, 1978).

Several case studies have shown that these changes (especially in the land consolidation areas) are fully experienced by a large part of the population, mostly in a negative way (Coeterier, 1979).

10.6 INTERNATIONAL ASPECTS

The background to agricultural intensification in the Netherlands and its effects on nature and landscape can be placed in an international context.

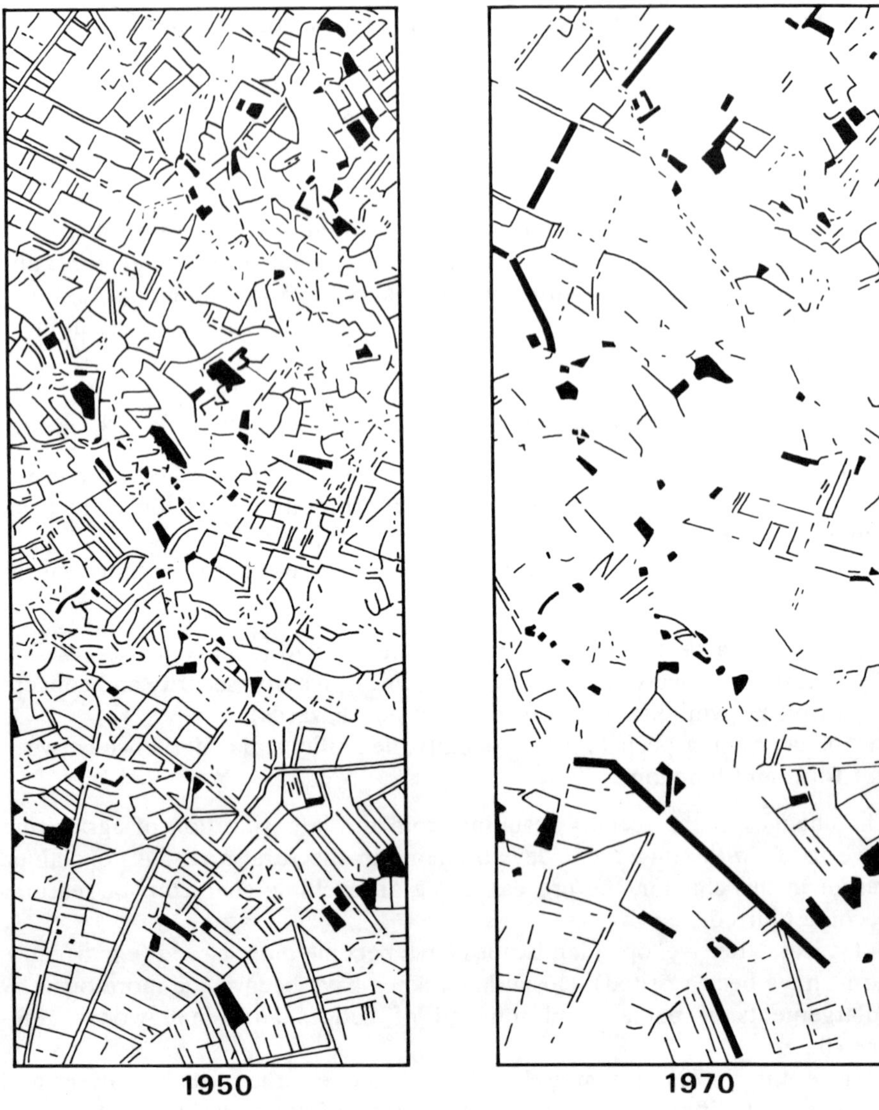

Figure 10.5 Small woods and hedges in 1950 and 1960 in a land consolidation area in 'de Achterhoek' (East Netherlands)

In the first place it has to be stressed that the development of Dutch agriculture is closely connected with the policy of the European Community. When considering the effects of agricultural intensification, we have to remember that some effects can cross frontiers. Also, Dutch nature and landscape are characterized by some internationally recognized specific qualities.

According to the Treaty of Rome, the agricultural policy of the European Community must fulfil the following tasks:

(1) deliver a reasonable income to farmers on well-managed enterprises;
(2) supply food to the European consumer at reasonable prices;
(3) not lose sight of the interests of trade with the rest of the world;
(4) help improve the farming structure by encouraging efficient management and sale.

For these purposes a Community marketing and prices policy was developed in the 1960s. In an attempt to keep pace with other economic interests, the Common Market introduced guaranteed prices above world market level. The difference between the guaranteed price and the market price is paid by the members of the Common Market. However, unrestricted production for these prices has caused structural surpluses in agricultural products (in the Netherlands especially in dairy products).

Faced with the fact that the organization of marketing and prices was not effective in solving problems in agriculture, in December 1968 the Commission introduced the '1980 Agricultural Memorandum', also known as the 'Mansholt Plan'. In the 1970s, this formed the basis of the policy on agricultural structure which was formulated in the following main directives of the Council of Ministers:

(1) the modernization of agriculture (Directive 72/159/EEC);
(2) promoting the closure of agricultural enterprises and the use of the land for the enlargement of other enterprises (72/160/EEC);
(3) the socio-economic enlightenment and education of people working in agriculture (72/161/EEC);
(4) compensation for permanent natural handicaps (hill-farming ruling; 75/268/EEC).

One of the results of this policy was that many farmers had to be phased out gradually, and some agricultural land had to be abandoned. In certain cases, farming on marginal land was subsidized for particular reasons (e.g. to ensure people remained in an area, for land conservation, etc.). In spite of these structural arrangements, the surplus in products has persisted.

Aware of the consequences of its policy for nature and landscape, in the 1970s the Commission initiated, as it were retrospectively, some studies on the ecological consequences of modern production techniques in agriculture. The results of these studies are still not traceable in the agricultural policy of the European Community.

The frontier-passing influence of intensification in agriculture is limited to just one of the migrating factors moving across Dutch frontiers: migrating birds. Because of its location the Netherlands cannot influence neighbouring countries by way of surface and groundwater, but can only be influenced by them. The influence of migrating birds cannot be quantified. Nevertheless, it is clear that given the important function of the Netherlands for international bird migration (as shown in Figure 10.3), the impact of agriculture on certain habitats will extend beyond the Netherlands. This is especially true for the species restricted to wetlands, e.g. waders and geese. As described earlier, lowering the water table is one of the major ways of improving physical conditions for agriculture.

Some Dutch landscapes are unique. This applies, for instance, to the low-lying polders and meres on abandoned turberies. This is, of course, true with respect to the scenery and the cultural and historical conditions, but also holds for the ecological conditions.

Intensification has also created some new landscapes (e.g. the polders claimed from the former Zuiderzee).

10.7 DEVELOPMENTS IN RURAL PLANNING POLICY

In this section attention is focused on the role of the national government, *vis à vis* developments in agriculture, the measures it has taken to stop the decrease in quality of the environment, and existing ideas about the possibilities of reconciling agricultural interests and the conservation of nature and landscape. Dutch government policy in rural areas was mainly directed at stimulating the intensification in agriculture to improve farmers' incomes. To achieve this, national government had two instruments at its disposal: the agricultural structural policy, and the marketing and prices policy.

The marketing and prices policy of the national government is, to a great extent, determined by the policy of the Common Market.

The agricultural structural policy of the national government fits in with the policy of the Common Market. For many years land consolidation formed the cornerstone of the agricultural structural policy. At first, land consolidation focused on agricultural interests only, but since the 1960s it has broadened its interests. Meanwhile, in agriculture there has been a shift from increasing production to increasing productivity. The multiple interests of land consolidation include improving outdoor recreation facilities, establishing nature reserves, and landscaping, as well as improving the physical conditions of agricultural production.

The promotion of other interests concerning nature and landscape is not only a matter of the preservation and rehabilitation of endangered flora, fauna and aspects of our cultural heritage, but also of the conservation of entire existing landscapes. To permit the latter, the government proposed the establishment of National Landscape Parks, i.e. areas characterized by a combination of nature reserves, traditional agricultural land and buildings or sites of historical and cultural importance. To prevent farm income in these areas dropping below the average because of the deliberate attempt to forgo agricultural rationalization and intensification, it was proposed to set up 'management areas' inside and outside NLPs. These would be areas of some 200 000 ha where farmers would voluntarily manage their farms under restrictions, for financial compensation. Thus, agriculture would be spared from modernization by the establishment of nature reserves on the one hand and by the introduction of management areas on the other hand. In this way, government policy would not only benefit interests other than agriculture, but would also conform to Common Market policy by discouraging farming on marginal agricultural land.

So far, these government proposals have not produced satisfactory results. There has been much resistance from the farmers to the concept of the National Landscape Park. They fear they will have to relinquish their independence and become park wardens instead. Only a few farmers have expressed willingness to enter into management arrangements in accordance with the governmental proposals. The nature and landscape conservation lobby are also less than enthusiastic about the government's proposals. They fear that saving areas in national landscape parks will be accompanied by further intensification of agriculture on the rest of agricultural land. Thus the establishment of landscape parks would lead to a deterioration in the rest of the landscape. It would also mean that preserved natural areas would remain like islands in a hostile sea. This would discourage interaction between organisms and lead to a further decline in ecological quality. The policy would lead to the spatial segregation of agriculture from nature and landscape conservation. Physical planners respond to the enlargement of scale in agriculture by saying that their policy in the rural areas would be to enlarge the scale of the conservation of nature and landscape concomitantly, by establishing large national landscape parks. But this is not a real change in agricultural policy.

The nature and landscape conservation lobby has become aware that preservation for the short term without changing the roots of agricultural development means even more decay in the long run. The strategy of the lobby has shifted from protection to active attack on the bottlenecks of agricultural policy. This strategy has found support among many farmers who are not satisfied with agricultural policy. They are aware that the policy is unable to solve the problems of surpluses, and in addition farmer redundancy is increasing because of the closure of farms that cannot fulfil the requirements for enlargement of scale.

The starting point for a new conservation strategy is the concept that a reasonable existence for farmers must, as far as possible, be in harmony with a warranted management of nature and landscape. It is essential to have a policy focused on production control. This could be achieved by a prices policy that would pay more for the first part of the production than for the following part. Production control would enable the total volume of dairy production to be fixed at farm level. This would mean that a further intensification of land use would be limited. It is also possible that in future the farmer could direct his skills to limiting costs and to enhancing the quality of the products as much as to increasing production. Only under these conditions will it be possible to integrate the production function with the management function in landscape.

10.8 CONCLUSIONS

As shown above, there always has been a strong relationship between landscape (and nature) and agriculture. In the past the agricultural system was strongly restricted by the environmental conditions and man was forced to adapt the land use system to local circumstances. The inability of man to manipulate these conditions resulted in a further fundamental differentiation of nature and landscape in space as well as in time. The factors man and environment were in equilibrium.

The intensification of agriculture as described earlier was made possible by technical progress. The landscape became modelled more and more according to rules for maximum food production. The subtle interaction between man and nature and landscape, which had existed for centuries, was disturbed. Man became dominant.

The consequences of this intensification of agriculture for nature, cultural and historical elements and structures in the landscape and for the scenery were enormous (as illustrated earlier). In other words, the production function of the landscape increased, but other functions that the landscape fulfilled for man and society decreased markedly: information functions (recreation, education, gene pool) and the natural regulation function.

10.9 REFERENCES AND BIBLIOGRAPHY

Algemene Milieustatistiek (1977/78) Centraal Bureau voor de Statistiek/ Staatsuitgeverij, 's-Gravenhage.

Coeterier, J. F. (1979) *De betekenis van de omgeving. VIII: Onderzoek naar het landschapsbeeld in Gieten en Heiloo*. Rapport Rijksinstituut voor onderzoek in de bos- en landschapsbouw De Dorschkamp, Wageningen, no. 217.

Dobben, W. H. van, Algra, S., Nijhoff, P., *et al*. (1980) Landbouw en natuurbehoud in Nederland: stellingen met discussies. *Landbouwkundig Tijdschrift*, **92**, (9), 346–354.

Eijerman, C. (1975) De problematiek in onze weidevogelgebieden. *Het Vogeljaar*, **23** (3), 105–107.

Glutz van Blotzheim, U. N., Bauer, K. M., and Bezzel, E. (1977) *Handbuch der Vögel Mitteleuropas*, Vol. 7. Charadriiformes, Wiesbaden.

Harms, W. B., and Kalkhoven, J. T. R. (1979) *Landschapsecologie en natuurbehoud in Midden-Brabant*. Rapport Rijksinstituut voor onderzoek in de bos- en landschapsbouw De Dorschkamp, Wageningen, no. 208; RIN-rapport 79/15, Leersum.

Jaarverslag Rijksplanologische Dienst (1974) 's-Gravenhage.

Landschapsbeeld. Algemeen Ruimtelijk Planningskader, deel 4 (1976) Rijksplanologische Dienst, 's-Gravenhage.

Maarel, E. van der, and Dauvellier, P. L. (1978) *Naar een Globaal Ecologisch Model voor de ruimtelijke ordening van Nederland*. Studierapport Rijksplanologische Dienst, 's-Gravenhage, no. 9.

Molenaar, J. G. de. (1980) *Bemesting, waterhuishouding en intensivering in de landbouw en het natuurlijk milieu*. RIN-rapport 80/6, Leersum.

Manten, A. A. (1975) Fifty years of rural landscape planning in the Netherlands. *Landscape Planning*, **2**, 197–217.

Selected Agri-figures of the EEC (1980) Ministerie van Landbouw en Visserij, 's-Gravenhage.

Veer, A. A. de, and Burrough, P. A. (1978) Physiognomic landscape mapping in the Netherlands. *Landscape Planning*, **5** (1), 45–62.

Verslag van de Werkgroep natuur- en landschapsbeheer door landbouwbedrijven (1979) Cie onderzoek Natuur- en Landschapsbeheer door landbouwbedrijven, 's-Gravenhage.

Vink, A. P. A. (1975) *Land Use in Advancing Agriculture*. Springer, Berlin.

Land Transformation in Agriculture
Edited by M. G. Wolman and F. G. A. Fournier

CHAPTER 11
Saline Seeps in the Northern Great Plains of the USA and the Southern Prairies of Canada

R. B. DANIELS

11.1 INTRODUCTION

In about 1945 saline seeps were recognized as a potential problem in non-irrigated agricultural areas of the semi-arid Northern Great Plains and Southern Prairies of the USA and Canada (Figure 11.1). Seeps were rare, however, in non-cultivated rangeland. During the last 10–15 years saline seeps have grown rapidly in cultivated areas and have been intensively investigated by numerous organizations (Van der Pluym, 1978). This report attempts to summarize and evaluate some of the work that has been published and is available to the reader in a large university library. Some of the works cited have not been read by the author (indicated by an asterisk) but are included because they have been extensively cited by individuals working in the area and indicate the large amount of literature that is now readily available.

Saline seeps are defined as areas of saline soils that have formed in non-irrigated areas within the last 30–40 years. This is to differentiate them from the saline soils that were present earlier, many before the areas were settled by Europeans. The seeps are wet some or all of the time, may or may not have a saline water discharge, and commonly have white salt crusts. Vegetative production ranges from abundant to almost nil, depending on the vegetation salt tolerance and the amount of salt collected at or below the surface (Miller *et al*., 1981). The seeps develop near the shoulders, backslopes and footslopes, usually accompanied by a distinct change in topography (Halverson and Black, 1974). The seeps do not always occur at the low point in the landscape.

The degradation produced by saline seeps, formerly called north slope

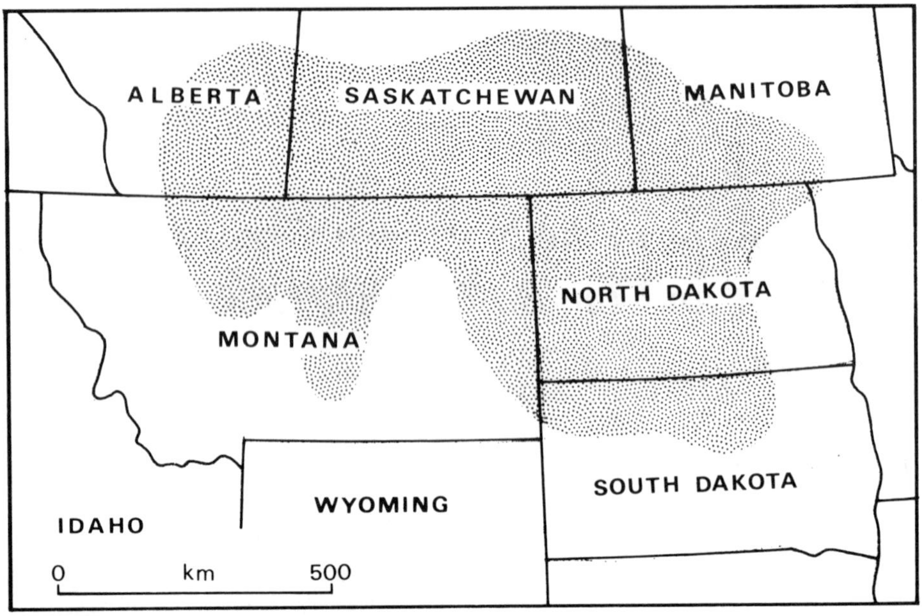

Figure 11.1 Areas (shaded) of potential saline seep development on the Northern Great Plains (Miller *et al.*, 1981)

alkali (Thacker, 1976), was hardly recognized in the early 1940s, but since then has taken approximately 0.8 million hectares out of crop production. Saline seeps and the factors leading to their development have degraded the surface and shallow groundwater supplies in many areas that are the only or primary source of potable water (Miller *et al.*, 1981). The high salt, particularly nitrogen, and metal content of the saline water is believed responsible for a number of livestock, wildlife and fish kills.

11.2 GEOLOGICAL SETTING

The dryland saline seeps in the prairie provinces of Canada and Great Plains of northern USA are underlain by dark, salt-laden, Cretaceous marine shales, and to a lesser extent the non-marine Tertiary sediments derived from the shales (Figure 11.2). The Colorado and Bearpaw shale in Montana and their equivalents elsewhere have a high amount of gypsum and other elements with lesser amounts of calcium. The Judith River–Claggett–Eagle (JCE) unit includes marine and non-marine units, but saline seeps are not widespread in its outcrop area. The groundwater of the JCE unit has a large salt load (Miller *et al.*, 1981). The soluble salt load contained within each unit is believed to be a reflection of the total dissolved solids (TDS) in the groundwater (Table 11.1).

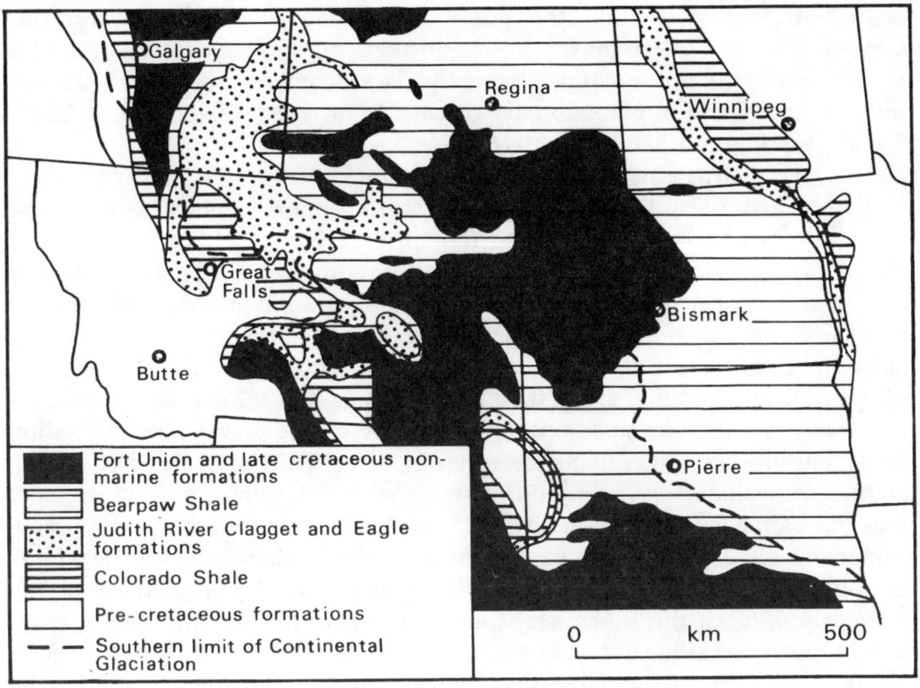

Figure 11.2 Generalized geological map of the Northern Great Plains Region (Miller *et al.*, 1981)

The bedrock is overlain by till in much of the saline seep area (Figure 11.2). In the Colorado or Bearpaw shale area, a weathered zone at the till–shale contact is 1–10 m thick. The underlying shale is dry, suggesting the shale is an aquiclude (Miller *et al.*, 1981). The Judith River–Claggett–Eagle unit has more silt and sand beds than the Colorado or Bearpaw units, but does have numerous bentonite, shale and mudstone beds that perch water. The widespread saline groundwater in the JCE unit suggests permeable materials and a

Table 11.1 Total dissolved solids (TDS) in major bedrock units (Miller *et al.*, 1978[a], 1981)

Geological unit	TDS (g/l)
Colorado and Bearpaw shale	20–50
Judith River–Claggett–Eagle	10–25
Fort Union	3–15

[a]Literature cited but not seen by author, but included to indicate the large amount of literature on the subject that is not readily available to the average reader.

deep flow system. Within the non-marine Fort Union unit, water that escapes the root zone moves down through permeable materials until it perches on the dense underclay beneath a coal seam. The perched water can move laterally and in places breaks out at the surface (Halverson and Black, 1974; Doering and Sandoval, 1976).

The glacial drift covering much of the saline seep area (Figures 11.1 and 11.2) has a swell-and-swale topography, with numerous potholes that pond water for extended periods during the spring months (Miller *et al.*, 1981). Post-glacial erosion has cut a drainage system into the drift plain, resulting in an increased local relief and increased local hydraulic gradient. The till has extensive vertical joints and fractures. The till and the underlying weathered shale have low horizontal permeability, so lateral drainage is slow (Miller *et al.*, 1981). In Montana, along the major streams, the till is 1–25 m thick but may exceed 100 m in Canada. Below 1 m it has abundant calcium and sodium salts. Laboratory data by Sanderegger *et al.* (1978) indicate that enough sodium is available in the aquifer systems within the upper 6 m to maintain existing saline seeps for the next 25–100 years; carbonate and sulphate minerals control the alkaline earth content of the leachate; and total leachable dissolved constituents make up 2–23% by weight of the original sample.

Nearly 95% of the saline seeps are in the glaciated region (Miller *et al.*, 1981), but several have developed in the unglaciated areas. The unglaciated areas in Montana with saline seeps have shallow soils developed from the Colorado, Bearpaw and Fort Union units (Figure 11.2). In both glaciated and certain unglaciated areas a large reserve of soluble salts is available for movement in the groundwater. Weathering and leaching during the Holocene era apparently have modified only the upper 1 or 2 metres.

Adding to the problem of high soluble salts in the till is the relatively high nitrogen content of the shale. Power *et al.* (1974) found very little nitrate nitrogen in the upper 2 m of the soil (apparently lowered by plant uptake), but from 2 to 10 m the NO_3^-N increased to 20–40 ppm (parts per million), with only 2–5 ppm NH_4^+N. Below 10 m the shale had 300–500 ppm N by Kjeldahl analyses, 10–40 ppm NH_4^+N exchangeable with 2N KCl and 150–300 ppm NH_4^+N released by HF treatment. Below 10 m, NO_3^-N was only 1–3 ppm. Power *et al.* believe the large store of NH_4^+N is readily nitrified when moisture, temperature and oxygen supply are favourable. In some shales, at least, a large reservoir of potentially available NO_3^-N exists to be added to the soluble sodium and calcium when conditions are right.

11.3 CLIMATIC FACTORS

Saline seeps occur in an arid to semi-arid climate with a precipitation range of 250–450 m a year. Winters are cold but snow cover is seldom more than 0.1 m. Strong winds usually remove snow from the cultivated upland areas and deposit it on the north and north-east facing slopes. About one-half the annual precipitation falls during April, May and June, when water use by crops is low. Crop yields are directly related to the spring rainfall. The largest outbreaks of saline seeps are in areas with more than 350 mm of annual precipitation, but there are many exceptions (Miller *et al*., 1981).

11.4 HYDROLOGY

Several authors have studied the hydrology of saline seeps, and while each area is somewhat different there is a general similarity (Miller *et al*., 1981; Halverson and Black, 1974; Doering and Sandoval, 1976a and b; Ballantyne, 1963; Luken, 1962). The till mantles an irregular eroded topography (Figure 11.3) and locally contains silt or sand lenses that are more permeable than the till matrix (Figure 11.4). A dense clay occurs at the top of the shale, or in other sediments permeable lignite or sandstone layers overlie dense clays (Halverson and Black, 1974). The aquitards perch water moving down below the root zone, and the water then moves towards the low areas of the buried landscape (Figure 11.5). Saline seeps develop wherever the saline groundwater comes within about 1.5 m or less of the surface, and evaporation during

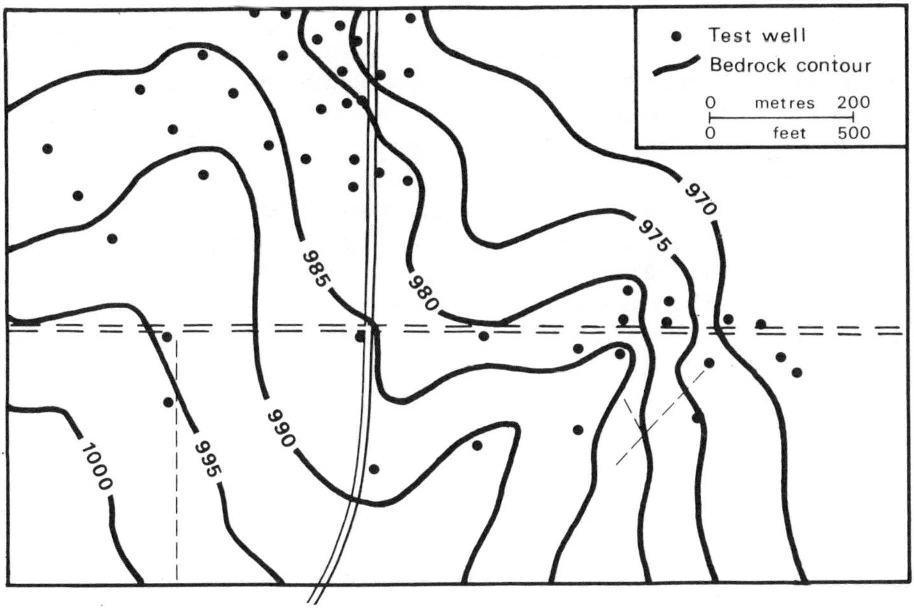

Figure 11.3 Configuration of eroded surface of the Colorado Shale underlying the Hanford–Bramlette test area (Miller *et al*., 1981)

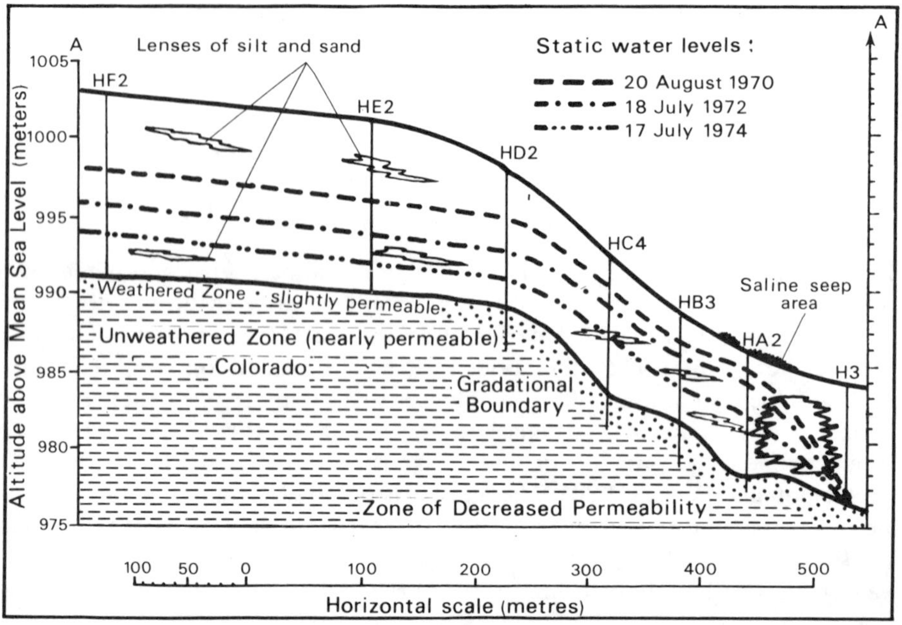

Figure 11.4 Geological section A–A′ across the Hanford–Bramlette test area showing change in static water level from 1970 to 1974

the dry summer months produces a net upward movement of water and salts (Rhoades and Halverson, 1976). Most of the seeps investigated have a local recharge area that is responsible for the salt load at the discharge (seep) area (Doering and Sandoval, 1976a and b; Halverson and Black, 1974; Miller *et al*., 1981). Paul Brown (personal communication) believes that the upland areas with slopes of 0–2% are the major recharge area for the adjacent seep.

Data from a test area in Montana (Figure 11.5) indicate transmissivity values of 0.52 m^2 a day and a seepage rate of 1.6 m^3/ha/day (Miller *et al*., 1981). Water in the weathered zone is commonly under artesian pressure in the discharge areas. Miller *et al*. estimate that 1–4% of the annual precipitation reaches the shallow groundwater flow system under native sod, but about 7–15% under an alternate crop–fallow system. Under 350 mm precipitation, 3.5–14 mm would be added to the shallow groundwater flow under sod and 2.45–52.5 mm under a crop–fallow system.

11.5 WATER QUALITY

Several investigations have quantified the salt load of the seepage and groundwaters in areas with saline seeps (Miller *et al*., 1981; Halverson and

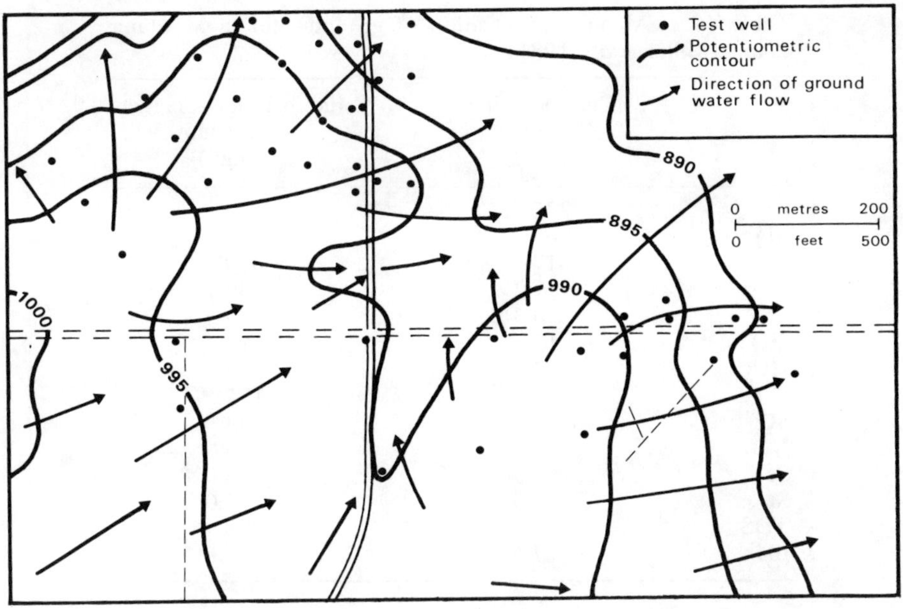

Figure 11.5 Configuration of the potentiometric surface and direction of groundwater flow on the Hanford–Bramlette test area (Miller *et al*., 1981)

Black, 1974; Doering and Sandoval, 1976a; Sanderegger *et al*., 1978; Power *et al*., 1974; Ballantyne (1963); Luken, 1962; Rhoades and Halverson, 1976). Sodium and magnesium are the major cations, and sulphate and bicarbonate are major anions (Tables 11.2 and 11.3). In Montana significant concentrations of trace elements occur, particularly selenium (Miller *et al*., 1981). Nitrate concentration is variable, but concentrations as high as 1829 ppm have been reported (Halverson and Black, 1974), and it is not uncommon for nitrate values in reservoirs to exceed 100 mg/l. The influence of bedrock on salt concentration is shown in Table 11.2, where the mean total dissolved solids (TDS) is greater than 9000 mg/l in the Colorado shale compared with about 3000 mg/l in the Judith River–Claggett–Eagle unit. Total dissolved solids greater than about 7000 mg/l usually result in fish kills. Detailed discussions of saline seep chemistry have been given by Oster and Halverson (1978) and St. Arnaud (1978).

Shallow groundwater in the saline seep areas is not a desirable source for either livestock or human consumption, and reservoirs are receiving an increased salt load. But the detailed historical data are lacking to quantify the effects of saline seep development on surface and groundwater. However, some authors believe there has been a significant deterioration in ground-

Table 11.2 Mean concentration of groundwater in Montana Plains (Miller *et al*., 1981)

	Colorado shale unit (mg/l)	Judith River–Claggett Eagle unit (mg/l)
Ca	275	120
Mg	866	71
Na	1317	670
K	17.7	6.1
Fe	0.3	0.3
SiO_2	12	12
HCO_3	534	722
Cl	141	177
SO_4	6041	1143
NO_3	57	6
Se	0.308	0.028
Sr	4.6	2.0
TDS	9262	2928
pH	7.62	7.89

water quality in the glaciated portions since the crop–fallow system of farming began (Miller *et al*., 1981).

In areas the shale contains relatively large amounts of NH_4 that can be oxidized to nitrate (Power *et al*., 1974), but most of the $NO_3^{-1}N$ occurs in the upper 10 m of the shale. Because the shallow-water flow system involves the upper weathered part of the shale, it is reasonable to assume that at least some—and possibly a large part—of the $NO_3^{-1}N$ is from the shale. Custer (1976) disagrees. In his study of a cultivated and fertilized and native sod area, Custer found that water below non-cultivated sod has no nitrate, but below a similar cultivated area the summer fallow land had 33–55 ppm

Table 11.3 Chemical analysis for water samples from five seeps in Hettinger County, North Dakota (Doering and Sandoval, 1976a)

Seep	EC (mmhos/cm)	Cations Ca, Mg, Na (meq/l)			SAR	NO_3	Anions Cl (meq/l)	Anions SO_4^2 (meq/l)
1	7.2	15	18	47	12	–	1	70
2	6.5	18	40	21	4	–	1	77
3	8.2	16	25	52	11	–	1	86
4	9.6	14	76	59	9	6	2	139
5	7.2	15	58	36	6	5	2	100

nitrate. The nitrate values were largest in the spring. Within the upper 7.6 m there is 3 times more nitrate below cultivated land than below non-cultivated sod. The nitrate profile suggests a source near the surface. All the fertilizer, rainfall and manure nitrogen added during the cultivation history of the site do not account for the excess nitrate below the cultivated land. Custer concluded that only oxidation of organic matter in the soil is quantitatively large enough to produce the excess nitrogen observed in his study area. Later Custer (1978) sampled the groundwater beneath a native sod and again the following year after the area had been cultivated for one season. About 68 kg/ha of 18% nitrogen was applied in fertilizer. The nitrogen in the groundwater after one season's cultivation increased nearly 7.7 times the amount of nitrogen available from the fertilizer, but was still considerably below the nitrogen in the groundwater of adjacent sites cultivated for at least 60 years. Apparently the oxidation of organic matter is releasing large amounts of nitrate nitrogen that escapes uptake by cultivated crops. However, the first year's cultivation of a native sod probably released more nitrogen than similar areas that have been cultivated for some time.

11.6 DETECTION OF SALINE SEEPS

Several methods, including visual, remote sensing and soil resistance measurements, have been used to detect saline seeps (Rhoades and Halverson, 1976; Brown, 1976; Horton and Moore, 1976). Early detection is important if counter-measures are to be effective. Brown (pp. 59–60) gives the following visual criteria:

(*a*) Prolonged soil surface wetness following a substantial rain. Surface wetness is easily observed in a fallow field but can also be observed on cropland, especially in late spring and summer. There will nearly always be uneven surface drying in a field following a rain but if wetness persists over a localized area for more than two or three days, it may indicate a shallow or perched water table. Sampling with a soil moisture probe may reveal that the soil is much wetter than field capacity and a film of free water may show on the end of the probe. Tractors 'bog down' or cut deep ruts in these areas. Water often seeps into the ruts and remains there for a period of time. Salt crystals form on the surface as the disturbed area dries.

(*b*) Rank wheat or barley growth is sometimes accompanied by lodging in an area that was normal in previous years. Farmers sometimes report that the crop of grain on these areas was not only their best but their last. The areas are sub-irrigated by a rising water table, however, the salt content was not yet high enough the first year to reduce growth. Following the bumper grain crop, the spot may become so wet and salty that the next crop either can't be seeded, or if seeded, doesn't survive. Such wet and saline areas usually expand rapidly.

(*c*) Scattered salt crystals on the soil surface or localized cloddiness are the other indications of a pending saline-seep outbreak.

(*d*) Rank *Kochia scoparia* growth after grain harvest in small areas where the soil would normally be too dry to support weed growth. Probing these areas may reveal that the surface and subsoils are wet.
(*e*) New foxtail barley (Hordeum jubatum) infestations indicate possible saline-seep outbreaks. The foxtail usually increases with time.
(*f*) Stunted or dying trees in a shelter belt or windbreak. Leaves may be light green or yellowish before stunting or death.
(*g*) Other early warning signs are delayed small grain germination, abnormally mellow soil and darker colored surface soil on lower slopes. The color change may be caused by dispersed organic matter or lignite fractions mixed with the surface soil. Another indication of saline seep is a sloughed hillside covered by native vegetation—and located next to a cultivated field.

After crop failure there is a typical sequence of vegetation (Custer, 1976):

(1) Kochia thrives vigorously for a year or two followed by stunted growth. Stunting results from salt build-up and lack of nitrogen.
(2) Kochia dies out and foxtail barley increases to a pure stand. Kochia probably is still present around seep perimeter.
(3) Bare soil, especially in the center of the seep, is too salty to support plant growth. Not all saline seeps reach this stage.

Rhoades and Halverson (1976) proposed the diagnostic criteria in Table 11.4 for distinguishing between unaffected, encroaching and developed saline seeps using extensive soil resistance measurements.

Other methods used to detect saline seeps are remotely sensed data (Horton and Moore, 1976) and rural assessment data used primarily for taxation but containing information on soil salinity on an individual quarter section basis (Crosson and Ballantyne, 1976).

11.7 DISTRIBUTION AND GROWTH RATES

The dryland saline seeps are most abundant in areas with greater than 350 mm of annual precipitation but occur within a precipitation range of

Table 11.4 Diagnostic criteria for distinguishing between unaffected soil sites, encroaching and developed saline seeps

Site type	Salt content	Water content
Unaffected	Low, increasing with depth	Low, increasing with depth
Encroaching saline seep	Intermediate, increasing to a peak at a relatively shallow depth, then decreasing with further depth	Moist surface becoming wet with depth
Saline seep	High, decreasing with depth	Relatively uniformly wet to the water table

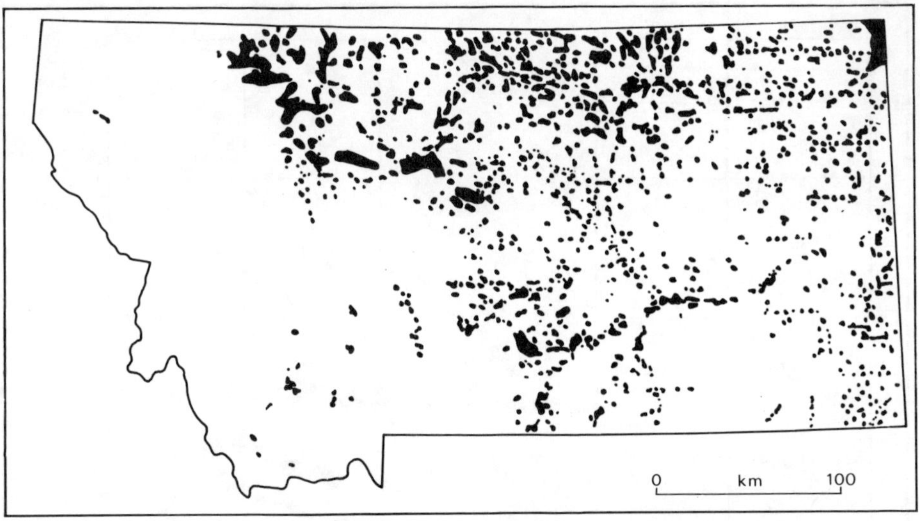

Figure 11.6 Distribution of saline seeps in Montana (1978)

250–450 mm. They are closely associated with Cretaceous saline shale bedrock and some of the Tertiary sediments derived from the saline shales (Figures 11.1 and 11.2). Within an area, the saline seeps are unevenly distributed (Figures 11.2 and 11.6), but regionally they occupy 0.8 million hectares or more (Van der Pluym, 1978). Within an affected area the seeps seldom develop in rangeland but are common in cultivated land. The seeps were essentially unknown before the early 1940s and apparently did not develop rapidly or attract much attention until the 1960s. Seeps expand very little during dry cycles but grow by 20% to 200% after wet cycles; a growth rate of about 10% is commonly cited (Miller *et al.*, 1976). Figure 11.7 is an example of how rapidly saline seeps can grow in an area. The growth rates indicated in Figure 11.7 cannot be maintained because a balance between recharge and discharge areas is maintained.

11.8 SOILS

The soils affected by dryland saline seeps have not been carefully described in the symposiums held on the problem. Conversations with pedologists from North Dakota suggest there may be some morphological clues that can be used to delineate potential problem areas, but published documentation cannot be offered. It is, in some ways, odd that hundreds of person hours have been spent on the hydrology, geology, chemistry and detection of dryland saline seeps, but apparently little time has been spent documenting the soil morphology of active or potential saline seeps. Eight recently published US

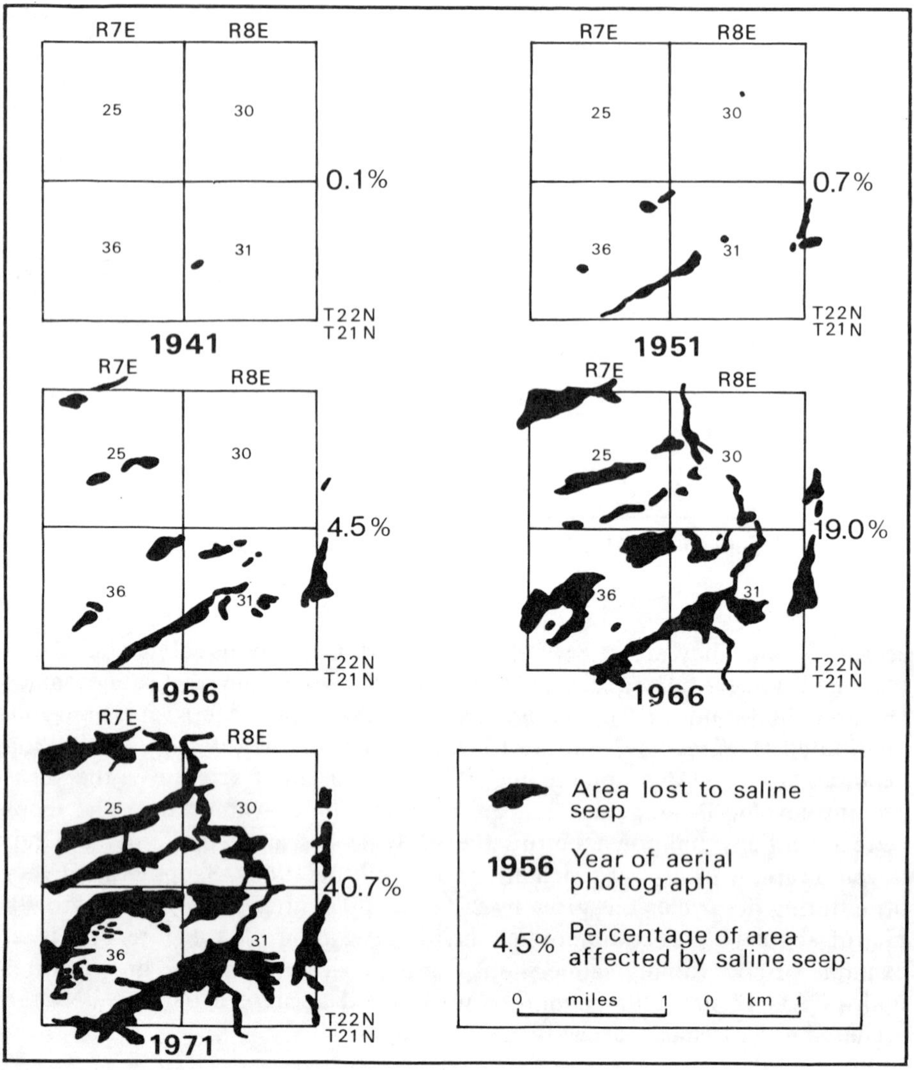

Figure 11.7 Rate of growth of a saline seep in a 10 km^2 area of Montana (1841 to 1971)

Department of Agriculture Soil Conservation Service soil surveys from North Dakota were examined (Benson, 1979; Emmons, 1980; Mercer, 1978; Oliver, 1975; Pierce, 1978; Pembina, 1977; and Slope, 1978). All counties were within the area of saline seeps (Figure 11.1) but no mention of saline seeps were noted. About one-half the maps of Benson County were examined for saline and wet spot symbols. Fifteen map units had a few too many saline spot symbols, but wet spot symbols were not found. There was no discussion either in the series or map unit descriptions of the significance of the saline spot symbols, although the need for intensive cropping and limiting fallow was mentioned in the saline phase descriptions. The other county surveys were similar to Benson County and little descriptive material that related to saline seeps could be found. Possibly saline seeps did not occur in these areas, but this seems unlikely in view of other evidence. Few modern soil surveys were found within the saline seep area of Montana. After spending several hours looking for information on the distribution of saline seeps and the properties of the affected soils, the author concluded that only someone well acquainted with the area could interpret the soil survey. Insufficient information is given for less-informed individuals.

It is highly unlikely that the soil survey party did not map saline seeps in the area and have considerable knowledge about the soil characteristics. Most of that knowledge, unfortunately, will remain with those individuals.

11.9 MAJOR FACTORS LEADING TO OUTBREAKS OF DRYLAND SALINE SEEPS

Most authors agree that saline seeps are produced when excess water over evapo-transpiration moves below the rooting zone through salt-laden material and eventually reappears at the surface. The area affected depends upon the local stratigraphy and geomorphology since the saline water must be within 1.5 m of the surface (Figures 11.3, 11.4, 11.5) (Rhoades and Halverson, 1976). Dryland saline seeps are rare under native sod and an estimated 1–4% of the annual precipitation moves below the rooting zone to the groundwater (Miller *et al*., 1981). Under an alternate crop–fallow system, 7–15% of the precipitation reaches the groundwater. Anything that allows water to be lost below the rooting zone, including farm ponds, water-filled potholes, accumulation of snow on north-east slopes (Miller *et al*., 1981; Thacker, 1976), or tillage techniques that increase snow cover and water intake (Greb *et al*., 1970; Sommerfeldt, 1976), can speed and aggravate the development of dryland saline seeps. The excellent evidence presented by Miller *et al*. (1981), Haas and Willis (1962), Fergerson (1963) and Doering and Sandoval (1976b), show that summer fallow is the major cause of dryland saline seep growth during and following the wet years of favourable precipitation of the 1940s and 1960s.

The dryland saline seep is a problem inherited from the geology and hydrology of the area. Marine shales high in gypsum and carbonate are overlain in large areas by a glacial till that also has a large amount of salt, derived primarily from the underlying marine shales. Groundwater becomes saline as it moves through the saline till below the rooting depth of the crop or native vegetation. The till–shale contact and the underclay of lignite beds are aquitards and perch water. Shallow valleys have been eroded in the till, so the water perched above the aquitard moves towards the valley under a local hydraulic gradient. Where the aquitards are close enough the saline groundwater appears at the surface. Evaporation concentrates the salt and a saline seep develops. Agriculture has increased the amount of water moving to the shallow groundwater system and increased the area of saline seeps by using a fallow–cropping system and building ponds. In some cases the increase in area of saline seeps has been dramatic.

Although agriculture has aggravated the situation, the potential for dryland saline seeps was always there, and even under natural conditions a slight change in factors that would allow more water to move to the shallow groundwater flow system would result in the development of saline seeps. Saline and saline–alkali soils are common in the saline seep area, and the same factors responsible for their characteristics and distribution are responsible for the saline areas produced by agricultural use. The problem and major causes have been clearly defined, but now what can be done to alleviate or control the situation?

11.10 SALINE SEEP CONTROL

Most saline seeps develop from local shallow groundwater systems and therefore their control can be local. Two major control methods are used. The agronomic approach attempts to use all or nearly all the precipitation in growing crops. By reducing or eliminating water loss to the groundwater in the recharge area, the amount of water moving to the seep is reduced. If groundwater tables can be lowered, many seep areas will disappear. The engineering approach attempts to intercept the saline groundwater before it reaches the surface. Both approaches have advantages and disadvantages.

The agronomic approach uses deep-rooted crops and several combinations of crops and cropping systems to reduce the volume of water in the rooting zone and still maintain enough for economical crop production. The amount of water lost below the rooting zone in the recharge area is kept to a minimum. By drying out the deeper subsoil a storage reservoir is established during the periods of higher rainfall or when crops that use less water are grown. The water use and rooting depths of various crops have been intensively studied (Brown *et al.*, 1976; Black *et al.*, 1981). Deep-rooted perennials such as alfalfa remove large amounts of water from the soil (Tables 11.5 and 11.6). The soil-water depletion and rooting depth over a 5-year period of several crops are shown in Figure 11.8. In one test, alfalfa grown on the recharge area lowered the water table 2.9 m in the discharge area and 2.0 m in the recharge area in six years (Brown and Miller, 1978). The alfalfa used current precipitation and reduced the volume of water in the deeper subsoil. At the seep area the deeper water table allowed leaching of the salts and

Table 11.5 Cumulative soil-water depletion and rooting depths for perennial crops and biennial sweetclover (Brown *et al*., 1976)

Crop	1971		1972		1973		1974		1975	
	Soil water used (in)	Root depth (ft)	Soil water used (in)	Root depth (ft)	Soil water used (in)	Root depth (ft)	Soil water used (in)	root depth (ft)	Soil water used (in)	root depth (ft)
Vernal alfalfa	7.0	7	11.3	10	20.2	14	22.2	18	30.9	19
Eski sainfoin	5.9	5	10.3	9	17.9	11	22.1	13	22.2	13
Cicer milkvetch	4.2	5	8.9	9	15.6	13	21.5	15	22.6	15
Russian wildrye	–	–	–	–	12.5	7	16.1	9	18.7	10
Tall wheat	7.6	6	11.1	7	14.0	8	17.1	9	17.3	9
Sweetclover	–	–	10.9[a]	6[a]	13.9[b]	8[b]	–	–	15.9[b]	9[b]

[a]First-year sweetclover.

[b]Second-year sweetclover.

Table 11.6 Soil water use and rooting depths for annual crops seeded on fallow or annually cropped (Brown *et al.*, 1976)

Crop	1971		1972		1973		1974		Average	
	Soil water used (in)	Root depth (ft)	Soil water used (in)	Root depth (ft)	Soil water used (in)	Root depth (ft)	Soil water used (in)	Root depth (ft)	Soil water used (in)	Root depth (ft)
Fallow–wheat	9.6	6	6.1	6	8.6	7	10.1	8	8.6	6.8
Fallow–barley	6.1	4	5.4	5	6.7	5	7.6	6	6.4	5.0
Proso millet	5.3	5	2.0	3	10.4	7	2.3	4	5.0	4.8
Corn	3.2	4	2.3	2	6.2	4	3.2	5	3.7	3.8
Kochia	–	–	8.2	5	7.7	4	4.6	5	6.8	4.7
Safflower	–	–	5.6	5	16.7	8	10.5	8	10.9	7.0
Sunflowers										
Mingren M	–	–	–	–	12.0	6	8.2	9	10.1	7.5
Mingren L	–	–	–	–	9.6	6	9.2	6	9.4	6.0
Krasnodarets	–	–	–	–	9.1	5	5.4	8	7.2	6.5
Peredovik	–	–	–	–	9.0	6	4.4	8	6.7	6.8
Annual cropping:										
Barley after WW	–	–	–	–	6.9	4	–	–	–	–
WW after Barley	–	–	–	–	–	–	8.3	8	–	–
									7.6	6.0

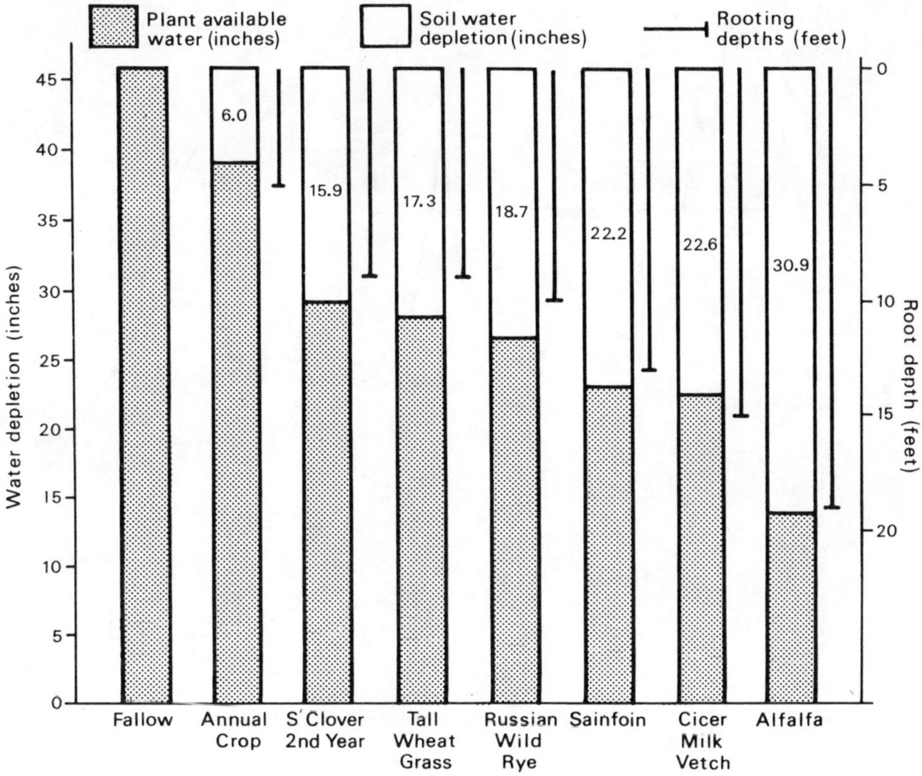

Figure 11.8 Soil-water depletion and rooting-depth measurements in 20 ft (6 m) soil profile after 5 years (1971 to 1975)

reduction of the surface soil electrical conductivity, enough for barley and winter wheat to produce 70% of the normal yield.

Halverson and Reule (1976) used alfalfa on about 80% of the recharge areas of two saline seeps and a 46 m wide alfalfa buffer strip (about 15% of the recharge area) to study the effect of deep-rooted crops on groundwater levels. Annual precipitation was above average for most of the three years of their study and for the two years before the study started. Groundwater levels decreased at the seep (J3) and the wells in the recharge area, J1 and J10 (Figure 11.9), where alfalfa occupied about 80% of the recharge area. The salinity at the seep decreased slightly, the salt crust disappeared, and farm equipment could cross the seep most of the year. Groundwater levels at the seep with only 15% alfalfa in the recharge area—buffer strip alfalfa—changed little if any and the seep was wet most of the time.

The success in controlling many saline seeps by agronomic means has spurred the development of flexible cropping systems to use most of the

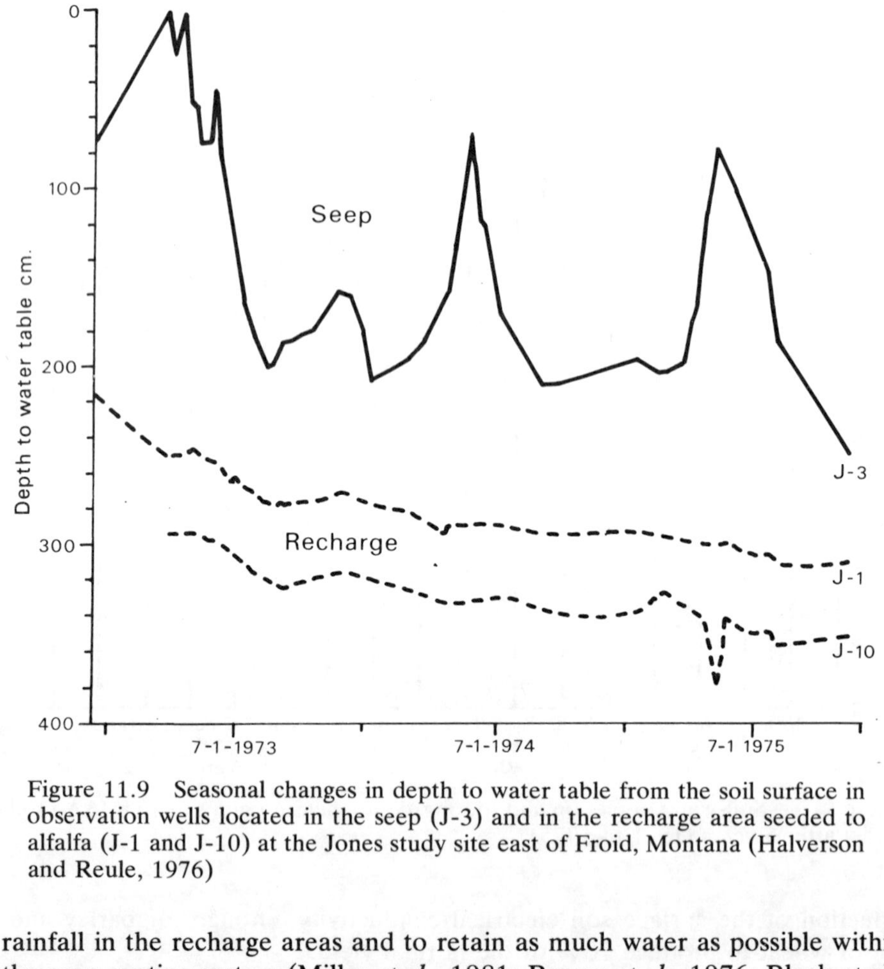

Figure 11.9 Seasonal changes in depth to water table from the soil surface in observation wells located in the seep (J-3) and in the recharge area seeded to alfalfa (J-1 and J-10) at the Jones study site east of Froid, Montana (Halverson and Reule, 1976)

rainfall in the recharge areas and to retain as much water as possible within the crop rooting system (Miller *et al*., 1981; Brown *et al*., 1976; Black *et al*., 1981; Brown and Cleary, 1978; Brown and Miller, 1978; Halverson and Reule, 1976; Jackson and Kroll, 1978; Duff, 1976; Black and Siddoway, 1976). In general the cropping systems eliminate fallow if possible, but do not rule out its use if stored moisture and expected precipitation are not large enough to insure a crop (Table 11.7). Although the dramatic growth of saline seeps influenced the move into flexible cropping systems, there was considerable earlier information that suggested fallow was not necessary in most years over much of the affected area (Greb *et al*., 1970; Haas and Willis, 1962; Fergerson, 1963; Ford and Kroll, 1979).

The engineering approach to saline seep control intercepts the saline water before it comes close enough to the surface to develop into a seep. Tile or mole drainage systems are installed, where possible, to intercept the water-

Table 11.7 Types of cropping strategies needed, based on effective soil root-zone depth, crop rooting, and precipitation (Black *et al*., 1981)

Precipitation (mm)	Cropping strategies for three effective depths of soil root zone		
	<1.5 m	1.5–3.0 m	3.0–6.0 m
250–380	Winter wheat Barley Winter wheat Fallow[b] or Spring wheat Winter wheat Barley Fallow[b]	Alfalfa (3 to 4 years)[a] Fallow[b] and/or and/or Barley Winter wheat Fallow[b] Barley Wheat[c] Fallow[b] Safflower or Barley Spring wheat Fallow[b] Winter wheat Fallow[b]	Alfalfa (4 to 5 years) Fallow[b] and/or and/or Wheat[c] Spring wheat Safflower Winter wheat Barley Fallow[b] Fallow[b]
380–500	Spring wheat Oilseed (sunflower) Barley or oats Fallow[b] or Barley Spring wheat Sunflower Barley or oats Fallow[b]	Alfalfa (3 to 4 years[a] Fallow[b] and/or Sunflower Spring wheat Barley or oats Fallow[b] Spring wheat Sunflower	Alfalfa (4 to 5 years) Fallow[b] and/or Spring wheat Sunflower Barley or spring wheat Fallow[b] Spring wheat Sunflower

[a]Sweet clover could be substituted for clover (2 years).
[b]Fallow is optional.
[c]Spring or winter wheat.

Table 11.8 Total drain outflow and dissolved salt load for drains A and B (Doering and Sandoval, 1978)

	Outflow		Dissolved salt	
	(ft^3)	(m^3)	(lb)	(kg)
1971[a]	110 410	3 127	63 500	28 800
1972	836 460	23 689	483 600	219 300
1973	205 860	5 830	101 400	46 000
1974	36 580	1 036	19 200	8 700
1975[b]	31 780	900	12 600	5 700
Totals	1 221 100	34 582	680 300	308 500

[a]September to December.
[b]January to June.

bearing strata (Doering and Sandoval, 1976b, 1978; Sommerfeldt *et al*., 1978). The combined flow and dissolved salt load from two drains in North Dakota are shown in Table 11.8. The probable recharge area for the seeps was about 125 acres (50 ha). Combined outflow from the drains was equivalent to 2.7 inches (68 mm) over the recharge area. About 1.8 inches (46 mm) of the drainage occurred in 1972. The dissolved salts (Table 11.8) amounted to about 5440 pounds per acre (6100 kg/ha) and about 44 pounds of nitrogen per acre (49 kg/ha) of recharge area. The chemical composition of water from seeps (Table 11.9) is somewhat different from the chemistry of the drainage

Table 11.9 Chemical composition of water samples from three seeps in southwestern North Dakota (Doering and Sandoval, 1978)

	#1	#2	#3
Electrical conductivity (mmhos/cm)	9.5	9.9	6.1
Calcium (meq/l)	11.4	25.3	21.9
Magnesium (meq/l)	22.4	42.8	38.2
Sodium (meq/l)	95.8	80.2	31.5
Ammonium (meq/l)	0.2	0.2	1.1
Sodium adsorption ratio[a]	23.3	13.7	5.7
pH	7.8	7.9	3.7
Bicarbonate (meq/l)	4.3	7.9	
Chloride (meq/l)	0.4	2.8	0.9
Nitrate (meq/l)	0.2	0.3	1.4
Sulphate[b] (meq/l)	125.0	138.0	90.0

[a]Sodium adsorption ratio = $Na/[(Ca + Mg)/2]^{\frac{1}{2}}$ when chemical compositions are in meq/l.
[b]Sulphate = (sum of cations) − (sum of determined anions).

water (Table 11.10) that has not yet passed through several feet of calcereous soil. The low pH of the drainage water is from the acidic nature of the lignite layers that conduct the water toward the seep area.

The mole drain systems (Sommerfeldt *et al.*, 1978) were designed to remove temporary excess water and control the water table at a safe depth through a combination of practices including plant use, surface and subsurface drainage, and natural drainage. The mole drains and associated practices appear to be functioning properly, but how long they will function is not known. However, mole drains have been used in England for the past 200 years and some have been operating in Canada for more than 20 years (Sommerfeldt, 1976b). Their low cost and ease of installation may give them a place in saline seep reclamation.

The most effective methods of controlling saline seeps are deep-rooted perennial crops, surface drainage of upland recharge areas, and flexible cropping systems (Miller *et al.*, 1981). Problems encountered with these methods are economic and cultural. Many farming operations have little use

Table 11.10 Average chemical composition of the drain waters and the standard deviation of the determinations for the intial drainout period and for the remainder of the study (Doering and Sandoval, 1978)

	Drain A				Drain B			
	9/71–4/73		4/73–6/75		9/71–4/73		4/73–6/75	
	Avg	σ	Avg	σ	Avg	σ	Avg	σ
	n = 19		n = 27		n = 20		n = 32	
Electrical conductivity (mmhos/cm)	8.1	0.9	7.3	0.3	10.1	1.0	9.0	0.3
Calcium (meq/l)	17.0	2.9	21.1	2.9	15.8	3.1	20.1	1.7
Magnesium (meq/l)	62.4	7.8	53.1	5.6	76.6	11.7	63.6	3.9
Sodium (meq/l)	42.5	6.5	36.1	1.8	64.4	8.1	52.0	2.5
Ammonium (meq/l)	0.8	0.3	0.6	0.3	1.3	0.3	1.5	0.3
Sodium absorption rainfall[a]	6.7	–	5.9	–	9.5	–	8.0	–
pH	4.6	0.4	4.8	0.7	3.6	0.1	3.8	0.2
Carbonate (meq/l)	0	–	0	–	0	–	0	–
Bicarbonate (meq/l)	0	–	0	–	0	–	0	–
Chloride (meq/l)	2.7	0.3	2.3	0.8	2.3	0.5	1.8	0.1
Nitrate (meq/l)	4.7	0.7	4.3	0.9	6.2	2.1	3.9	1.2
Sulphate[b] (meq/l)	115.0	15.0	104.0	6.0	150.0	18.0	132.0	5.0

[a]Sodium adsorption ratio = $Na/[(Ca + Mg)/2]^{\frac{1}{2}}$ when chemical compositions are in meq/l.
[b]Sulphate = (sum of cations) − (sum of determined anions).

for alfalfa without radical changes; the recharge areas do not always fall within one operating unit, and the needed degree of water control in the recharge area cannot be exercised by the operator of the affected area; the recharge area can be dried out by deep-rooted crops, but a crop–fallow rotation soon refills the soil moisture reservoir reduced by the deep rooted crops; the deep-rooted crops may concentrate salts in the soil profile over a period of time; a high level of management is required to monitor soil moisture levels, so changes in cropping sequence can be timely; and weed, fertility and mulch control also require a high level of management (Jackson and Kroll, 1978; Duff, 1976).

Drainage is a necessary part of the control of saline seeps, especially surface drainage of potholes and some farm ponds. Again problems arise: the potholes are excellent wildlife breeding areas; disposal of saline waters from the subsurface is difficult because high nitrate and salt contents contaminate surface waters and are marginal if not dangerous drinking water for livestock; and fish kills may result from allowing drainage waters to contaminate surface reservoirs.

11.11 SUMMARY

The work conducted by numerous investigators on the origin and control of dryland saline seeps is, in general, excellent. The geology and hydrology of saline seeps have been well documented, and the most probable causes are now understood. As a result methods to control the seeps have been developed that alleviate the primary cause of seeps—water moving below the rooting zone to a shallow groundwater flow system. The effects of various cropping systems on the control and reclamation of saline seeps has been well documented, and some authors believe that 60–70% of the seeps can be controlled and reclaimed by agronomic practices alone. The remaining 30–40% will require a combination of agronomic and engineering practices. The reclamation of saline seeps can be relatively rapid; within 5 years wheat and barley production have reached 70% of normal and was 100% within 7 years using agronomic techniques. The reclamation process is not without problems, but considerable progress has been made on overcoming the difficulties of various cropping systems and the educational needs to put the systems into effect. There are, however, a few areas of investigation or interpretations that seem questionable.

Published reports by pedologists on the morphology of saline seep areas are rare except for those of the early Canadian workers. Soil scientists, including pedologists, have published on the properties of saline seeps, but a recent strictly pedological study has not been found. Although other disciplines within the broad fields of soil science, engineering, agronomy and geology were used in the multitude of investigations, the one apparently missing team member has been an experienced field pedologist. One strongly suspects that much more is known about the morphology of soils in potential and active seep areas than can be found in published literature, including published soil surveys. The pedologist, working with other scientists, has an opportunity to delineate potential seep areas if there have been past events of saline seeps that left some morphological clue. Yet the published literature leads one to believe that each area must be investigated by deep drill holes, water table studies, and detailed field examination using various techniques that can be described as research tools, not production mapping tools. It would be an exception if a soil survey party leader in a county did not quickly develop a 'feel' for the areas where potential saline seeps may occur and their maximum size based upon geologic, morphologic, and topographic inference. But this information is not readily available in the published literature, and apparently not even in recently published soil surveys from saline seep areas. Part of the reason is that the drive for uniformity in published soil surveys have forced the party leader to publish in a standard format that may almost preclude any direct discussion of saline seeps. Many of seeps initially occupy relatively small areas, and spot symbols are very easy for high authority to eliminate. The soil survey party seldom publishes their findings in national or local journals because this is not always encouraged or even deemed to be of any importance by their immediate supervisors. The journal reviewers usually are dominated by laboratory-oriented scientists who frequently frown on publication of field observations. These reviewers are used to volumes of laboratory data from one or two points, not hundreds or thousands of field observations that describe the relations of several natural bodies. Frequently, the 'real world' does not fit the provincial theories developed by laboratory-oriented scientists.

The inference is strong in many published articles on saline seeps that the seeps are a first-time phenomenon produced by cultivation. The glacial till in much of the saline seep area (Figure 11.1) apparently was deposited about 20 000 years ago, and since that time there have been several climatic fluctuations. Unless the landscape of the saline seep spots were developed only in the mid- to late-Holocene, there may have been several periods of saline seep development even under the native vegetation. Even the late-Holocene may have had short periods favourable for seep development.

Another inference from published papers is that agriculture has polluted the shallow groundwater supplies by allowing more water to escape to the groundwater table (Miller *et al*., 1981; Doering and Sandoval, 1976a, 1978; Ballantyne, 1963; Black *et al*., 1981; Botz, 1976). There can be little doubt that agriculture has contributed to the nitrate loading of the groundwater, but how has the non-irrigated agriculture contributed to groundwater pollution? Any water moving below the root zone must pass through salt-rich till or shale to reach the groundwater. The leachable salt content of the material above the groundwater reservoir is from 2.2% to 23.5% (average of 8.1%) by weight. How could the water moving below the rooting zone in pre-agriculture times, estimated at 1–4% of precipitation under current grassland conditions, have a lower salt load than water now moving through the same material? Has there been an interval in pre-agricultural time when the water-soluble salt content of the till and shale was less than present, or when there was no water escaping the rooting zone? If the latter is true, is the present salt load a flush of soluble salt released by earlier weathering but not moved into the groundwater? These questions need to be answered before an increase in the groundwater salt load on a unit volume basis (a possible exception is nitrogen) can be attributed to agriculture.

Although there is considerable evidence supporting the statements that agronomic and engineering methods can control and reclaim saline seeps, will they be controlled? The control methods require a high level of management using techniques new to the land operators. Will the land operators apply the needed technology to control saline seeps? Can they afford to control the seeps under the future economic conditions, especially when control measures call for crops such as alfalfa on a grain farm? Will the control of saline seeps be like the control of erosion—the control technology is available but for several reasons it is seldom applied. Time will tell.

11.12 REFERENCES

Ballantyne, A. K. (1963) Recent accumulation of salts in the soils of south-eastern Sackatchewan. *Can. J. Soil Sci.*, **43**, 52–58.

Black, A. L., Brown, P. L., and Siddoway, F. H. (1981) Dryland cropping strategies for efficient water-use to control saline seeps in the northern Great Plains, U.S.A. *Agricultural Water Management*, **4**, 295–311.

Black, A. L., and Siddoway, F. H. (1976) Dryland cropping sequence within a tall wheatgrass barrier system. *Regional Saline Seep Control Symposium Proceedings*, Montana State Univ., Coop. Ext. Serv. Bul. 1132, 154–163.

Botz, M. K. (1976) Salinity in hydrological systems in Montana. *Regional Saline Seep Control Symposium Proceedings*, Montana State Univ., Coop. Ext. Service Bul. 1132, 91–98.

Brown, P. L. (1976) Saline-seep detection by visual observations. *Regional Saline Seep Control Symposium Proceedings*, Mont. St. Univ., Cooperative Extension Service. Bul. 1132, 59–61.

Brown, P. L., and Cleary, E. C. (1978) Water use and rooting depths of crops for Saline Seep Control. *Dryland-Saline-Seep Control Proceedings. Meeting of the subcommission on salt-affected soils. 11th International Soil Science Society Cong.*, 7-1 to 7-7.

Brown, P. L., Cleary, E. C., and Miller, M. R. (1976) Water use and rooting depth of crops for saline seep control. *Regional Saline Seep Control Symposium Proceedings*. Montana State Univ., Coop. Ext. Serv. Bul. 1132, 125–136.

Brown, P. L., and Miller, M. R. (1978) Soil and crop management practices to control saline seeps in the U.S. Northern Plains. *Dryland-Saline-Seep Control Proceedings. Meeting of the subcommission on salt affected soils. 11th International Soil Science Society Cong.*, 7-9 to 7-15.

Crosson, L. S., and Ballantyne, A. K. (1976) Soil salinity studies using rural land assessment data. *Regional Saline Seep Control Symposium Proceedings*, Montana State Univ. Cooperative Ext. Service Bul. 1132, 47–58.

Custer, S. (1976) The nitrate problem in areas of saline-seep—A case study. *Regional Saline Seep Control Symposium Proceedings*, Montana State Univ., Cooperative Extension Service Bul. 1132, 63–85.

Custer, S. G. (1978) Shallow groundwater nitrate one season after initial cultivation in Stillwater County, Montana. *Dryland Saline Seep Control Proceedings. Meeting of the subcommission on salt affected soils. 11th Int. Soil Sci. Soc. Cong.*, 2-1 to 2-6.

Doering, E. J., and Sandoval, F. M. (1976a) *Saline-Seep Development on Upland Sites in the Northern Great Plains*. US Dept. Agr., ARS NC-32.

Doering, E. J., and Sandoval, F. M. (1976b) Hydrology of saline seeps in the Northern Great Plains. *Trans. Am. Soc. Ag. Eng.*, **19**, 856–861, 865.

Doering, E. J., and Sandoval, F. M. (1978) Chemistry of seep drainage in southwestern North Dakota. *Dryland-Saline-Seep Control Proceedings, Meeting of the subcommission on salt-affected soils. 11th International Soil Science Society Cong.*, 4-1 to 4-14.

Duff, A. L. (1976) Dryland cropping systems in Central Montana. *Regional Saline Seep Control Symposium Proceedings*, Montana State Univ. Coop. Ext. Serv. Bul. 1132, 145–153.

Fergerson, W. S. (1963) Effect of intensity of cropping on the efficiency of water use. *Canadian Journal Soil Sci.*, **43**, 156–165.

Ford, G. L., and Kroll, J. L. (1979) *The History of Summer Fallow in Montana*. Montana Ag. Exp. Sta. Bul. 704.

Greb, B. W., Smika, D. E., and Black, A. L. (1970) Water conservation with stubble mulch fallow. *J. Soil and Water Conserv.*, **25**, 58–62.

Haas, H. J., and Willis, W. O. (1962) Moisture storage and use by dryland spring wheat cropping systems. *Soil Sci. Soc. Amer. Proc.*, **26**, 506–509.

Halverson, A. D., and Black, A. L. (1974) Saline seep development in dryland soils of north-eastern Montana. *J. Soil and Water Conserv.*, **29**, 77–81.

Halverson, A. O., and Reule, C. A. (1976) Controlling saline seeps by intense cropping and recharge area. *Regional Saline Seep Control Symposium Proceedings*, Montana State Univ., Coop. Ext. Ser. Bul. 1132, 115–124.

Horton, M. L. and Moore, D. G. (1976) Remote sensing as a means of detection of saline seeps. *Regional Saline Seep Control Symposium Proceedings*, Montana St. Univ., Cooperative Extension Service. Bul. 1132, 41–45.

Jackson, G. D., and Kroll, J. L. (1978) The flexible method of recropping. *Dryland-Saline-Seep Control Proceedings. Meeting of the subcommission on salt affected soils. 11th International Soil Science Society*, 7-23 to 7-28.

Luken, H. (1962) Saline soils under dryland agriculture in southeastern Sackatchewan (Canada) and possibilities for their improvement. I. Distribution and composition of water soluble salts in soils in relation to physiographic features and plant growth. *Plant and Soil*, **17**, 1–25.

Miller, M. R., Bergantino, R. N., Bermel, W. U., Schmidt, F. A., and Baltz, U. K. (1978)* *Regional Assessment of the Saline-Seep Problem and a Water Quality Inventory of the Montana Plains*. Montana Bureau of Mines and Geology. Open File Report 42.

Miller, M. R., Brown, P. R., Donovan, J. J., Bergatino, R. N., Sanderegger, J. L., and Schmidt, F. A. (1981) Saline seep development and control in the North American Great Plains—Hydrogeological Aspects. *Agricultural Water Management*, **4**, 115–141.

Miller, M. R., Vander Pluym, H., Holm, V. M., Vasey, E. H., Adams, B. P., and Bahls,

L. H. (1976) An overview of saline-seep programs in the states and provinces of the Great Plains. *Regional Saline Seep Control Symposium Proceedings*, Montana State Univ., Coop. Ext. Service Bul. 1132, 4–18.

Oster, J. D., and Halversen, A. D. (1978) Saline seep chemistry. *Dryland-Saline-Seep-Control Proceedings, Meeting of the Subcommission on salt-affected soils. 11th International Soil Sci. Cong*., 2–7 to 2–29.

Power, J. F., Bond, J. J., Sandoval, F. M., and Willis, W. O. (1974) Nitrification in Paleocene shale. *Sci*., **183**, 1077–1079.

Rhoades, J. D., and Halverson, A. D. (1976) Detecting and delineating saline seeps with soil resistance measurements. *Regional Saline Seep Control Symposium Proceedings*, Montana State University, Cooperative Extension Service Bul. 1132, 19–34.

St. Arnaud, R. J. (1978) The relationship between salinity and secondary soil carbonates. *Dryland-Saline-Seep Control Proceedings. Meeting of the subcommission on salt-affected soils. 11th International Soil Sci. Cong*., 2–30 to 2–39.

Sanderegger, J. L., Donovan, J. J., Miller, M. R., and Schmidt, F. A. (1978) *Investigations of Soluble Salt Loads, Controlling Mineralogy and Some Factors Affecting the Rates and Amounts of Leached Salts*. Saline Seep Program Report. Montana Bureau of Mines and Geology Open-File Report 30.

Sommerfeldt, T. G. (1976a) Snow trapping by windbreaks. *Regional Saline Seep Control Symposium Proceedings*, Montana State Univ., Coop. Ext. Service Bul. 1132, 87–90.

Sommerfeldt, T. G. (1976b) Mole drain for saline-seep control. *Regional Saline Seep Control Symposium Proceedings*, Montana State Univ., Coop. Ext. Ser. Bul. 1132, 296–302.

Sommerfeldt, T. G., Vander Pluym, H., and Christie, H. (1978) Drainage of dryland saline seeps in Alberta. *Dryland-Saline-Seep Control Proceedings, Meeting of the subcommission on salt-affected soil. 11th International Soil Science Society*, 4-15 to 4-23.

Thacker, W. (1976) An overview of saline seep. *Regional Saline Seep Control Symposium Proceedings*, Montana State University, Cooperative Extension Service Bul. 1132, 1–3.

Van der Pluym (1978) Extent, causes and control of dryland saline seepage in the Northern Great Plains Region of North America. pp. 1–48 to 1–58. *Dryland-Saline-Seep Control Proceedings, Meeting of the subcommission on salt-affected soils. 11th Int. Soil Sci. Soc. Congr*., 1–48 to 1–58.

Land Transformation in Agriculture
Edited by M. G. Wolman and F. G. A. Fournier

CHAPTER 12

Soil Erosion and Degradation in the Southern Piedmont of the USA

R. B. DANIELS

12.1 INTRODUCTION

S. W. Trimble in 1974 and 1975 published two important papers on soil erosion in the southern Piedmont of the United States. His publications form the basis for an analysis of soil degradation and its influence on soil productivity. Trimble traced the erosive land use of the southern Piedmont (VA, NC, SC, GA and ALA) from 1700 to 1970. He quantified the amount of erosion in space and time by using a variety of data and techniques. The four major objectives of the study were (i) to ascertain if soil erosion was occurring before European settlement, (ii) to show that erosion increased with the accelerating erosive activities of European man, (iii) to describe the spatial and chronological pattern of erosive land use and consequent erosion, and (iv) to show that erosive land use in the Piedmont has subsided and only a few vestiges remain. A second study was designed to quantify the volume of erosion by comparing present soils with those in the virgin condition and extrapolating the results by integrating modern soil surveys and erosion surveys made about 35 years earlier.

The southern Piedmont is an erosional landscape about 750 miles (1200 km) long and 150 to nearly 200 miles (240–320 km) wide. It has a gently to strongly rolling topography with a few broad plateaux. The normal topography is a gently rolling upland with moderate to steep valley slopes that grade to the adjacent stream system. The soils are in residuum from acid and basic igneous rocks, metavolcanics, slates, sandstones and Triassic sediments. The saprolite is several feet thick in most places and is loamy or sandy. Many of the weatherable minerals originally in the rock are absent in the upper part of the saprolite and the nutrient content is low (Calvert *et al*., 1980; Cady, 1950). The saprolite is easily dug and eroded, has a bulk density of 1.3–1.4 in the upper part, and has moderate to low permeability (O'Brien and Buol,

1984). The major soils of the Piedmont are Ultisols that have low cation exchange capacity and base saturation of <35% (*Agriculture Handbook*). Alfisols are common in saprolite from more basic rock and their nutrient status is somewhat better. The surface horizons of the Ultisols from acid igneous rocks are loamy sands to sandy loams and under mature forests are relatively thin—less than 3 or 4 inches (75–100 mm) (Davis *et al*., 1931). The B horizons are clay in all but a few soils. Most upland Piedmont soils are well to moderately well drained.

12.2 METHODS

Trimble's studies (1974, 1975) were based upon historical records of land use and population migration. Data were extracted from many sources. The

Table 12.1 Relative rates of erosion under various types of cover and cover conditions (Trimble, 1974)[c]

Type and cover condition	Percentage of row crop erosion rate[a]
Row crop—poor or no rotation	0.80–1.00
Small grains (autumn planted)	0.30–0.40
Small grains (spring planted)	0.40–0.50
Grasses and legumes in rotation	0.10–0.20
Pasture (excellent cover)[b]	0.01
Pasture (good cover)	0.03
Pasture (fair cover)	0.07
Pasture (poor cover)	0.15
Pasture (very poor cover or idle)	0.30
Woods (excellent cover)	0.002–0.005
Woods (good cover)	0.01
Woods (fair cover)	0.03
Woods (poor cover)	0.07

[a]Factors can be raised or lowered based on field conditions. For example: pasture fair cover could range from 0.04 to 0.14.

[b]Cover density guide (percentage ground cover including litter:

Excellent	90–100
Good	70–89
Fair	50–69
Poor	30–49
Very poor	15–29

[c]Primary source: USDA Soil Conservation Service, *Guide to Sedimentation Investigations, South Regional Technical Service Area*, Engineering and Watershed Planning Unit *Technical Guide*, No. 12, 1968, p. VII-8. These values are for use in the Musgrave Equation and are not to be confused with those for the Universal Soil Loss Equation which is now being used by the Soil Conservation Service.

history of land use was converted to a composite index of erosional significance or intensity called erosive land use (ELU). The ELU is based primarily on the equations of Musgrave (1947) and Wischmeier and Smith (1965). A relative rate of erosion was assigned to different crops or vegetation types (Table 12.1). This index of erosivity, called the cover factor, is a negative expression of the effectiveness of a particular ground cover in preventing erosion. A composite cover factor was calculated for large areas having more than one land use by weighing the individual cover factors according to the proportion of area in a particular use. This composite ELU intensity permits analysis on a broad scale.

Trimble constructed a map of the Piedmont showing the areal distribution of erosive land use in the Piedmont for 1700, 1770, 1810, 1860, 1920, and 1967 (Figure 12.1). For example, the mean intensity of ELU in 1860 was 38% and in 1920 it was 42%. This means that in 1860 the equivalent of 38% of the area was in row crop and in 1920 it had increased to 48%. The data used were the US Census of Agriculture where possible. The 1967 data were from the conservation needs inventory published by each Soil Conservation Service State Office. Where US Census data were not suitable, Trimble used

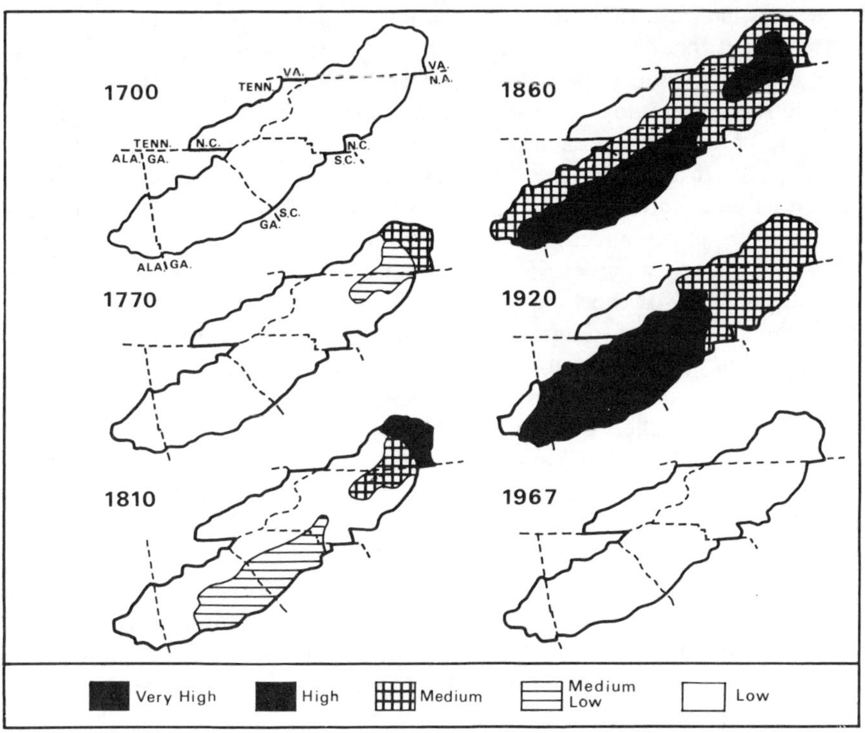

Figure 12.1 Erosive land use, 1700–1967 (Trimble, 1974)

the relationship between ELU population variables

$$\text{ELU} = 17.6 + 0.9\,(\text{density of slave population}) + 0.1(\text{density of non-slaves})$$

to estimate the ELUs before 1860. Trimble (1974) gave a detailed account of how ELUs were calculated and the problems with various parts of the data.

Trimble also estimated the volume of material removed from soils of the Piedmont by a combination of methods. He used the USDA Physical Land Surveys made in the 1930s and 1940s and modern soil surveys published between 1949 and 1967 as the basic data. The 34 surveys used were made at a scale of 1:20 000 or larger and covered 39 160 km^2 of the 142 515 km^2 of the Piedmont.

Average soil-removal estimates were assigned to each erosion class and soil mapping unit by soil scientists experienced in the general area of each survey or those soil scientists who had taken part in the surveys. Estimates were made for the mapping unit in each survey area by two or more soil scientists. It was recognized by Trimble that the numbers assigned are estimates or value judgements but these judgements are based upon thousands of field measurements and observations. Trimble believed the values begin to assume the characteristics of statistical means.

The average depth of soil removed from each survey area was computed from the values assigned to each mapping unit within a survey area. The process was repeated for each of the 34 large-scale surveys used. The information was then expanded to the entire Piedmont for each county by using the Reconnaissance Erosion Surveys completed in 1934 (Lee *et al.*) and 1935. The percentage of each of six erosion classes for the county were used as coefficients in an equation for determining soil loss. The equations were solved for the unknowns by using the estimates of soil loss from each erosion class made by experienced soil scientists. The predicted losses were within 27% of the observed values 65% of the time and within 54% of the observed values about 95% of the time.

By using the above data Trimble calculated the average depth of soil erosion for the entire Piedmont by counties (Figure 12.2). Maps showing the areal distribution of six erosion classes from 1.9 to 12.3 inches (48–312 mm) soil loss were published in both studies.

Trimble used a variety of evidence—such as clear streams, very low ELUs (0.2), and large areas of undisturbed vegetation—as evidence that soil erosion was minor in the Piedmont before European settlement. He probably was correct in his conclusion, but it should be recognized that the Southern Piedmont is an old land mass that has been eroding for several million years. Several geologists, using different techniques, have estimated the average rate of erosion covering several million years for various parts of the United States (Hack, 1978; Judson and Ritter, 1964; Matthews, 1975; Menard, 1961). Most of the estimates are within a range of 4–5 cm per 1000 years for the

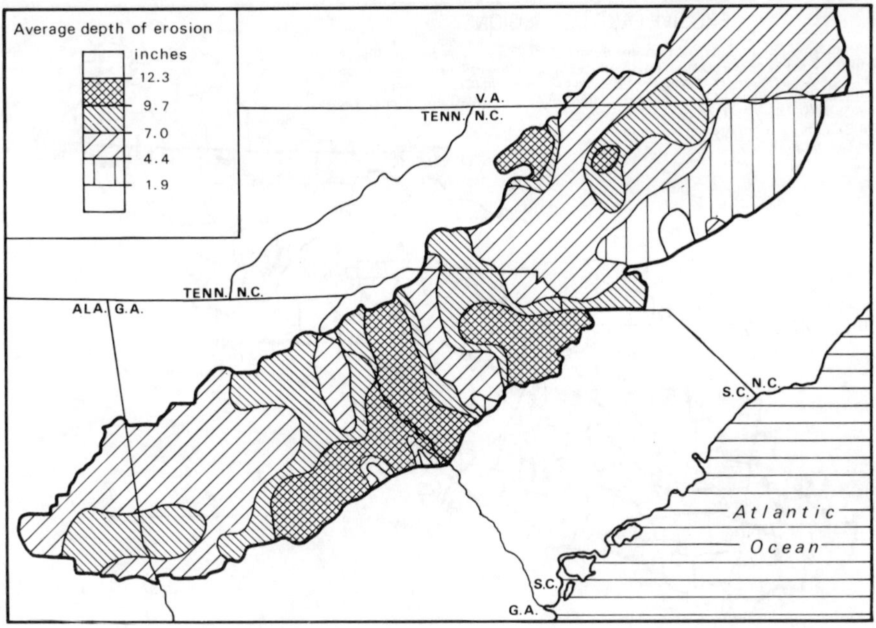

Figure 12.2 Average depths of total erosion (Trimble, 1974)

Piedmont and Applachian region. The low estimates are 0.9–0.4 cm/1000 years for modern erosion of some small watersheds. These rates are the equivalent of 0.75–0.60 t/ha/year, with the low estimates equal to 0.126–0.05 t/ha/year. The high estimates would be difficult to measure by most monitoring techniques and could easily be dismissed as being of no consequence. Yet at 0.75 t/ha/year the land mass of the United States lying above sea level could be moved to the ocean in 11–12 million years (Judson and Ritter, 1964)—only a short part of the time the Southern Piedmont has been exposed to erosion. Other evidence of erosion and deposition in the Piedmont that predates European settlement are the buried soils in alluvium or colluvium in South Carolina that have more northern vegetation pollen, and the radiocarbon dates from wood in alluvium in Georgia that range from 8725 to 425 years before the present (Cain, 1944; Staheli *et al.*, 1974, 1977), with sedimentation rates equal to 0.46–2.47 m/1000 years. The problems with average rates of erosion or sedimentation are that we have no idea of the rate during the active cycle, nor do we know how long the cycles lasted.

European settlement started in the Virginia Piedmont in about the year 1700 and was completed in the Alabama Piedmont during the 1830s. The ELU of highest intensity shifted from the north-east to the south-west Piedmont where plantation agriculture, cash crops and slavery predominated.

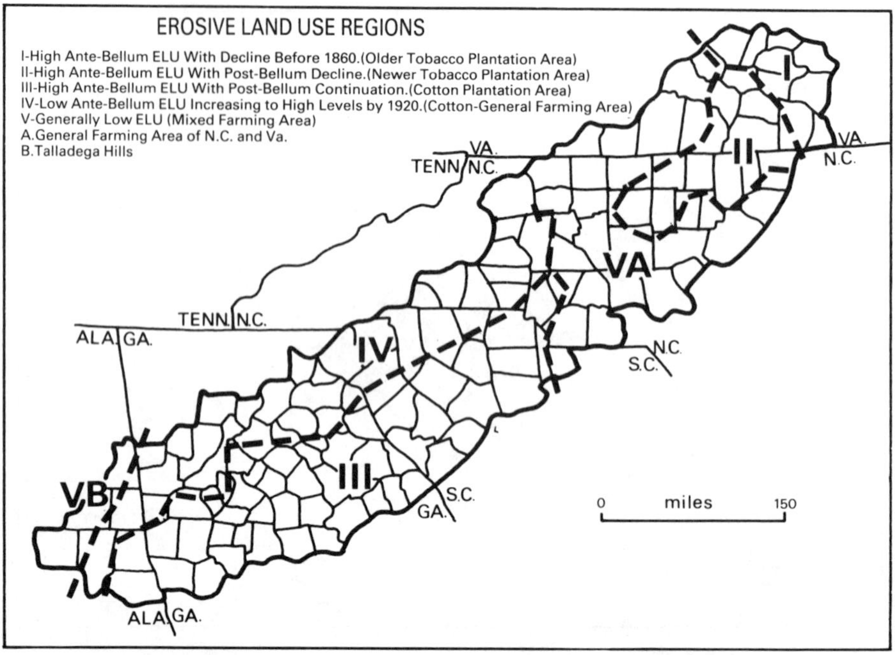

Figure 12.3 Erosive land use regions (Trimble, 1974)

The shifting intensity allowed Trimble to subdivide the Piedmont into five regions (Figure 12.3) that had similar ELU trends through time. These regions are:

Region 1—Old tobacco plantation area: probable high early antebellum ELU with late antebellum decline. 1860 ELU intensity less than 38%.

Region 2—Newer tobacco plantation area: high antebellum ELU (>38%) with postbellum decline (<43%).

Region 3—Cotton plantation area: high antebellum ELU intensity with postbellum continuation (1920 ELU >43%).

Region 4—Cotton—general farming area: low antebellum ELU (<38%) with significant postbellum increase (1920 ELU >43%).

Region 5—Mixed farming area: no definite trends, but with ELU intensities generally remaining at levels below the mean.

Region 1 declined agriculturally before 1860, and by 1860 the greatest ELU was in region 3, the cotton plantation area. Between 1860 and 1920, ELU was medium to high over the Piedmont, although there was a decrease in regions 1 and 2, and a sharp increase in region 4 where cotton cultivation expanded rapidly. High farm tenancy increased the ELU locally. After 1920 the ELU decreased significantly from 22% to 55% and everywhere on the Piedmont was <13%. The decrease in ELU was the result of agricultural decline, change of crop land to forest and pasture, and widespread application of soil conservation practices. Not all effects have been good. The alluvial fills in the headwater valleys are becoming unstable and degrading since the sediment load has decreased (Trimble, 1970).

12.3 CRITIQUE

Trimble used data of variable credibility, completeness and age in reconstructing the erosive land use of the Southern Piedmont from 1700 to 1970. He discussed the problems with these data and assumptions. It is an excellent piece of detective work and synthesis that required imagination and patience to assemble. With present knowledge and data there probably can be few changes of any significance. But as in all studies there are some assumptions that need to be re-examined. The major assumption involves the validity of soil scientists estimating the amount of truncation of a cultivated soil by comparing it to a virgin profile on similar topography. The second assumption is that the 1935 reconnaissance maps accurately represented soil erosion in an area.

The question is not Trimble's approach to measuring erosion, but how accurately can soil scientists estimate the average removal of soil from a mapping unit by comparing it to a virgin or assumed virgin profile? Within the last 30 years it has been difficult to find virgin sites that have much areal extent, so even those few virgin sites remaining may not be representative of the former undisturbed mapping unit. The uneven topography of the Piedmont is another factor, and the surface thickness and textural variations within a virgin area mapping unit have not been documented. It is unusual even on a nearly level Coastal Plain surface to have A horizons (A1 and A2) of uniform thickness across an area because the local soil environment is variable (Daniels, 1967). It would be even more unusual to expect surface thickness and textural uniformity in a gently to moderately rolling landscape shaped by geologic erosion. Local areas of convexity, concavity, runoff and runon on the uplands and valley slopes of the Piedmont should result in localized areas of erosion and deposition even in the undisturbed state. Thus, there is reason to doubt the accuracy of attempts to estimate average soil removal unless considerable detailed work on variations within a mapping unit has been completed. Trimble did not give the documentation, nor can the necessary documentation on surface properties be obtained from the literature.

Although there is ample reason to question the accuracy of an estimated soil loss from a mapping unit by experienced soil scientists, there probably is no better estimate available. Documentation either for or against an accurate estimate is missing. It appears that neither Trimble's approach nor the soil scientists' estimates can be improved upon at the present time. But caution should be exercised in interpreting such data and it should be considered only as a reasonable but rough approximation until other data are available. Authors using these techniques, however, should warn their readers of the possible or probable shortcomings of the approach.

The second assumption—that the reconnaissance and later soil erosion maps accurately represent soil erosion in an area—has been questioned by several, but some authors (Held and Clawson, 1965) feel that these maps are reasonable representatives of the conditions in the field. The Soil Conservation Service completed a reconnaissance soil erosion survey of several states in the early 1930s. No text accompanied these maps and most were published at a scale of 1:500 000. Later the Soil Conservation Service made erosion maps of counties, watersheds and selected project areas (e.g. Stevens *et al.*, 1938). A text accompanied the later maps, and soil series as well as erosion were delineated. On both sets of maps the degree of erosion was determined by comparing the soil with a virgin profile on the same topography, or if none were available, then a virgin profile was estimated. A detailed discussion of the criteria used was not given in the later publications, nor were the local variations in surface horizon thickness for each mapping unit. The virgin or assumed virgin profile used to determine the amount of erosion on cultivated soils was not described, so the original thickness of the A horizon is unknown.

The lack of descriptive detail on how the erosion was measured and the absence of documentation of what was used for the uneroded soil of each mapping unit is unfortunate. Today it is nearly impossible for workers to reconstruct or evaluate the changes in an area since the original maps were made and their value has been decreased considerably. Apparently there was considerable editing of the original manuscripts compiled by the field personnel. Stevens (oral communication, 1981) states that detailed descriptions of each map unit were sent in with the maps, both on the original reconnaissance erosion survey and the later surveys. Enough detail was given so that soil differences of slope groups within a series were described. This documentation was not published in the surveys examined; apparently it was eliminated by higher authority during the compilation and publishing process. For many reasons, it is very difficult to publish the basic data collected, and much data giving a detailed insight into the knowledge of the field party is eliminated before publication. (Modern soil survey reports are little different. A map is published, but information needed to help interpret the map 10–30 years after publication is often eliminated or never written.)

The question of how accurately soil scientists can map erosion needs to be discussed because soil scientists operate under some very real restrictions. Modern soil maps are made at scales of 1:20 000 and 1:24 000 and minimum size delineations are about 2 ha (5 acres). Even in the 1:15 840 map scales the minimum delineation is about 0.8 ha. Smaller delineations are discouraged because the symbols must fit within the boundaries of the mapped unit and spot symbols can be lost, eliminated or misplaced during the correlation and compilation process. Estimates of the percentage of inclusions of similar and contrasting soils within a mapping unit are made, but most soil surveys in the United States say little about the variations to be found within an erosional phase. There also has been little published work on the accuracy of mapping erosion phases. Powell and Springer (1965), in a limited study in the Southern Piedmont and using established criteria, stated that erosion was mapped correctly at 77% of their sites investigated. If the work of Powell and Springer is typical of the Piedmont mapping, then the soil scientists are doing an exceptional job. Additional checking is needed to verify this earlier work because most studies of soil variability indicate much more variation in soil properties among morphologically similar pedons than was originally thought (McCormack and Wilding, 1969; Amas and Whiteside, 1975; Bracewell *et al*., 1979; Beckett and Webster, 1971).

One reason to doubt the accuracy of mapping erosion is the fact that most Piedmont soil mapping units subject to erosion have an uneven topography and contain both erosional and depositional slopes or segments. Delineating a reasonably pure unit is very difficult, especially at the smaller scales. Unpublished data by Stone (1981) indicates the variation in surface characteristics within the same mapping unit in the Piedmont of North Carolina. The sites chosen were mapped as severely eroded Cecil, a Typic Hapludult, yet surface textures of the Ap within the delineations ranged from clay loam 75–100 mm thick to sandy loam up to about 450 mm thick. Corn yields in 1981 appear to be directly related to plow layer characteristics and micro landscape position. The lowest yields were on thin clay loam Ap horizons on convex ridge crests and noses, and the highest yields on sandy loam Ap horizons less than 380 mm thick on concave valley slopes. The yield variation in bushels per hectare per plot within fields composed of one mapping unit was as shown in Table 12.2.

Langdale *et al*. (1979) published data supporting the yield variations in eroded mapping units found by Stone. The test site was a 1.3 ha (3.2 acres) watershed mapped as eroded Cecil, but the test plots were eroded and moderately eroded Cecil plus a soil in local alluvium. Corn yields on severely and moderately eroded Cecil and the local alluvium were, respectively, 2226, 4674 and 6429 kg/ha (or 36, 75 and 92 bushels per acre) over three growing seasons. Corn grain and dry matter production were related positively to soil depth above the Bt horizon. Yields increased until the Bt horizon was about 40 cm from the land surface. A 40% yield reduction occurred when 15 cm of topsoil was eroded or the depth to the Bt was reduced from 40 to 25 cm. Similar results were found by Adams (1949). The only problem with trying to

Table 12.2 Corn yields in Stone's study

Location	Average yield (bu/acre)	Yield range (bu/acre)	Standard deviation (bu/acre)	Growing season rainfall (mm)	Rainfall (% >25 mm/hour)
Wake Co.	42	22–77	12	152	20
Rockingham Co.	76	35–139	24	254	50
Davidson Co.	106	81–131	14	178	50
Yadkin Co.					
1	156	71–200	27	203	28
2	153	100–199	25	203	32

1 US bushel = 35.2391 dm^3.
1 acre = 0.405 ha.

extrapolate Langdale *et al.*'s data is that the plots were not all Cecil soil nor did each plot have the same landscape position and configuration. Only one plot was moderately eroded and three were in local alluvium. The micro-environment was not the same on all plots, especially the plots on the concave sites on the lower slopes and those in local alluvium. But the variable yields found by Langdale *et al.* are probably typical for the gently sloping Cecil uplands and mapping units. Within the uplands the more productive sites are usually too small for a soil surveyor to delineate the different soils. At scales of 1:20 000 the entire watershed studied by Langdale *et al.* has less area than the smallest delineation allowed by modern surveys.

The data of Stone and Langdale *et al.* are indicative of the probable variations within a soil mapping unit in the Piedmont. The areal extent of the inclusion of other erosional phases within the mapping units and their influence on map unit production needs to be carefully documented.

12.4 PRESENT SOIL PRODUCTIVITY

Many kinds of soils are found in the Southern Piedmont, but one of the most common is the Cecil series derived from acid rocks, usually schists and gneiss, and its associate the Appling series, usually found on granites. These two soil series are still used for agriculture in the Piedmont and their present productivity can give an indication of changes in productivity over time.

The US agricultural censuses have been reasonably accurate since about 1860 (Trimble, 1974), but yields per acre are available only since about 1880 (Table 12.3). The average yields of corn harvested for grain in several southern Piedmont counties are given in Table 12.3. The average volume of soil lost by erosion as estimated by Trimble (1975) is given for each county. If soil erosion affects soil productivity, then we should expect lower production from severely eroded counties than from less severely eroded counties in the same state. The productivity loss should be large in South Carolina and Georgia between 1860 and 1920 because these are the areas of most erosive land use during these periods (Trimble, 1974: Figures 1, 3). But average corn yields in all counties remained nearly constant from 1879 to about 1954 in Georgia and South Carolina. Table 12.3 suggests that yields in North Carolina did not increase much until about 1929, although annual yield data (Table 12.4 and Figure 12.4) show that corn yields increased from a range of 11–14 bushels per acre starting about 1900 to a range of 18–20 bushels per acre by 1920. The increase was a result of 4-H Corn Clubs and educational programmes (Krantz and Chandler, 1954). Data in Table 12.4 show a large increase in North Carolina corn yields in the 1950s, probably as a result of suitable hybrids and an intensive educational programme that increased fertilizer use and helped install other practices (*ibid*). The increase in average corn yields in South Carolina and Georgia counties did not take place until the 1960s (Table 12.3).

If one looks only at average county yields, there apparently is no loss in productivity even in the severely eroded counties. Technology has increased the average yield per acre considerably during the period of record. But there has been a large decrease in corn acreage in all counties (Table 12.3), and the decrease in corn acreage in the Georgia counties could be interpreted as little more than transfer of corn acreage to the better soils.

Soil Conservation Service soil scientists have estimated average crop yields under a high level of management for a number of years. The yields are not measured but are determined for each mapping unit through discussions with land owners and operators, the experience of the local conservationist and extension agent. For the time collected, the estimated yields are probably reasonable for the soil mapping unit in the county. The estimates are quickly outdated by changes in technology, although the relative yield differences among mapping units probably hold for a few years. Table 12.5 lists the soil survey report's estimated yields for Cecil and Appling mapping units, slope and surface texture phases, in four counties. The sandy loam phases are the least eroded, or at least have the highest predicted yields, and the clay loam phases usually have lost the A horizon and the underlying B is now the plough layer. In most counties the yields on eroded phases, or the soils with the finest textured plough layer, were estimated to be 20–25% less than similar soils with a coarser textured surface. Although subject to criticism, these data indicate that mapping units with B horizon for the plough layer have lower yields than similar mapping units with some thickness of sandy loam surface remaining. The yield reductions for the mapping unit as estimated in soil survey reports are about one-half the reduction suggested by Langdale *et al.* (1979) and Adams (1949) for similar soils. The researchers were looking at plot data and the estimates are for mapping units that usually have inclusions.

Unpublished data compiled in 1964 by W. W. Stevens, former State

Table 12.3 Corn yields in the Southern Piedmont from 1879 to 1978

State:	Georgia				South Carolina						North Carolina					
County:	Baldwin		Harris		Edgefield		Fairfield		Greenville		Catawba		Person		Wilkes	
	A[a]	B[b]	A	B	A	B	A	B	A	B	A	B	A	B	A	B
Average depth erosion (cm)[c]	>31.2		11.4–17.8		>31.2		>31.2		11.4–17.8		11.4–17.8		4.8–11.2		>31.2	
Year																
1879	17.6	7	26.9	9	67.8	8	40.3	9	52.6	11	21.2	17	19.4	12	34.9	14
1890	13.3	10	24.3	11	68.0	10	34.1	9	53.5	12	24.3	13	18.8	12	40.8	13
1899	22.6	7	29.5	11	38.3	8	40.4	20	63.5	10	27.8	14	21.3	16	44.5	14
1909	18.6	10	26.0	10	36.7	11	35.1	9	57.2	11	30.9	14	20.9	15	44.3	14
1919	23.3	10	28.9	10	33.8	15	36.2	11	58.0	12	25.9	15	20.7	15	40.7	16
1924	16.9	9	21.5	12	27.7	12	25.5	10	50.1	15	19.4	14	16.5	11	38.5	15
1929	15.9	–	12.1	15	22.6	10	20.2	10	39.9	7	19.1	19	19.1	18	42.5	21
1934	21.6	7	25.2	9	29.0	8	22.5	6	54.9	12	21.8	18	19.2	16	37.6	18
1940	19.4	8	20.7	7	23.4	11	22.0	10	46.1	16	19.4	20	23.1	21	31.5	22
1945	16.3	11	11	13	22.5	17	16.7	14	46.2	18	16.8	26	20.1	26	27.9	27
1949	8.5	16	8.3	13	19.0	16	11.2	14	29.2	20	13.3	30	18.3	31	19.3	35
1954	8.7	10	4.9	10	12.9	10	4.8	9	14.2	11	8.1	20	17.5	20	11.3	22
1959	5.0	24	2.9	18	7.1	22	3.1	20	7.3	25	7.8	36	13.5	30	8.4	37
1964	0.7	31	0.7	30	3.3	24	1.2	31	4.3	38	4.5	49	10.0	42	2.7	47
1969	1.6	41	0.5	36	1.4	40	0.4	31	1.6	35	4.7	53	7.1	54	4.3	63
1974	2.6	34	1.6	47	0.8	33	0.2	56	2.9	54	5.8	63	8.2	60	4.3	77
1978	2.5	41	0.7	32	0.7	23	0.3	33	4.4	44	4.5	57	7.1	53	5.2	73

[a] Acres ×1000 (1000 acres = 405 ha).
[b] Yield in bu/acre (1 US bushel = 35.24 dm^3).
[c] From Trimble (1975).

Table 12.4 Yield (in bushels/acre) and acreage of corn harvested for grain in three North Carolina counties[a]

	Catawba		Person		Wilkes	
Year	Acres	Yield	Acres	Yield	Acres	Yield
1879	21 248	17	19 372	12	34 865	14
1890	24 275	13	18 776	12	40 760	13
1899	27 837	14	21 292	16	44 466	14
1909	30 936	14	20 897	15	44 281	14
1919	25 917	15	20 724	15	40 711	16
1925	20 250	14	17 260	11	35 970	15
1926	20 010	18	21 350	18	37 230	20
1927	20 340	18	21 650	19	32 850	18
1928	20 160	17	21 470	16	32 910	19
1929	18 800	18	20 950	19	33 680	21
1930	21 460	17	22 170	15	38 480	11
1931	21 810	18	26 060	20	41 790	19
1932	20 400	12	22 210	12	37 830	18
1933	20 680	13	24 180	16	40 300	18
1934	21 640	18	21 150	16	36 760	18
1935	22 860	19	21 360	16	35 740	19
1936	20 560	17	20 540	19	32 360	20
1937	18 310	20	19 970	20	30 340	20
1938	19 090	20	21 660	19	31 880	22
1939	19 760	20	21 350	19	35 070	21
1940	18 860	20	21 640	21	30 910	22
1941	18 490	21	20 690	21	30 150	20
1942	17 010	23	21 080	21	26 060	19
1943	17 950	21	22 020	20	27 180	25
1944	17 500	26	21 000	20	26 900	27
1945	16 100	26	19 500	26	21 800	27
1946	1 500	29	20 000	26	2 300	27
1947	14 900	31	18 600	28	20 700	33
1948	16 430	30.5	19 290	32.3	21 800	36
1949	15 230	30.4	18 380	30.9	19 500	35.1
1950	13 890	31.3	19 090	35.2	17 540	36.7
1951	13 730	31.3	18 140	28.3	15 980	37.5
1952	13 400	21.7	18 300	22.5	15 300	23.4
1953	12 400	21.6	18 100	17.4	13 800	27.9
1954	18 800	20.0	17 700	20.6	11 600	23.2
1955	9 900	34.1	16 200	28.7	11 600	41.1
1956	9 400	27.0	15 400	34.0	9 200	37.8
1957	7 400	32.4	15 100	21.0	9 100	43.3
1958	6 700	37.5	14 700	36.0	9 200	43.0
1959	8 650	36.0	14 400	29.6	9 200	36.7
1960	7 900	41	14 400	36	7 650	49
1961	7 500	40	10 650	37	7 900	45
1962	6 650	37	9 200	40	5 850	49
1963	8 050	40	11 300	35	5 950	46
1964	8 350	49	11 000	42	5 550	47

Table 12.4 (*contd*)

	Catawba		Person		Wilkes	
Year	Acres	Yield	Acres	Yield	Acres	Yield
1965	8 450	55	10 500	45	5 900	60
1966	9 600	40	11 100	25	6 350	45
1967	9 100	55	11 500	55	7 300	65
1968	8 700	50	10 950	45	7 350	60
1969	7 600	53	8 750	54	7 000	63
1970	5 750	26	9 650	41	7 750	48
1971	5 250	44	8 950	36	6 250	60
1972	4 800	53	7 800	58	2 650	62
1973	5 800	64	8 600	61	2 900	78
1974	7 050	63	9 300	60	4 000	77
1975	7 070	58	9 510	60	5 540	70
1976	6 050	58	11 000	46	7 650	75
1977	7 000	18	9 500	25	6 000	44
1978	6 000	57	9 000	53	5 100	73
1979	5 700	82	8 300	60	5 920	78
1980	5 650	63	9 300	44	6 250	37

[a]From NC Dept. Agriculture, Division of Agricultural Statistics, Raleigh, NC, and US Agricultural Censuses.

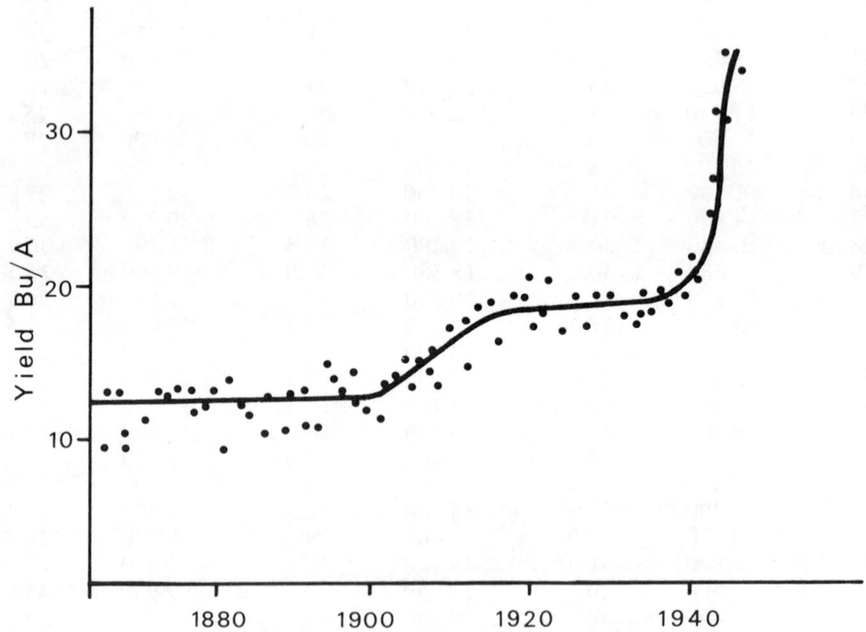

Figure 12.4 Corn yields in North Carolina (Krantz and Chandler, 1954)

Table 12.5 Corn yield estimates for Piedmont soils

Series and phase	Catawba, NC, 1975		Wake, NC, 1970		Greenville SC, 1975		Baldwin, GA, 1976	
	Y[a]	A[b]	Y	A	Y	A	Y	A
Appling								
2–6% sandy loam	90	4.9	73	3.2	80	12.1	–	–
Sandy loam eroded	–	–	68	7.1	–	–	–	–
6–10% sandy loam	–	–	65	4.6	70	2.5	–	–
Sandy loam eroded	80	2.8	60	8.1	–	–	–	–
Cecil								
2–6% sandy loam	–	–	–	–	90	56.4	95	7.9
Sandy loam eroded	90	37.7	–	–	–	–	–	–
Clay and sandy clay	–	–	–	–	–	–	–	–
loam	70	1.5	–	–	65	2.6	70	2.6
6–10% sandy loam	–	–	–	–	80	64.0	90	5.2
Sandy loam eroded	80	36.6	–	–	–	–	–	–
Clay and sandy clay	–	–	–	–	–	–	–	–
loam	60	6.3	–	–	55	11.0	60	8.0

[a]Yield in bu/acre (1 US bushel = 35.24 dm^3).
[b]Acres ×1000 (1000 acres = 504 ha).

Resource Conservationist, Soil Conservation Service, Raleigh, North Carolina, show the effect of surface thickness on stand (Table 12.6). Only a small part of the data are shown to illustrate the relationships. The stands were measured for 100 feet (30.5 m) of row by experienced agronomists. The soils were identified by soil scientists or other qualified personnel.

The relationship between stand (Y) and topsoil thickness (X) for Cecil is $R^2 = 0.6616$ ($R = 0.8195$) and for Appling $R^2 = 0.5449$ ($R = 0.7429$); both correlations are statistically significant. Although there are several criticisms of these data that can be raised, they clearly illustrate the difficulty in 1964 of obtaining a satisfactory stand on soils with little or no coarse textured surface horizon. Technology still has not solved the problem of producing a uniform seed bed in the variable surface textures of the Piedmont and is one of the problems restricting corn yields in the area (oral communication, G. Naderman, 1982). The yields on the plots with small plant populations would be low even if hybrids producing 2 or 3 ears per stalk were grown under high fertility levels (Kamprath *et al*., 1973). Present recommendations are for populations of 20 000 plants per acre for modern hybrids that produce only one ear.

If there is any region in the United States where soil erosion should have a maximum effect on soil productivity it would be in the Southern Piedmont. The soils (Ultisols), or most of them, were and are acid, nutrient-poor mineral soils formed from saprolite that has lost most of its weatherable minerals. The

Table 12.6 Relationship among surface thickness and corn stand (unpublished data from W. W. Stephens)

	Topsoil (plough layer) thickness (inches)				
	0	0–2	2–4	4–6	6–8
Cecil					
No. of plots	8	21	35	27	12
Plants/acre					
r	124–5977	1084–6973	996–7845	5479–10 958	7845–12 453
$\bar{X}$	2241	2864	4358	7721	10 585
Appling					
No. of plots	4	16	36	38	43
Plants/acre					
r	0–3985	0–5479	0–7222	2247–10 460	4483–12 826
$\bar{X}$	1120	1618	3985	6102	7721
White Store					
No. of plots	0	11	21	12	11
Plants/acre					
r		1245–5728	2490–6849	4109–7721	5479–8966
$\bar{X}$		3362	4732	5603	6724
Creedmore					
No. of plots	5	24	32	31	35
Plants/acre					
r	747–2490	373–12 951	2739–10 460	1743–10 958	4732–12 453
$\bar{X}$	1245	4607	6102	6973	7596

r = range; $\bar{X}$ = mean.

A1 horizons were thin. The underlying B horizons are usually clay, most are acid and have kaolinitic or mixed 2:1–1:1 clay minerals. The clay B horizons with the most favourable physical properties are always described as sticky or slightly sticky, plastic or slightly plastic when wet. In soils developed from basic rock, the B horizons are sticky and plastic and it is a common statement that you may have 30 minutes or an hour when ploughing is favourable; otherwise it is too wet or too dry. The plastic B horizons do not produce good seed beds under most conditions.

The soils of the Southern Piedmont have been farmed for at least 200 years and many have been severely eroded, or at least the B horizon is exposed in some areas. If Trimble's 1975 estimate of the volume of soil removed is correct, many areas have eroded well into the B horizon. Yet crop yields are improving, and since about 1870 or 1880 held at rather constant levels during periods of erosive land use (Tables 12.3, 12.4, 12.5 and Figures 12.1, 12.2, 12.4). Improved technology can account for the increase in average yields, although soils with thin plough layers of fine-textured material still produce less than the thicker slightly coarser surface horizons (Tables 12.5 and 12.6). Similar relationships occur in other southern Piedmont counties with published soil surveys during the last 10 years.

Trimble's 1975 data suggest that in severely eroded areas such as Baldwin County, Georgia, most of the soils should have clay loam plough layers because more than 30 cm of soil has been removed by erosion (Figure 12.2). The 1976 Baldwin County, Georgia, soil survey report suggests otherwise (Table 12.5). Similar relationships occur in other counties with published soil surveys during the last 10–20 years (Tables 12.7 and 12.8). In the Southern Piedmont surveys examined, about 75% of the Appling, Cecil and Madison soils on slopes from 2% to 10% still have sandy loam surfaces. In both Abbeville and Edgefield Counties, South Carolina, where Trimble predicted >31.2 to 24.6 cm of erosion, the majority of the soils have sandy loam surfaces (Table 12.8). Only Morgan and Walton Counties, Georgia, have more area of sandy clay loam or clay loam surfaces in the 2–10% slope groups of Cecil, Appling and Madison than sandy loam (Table 12.8). Walton County

Table 12.7 Total acreage (×1000 acres) of Appling, Cecil and Madison soil series with sandy clay loam or clay loam surface[a]

Soil series and phase Number of counties	ALA 3	GA 18	SC 9	NC 8	VA 3	Totals 41
Appling						
SL 2–6[b]	9.5	130.5	81.7	140.1	20.8	382.6
SL 6–10	32.0	97.5	30.0	65.4	12.1	237.0
SCL 6–10	16.7	17.3	–	0.3	1.3	35.6
Cecil						
SL 2–6	4.8	301.0	353.5	140.4	31.3	831.0
SL 6–10	13.2	228.3	273.7	140.1	23.4	678.7
SCL-CL 2–6	2.8	84.6	35.6	122.9	4.8	250.7
SCL-CL 6-10	30.1	287.3	77.1	55.2	18.8	468.5
Madison						
SL 2–6	4.2	52.9	25.6	8.7	1.9	93.3
SL 6–10	36.3	84.8	45.7	16.6	1.7	185.1
SCL 2–6	1.1	7.1	2.1	0.4	0.9	11.6
SCL 6–10	36.5	51.1	16.0	2.3	2.2	108.1
Totals	187.2	1342.4	941.0	692.4	119.2	3282.2

[a]From soil surveys published since 1959.
[b]Slope groups.
SL = sandy loam; SCL = sandy clay loam; CL = clay loam.
1000 acres = 405 ha.

Table 12.8 Percentage of Appling, Cecil and Madison soils on 2–6% and 6–10% slopes with sandy loam and sandy clay loam or clay loam surface horizons

Location	Approx. average soil loss (cm)[a]	Total acres (×1000)	Sandy loam (%)	Sandy clay loam and clay loam (%)
Georgia				
Banks, 1971	24.4–18.0	33.3	55.0	45.0
Barrow, 1977	24.4–18.0	59.9	77.1	22.9
Clarke, 1968	24.4–11.4	25.6	82.0	18.0
Clayton, 1979	17.8–11.4	43.0	85.1	14.9
Coweta, 1980	17.8–11.4	137.9	79.6	20.4
Fayette, 1979	17.8–11.4	93.5	72.5	27.5
Gwinett, 1967	17.8–11.4	74.4	89.1	10.9
Hall, 1977	24.4–18.0	46.6	53.2	46.8
Heard, 1980	17.8–11.4	45.3	68.9	21.1
Henry, 1979	17.8–11.4	126.2	72.8	27.2
Jackson, 1977	24.4–18.0	113.7	50.6	49.4
Meriwether, 1965	24.4–11.4	112.1	53.3	46.7
Morgan, 1962	31.2–18.0	72.6	37.2	62.8
O'Conee, 1968	24.4–11.4	42.1	83.6	16.4
Spalding, 1964	17.8–11.4	80.6	52.3	47.7
Stephens, 1971	24.4–11.4	17.0	54.1	45.9
Troup, 1980	24.4–18.0	100.1	71.1	28.9
Walton, 1964	17.8–11.4	118.5	43.0	57.0
Alabama				
Celeburne, 1979	17.8–11.4	4.8	100.0	0.0
Chambers, 1959	24.4–11.4	107.8	48.0	52.0
Randolph, 1967	17.8–11.4	74.6	58.3	41.7
Virginia				
Charlotte, 1974	17.8–11.4	101.2	73.2	26.8
Chesterfield, 1978	17.8–11.4	3.8	100.0	0.0
Henrico, 1975	17.8–11.4	14.2	93.7	6.3
South Carolina				
Abbeville, 1980	>31.2–24.6	91.3	91.4	8.6
Anderson, 1979	–	265.2	96.6	3.4
Edgefield, 1981	>31.2–24.6	49.9	99.5	0.5
Greenville, 1975	24.4–11.4	159.4	90.3	9.7
Greenwood, 1980	31.2–18.0	63.3	77.4	22.6
Laurens, 1975	24.4–11.4	167.6	78.1	21.9
McCormack, 1980	31.2–24.6	14.2	97.2	2.8
Union, 1975	>31.2–24.6	63.9	99.5	0.5
York, 1965	24.4–11.4	66.2	58.0	42.0
North Carolina				
Catawba, 1975	17.8–11.4	102.2	92.3	7.7
Durham, 1976	11.2– 4.8	6.5	100.0	0.0

Table 12.8 (*contd*)

Location	Approx. average soil loss (cm)[a]	Total acres (×1000)	Sandy loam (%)	Sandy clay loam and clay loam (%)
Forsyth, 1976	31.2–11.4	57.0	90.0	10.0
Iredell, 1964	22.9–11.4	131.4	88.4	11.6
Mecklenburg, 1980	17.8–11.4	122.2	3.4	96.6
Orange, 1977	17.8– 4.8	22.5	100.0	0.0
Vance, 1980	11.2– 4.8	71.1	57.9	42.1
Wake, 1970	11.2– 4.8	179.5	97.5	2.5

was predicted to have lost 17.8–11.43 cm and Morgan County 31.2–18.0 cm of soil by Trimble (1975).

Data in Tables 12.7 and 12.8 seemingly contradict the data of Trimble's that large volumes of soils have been eroded from the soils of the southern Piedmont. Large areas still retain a sandy loam surface that is at least grossly similar to the original surface. Trimble's estimate of soil loss is probably the best available using current information. Therefore, we must ask whether modern cultural practices are developing sandy loam plough layers from the underlying or upslope clay B horizons at a rate approximately equal to the loss of surface, or whether our interpretations of the Universal Soil Loss Equation grossly over-estimate the amount of material removed from a field. The upland Piedmont landscape is a combination of gentle concave and convex slopes, so erosion and deposition of material can occur within a few square metres. Modern cultural practices expose the B horizon at the surface where it can be dispersed by raindrop impact and aggregates moved downslope. The large and heavy equipment now in use, too, is capable of mixing the sandy surface and the clayey B horizon to a much greater extent than the light mule-drawn equipment of a few years ago.

Is it possible that erosional and depositional processes in cultivated fields in the Piedmont have reached a pseudo-equilibrium so that continued erosion does not change the present field relations much even though it may remove the original Ap horizon and a large part of the B horizon? Are the areas of Cecil and similar soils with clay loam surfaces the zone of most active removal and the adjacent lower areas with sandy clay loam to sandy loam surfaces the zone of transport and deposition? Are the gentle interfluves of the upland maintaining a sandy loam surface because the fines exposed by ploughing are being partially dispersed by raindrop impact and removed by runoff, as well as clayey aggregates being moved downslope as bed-load?

The idea that erosion will remove all of an A horizon, and leave only a fine textured B horizon at the surface, should be carefully examined in light of estimates of erosion and the conditions that actually exist. The variations in colour of a freshly ploughed Cecil soil that has been cultivated for nearly 200 years is convincing evidence that we either over-estimate the amount of erosion, or other processes are maintaining a sandy loam surface over much of the upland landscape developed from acid igneous rocks.

Is it also possible that we are over-predicting the amount of material removed from a field when we interpret erosion by the Universal Soil Loss Equation? This equation predicts the amount of erosion, but does not account for sorting and local deposition. It is an estimate of the amount of material moved, but not necessarily from the field. In most interpretations an annual soil loss of 10 tons per acre (25 t/ha) is interpreted as meaning the movement of that much material from the field. This claim is not made for the equation.

12.5 SUMMARY AND SPECULATION

There is little doubt that erosion since European settlement has changed the soils of the Southern Piedmont. These changes have resulted in lower soil productivity using the same technology, yet with modern technology the loss in productivity from erosion is minor compared with the large advances in yields produced by better hybrids, cultural practices and fertilizer applications. Productivity today, as measured by corn yields, is larger than at any time since European settlement. Problems—such as poor stands, as a result of unfavourable seed beds—exist but there appears to be little evidence that suggests continued erosion will reduce soil productivity to zero or even much below present levels. In most of the Southern Piedmont the saprolite is several feet thick, and except for a lower permeability to water it probably has physical properties more conducive to high soil productivity using present technology than the clay B horizons of the soils. There is also a suggestion that erosional and cultural processes may help maintain a relatively favourable surface horizon over much of the area on slopes of 10% or less. When one considers the low natural fertility of most Piedmont soils, the maintenance of a favourable rooting zone may be more important than maintaining the original surface layer of low fertility.

If it becomes necessary to again cultivate large areas of the Southern Piedmont, present technology should maintain productivity at a higher level than was possible at any time in the past. Erosion since European settlement has damaged soil productivity but not irreversibly. Without modern technology, however, soil productivity would soon decrease somewhere near the levels of the late 1800s or early 1900s, largely because the soil and underlying materials usually have low inherent fertility.

12.6 REFERENCES

Adams, W. E. (1949) Loss of topsoil reduces crop yields. *J. Soil and Water Conservation*, **4**, 130–133.

Agriculture Handbook no. 436. Soil Taxonomy: A Basic System of Soil Classification for Making and Interpreting Soil Surveys. Soil Survey Staff, US Dept. Agriculture Soil Conservation Service.

Amos, D. F., and Whiteside, E. P. (1975) Mapping accuracy of a contemporary soil survey in an urbanizing area. *Soil Sci. Soc. Amer. Proc.*, **39**, 937–942.

Beckett, P. H. T. and Webster, R. (1971) Soil variability: a review. *Soils and Fertil.*, **34**, 1–15.

Bracewell, J. M., Robertson, G. W., and Logan, J. (1979) Variability of organic matter and exchangeable cations within the A_2 horizon of an iron podzol. *J. Soil Sci.*, **30**, 327–332.

Cady, J. G. (1950) Rock weathering and soil formation in the North Carolina Piedmont region. *Soil Sci. Soc. Amer. Proc.*, **15**, 337–342.

Cain, S. A. (1944) Pollen analysis of some buried soils, Spartanburg County, South Carolina. *Bull. Torrey Bot. Club*, **71**, 11–22.

Calvert, C. S., Buol, S. W., and Weed, S. B. (1980) Mineralogy characteristics and transformations of a vertical rock-saprolite-soil sequence in the North Carolina Piedmont. I: Profile morphology, chemical composition, and mineralogy; II: Feldspar alteration products—their transformation through the profile. *Soil Sci. Soc. Am. J.*, **44**, 1096–1112.

Daniels, R. B., Gamble, E. E., and Nelson, L. A. (1967) Relation between A2 horizon characteristics and drainage in some fine loamy ultisols. *Soil Sci.*, **104**, 364–369.

Davis, W. A., Goldston, E. F., and Warner, C. H. (1931) Soil survey of Franklin County, North Carolina. US Dept. Agriculture, Bureau of Chemistry and Soils, Series 1931, no. 21 (24p plus map).

Hack, J. T. (1978) *Rock Control and Tectonism—their Importance in Shaping the Appalachian Highlands*. US Geol. Survey, Open file report no. 78403, and Prof. Paper 1126B.

Held, R. B., and Clawson, M. (1965) *Soil Conservation in Perspective*. Resources for the Future Inc., Baltimore, Md.

Judson, S., and Ritter, D. F. (1964) Rates of regional denudation in the United States. *J. Geog. Res.*, **69**, 3305–3401.

Kamprath, E. J., Broome, S. W., Raja, M. E., Tonapo, S., Baird, J. V., and Rice, J. C. (1973) *Nitrogen Management, Plant Population and Row Width Studies with Corn*. N. C. Agr. Exp. Sta., Tech. Bull. 217.

Krantz, B. A., and Chandler, M. V. (1954) *Fertilize Corn for Higher Yields*. N. C. Agr. Exp. Sta. Bull. 366.

Langdale, G. W., Box, J. E., Leonard, R. A., Barnett, A. P., and Fleming, W. G. (1979) Corn yield reduction on southern Piedmont soils. *J. Soil and Water Conservation*, **34**, 226–228.

Lee, W. D., Stevens, W. W., and Patrick, E. P. (1934) *Reconnaissance Erosion Survey of the State of North Carolina*. US Dept. Agr. Soil Conservation Service. Map Scale 1:500 000 (legend, no text).

Matthews, W. H. (1975) Cenozoic erosion and erosion surfaces of Eastern North America. *Am. J. Sci.*, **275**, 818–824.

McCormack, O. D., and Wilding, L. P. (1969) Variation of soil properties within map units of soils with contrasting substrata in N.E. Ohio. *Soil Sci. Soc. Amer. Proc.*, **33**, 587–593.

Menard, H. W. (1961) Some rates of regional erosion. *J. Geol.*, **69**, 155–161.
Musgrave, G. W. (1947) The quantitative evaluation of factors in water erosion, a first approximation. *J. Soil and Water Conser.*, **2**, 133–318.
O'Brien, E. L., and Buol, S. W. (1984) Physical transformation in a vertical rock–saprolite–soil sequence. *Soil Sci. Soc. Amer. J.*, **48**, 354–357.
Powell, J. C., and Springer, M. E. (1965) Composition and precision of classification of several mapping units of Appling, Cecil and Lloyd series in Walton County, Georgia. *Soil Sci. Soc. Amer. Proc.*, **29**, 454–458.
Staheli, A. C., Ogren, D. E., and Whorton, C. H. (1974) Age of swamps in the Alcovy River drainage basin. *S. E. Geol.*, **16**, 103–106.
Staheli, A. C., Ogren, D. E., and Whorton, C. H. (1977) Age of swamps in the Alcovy River drainage basin, a reply. *S. E. Geol.*, **18**, 195–198.
Stevens, W. W., Bragg, H. V., Sease, E. C., and Lewis, O. C. (1938) *Erosion and Related Land Use Conditions on the Reedy Fork Demonstration Area*. North Carolina USDA Soil Conservation Service, Washington. US Govt. Printing Office.
Stone, J. (1981) Unpublished data, Soil Sci. Dept., NCSU, Raleigh, NC.
Trimble, S. W. (1970) The Alcovy River swamps: the result of culturally accelerated erosion. *Georgia Academy of Science Bulletin*, **28**, 131–141.
Trimble, S. W. (1974) *Non-induced Soil Erosion on the Southern Piedmont*. Soil Conservation Society of America.
Trimble, S. W. (1975) *A Volumetric Estimate of Man-induced Erosion on the Southern Piedmont*. USDA, Agricultural Research Service Pub. S40, 142–145.
Wischmeier, W. H., and Smith, D. D. (1965) Predicting rainfall erosion losses from cropland east of the Rocky Mountains. *Agr. Handbook no. 282*. US Dept. Agr., Washington, DC.

Land Transformation in Agriculture
Edited by M. G. Wolman and F. G. A. Fournier

CHAPTER 13
USDA Soil Depletion Study of the Southern Iowa River Basin, USA

R. B. DANIELS

13.1 INTRODUCTION

The soil depletion study (Rosenberry *et al.*, 1980) used soil mapping units of the Soil Survey as the basis for extrapolation and the 1967 Iowa Conservation Needs Inventory as the initial data source. Three erosion phases were recognized: Phase 1 had no mixing of the surface and subsoils in the plough layer, phase 2 had some subsoil mixing in the plough layer, and phase 3 is where the plough layer is predominantly subsoil material. The sample data were sorted by county, land resource area, sub-basin and soil mapping unit. Crop land area was summed across all soil mapping units in each county and the totals were then adjusted to the crop areas reported in the 1974 Agricultural Statistics. Four land uses were reported: corn for grain, soy-beans, oats (includes all small grains) and hay (includes rotation pastures). The percentage of each soil mapping unit area in each land use was assumed to represent the rotation, i.e. 66% corn and 33% soy-beans represented a corn–corn–soy-beans rotation.

From the above information the gross erosion per unit area was computed for each production alternative by using the Universal Soil Loss Equation (Wischmeier and Smith, 1978). The individual soil losses were weighted to give an average soil loss for each mapping unit.

The tonnage of soil loss for each mapping unit was converted to depths of soil after adjusting for the bulk density of a partially compacted plough layer for each mapping unit.

The soil depletion study of the Southern Iowa Conservancy District covers 5.3 million acres (13.1 million hectares), with 3.5 million acres of crop land. The area is predominantly general grain and livestock farms and the topography is gently to strongly rolling with Wisconsinan loesses mantling earlier

Pleistocene tills. Soils are dominantly Mollisols with small areas of Alfisols. Surface horizons are silt loam, silty clay loam and loam, and the B horizons, if present, are fine silty, fine loamy and fine textures (Rosenberry *et al*., 1980).

The study was made to 'predict the effect that current levels of soil erosion, if continued, will have on individual soils in the Southern Iowa Conservancy District by the year 2020'. Production inputs and yields were estimated for each erosion phase of the soil. The study estimated the impact on productive potential and compared the productivity capacity in 1974 and 2020. The extra inputs required to keep yields at levels consistent with soil-test recommendations were estimated. A secondary goal was to examine six alternative methods of controlling erosion at tolerable levels and to determine the impacts of alternatives on land use and income. The authors defined tolerable erosion as: 'that amount of annual soil loss in tons per acre a soil mapping unit can lose and still be able to maintain its natural productivity and top soil over time'. No definition was given for natural productivity.

The logic used in computing soil degradation assumed a constant soil loss for each alternative between 1965 and 2020. The total tonnage eroded was converted to inches of soil and then the percentage of the soil mapping unit that degraded one erosional phase was estimated from a graph. The graph assumed an equal distribution of topsoil across each point on the scale from maximum to minimum topsoil for each phase. For example, 10% of the soil mapping unit would have the amount of topsoil allowed in the upper 10% of the erosional phase and 10% would have the amount allowed for the lower 10% of that phase. The other 80% would be evenly distributed throughout the remainder of the mapping unit.

Fuel requirements, fertilizer, pesticide and terrace costs were estimated for each mapping unit and adjusted when a soil changed from one erosional phase to another.

The conservation practices considered were rotations, contouring, terracing and residue management. The practices were combined so that soil loss was less than or equal to the tolerable soil loss that was assumed to prevent soil depletion. A total of six alternatives were examined:

(1) rotations alone with up-and-down the slope cropping and residue incorporated;
(2) rotations alone with up-and-down the slope cropping but 2500 pounds (1134 kg) of corn and 1000 pounds (454 kg) of bean residue left on the surface after planting;
(3) rotations and contouring, residues incorporated;
(4) rotations, contouring and residues on surface after planting;
(5) rotations and terracing with residues incorporated;
(6) rotations, terracing, and residue left on surface after planting.

Table 13.1 Harvested crop land (×1000 acres) by erosion phase, projected over time (Report Table 5)

Erosion phase[a]	1974	2000	2020
1	683	569	501
2	1217	1007	768
3	195	519	826
Totals	2095	2095	2095

[a]Erosion phase 1: no apparent or slight erosion, > 10 inches to B horizon: phase 2: moderate erosion, Ap may contain a mixture of A and B horizons, depth to B ranges from about 6 to 10 inches; phase 3: severely eroded, <3 inches of A horizon, the Ap is 50–100% B material.

13.2 RESULTS

About 1.4 million acres of crop land, nearly 40%, were within the tolerable soil loss limits in 1974. The authors assumed these soils would maintain their topsoil and their productivity. The other 60% of the soils were above the tolerable soil loss so, by the assumptions made, they would be degraded over time (Table 13.1).

The projected change in erosion classes requires 30% additional fertilizer to help counteract the depletion from erosion (Table 13.2). Yields are assumed to decrease as the soil is eroded from one phase to another (Table 13.3).

With the exception of alternative rotation (6), terracing and residue, the major impact of the alternative is to shift land from row crops to small grain and hay to bring all soils within the tolerable erosion levels. The decrease in projected income, averaged across all soils now exceeding the tolerable soil loss, is proportional to the area of row crops being taken out of production (Table 13.4).

Table 13.2 Increases in fertilizer needs (pounds/acre) as soil is depleted (Report Table 7)[a]

Change in erosion phase	Corn			Soy-beans			Oats		
	N	P_2O_5	K_2O	N	P_2O_5	K_2O	N	P_2O_5	K_2O
1–2	10	2	6		1	5	5	2	6
2–3	30	1	7		1	7	15	1	7

[a]From unpublished data, Iowa Agricultural Station, Ames, Iowa.
1 pound/acre = 1.12 kg/ha.

Table 13.3 Reduction in yield (per acre) as soil is depleted (Report Table 8)[a]

Change in erosion phase	Corn	Soy-beans	Oats	Hay
1–2	16	5	9 bu	0.6 ton
2–3	7	3 bu	4 bu	0.5 ton

[a]From Fenton *et al.* (1971); Soil Conservation Service, Iowa 2, US Dept. Agric., Feb 1969.
1 US bushel = 32.24 dm^3; 1 acre = 0.405 ha; 1 ton = 1.016 t.

13.3 CONCLUSIONS

The authors of the study concluded that the cost of reducing soil erosion to tolerable levels is three times as expensive as benefits. Farmers will be hard-pressed to finance erosion control without some form of assistance. The authors believed that the used surface residues with rotations and contouring can be an important and economically feasible method of controlling erosion. They also concluded that fertilizer can offset yield reductions for a short time, but cannot prevent a decrease in potential productivity.

13.4 CRITIQUE

The study of erosion's impact on the Southern Iowa Conservancy District used the best available information in its computations. Some refinement probably could be made, but it is doubtful if the results or conclusions would be changed significantly. We need to look at the basic assumptions of the study, however, some of which were:

(1) The Universal Soil Loss Equation is applicable to the study and can be used to predict rates of erosion and interpret the amount of soil degradation.
(2) There apparently was no allowance made for regeneration of the topsoil, although mixing of topsoil and B horizon were recognized.
(3) The tolerable level of erosion is the T value set for each mapping unit. This level of erosion, if not exceeded, will maintain the productivity of the soil, but degradation will start when the T value is exceeded. In other words, surface horizons are generated at a rate equal to the T value. If the tolerable annual soil loss is 5 tons per acre (12.6 t/ha), then 1 inch (25 mm) of surface horizon is produced every 30 years. Assumptions 2 and 3 are contradictory.
(4) The mapping or identification of erosion phases is accurate.
(5) The reduction in productivity with increasing erosion applies equally to all soils.

Table 13.4 Comparison of alternative methods of controlling soil erosion from water runoff (Report Table 10)

Alternative	Row crop	Close grown	Hay	Residue	Terraced	Net Income (lost dollars)
			(×1000 acres)			
(1) Rotation control	−1190	274	911	0	0	97 260
(2) Rotation and residue	−881	210	679	1727	0	66 043
(3) Rotation and contouring	−809	206m	603	0	0	59 388
(4) Rotation, contouring and residue	−623	244	379	1725	0	34 478
(5) Rotation and terracing	−243	128	116	0	2009	99 216
(6) Rotation, terracing and residue	401	−168	−233	1586	2009	49 473

The Universal Soil Loss Equation was specifically designed and field tested to predict the average annual soil movement from a given field slope under specified land use and management conditions and to estimate the reduction in soil loss from various changes in cropping systems and cultural practices (4). The question is whether the equation can be used effectively in estimating soil loss from county-wide mapping units, since it was designed primarily for a single field application. Rosenberry *et al.* defined *L*, the slope-length factor, as the distance from the top of the hill to the sediment fan at the bottom of the slope. Yet Wischmeier and Smith (1965) defined effective slope length as: 'the distance from the point of origin of overland flow to the point where either the slope decreases enough that deposition begins or runoff water enters a well defined channel'. Few details were given by Rosenberry *et al.* on how an average slope length or percentage slope was derived for each mapping unit. However, they stated that if two mapping units were on one slope, the soil at the bottom would have the entire slope length for *L*.

Most mapping units subject to erosion have an uneven topography that varies from convex to concave, and most have first- and probably second-order stream channels to route the runoff. The delineations are irregular in shape and a sloping unit may occupy narrow spur ridges as well as valley slopes. The uneven or irregular slopes need to be divided into relatively uniform segments but may or may not be treated as independent units (Wischmeier, 1977). Establishing an average slope length and percentage slope would require many measurements for each mapping unit. The slope length from the ridge crest to depositional footslope ignores the converging water flow in the concave areas, potential depositional areas, and the diverging or straight flow, usually eroding segments, in other areas. It is also possible for slope length to be made longer and shorter than normal by cultivation patterns. Thus, the slope-length factor, which is very important in determining soil loss for slopes above 3% (Williams and Berndt, 1977), is probably overestimated by Rosenberry *et al.* on slopes over 3% where most of the erosion occurs. The question that needs to be carefully examined is how one applies the Universal Soil Loss Equation to predict erosion from large areas when each delineation is a special problem unto itself.

Although it was not directly stated, apparently no allowance was made for regeneration of the surface or A horizon in mapping units where erosion exceeded the tolerable limit. If topsoil and productivity can be maintained at a *T* value of 5 tons per acre per year, with the inference that this is the rate of regeneration (McCormack *et al.*, 19??), why should not 6 or 8 tons per acre per year erosion be the equivalent loss of only 1 or 3 tons when computing topsoil or productivity losses for that soil? A 6-ton soil loss should be an effective loss of 1 inch of surface soil in 150 years. This means that a 6-inch plough layer or surface horizon would still be 5 inches thick at the end of 150 years, not lost completely at the end of 150 years (based on 150 tons/acre/

Table 13.5 Guide for assigning soil-loss tolerance values *T* to soils having different rooting depths

Rooting depth (inches)	(cm)	Soil-loss tolerance values (Annual soil loss, tons/acre) Favourable substrata[a]	Unfavourable substrata[b]
0–10	0–25	1	1
10–20	25–51	2	1
20–40	51–102	3	2
40–60	102–152	4	3
60+	152+	5	5

[a]Can be renewed by cultural and management practices.
[b]Substrata such as rock or soft rock that cannot be renewed by economical means.
1 ton/acre = 2.51 t/ha.

inch) as apparently as assumed by Rosenberry *et al*. If any consideration is given to regeneration of the A horizon it would decrease the rate of topsoil loss so the interval between changes from one erosion class to another would increase and the computed loss of soil productivity would be much less. There, of course, must be an upper limit; i.e. if erosion is 40 tons per acre per year it is doubtful there would be any regeneration of the surface horizon.

Soil-loss tolerance is defined as the maximum rate of annual soil erosion that will permit an economically high level of crop productivity to be maintained indefinitely (McCormack *et al.*, 1982). Guidelines for soil-loss tolerance were formulated in the early 1960s after years of discussion by soil scientists, agronomists and others. During 1961 and 1962 the Soil Conservation Service held regional workshops where the guidelines for establishing soil-loss tolerances and the *T* values were set for the major soils. A maximum of 5 tons/acre (12.6 t/ha) per year was established. Current Soil Conservation Service guidelines are (*Guide*, 1973):

(1) An adequate rooting depth must be maintained.
(2) Soils that have significant yield reduction when the surface layer is removed by erosion are given lower soil-loss tolerance values than those where erosion effects yields very little.

A maximum of 5 tons soil loss per acre per year was selected for use with the Universal Soil Loss Equation for the following reasons:

(1) Soil loss in excess of this value affects the maintenance, cost and effectiveness of water control structures affected by sediment.

(2) Gully erosion accompanies excessive sheet erosion in many places.
(3) The loss of plant nutrients is excessive with greater soil losses.
(4) Numerous practices can keep soil losses below this value.

The prevention of off-site damages by sediment was also used to limit the maximum soil loss to 5 tons per acre per year. The guidelines used by the Soil Conservation Service do not document the supporting data for establishing the allowable soil loss. Such documentation is difficult to find, and when presented is often of questionable value. For example, Browning *et al*. (1948, 1949) were among the first papers to establish a maximum allowable soil loss. They stated that: 'all known conservation practices may be needed if soil losses are to be reduced to an allowed minimum and the productivity is to be maintained over a period of time'. In a paper published while the 1948 paper was in press, Browning *et al*. stated that soil loss should be reduced so that productivity will be maintained over a period of time. To accomplish the above, gully formation must be prevented and loss of fertility by erosion, leaching and crop removal should not exceed that which is being replaced or built up from the lower soil by management practices. They developed a table showing permissible soil loss for a number of Iowa soils but gave no details in either paper on criteria used and what evidence supported the figures given.

Smith and Stamey (1965), in a discussion on determining tolerable erosion rates, quoted Chamberlin's remark (1908) that a foot (30.5 cm) of soil developed in 10 000 years as a mean rate of soil formation. Chamberlin's ideas were based on observations of the Mid-west Pleistocene in the period before radio-carbon dating drastically shortened the time intervals between deposits and increased the rates of soil formation. After citing Jenny's (1941) review of rates of soil formation, Smith and Stamey (1965) used a renewal rate of 0.2 tons for weathering of the average rock and between 0.1 and 0.6 tons from adjusted erosion under close-growing vegetation if a balance were assumed between erosion and renewal. These renewal rates from Smith and Stamey are for weathering of parent material, not development of surface horizons from B horizon material.

Van Doren and Bartelli (1956) maintained that the rate of weathering of soil material, the soil-building potential of the material under the surface soil, the texture of the subsoil, the climate, and a host of other factors should be considered when assigning a permissible soil-loss value. These authors used crop yield data and its variation with surface depth to help establish permissible soil-loss rates. The assumption was made that the Tama soil could lose 4.5 tons per acre (11.3 t/ha) per year without materially affecting crop yields. The 1 inch (25 cm) of soil lost about every 35 years would be compensated for by the management practices used so that crop production would be maintained. There was no evidence presented that supported the assumption that 1 inch of topsoil could be formed every 35 years, but it probably came from

Bennett's (1939, pp. 8, 95) ideas on how fast topsoil forms under natural conditions and Hudson's (1971, p. 36) inference that modern agriculture could build surface soil at about this rate.

Although the allowable erosion rates have been used for many years, the rates apparently are based upon the experience of a large number of individuals and have been arrived at more or less by committee vote. Hard evidence to support the decision is scarce. As Smith and Wischmeier (1962) stated: 'establishing the tolerance values (for allowable soil loss) has been largely a matter of judgment'. It remains so today. But no better alternative exists because research has not established the hard data, nor is it likely to in the near future.

Major criticism of attempts to establish allowable erosion rates by most authors or organizations were not documented, or the documentation is difficult to obtain so that someone else can logically reconstruct the information used and arrive at a similar conclusion.

The unstated assumption by Rosenberry *et al.*, that soil erosion maps are accurate, must also be challenged. There are few uniform areas of moderate or severe erosion. Most delineations of erosional map units, especially at scales of 1:20 000 to 1:24 000, have areas of phase 2 and 3 erosion because the landscape within the delineation is not uniform and has areas of erosion and deposition. The minimum size delineation on the small-scale maps is about 5 to 10 acres, so micro landscapes can occur within each delineation.

Dideriksen (1966) summarized the plot data from various mapping units in Iowa and evaluated the correctness of mapping series, slope, and erosion based upon concepts and criteria used at that time. These data are summarized by series and slope group in Table 13.6. The plots were drawn randomly from a 2% statistical study used in the Conservation Needs study. A point was located by drawing random numbers to determine the corner of the quarter section and distance along the X and Y axes. From Table 13.6 it is evident that erosion was mapped correctly most of the time, especially on the lesser slopes. These data are somewhat surprising because most studies have indicated wide variability within mapping units (McCormack and Wilding, 1969; Amos and Whiteside, 1975). When only phase-3 map units are considered, erosion was mapped correctly on 28 out of 34 plots studied, or about the same percentage as shown for all erosion phases in Table 13.6. But it must be recognized that these data are from loess soils and the uniformity under these conditions is probably much greater than in more complex parent materials and landscapes.

If Dideriksen's summary is correct, erosion on about 70–80% of the severely eroded map units is correctly classified. But this also means that about 20–30% of the map unit is less severely eroded and can produce from 7 to 23 bushels more corn per acre (Table 13.3). At a minimum this is a productivity under-estimation of 1.5–3 bushels of corn per acre if yields are

Table 13.6 Summary of corn yield sites—testing series slope, erosion groups (Dideriksen, 1966)

Series	Slope (%)	Number of sites	Average percentage correctly mapped		
			Series	Slope	Erosion
Ida	5–9	12	91	100	91
	9–14	10	80	70	60
	14–20	7	71	100	86
Monona	0–2	3	66	100	100
	2–5	10	60	100	90
	5–9	13	69	100	69
9–14	9–14	16	100	100	94
	14–20	7	57	86	71
Marshall	0–2	5	76	98	84
	2–5	12	83	66	83
	5–9	14	79	100	71
	9–14	10	70	80	80
Sharpsburg	0–2	2	100	100	100
	2–5	9	63	100	78
	5–9	11	55	90	73
Tama	0–2	2	100	100	100
	2–5	2	100	100	100
	5–9	4	100	100	75
Shelby	5–9	1	100	100	100
	9–14	9	77	89	100
	14–18	2	75	92	100

near 100 bushels per acre, or well within the experimental error. But the apparent high accuracy of mapping erosion shown by Dideriksen's data may be an artifact of a small number of plots located in the more complex landscapes of the severely eroded phase-3 units. Final judgement should be withheld until yield data for individual delineations are available, especially in the more complex landscapes.

The areal extent of other erosion phases within a mapping unit needs to be known to predict the changes in productivity with continued erosion. For example, if all convex areas in the mapping unit have the B horizon as the plough layer, will continued erosion increase the area of severely eroded soils significantly? Will continued erosion remove the surface horizon from the straight and concave slopes so the entire map unit is severely eroded? If the area of severely eroded soil within the map unit does not expand significantly with continued erosion, productivity should stabilize, providing unfavourable horizons are not exposed on the eroding areas.

Rosenberry *et al*. assumed the same reduction in yield for all soils when changing from one erosion phase to another (Table 13.3). Yet in the original publication, Fenton *et al*. (1971) recognized that the decrease in yield was greater for some soils than others as erosion phases changed. The use of a constant decrease in yield for each phase change may have changed the overall results very little, but for the individual mapping units the change can be considerable.

Much of the work cited by Rosenberry *et al*. to support their productivity analyses is unpublished yield data from Iowa State University. These data were collected from plot studies where each plot was classified by an experienced soil scientist. Thus the change in productivity with increasing erosion shown in Table 13.3 should be an accurate prediction for the area of a mapping unit within a specific erosion phase. But if more than one erosion phase is included within a mapping unit, the map units' productivity would be over-estimated or under-estimated by assuming that the area was all one erosion phase. Considering the normal inclusions within various mapping units, it is probable that Rosenberry *et al*.'s analysis over-estimated the decrease in productivity with continuing erosion.

13.5 SUMMARY

The study by Rosenberry *et al*. used the best available information to compute the impact of erosion on productivity. Procedural changes are probably not warranted in most cases, except the method for determining slope length, and some allowance should be made for regeneration of the surface horizons. The major problem is that the assumptions used—such as (i) that regional erosion can be computed by the Universal Soil Loss Equation, (ii) that erosion rates equal to or less than the *T* values will permit long-time production by maintaining the surface horizon, or if the *T* value is exceeded production will decrease; (iii) that A horizons or surface soil can be generated at a rate of 1 inch per 30–35 years under modern cultural practices—are not supported by adequate data or rigorous field testing. The stated and unstated assumptions used in the study have been accepted and used for 20 years or more. While unsupported by data in many cases, the judgements were made by knowledgeable individuals using the best available information at that time. It is doubtful that we could do any better if it had to be done today with current information. It is also doubtful that additional research could give better data for a considerable period because how do you approach the problem except through inference based on past history, including the Holocene? However, this still does not excuse us for blindly accepting assumptions that have questionable validity, or for not documenting how decisions were reached so that others can evaluate their validity as new evidence becomes available.

13.6 CHANGES IN SOIL PROPERTIES PRODUCED BY CULTIVATION

Soil degradation, if defined as the changes from a virgin profile, is easy to prove in Mollisols or other soils with thick dark A horizons, because a large

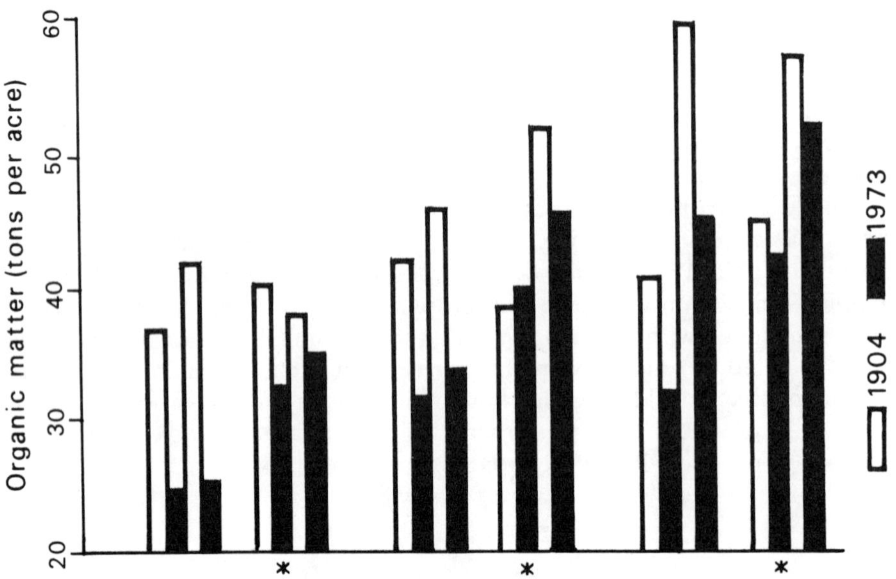

Figure 13.1 Organic matter content in the surface soils of selected sub-plots of the Morrow Plots in 1904 and 1973 (Welch, 1981, unpublished data)

decrease in organic matter usually occurs within a short time (Guernsey *et al*., 1969). Other changes are more difficult to document because many of the measurements required were not developed until recently, or the data are not available on virgin pedons.

Probably the best source for changes in Mollisols is the Morrow Plots in Illinois. These plots were established in 1876 on a nearly level Flannigan silt loam, a fine, montmorillonitic, mesic aquic Argiudoll, with about 18 inches (46 cm) or more of A1 horizon. Cultivation has produced some marked changes depending upon the treatment (Figure 13.1). The amount of organic matter has decreased from 1904 to 1973 in all plots, although at different rates (Figure 13.2—sample dates were 1904, 1913, 1923, 1933, 1944, 1953, 1955, 1961, 1967 and 1973). The two interesting points about Figures 13.1 and 13.2 are that a rotation of corn–oats–clover retained a relatively high organic matter status, and lime, nitrogen, phosphorus and potassium additions on continuous corn after 1954 increased the organic matter content by 1973 by about 5 tons per acre. The rotation corn–oats–clover rotation with manure, lime and phosphate was able to retain >85% of the original organic matter and nitrogen (Table 13.7).

Guernsey *et al*. (1969) found a higher total root weight of corn in the corn–oats–clover rotation plot than in the untreated continuous corn sub-plot. The A horizon averaged about 8 inches (20 cm) thicker and was more

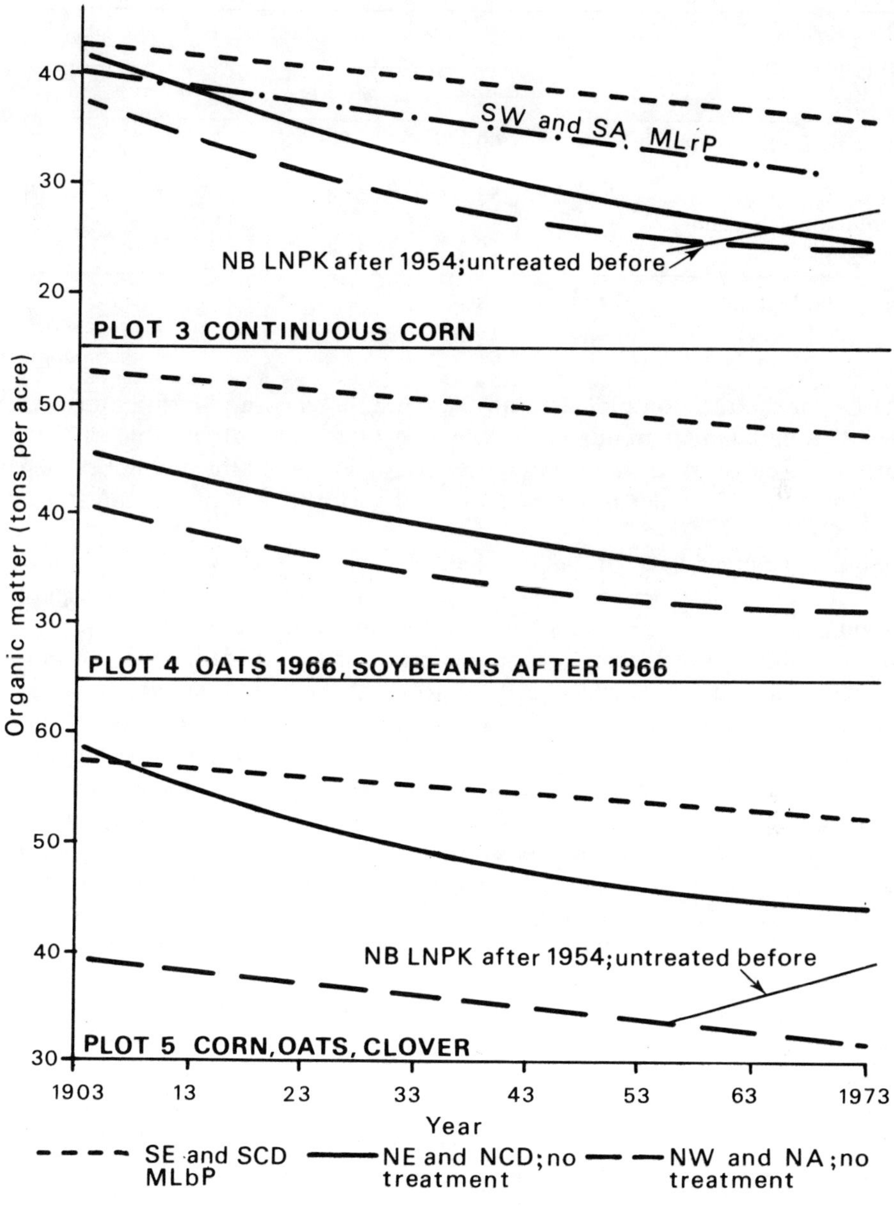

Figure 13.2 Trends in organic matter content of surface soils on the Morrow Plots, 1904–73 (Welch, 1981, unpublished data)

Table 13.7 Data from the Morrow Plots (see Figures 13.1, 13.2 and the text) (Welch, 1981, unpublished data)

Crop history and soil treatment	Organic matter (tons/acre)	Organic matter (%)	Nitrogen (lb/acre)	Nitrogen (%)
Sod border; no treatment	60.1	100	5237	100
Corn–oats–clover; MLP treatment	53.7	89	4550	87
Continuous corn; no treatment	28.7	48	2640	50

From L. F. Welch, 1981, Univ. Ill.—unpublished data.

porous on the rotation plot than on the continuous corn plot. Stauffer *et al.* (1940) found similar trends in the Morrow Plots in 1940 and that only the corn–oats–clover fertilized rotation materially increased the amount of large aggregates in the upper 6 inches compared with the continuous unfertilized corn plots.

Similar changes have probably taken place in other Mollisols. Very little data are available from specific plots on the changes in bulk density, aggregation, water-release curves and moisture retention produced by cultivating a virgin Mollisol. Most comparisons are made from adjacent fence rows or 'similar' soils. Such data will be difficult to obtain because there are few virgin sites left—probably not even enough to characterize the natural variability within the same soil. But comparing soil properties on virgin and cultivated soils is, in part, a negative academic exercise. It is nearly impossible using present technology not to 'degrade' or change some soil property. Maintaining a 'virgin' level of organic matter in Mollisols is not economically feasible or even desirable in most cultivated soils. The very act of cultivation to produce the desired seed-bed will areate the soil, oxidize more organic matter, and change the aggregation. Drainage, necessary to produce crops on wet soils, also increases the rate of organic matter oxidation. What is important is what has happened to the productivity of the soil using present technology—has it decreased, remained the same or increased? What can be done with what soil is remaining?

The best measure of productivity is yield, and in Iowa that means corn yields. Tables 13.8 and 13.9 are the average corn yields for counties within the Southern Iowa Conservancy district. Adair County is in the Shelby–Sharpsburg–Macksburg soil association area, Page County is about evenly split between the Marshall and Shelby–Sharpsburg–Macksburg soil association areas, and Wayne County is in the Adair–Grundy–Haig soil association (Fenton *et al.*, 1971). Estimated obtainable yields for the major

Table 13.8 Corn for grain: areas harvested and yield[a]

	Adair		Page		Wayne	
Year	(acres)	(bu/acre)	(acres)	(bu/acre)	(acres)	bu/acre)
1879	80 008	39.4	133 631	47.1	88 081	39.5
1890	91 581	40.9	112 591	47.8	67 560	40.8
1899	112 723	42.7	132 601	43.2	71 908	26.9
1909	98 366	33.3	109 725	38.5	76 911	25.3
1919	101 898	41.0	96 646	36.8	69 267	31.0
1924	90 838	25.6	108 180	31.5	67 186	24.8
1925	98 050	34.9	102 165	36.0	66 338	27.6
1934	5 616	5.3	14 805	7.3	6 794	4.5
1940	89 010	39.9	96 411	43.9	45 651	36.1
1945	92 941	37.4	115 039	46.4	77 836	24.7
1954	87 574	42.3	95 333	28.9	54 178	26.3
1959	94 783	62.6	105 404	56.8	57 239	47.8
1964	82 554	73.6	85 124	64.1	46 820	67.3
1969	75 405	87.9	86 155	90.7	45 504	75.6
1974	97 636	57.3	98 864	44.1	–	69.4
1978	102 997	103.7	103 380	97.2	57 227	100.9

[a]Source: US Census of Agriculture.
1 US bushel = 35.24 dm^3; 1 acre = 0.405 ha.

soils for each county are given in Table 13.10. The corn yields were relatively steady from about 1879 until 1900 when a slight decrease occurred from 1900 to 1929. The changes may not be statistically significant. The abrupt drop in areas harvested and yield in 1934 is the result of an extremely hot and dry summer. In most counties the area harvested for grain remained relatively high with the maximum area being in the late 1800s. Starting in the mid 1950s, the corn yields increased dramatically, and while fluctuating considerably have continued to increase until 1980. In 1979, the county average yields in all three counties were higher than the estimated attainable yield predicted by Fenton *et al*. in 1971. Page County, with an average yield of 133.4 bushels per acre in 1979, was about 20 bushels greater than the predicted attainable yield for the major soil (Table 13.10). Nineteen seventy-nine was an almost perfect year for corn production, but the average yields obtained in the above counties (Table 13.9) illustrates how rapidly corn yields have been increasing.

What has been responsible for the increase in corn yields? Thompson (1969) and Thompson and Troeh (1978) believe that the increases in corn yields have been due to improved technology (increased fertility level, better varieties, hybrid corn, higher plant populations, better weed control, and timely operations) and favourable weather. The cooler-than-normal weather and above-normal precipitation favoured high corn yields during the 1960s

Table 13.9 Corn yields (bushels/acre), Southern Iowa Conservancy District, 1942–80[a]

Year	County			Year	County		
	Adair	Page	Wayne		Adair	Page	Wayne
1942	51.4	47.6	48.1	1962	74.6	71.3	57.7
1943	50.6	52.3	35.4	1963	83.4	84.0	73.0
1944	43.0	49.4	33.9	1964	73.6	64.1	67.3
1945	37.4	46.4	24.7	1965	76.0	82.8	68.2
1946	56.8	51.9	43.3	1966	82.8	83.5	79.6
1947	20.8	31.9	15.7	1967	78.0	80.4	67.9
1948	59.3	54.3	46.1	1968	83.2	59.3	84.6
1949	45.0	45.3	44.0	1969	87.9	90.7	75.6
1950	49.4	55.0	42.0	1970	83.9	75.6	79.1
1951	36.4	34.1	29.1	1971	95.0	97.0	94.0
1952	58.2	56.7	46.8	1972	104.0	106.0	104.0
1953	46.1	50.0	37.1	1973	104.1	102.1	94.9
1954	42.3	28.9	26.3	1974	57.3	44.1	69.4
1955	40.8	32.8	36.5	1975	65.6	70.7	75.2
1956	50.4	42.2	46.3	1976	99.0	98.5	93.2
1957	52.2	50.8	48.0	1977	50.1	66.6	53.7
1958	64.6	61.6	54.3	1978	103.7	97.2	100.9
1959	62.6	56.8	47.8	1979	114.7	133.4	111.5
1960	62.0	63.3	49.6	1980	89.7	73.5	95.3
1961	69.1	64.7	57.3				

[a]Source: Iowa Statistical Reporting Service.
1 US bushel = 35.24 dm^3; 1 acre = 0.405 ha.

Table 13.10 Estimated obtainable yields (bushels/acre) for corn (Fenton *et al*., 1971)

Adair County		Page County		Wayne County	
Soil series	Corn yield	Soil series	Corn yield	Soil series	Corn yield
Shelby	81–90[a]	Marshal	99–107	Adair	65
Sharpsburg	113–105	Sharpsburg	113–105	Grundy	87–107
Macksburg	121	Shelby	81–90	Haig	105

[a]Range for slope and erosion phases.
1 US bushel = 35.24 dm^3; 1 acre = 0.405 ha.

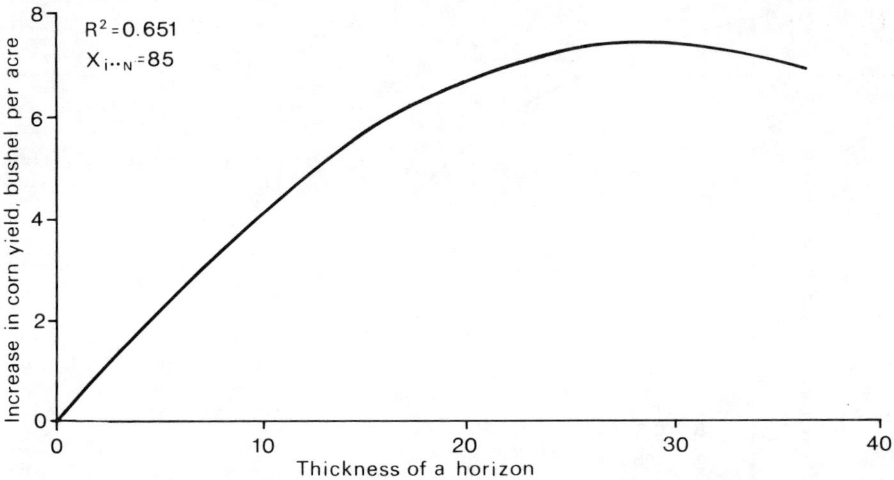

Figure 13.3 Correlation between the thickness of A horizon and increase in corn yield in Iowa (Dumenil, 1976, unpublished data)

(Thompson and Troeh, 1978; Shaw and Duorset, 1965; Hendricks and Scholl, 1943). Thompson (1973) also believes that heating and cooling cycles last about 10 years, i.e. cool summers in the 1940s and 60s, and warm summers in the 1930s, 50s and 70s relate to corn yields. Some of the yield variability, especially in the 1970s (Table 13.9), may be explained by temperature and moisture variations.

The high corn yields have been produced regardless of the degradation or changes in soils since they were first cultivated. Unpublished work by Dumenil at Iowa State University (Figures 13.3, 13.4, 13.5) show the correlation between thickness of A horizon, erosion class, the amount of organic carbon and corn yields in Iowa. The figures shown are summarized from several years' plot data taken in farmers' fields on some of the major soils in Iowa. Even under improved technology and rising yields, any decrease in organic carbon below about 4%, or decreased thickness of the A horizon below about 25 inches (63 cm), results in a loss of corn yield. The same applies when erosion class is changed to one of greater severity.

Dumenil's data require a cautious interpretation because there is not a simple relation between corn yield and A horizon thickness: several interactions need to be considered. Not all soils initially had the same thickness and organic carbon content of the A1 horizon and the same productivity capacity. The micro-environment such as drainage, runon or runoff, aspect, slope gradient, etc., can have a considerable effect on yields in any year. The temperature regimes and moisture-supplying capacity of the soil during the

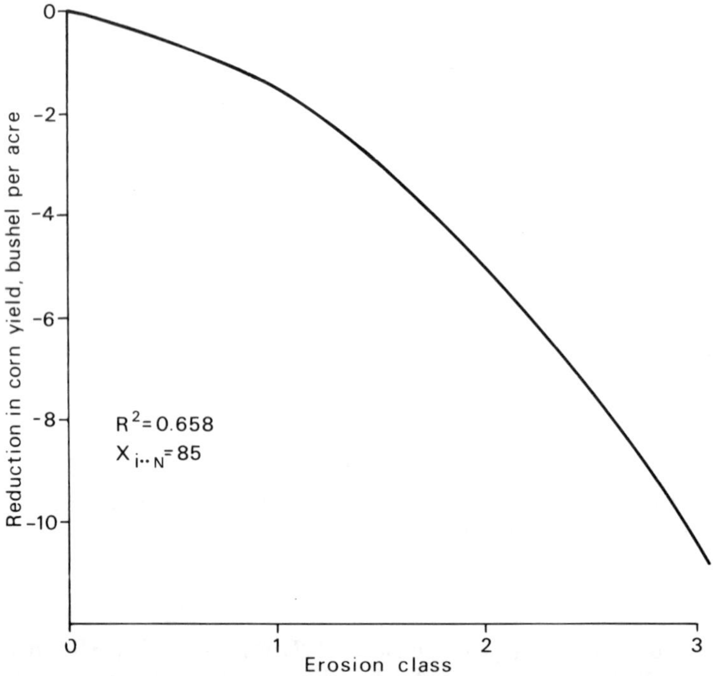

Figure 13.4 Correlation between erosion class and reduction in corn yield in Iowa (Dumenil, 1976, unpublished data)

critical periods of crop growth are very important, especially in high-management rain-fed agriculture. A soil on a level surface or a slight depressional area with 25 inches of A1 horizon has a much different environment from a soil in similar materials on a 10% convex nose slope with only 5 or 6 inches of A1 horizons. The productivity capacity also has to be different, especially when moisture and temperature regimes are different. To illustrate, Sopher *et al.*'s (1973) data from North Carolina indicate that corn yields on wet soils are improved by spring drought but decreased on well-drained soils. Cold, wet springs favoured yields on the well-drained soils compared with their wetter counterparts. Corn yields on wet soils commonly are reduced under conservation or no-till compared to ploughed soil, apparently because lower temperatures and damp conditions lead to a wide variety of processes unfavourable to maximum yields (Van Dorin *et al.*, 1976; Griffith *et al.*, 1973; Unger and McCalls, 1980). There is a lot more to high yields than organic matter or A1 horizon thickness, although this is part of it. Simple correlations of yields with organic matter combine several micro-environmental features that are equally as important as the organic matter.

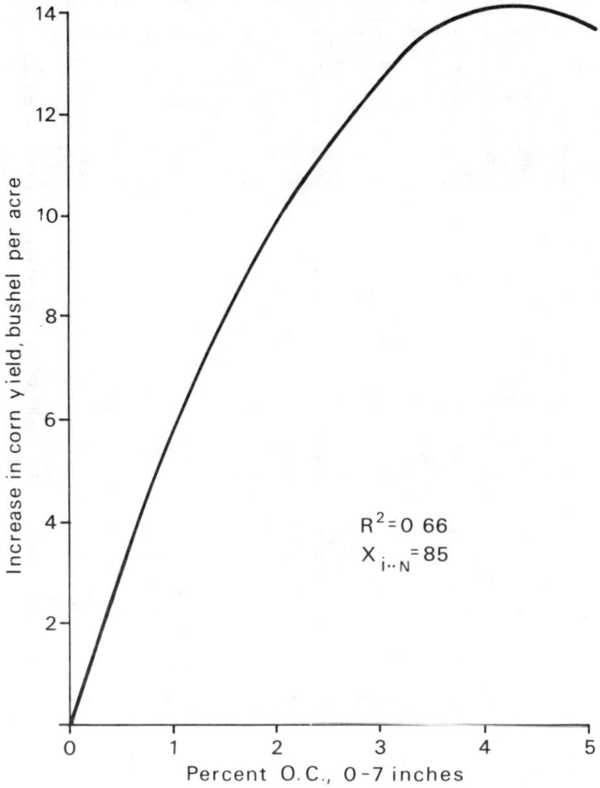

Figure 13.5 Correlation between amount of organic carbon and increase in corn yield in Iowa (Dumenil, 1976, unpublished data)

Every manipulation of the soil changes some soil property, but not always for the worse. The carbon content of a soil in the Morrow Plots that had been under continuous corn for nearly 100 years increased considerably when lime, nitrogen, phosphorus and potash were added (Moldenhauer, *et al*., 1967; Barnhart, *et al*., 1978; Figure 13.2). The increase was about five tons of organic matter in about 20 years. Tables 13.11 and 13.12 are from the Morrow Plots in Illinois and show the changes in treated and untreated plots across three rotations. The base data are not available for the condition of the plots in 1876, but Tables 13.11 and 13.12 do indicate that management can easily alter the nutrient status and pH of the soil surface. There is also some unpublished data indicating that the nutrient status of the B horizons of many soils are increasing with continued cultivation and addition of amendments.

Although certain changes have taken place in the soils of the Southern

Table 13.11 Summary of surface soil tests for the Morrow Plots during selected years, 1955–77[a] (Welch, 1981, unpublished data)

Plot	Cropping system	Soil treatment	pH: No. of obs.	pH: Mean (lb/acre)	pH: R^2	pH: Coef. of var. (%)	pH: Intercept a	pH: Reg. coef. b	P_1: No. of obs.	P_1: Mean (lb/acre)	P_1: R^2	P_1: Coef. of var. (%)	P_1: Intercept a	P_1: Reg. coef. b	K: No. of obs.	K: Mean (lb/acre)
3NACD	Continuous corn	Not treated	7	5.2	0.947	1.3	4.8	0.030[e]	7	15	0.785	12.0	19.1	−0.373[e]	7	206
3NB[b]	Continuous corn	LNPK after 1954; untreated before	6	6.0					6	40					6	246
3SCD	Continuous corn	MLbP	7	6.5					7	79					7	273
3SB[b]	Continuous corn	MLrP 1904–54 + LNPK after 1954	7	6.3					6	55					6	271
3SA[c]	Continuous corn	MLrP 1904–66	3	6.6					3	73					3	277
3SA[d]	Continuous corn	MLrP 1904–66 + Hi LNPK after 1966	4	6.2					4	108					4	346
4NACD	Corn Oats through 1966 S.B. after 1966	Not treated	7	5.3	0.788	2.7	5.0	0.030[e]	7	13	0.910	9.0	17.9	−0.416[e]	7	206
4NB[b]	Corn ” ” ”	LNPK after 1954; untreated before	6	6.1					6	50					6	251
4SCD	Corn ” ” ”	MLbP	7	6.4					7	47					7	274
4SB[b]	Corn ” ” ”	MLrP 1904–54 + LNPK after 1954	7	6.4					6	51					6	297
4SA[c]	Corn–oats	MLrP 1904–66	3	6.6					3	41					3	252
4SA[d]	Corn-soybeans	MLrP 1904–66 + Hi LNPK after 1966	4	6.4					4	112					4	362
5NACD	Corn–oats–clover	Not treated	7	5.2	0.784	2.5	4.9	0.026[e]	7	10	0.913	15.0	16.4	−0.549[e]	7	214
5NB[b]	Corn–oats–clover	LNPK after 1954; untreated before	6	6.0					6	39					6	246
5SCD	Corn–oats–clover	MLbP	7	6.5					7	45					7	234
5SB[b]	Corn–oats–clover	MLrP 1904–54 + LNPK after 1954	7	6.6					6	90					6	280
5SA[c]	Corn–oats–clover	MLrP 1904–66	3	6.8					3	35					3	236
5SA[a]	Corn–oats–clover	MLrP 1904–66 + LNPK after 1966	4	6.7					4	96					4	321

[a]Soil tests were studied for the years 1955, 1957, 1961, 1967, 1970, 1974 and 1977.
[b]Soil samples were collected in April 1955 before new treatments were applied in May 1955 on sub-plots 3, 4 and 5NB and 3, 4 and 5SB. Therefore, the results of 1955 soil tests for plots 3, 4 and 5NB and the P_1 and K tests in 1955 for sub-plots 3, 4 and 5SB were not included in the analysis.
[c]1966 and before.
[d]After 1966.
[e]Highly significant.

Table 13.12 Interpretation of soil tests for the Morrow Plots (Welch, 1981, unpublished data)

		Available nutrients (lb/acre)	
pH	Soil reaction	P_1	K
5.1–5.5	Strongly acid	0–12 (very low)	0–90
5.6–6.0	Medium acid	12–20 (low)	90–135
		20–30 (slight)	135–180
6.1–6.5	Slightly acid		
6.6–7.3	Neutral	30–45 (medium)	180–300
		45–65 (high)	300–400
7.4–7.8	Mildly alkaline	>65 (very high)	>400

Conservancy District that have lowered corn yields under specific levels of management and weather conditions, improved technology has increased the total yield per acre considerably. There are many unanswered questions; but Figures 13.3 and 13.4 indicate that, even under the most severe conditions of removal or loss of the A horizon, the decrease in production will be about 10 bushels per acre, or about 10–12% of the maximum 'attainable' yield for the soils tested. Ninety plus bushels of corn per acre is not as good as 100 plus, but it is still much better than the 40 bushels or less grown during the last part of the 1800s when the soils were first broken from the prairie sod (Table 13.9). It is difficult to argue with success when the apparent productivity is still rising. However, 30 years is a short time and it has not been proven that technology can economically increase or even hold the current level of yields. Fifty years from now we may have some answers.

If the changes in soil properties and the decrease in productivity potential produced by cultivation and erosion are thought to indicate a steadily worsening condition, the reader should look to the geological past for an indication of the recuperative powers of the soil system. Data from western and south-western Iowa show that, within the last 1500 to 500 years, so-called 'geological' erosion has removed, at a minimum, three feet (>90 cm—the entire soil profile) from the surrounding landscape (Ruhe and Daniels, 1963). Yet these geologically eroded areas recuperated rapidly enough, so they became the 'rich' virgin soils that were first cultivated about 130 years ago. There will always be a soil, but within the next few hundred years what kind of soil exists will be strongly influenced by the manipulations of man. The Southern Iowa Conservancy study and the above critique show that erosion does not have the same effect under all conditions. The deep, favourable soil material in much of the conservancy district lowers the impact of erosion on the soils of the area, but care must be used when extrapolating to other areas.

13.7 REFERENCES

Amos, D. F., and Whiteside, E. P. (1975) Mapping accuracy of a contemporary soil survey in an urbanizing area. *SSSA Proc.*, **39**, 937–942.

Barnhart, S. L., Shrader, W. D., and Webb, J. R. (1978) Comparison of soil properties under continuous corn grain and silage cropping systems. *Agron. J.*, **70**, 835–837.

Bennett, H. H. (1939) *Soil Conservation*. McGraw-Hill, New York.

Browning, G. M., Parish, C. L., and Gloss, J. (1949) A method for determining the use and limitations of rotations and conservation practices in the control of soil erosion in Iowa. *Amer. Soc. Agron. J.*, **39**, 65–73.

Browning, G. M., Horton, R. A., McCall, A. G., and Bell, F. G. (1948) *Investigation in Erosion Control and the Reclamation of Eroded Land*. US Dept. Agr. Tech. Bul. 959.

Hudson, N. (1971) *Soil Conservation*. Cornell University Press.

Chamberlin, T. C. (1908) Soil wastage. In *Conference of the Governors of the United States*. White House, Washington, DC, Govt. Printing Office, 75–83.

Dideriksen, R. I. (1966) *An Evaluation of Soil Mapping Units by Erosion Groups in Iowa*. Unpublished Soil Conservation Service Report.

Fenton, T. E., Duncan, E. R., Shrader, W. D., and Dumenil, L. C. (1971) *Productivity Levels of Some Iowa Soils*. Ag. and Home Econ. Exp. Sta. and Crop Ext. Service., Iowa State University, Spec. Rept. #66.

Griffith, D. R., Mannering, J. V., Galloway, H. M., Parsons, S. D., and Riley, C. B. (1973) Effect of eight tillage-planting systems on soil temperature, percent stand, plant growth, and yield of corn on five Indiana soils. *Agron. J.*, **65**, 321–326.

Guernsey, J. B., Fehrenbacher, X. X., Wray, B. W., and Miller, L. B. (1969) Corn yields, root volumes, and soil changes on the Morrow Plots. *J. Soil and Water Conservation*, **24**, 101–104.

Guide for Developing the Soil Loss Tolerance Factor (T) in the Universal Soil Loss Equation. US Dept. Agr. Soil Conservation Service. Attachment to *Advisory Soils 6*, 6 Feb. 1973.

Hendricks, W. A., and Scholl, J. C. (1943) *Techniques in Measuring Joint Relationships: The Joint Effects of Temperature and Precipitation on Corn Yields*. NC Ag. Exp. Sta., Tech. Bul. 74.

Jenny, H. (1941) *Factors of Soil Formation*. McGraw-Hill, New York.

McCormack, D. E., and Wilding, L. P. (1969) Variation of soil properties with map units of soils with contrasting substrata in N.E. Ohio. *SSSA Proc.*, **33**, 587–593.

McCormack, D. E., Young, K. K., and Kimberlin, L. W. (1982) Current criteria for determining soil loss tolerance. In Press. Soil Science Soc. Amer. Spec. Pub. 45:95–111.

Moldenhauer, W. C., Wischneier, W. H., and Parker, O. T. (1967) The influence of crop management on runoff, erosion, and soil properties of a Marshall silty clay loam. *Soil Sci. Soc. Amer. Proc.*, **31**, 541–546.

Rosenberry *et al.* (1980)—listed the major soils in the Southern Iowa River Basin (Table 1, p. 5). Textural characteristics of the soils listed were from classification of soil series of the United States, Puerto Rico and the Virgin Islands, October 1980. US Dept. Agric. Soil Conservation Service, US Govt. Printing Office. (Two parts: part 1, 182pp; part 2, 149pp).

Ruhe, R. V., and Daniels, R. B. (1963) Landscape erosion, geologic and historic. *J. Soil and Water Conser.*, **20**, 52–57.

Shaw, L. H., and Duorset, D. D. (1965) *The Effects of Weather and Technology on Corn Yields in the Corn Belt, 1929–1962*. US Dept. Agr. Econ. Report no. 80.

Smith, D. D., and Wischmeier, W. H. (1962) Rainfall erosion. *Adv. in Agron*. **14**, 109–148.

Smith, R. M., and Stamey, W. L. (1965) Determining the range of tolerable erosion. *Soil Sci*., **100**, 414–424.

Sopher, C. D., McCracken, R. J., and Mason, D. D. (1973) Relationships between drought and corn yields on selected South Atlantic Coastal Plain Soils. *Agron. J*., **65**, 351–354.

Stauffer, R. S., Muckenhiern, R. J., and O'Dell, R. T. (1940) Organic carbon, pH, and aggregation of the soil of the Morrow Plots as affected by type of cropping and manual addition. *J. Am. Soc. Agron*., **32**, 819–832.

Thompson, L. M. (1969) Weather and technology in the production of corn in the US corn belt. *J. Am. Soc. Agron*., **61**, 453–456.

Thompson, L. M. (1973) Cyclical weather patterns in the middle latitudes. *J. Soil and Water Conserv*., **28**, 87–89.

Thompson, L. M., and Troeh, F. R. (1978) *Soils and Soil Fertility* (4th edn). McGraw-Hill.

Unger, P. W., and McCalls, T. M. (1980) Conservation tillage systems. *Adv. in Agron*., **33**, 1–58.

Van Doren, C. A., and Bartelli, L. J. (1956) A method of forecasting soil loss. *Ag. Engineering*, **37**, 335–341.

Van Dorin, D. M., Triplett, G. B., and Henry, J. E. (1976) Influence of long term tillage, crop rotation and soil type combination on corn yields. *Soil Sci. Soc. Amer. J*., **40**, 100–105.

Williams, J. R., and Berndt, H. D. (1977) Determining the Universal Soil Loss Equation's length-slope factor for watersheds. In *Soil Erosion: Prediction and Control*. Soil Conservation Society of America Spec. Pub. no. 21, 217–225.

Wischmeier, W. H. (1977) Use and misuse of the Universal Soil Loss Equation. In: *Soil Erosion: Prediction and Control*. Soil Conservation Society of America Spec. Pub. no. 21, 371–378.

Wischmeier, W. H., and Smith, D. D. (1965) *Predicting Rainfall Erosion Losses from Cropland East of the Rocky Mountains*. Agr. Handbook no. 282. US Dept. Agr., Washington, DC.

Wischmeier, W. H., and Smith, D. D. (1978) Predicting rainfall-erosion losses—a guide to conservation planning. Agricultural Research Service, US Dept. of Agric. Ag. Handbook no. 537.

Land Transformation in Agriculture
Edited by M. G. Wolman and F. G. A. Fournier

CHAPTER 14
Reclamation of Areas Affected by Open-cast Mining in the North Bohemian Brown Coal Basin, Czechoslovakia

S. Štýs

14.1 CHARACTERISTICS OF THE AREA

The North Bohemian basin is a significant fault enclosed to the north by the southern slopes of the mountains, to the south-east by the Czech Central Mountain Range, to the east by the river Elbe, and to the west by the strato-volcanic Doupovské hills. It is only open in the south-western direction, into the Žatec plateau. The valley character is given mainly by the significant elevation between the valley basin and the mountainous part of the Krušné Hory massif (250–900 m above sea level), whose northern slopes have a mild gradient while the eastern slopes which run towards the basin are very steep.

The dynamics of the basin relief are mild and correspond to its sedimentary origins in the Tertiary period, during which the major part of the basin was formed. The Quarternary formation of the terrain is only manifest at the foothills of the Krušné Hory Mountains (by sediments of dejection cones and slope debris) and in the south-western part of the basin where powerful layers of eolic sediments have been formed. From the geomorphological point of view, the destructive activity of the hydrographic system has only been manifest very mildly.

The area of outcrops of the coal seams is about 680 km^2, the orographic area covers about 1000 km^2, and the productive area of the basin covers 1000–1200 km^2.

14.1.1 Geological Characteristics

The geological development and formation of the basin started in the Lower Oligocene by subsidence of Saxonian tectonics on to the Neocene platform of the Czech Massif. The voluminous sedimentary area of the basin was formed especially in the Miocene by bulging out of the Krušné Hory massif, the formation of the Czech Central Mountains and the Coupovské Mountains, with the concurrent subsidence of the bottom of the basin which began filling with sediments. Following the period of sedimentation during the Miocene, a tectonic disintegration took place in the whole area, which was in turn followed by denudation.

As for the stratigraphy of rocks in the basin, there are crystalline rocks, calcareous, Tertiary rocks, i.e. of the basal series, the detrital vulcanite series, the sub-grade series of strata of sands and loam, the series of strata of brown coal seams, the series of strata of overburden rocks, with a prevalence of Miocene loams and Quarternary rocks.

Crystalline rocks in the basin mainly comprise gneisses which, in contact with the sediments of the basin, are strongly kaolinitic. The oldest developmental stage of the sedimentation is the chalk stage of the Mesozoic era, represented by sandstone, conglomerates, quartzites, argilites and marls.

The basal series of Tertiary sediments was formed in the Upper Eocene and Lower Oligocene. It is mostly formed of conglomerates, sandstones and argilites. The vulcanic detrital series was mostly formed in the upper Oligocene during increased volcanic and denudation activity. This was followed by the sedimentation of sandstone and argilite sediments, forming the direct sub-grade of the coal seam.

The most favourable conditions for the growth of vegetation existed in the period of the Upper Burdigal, Helvetium and Lower Torton eras. This was the period of formation of the series of strata of brown coal seams. The cleavage of the seams, especially in the western part of the basin, is the result of delta-shaped flows into the basin and denudation in the area of the Krušné Hory Mountain massif during the discontinuation of sedimentation of the organic substrate. The intermediate strata of coal seams are formed by clay, sandy clay and sand.

The Miocene sedimentation formed a powerful overburden in which clays and argilites are prevalent. Locally the series of strata which forms the overburden is interrupted by semi-siderite parts in the direction to the Žatec face and in the boundary areas, namely in the Duchcov–Bílina part of the basin where there are thick layers of sand and sandy clays.

Quarternary rocks cover almost the entire Tertiary fill of the basin. They are formed by substrates which differ widely in structure: ranging from clayey weathered tuffs to weathered eluvial rocks to boulders deposited at the Krušné Hory foothills, to stone seas and sandy gravels formed mainly by muscovite biotic orthogneiss.

14.1.2 Climatic Conditions

The climatic conditions of the area are characterized by low precipitation, and a relatively high atmospheric temperature.

The highest average total annual precipitation is reached in the northern part of the basin (670 mm); in the central part of the basin the average drops to 517 mm and in the south-western part to 470 mm. The average annual temperatures are extremely high in the last-mentioned area and reach an average of 8.4–8.8°C (Kadaň 8.09°C, Most 9.0°C, Ústí nad Labem 0.0°C). The average temperatures during the vegetation period reach 13–14°C. Atmospheric moisture is 72% (long-term average). The direction of the wind conforms to the relief, and therefore the wind mostly blows along the southern slopes of the Krušné Hory Mountains. In the whole basin the prevalent winds are those of the western quadrant, i.e. in Chomutov 305°, Litvínov 288°, Bílina 260°, Teplice 268°, in Ústí nad Labem 255°.

The climate of the basin is significantly influenced by the orographic situation in relation to the neighbouring orographic units, mainly the vertical topography of the terrain. With this is connected the considerable attenuation of flow, the frequent formation of a very stable boundary layer with poor conditions for dispersion, and the frequent occurrence of solar radiation. The annual number of foggy days is extremely high: in Chomutov 70, Teplice 87, Bílina 108, Litvínov 126, Ústí nad Labem 141.

The frequent occurrence of inversion prevents the dispersion of emissions from the industrial area. The relatively stable layer of cool air in this situation preserves the major part of industrial aerosols in the basin area.

14.1.3 Hydrological Conditions and Water Management

The hydrographic system of the basin is influenced by the fact that water from the whole area is diverted by the river Bílina, and partly by the river Ohře. Streams flow from the Krušné Hory Mountains and their upstream beds are sharply cut into the terrain. Water reservoirs mostly have the character and function of water-retention basins (accumulating industrial waters); pond systems are practically non-existent.

In the tenth century the Komořany Lake existed here and was the biggest basin lake in the country. It was described by the Arabian merchant Ibrahim Ibn Jakub. The Komořany lake covered an area of 60–70 km^2 and reached a depth of 40 m. It was fed by the waters of the Bílina river and several smaller tributaries and mountain streams. Archaeological surveys have shown that human settlements existed on the banks of this lake 6000 years ago and continued to exist throughout the Slavonic period. The Komořany lake was continually silted and its area and depth decreased. The last remnant of the lake, having an area of 165 ha, was drained and dried by the owners of the estate of Jezeří in 1835.

The hydrological conditions are characterized by the low efficiency of crystalline slates, permo-carboniferous localities, with the exception of the Teplice quartz porphyrite which is the bearer of the thermal and mineral waters in the Teplice area, of the Tertiary sub-grade and overburden clays and argilites, and of the Quarternary physically heavy loess and slope clays. Hydrogeologically effective rocks include calcareous sandstone: of the Tertiary, Oligocene sandstones, overburden sandstones, porcelainites and, mainly, the brown coal seam; of the Quarternary, the powerful strata of slope debris and dejection cones at the foothills of the Krušné Hory Mountains.

The groundwaters in the basin are considerably mineralized. The waters of the sub-grade strata series are mostly of the MgNa–HCO_3 or Na–HCO_3 types, their mineralization in the range 300–1400 mg/l. The waters of the coal seam strata are mainly of the CaMg–SO_4HCO_3 type, their mineralization in the range 900–1400 mg/l. In the Quarternary of the overburden the water is mainly of the CaMg–SO_4HCO_3 type, with a mineralization of approximately 1500 mg/l. The streams in the mountainous parts of the Krušné Hory have an average mineralization of 95–140 mg/l.

A very serious hydrological problem of the area is to secure the operation of the fourteen curative springs of the Teplice and Šanov line. With the development of mining their abundance has declined and seven springs have ceased to exist. In 1818 the Šanov springs had an abundance of 1262 m^3. The current unfavourable situation is being successfully resolved by the deep withdrawal of Tertiary waters, which secures not only a greater abundance and quality of the thermal springs but also the exploitation of coal in the vicinity of Teplice.

Water management in the basin is ruled by the necessity to economize the use of water for ecological, health, economic, cultural and aesthetic reasons. For water to fulfil all the required functions, it has to be available in adequate quantities and of appropriate quality. The need for potable and utility water in the basin is increasing significantly. The natural resources are very limited and there is an increasing trend to use surface waters instead of groundwaters. The surface waters are, however, highly polluted: in 1975 the river Bílina, for example, was polluted along its entire length by ammonia, heavy loadings of BOD, COD phenols and other toxic substances. The biggest polluters are the ČSSP Chemical Works in Litvínov, the towns of Litvínov, Most, Duchcov, Bílina, Teplice and Ústí nad Labem, and the Spolek works in Ústí nad Labem.

A supply of drinking water is increasingly being secured by the building of dams in the Krušné Hory Mountains, and utility water by the diversion of the Ohře and Elbe rivers to the Bílina catchment. However, these supplies of water are negatively affected by the considerable pollution of flows owing to the increasing level of emissions. The water is depleted by the occurrence of substances which in abnormal concentrations have unpleasant, harmful and toxic effects. In that part of the Krušné Hory Mountains where forests have been damaged by emissions, deforestation has altered the water management function of the forests; this is manifest in waterlogging, the increase of surface runoff and the reduction of precipitation infiltrated into the soil surface, the intensification of erosion processes and erosion phenomena, and an increase in the fluctuation of flow rates in water-courses.

The natural conditions in the North Bohemian brown coal basin are significantly affected by mining activities. In the preparation of the area for mining the routes of water flows are diverted and the surfaces exposed for mining are drained by wells. Annually, 30–40 million cubic metres of water are drained in the basin, and this significantly influences the groundwater regime.

14.1.4 Soil Conditions

The soil conditions in the basin are influenced not only by soil formation factors but also by anthropogenic factors: i.e. negatively by mining activity and positively by reclamation.

The texture of the soil is characterized by almost equal quantities of heavy and medium soils, with some light soils (ratio: 40:40:20). Prevalent are deep soils (75%), with 20% of soils of medium depth and 5% of shallow soils.

The dynamic development of soils in the basin is shown by the prevalence of climatogenic soils, mainly brown soils, chernozem, and rendzina. Meadow soils, allenral soils, mildly podzolized, riparian, meadow gleyic, woodland meadows, illimerized, pseudo-gleyic soils, gleisoil and peat soils are complementary.

In the area, reclamation is gradually taking place of all soils devastated by mining. This should return the soil to agriculture and forestry. Forest reclamation is carried out almost entirely by direct planting into dump substrates, which are thus influenced by soil formation factors and by reclamation culture. In this study we are almost entirely concerned with undeveloped waste-dump forest soils.

In agricultural reclamation there are two alternatives: with made-up ground (top humus profiles) or without made-up ground. In the former case the quality of the newly made-up soil is related to the quality and thickness of the humus, the character of the surface earths of the made-up ground, and the intensity of a biotechnical reclamation. The made-up ground usually has a thickness of 0.5 m, which after compaction forms a 0.4 m humus horizon. If the sub-topsoil layers of the made-up layer are excessively permeable, there results an excessive seepage of fine-grained parts, nutrients and water. In the North Bohemian brown coal basin, however, there is a prevalence on spoil banks and waste dumps of overburden loams, with a very low hydraulic conductivity; the situation therefore more often arises where the made-up layer is oversaturated with water. The pedogenetic development of these waste dump soils in such cases results in gleisation.

When a waste dump is reclaimed without the use of humus cover, by using land improvement seeding processes, the pedogenetic development of the new agricultural land is mainly affected by the nature of the waste dump earths, climatic factors and the nature of the reclamation. This type of reclamation mainly concerns clayey substrates, especially loess, loess clay and slope debris. In cases where there is an adequate calcium content conditions are good for the formation on these waste dumps of soils similar to soils formed in this region on soil-forming substrates, i.e. brown soils, brown earths and chernozem.

14.1.5. Phytogeographical Conditions and Pollution

The phytogeographical conditions of the basin are also mainly of a climatogenic nature. Reconstructed natural vegetation expresses the basic ecological potential of the landscape. The major part of the basin is thus covered by sub-xerophilous oak-hornbeam forests. On the more moist and heavy soils of the basin, formed on Tertiary loams, conglomerates and heavy slope clays, there are the hydric associations of the beech and alder woods, and on the wet alluvial stands there are the willow and poplar woods. At the Krušné Hory foothills there are acid oak woods, which in the northern direction are replaced by beech groves.

Man's activity has, however, changed the whole area: first, agricultural colonization formed a cultural steppe, and with the development of mining man turned the region into a restricted floristic area characterized by reduced forms of vegetation.

Currently the whole area is characterized by a low rate of afforestation (only 4% of the area is covered with forests); but this proportion is being increased by reclamation.

Prevalent is the agricultural use of the soil. Sugar-beet farms cover approximately 70% of the total agricultural area. The most frequent sub-types are beet–wheat farms on heavy soils; in the Duchcov area the beet–barley farms; and in small localities in the western part of the basin the beet–rye farms. The rest of the basin is mainly given over to potato or potato and rye farms (restricted to a narrow strip at the foot of the Krušné Hory mountains) and the potato and wheat farms which are located at the foot of the Czech Central Mountain range and the Doupovské Mountains. Pastures are only found in small localities on the boundary of the basin.

As a result of the development of industry, mainly in connection with the significant increase in power production by thermal power plants which burn considerable amounts of coal with a high sulphur and ash content, emission levels have considerably increased throughout the North Bohemain brown coal basin, including the adjacent landscape areas. The harmful effects are felt by all the physical and organic components of the landscape.

From the ecological point of view there is not only the significant *direct* impact of harmful emissions on plants, but also their *indirect* impact through the soil. The emissions enter the soil by direct fallout, by precipitation, by direct sorption through the soil surface, and by way of wastes of intoxicated organs, mainly of leaves of affected plants. The direct effect of separation by vegetation, which depends on the intensity of sorption, the accumulation capacity of plants and on the amount of produced organic mass, is small to counter the impact of the said emissions. As a result the concentration of emissions throughout the basin areas is increasing in the whole soil profile. This mainly concerns sulphur compounds which appear in the form of free sulphuric acid or sulphurous acid. In some parts of the area the pH of precipitation drops to 3.5 (in some cases even less). Free acid causes the washing out of a number of substances from the soil profile, including the plant nutrients calcium, magnesium, potassium and iron. Without human interference this is an irreversible long-term process: the changes have an cumulative character and act for a long time after the emissions have ceased. There is a negative impact on agricultural cultures whose vital functions are disturbed; the intensity of phytosynthetic assimilation is reduced and yields decline.

The regional agricultural station in Ůstí nad Labem reported in 1975 that agricultural yields had significantly decreased as a result of the increased emission levels: wheat by 9–11%, oats by 12–14%, barley by 14–18%, silage maize by 25–38%, sugar-beet by 10–15%, perennial fodder by 17–32%, annual fodder by 12–14%.

14.2 BROWN COAL MINING IN THE BASIN

In 1860, 500 000 t of coal was mined in the basin, and by the time of the outbreak of the First World War production reached 18.6 million tonnes. This level was not reached again in the period between the two world wars, but the German war-needs stimulated development, which in 1943 reached 20.2 million tonnes. An upsurge of production was, however, only recorded in the post-war years, as shown in Table 14.1. Current exploitation shows a continuously rising trend, accompanied by an increasing volume of stripping of overburden rocks which has been significant especially in the second half of this century. The ratio of stripping to coal production is expected to reach 4.5 by the year 2000, owing to the increasing depths of coal seams.

The brown coal basin was originally divided into a large number of mining units by past entrepreneurial interests. There were 1639 deep mines and 186 open-cast mines. The basic aim of the private entrepreneurs was to minimize investments and to maximize profits and yields, and this was reflected in

Table 14.1 Production and stripping data

Year	Production ($\times 10^6$ t)	Stripping of overburden ($\times 10^6$ m^3)	Ratio of stripping to production (m^3/t)
1945	11.1	10.3	0.93
1950	19.8	27.4	1.38
1960	39.6	97.2	2.45
1970	55.0	106.0	1.93
1975	60.7	159.5	2.63
1980	66.7	152.2	2.28
⋮			
2000			4.50 (projected)

uncoordinated mining which disregarded the principles of good mining practice.

The current development of the basin is characterized by the intensive development of quarry exploitation which is concentrated in a small number of mines with a productive and efficient technology. Mining is advancing to greater depths in an area with increasing stripping ratios. This results not only in taking up plots, but also in intensified devastation of the natural and cultural components of the landscape, with significant growth of the volume of overburden stripping and the resulting elevated sandfills and waste dumps. It is significant for the concept of reclamation that spoil banks are established in a more or less exploited area; for complete exploitation of the area they will later be quarried, and so the reclamation of the area is only temporary. The intensive development of quarrying is made possible mainly by technology orientated to more efficient practices.

14.3 IMPACTS OF OPEN-CAST MINING

In 1980, 92% of the coal mining in the North Bohemia brown coal mining concern was of the open-cast type. This is very evident in the area (Figure 14.1). We shall therefore devote our attention solely to the impacts of such mining on the natural components of the area.

Open-cast mining technologies influence mainly the powerful profile of the *lithosphere* by transformation of the rock medium and the change in relief. They also have an impact on and contaminate the micro- and meso-climatic factors of the *atmosphere*, deform the character and regime of the *hydrosphere*, devastate and destroy the whole *pedosphere*, contaminate or liquidate the phytocenoses, zoocenoses and microcenoses, thereby significantly

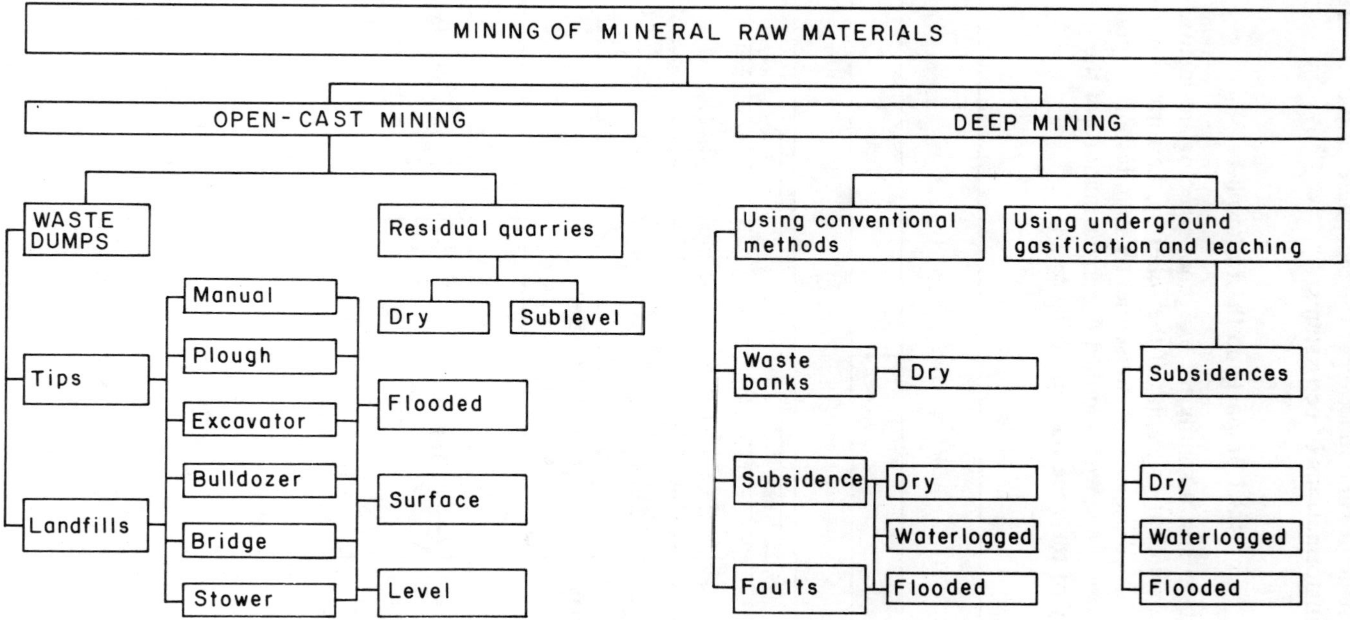

Figure 14.1 Basic forms of impacts of mining on the landscape

influencing the structure and function of the biosphere, and thus participate in the extreme transformations of ecosystems.

14.3.1 Impacts on the Lithosphere

Montane activity is still a significant exogenous geological factor; it usually creates a more explicit topography of the relief by the formation of quarries and elevated waste dumps. It is subsequently manifested by genetical and morphological processes, with significant transformations of the petrography and stratigraphy of the rock medium (Figure 14.2).

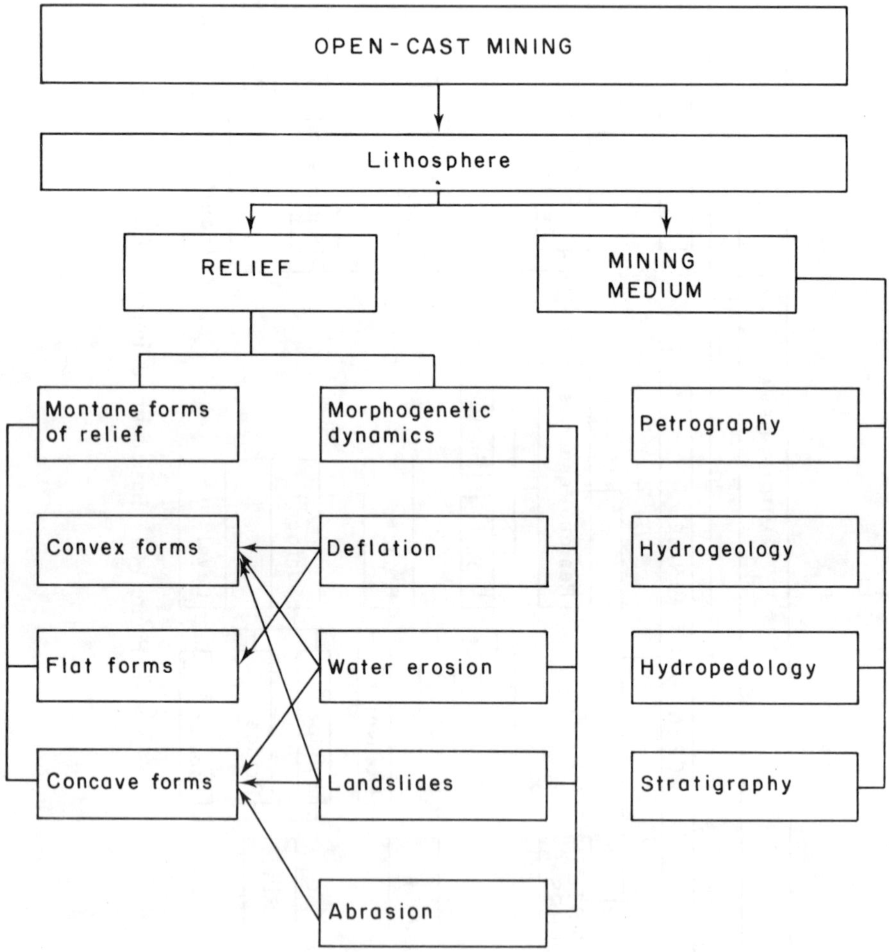

Figure 14.2 Structure of impacts of open-cast mining on the lithosphere

The convex forms of the relief are mainly due to spoil banks and waste dumps. Flat forms are due to landfills (filled or sluiced) and residual quarries. A concave relief is formed by underground landfills.

Morphogenetic processes result in the convex forms being modelled by deflation, water erosion and landslides. Flat reliefs are modelled mainly by deflation, and concave forms are transformed by landslides, mainly the solifluxion of the boundary slopes of residual quarries and by the abrasion of the banks of quarry lakes.

Open-cast mining models the relief in height, exposure and incline. Waste dumps and tips have the form of terraces, cones, plateaux, ridges and hybrid forms. In the process of overburden stripping, transport and stowing, a significant destruction occurs of the initial rock medium, as do significant petrographic, stratigraphic hydrogeological and hydropedological transformations. During the overburden stripping the devastated area is affected by mechanical, chemical, physico-chemical and mineralogical changes in the structure or rocks, which gives the area an atypical and highly specific petrographic character. The stratigraphy of the waste dump area also has a significantly anthropogenic character and does not correspond to any type of geological development. As a result of these processes the area assumes a specific *hydrogeological and hydropedological character*, with specific hydrogeological and hydropedological conditions.

The relief and the rock medium are significant ecological components of the natural environment and as such have great significance in the optimization of the reclamation of the landscape. They must therefore be considered *during* exploitation, and during the period of reclamation.

14.3.2 Impacts on the Atmosphere

Open-cast mining causes significant changes in classical climatic factors and has an impact on the quality of the air (Figure 14.3). Micro-climatic and meso-climatic changes are mainly caused by transformations of the relief, altitude, topography, its exposure and incline, colour, moisture conditions and vegetation cover. In some localities this is accompanied by thermal emissions from exothermal reactions arising during mine fires and gob fires.

The most significant changes take place in all basic micro-climatic factors in the area of the mines and quarries and in the area of barren waste dumps with no vegetation cover. The surface of these areas is exposed to more intensive solar radiation. Ecologically significant is the considerable overheating of the ground layers of the atmosphere on slopes exposed to the south, especially in areas with dark substrates and excessive albedo values; here there is increased evaporation and reduced moisture content of the ground layers of the atmosphere.

The concave parts of the relief are susceptible to inversion situations,

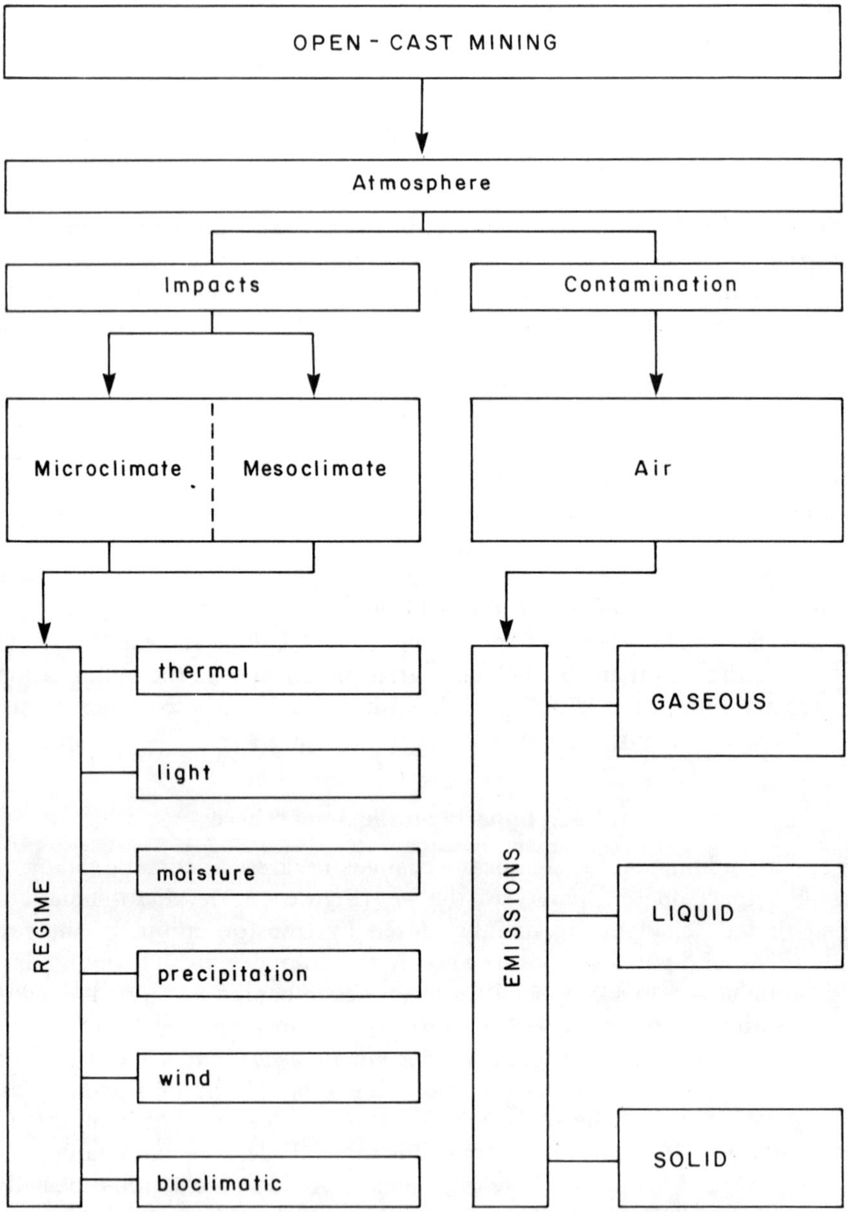

Figure 14.3 Structure of impacts of open-cast mining on the atmosphere

characterized by a relatively stable temperature stratification which is manifested by more frequent ground frosts and, in areas with a contaminated atmosphere, by considerable disturbances in the bioclimate.

The technological processes of mining usually cause contamination of the atmosphere by emissions resulting from mine fires and gob fires in quarries and waste dumps, during the liquidation of waste dumps for further mining, during coal crushing and the crushing of bituminous rocks during mining, and during blasting and transportation. Harmful gaseous substances—namely the products of the burning of the coal substance and combustion processes taking place in combustion engines—have more intensive effects than *dust* generated by the mining operation and by deflation occurring on barren waste dumps and quarries.

Brown coal basins of Tertiary origin, by virtue of their position in relation to the adjacent orographic entities, have a natural tendency to create inversion states. This situation is made worse by increasing in the depth of open-cast mines, whose slopes are made as steep as possible with the intention of minimizing overburden stripping. From this ensues the danger of a considerable concentration of biotic emissions in the area of the mine.

14.3.3 Impacts on the Hydrosphere

In the system of the natural components of the landscape in the North Bohemia brown coal basin the hydrosphere has an infrastructural character and penetrates the whole ecosphere. For practical reasons in this study we shall restrict ourselves to the groundwaters and to the surface parts of the hydrographic system.

The open-cast mining of mineral raw materials has an impact on the quantity and quality of water—usually by disturbing the regime, and by contamination. These changes may be negative and/or positive (Figure 14.4).

Open-cast mining may have the following *negative* impacts on the water regime:

(1) The mine functions as a drainage system; it uncovers the aquifers, reduces the ground water table and desiccates the whole area.
(2) There is artificial drainage of areas designed for mining, through restriction of surface inflow and the acceleration of outflow by hydrotechnical means (e.g. by deep drainage), where a system of hydrodynamic barriers is realized by a system of hydrogeological wells, mainly in those parts of the basin where, in the sub-grade, coal deposit or overburden, there are hydrogeologically efficient rocks with an artesian water regime.
(3) There may be liquidation or diversion of existing water-courses and water reservoirs. The hydrogeological effect of such action mainly consists in the reduction of infiltration owing to the impermeable lining of the beds.

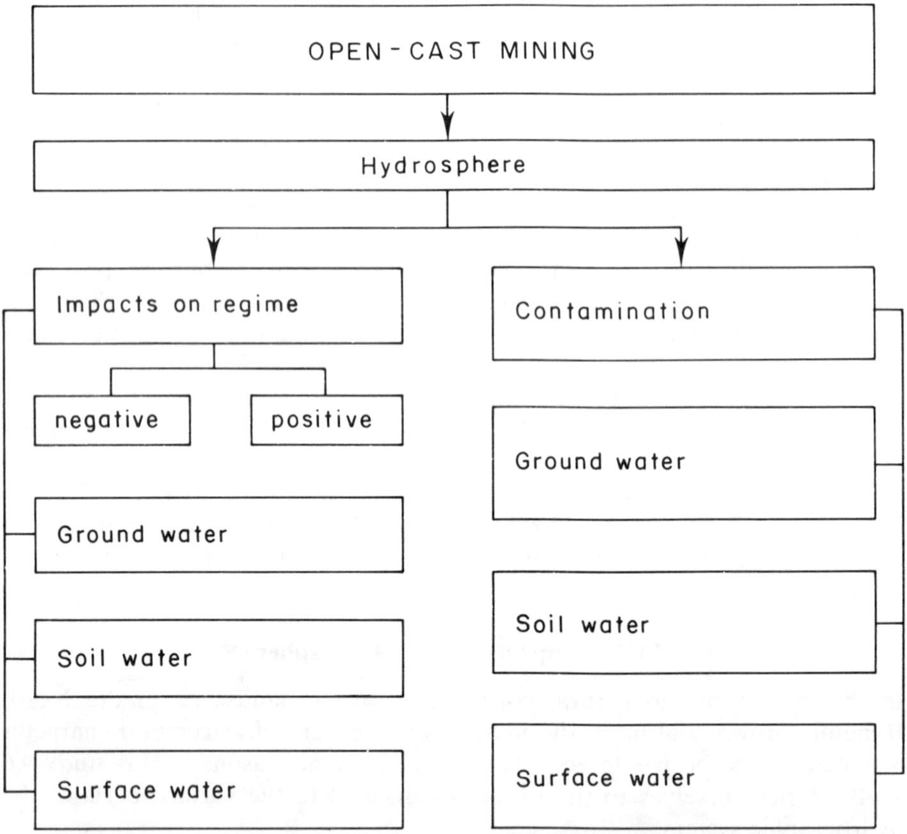

Figure 14.4 Structure of impacts of open-cast mining on the hydrosphere

(4) There may be changes in the rock medium. Significant deformations of the water regime occur in cases where the surface of the waste dumps is formed by rocks with extreme mechanical and physical properties.
(5) There may be changes in runoff conditions owing to the new relief of the exploited area.
(6) Unsuitable diversion of mine waters may cause uncontrolled waterlogging of adjacent plots.
(7) There may be an excessive rise of the groundwater table owing to uncontrolled accumulation of water in residual quarries, which is reflected in undesirable impacts on the environment (e.g. the expulsion of groundwater from the sub-grade of waste dumps).

Open-cast mining in some cases has *positive* impacts on the water regime:

(1) It increases the accumulation capacity of the exploited area (inundated residual quarries, other water reservoirs).

(2) It increases the flow rate in water-courses by the discharge of mine waters.
(3) Mine waters may be used as utility or potable water.
(4) There may be a rise of deeply sunk groundwater tables with previously low ecological effectiveness.
(5) There may be a lowering of excessively high groundwater tables, which is beneficial for agricultural and forest ecosystems.
(6) The formation of recreational water spaces may, in association with vegetation, be suitable for leisure-time activities.

From the point of view of *contamination*, the mine waters of open-cast brown coal mines in North Bohemia are usually characterized by low pH values, the most frequent cause being iron sulphides, mainly pyrite. Also contributory to the reduction of pH values is trivalent iron. The reactions are:

$$FeS_2 + 2H_2O + 5O_2 \longrightarrow FeSO_4 + 2H_2SO_4$$
$$4FeSO_4 + O_2 + 2H_2SO_4 \longrightarrow 2Fe_2(SO_4)_3 + 2H_2O$$
$$Fe_2(SO_4)_3 + 6H_2O \longrightarrow 2Fe(OH)_3 + 3H_2SO_4$$
$$FeSO_2 + 14Fe^3 + 8H_2O \longrightarrow 15Fe^2 + 2SO_4^2 + 16H^+$$

The excessive occurrence of iron sulphides has the same negative effect on the water of the pedosphere of certain waste dumps, which in turn contributes to the long-term biotoxicity of all components of the waste dump soil-formation substrate (phytotoxic earths).

The mine waters also have a high level of hardness. Groundwater in the area of the coal seam is influenced by the oxidation of sulphides and the ensuing sulphate mineralization. In the North Bohemia brown coal basin the prevalent chemical types are CaNa—SO_4HCO_3, CaMg—SO_4.NaCa—SO_4, possibly Ca—HCO_3SO_4 or CaMg—HCO_3.

There is, too, a high content of metal ions. The released mineral acids, mainly H_2SO_4, act on silicate minerals—of which the ions of Al, Fe, Na, Zn, Cu, Pb and other metals may be released into mine waters.

Finally, the mine waters have a high concentration of dissolved substances and suspensions, but extremely low levels of organic substances.

14.3.4 Impacts on the Pedosphere

Open-cast mining will cause degradation or destruction of the soil as the basic ecological factor, the basic means of agricultural and forestry production, and as the space for recreation and for building (Figure 14.5).

Degradation of the soil is caused by excessive waterlogging or desiccation in the area. Contamination of the pedosphere is usually the result of pollution as well as of the negative impacts of the emission background on the whole mining area, where emissions are caused not only by the mines themselves but

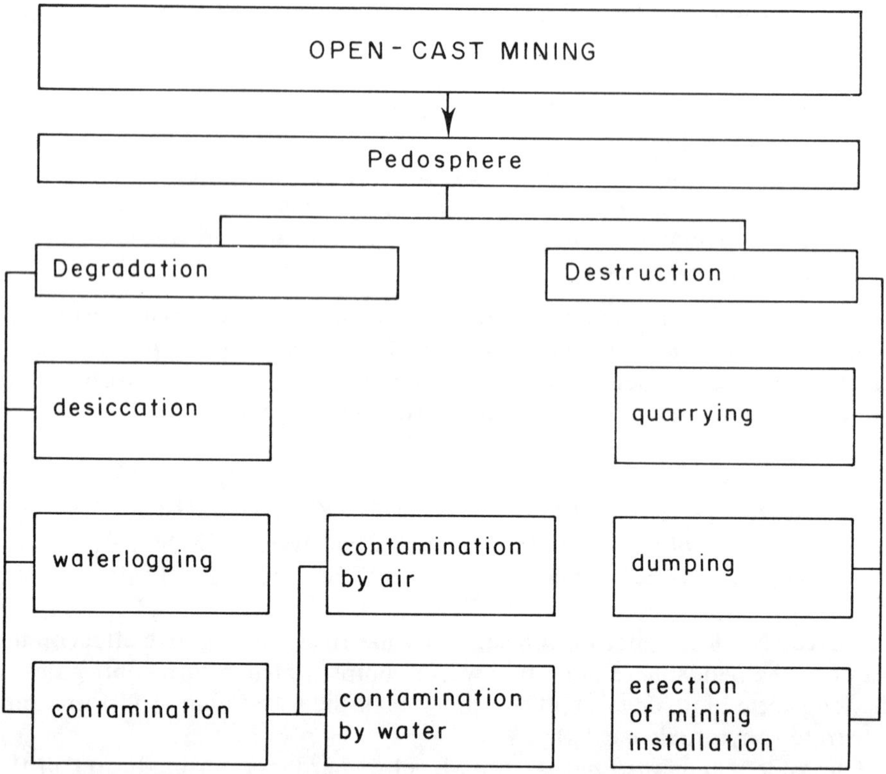

Figure 14.5 Structure of impacts of open-cast mining on the pedosphere

also by other sources (mainly by thermal power plants and other industrial furnaces).

Destruction of the pedosphere eventually occurs inevitably in the whole area of the mine, and gradually in the whole mining space, including waste dumps and areas used for the deployment of mining equipment. Selective stripping of the topsoil (humus profile), which is the most valuable component of the soil, and especially its use for reclamation, may minimize the lasting damage.

14.3.5 Impacts on the Biosphere

Open-cast mining affects the organic components of nature *indirectly* by the degradation and destruction of the environment of organisms, of the ecotope which is formed by the rock medium, relief, atmosphere, hydrosphere,

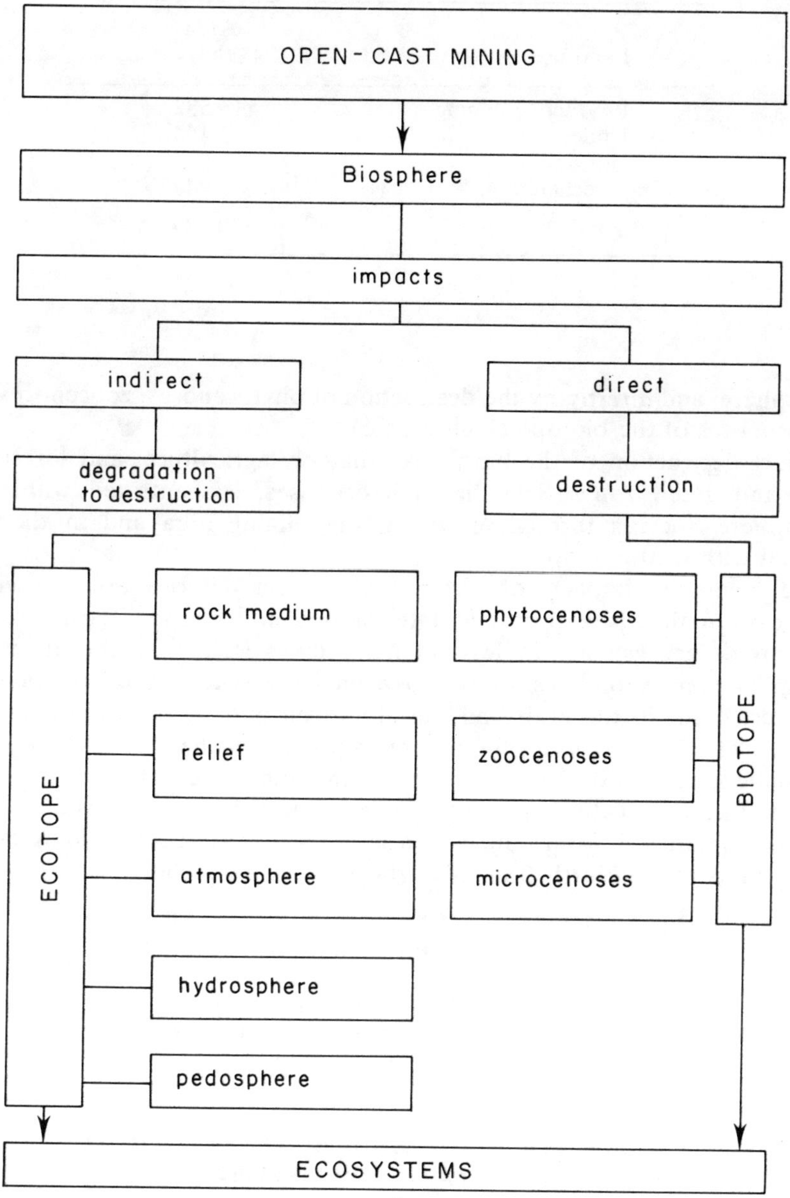

Figure 14.6 Structure of impacts of open-cast mining on the biosphere

Table 14.2 How the mining concern used the available land in 1980

Land use	Area (ha)
Productive mining	15 582
Under reclamation	3 454
Not under reclamation	1 048
Total	20 084

pedosphere, and *directly* by the destruction of phytocenoses, zoocenoces and microcenoses of the biotope (Figure 14.6).

Direct destruction of the biosphere—namely agricultural and forest cultures and green zones, including microcenoses, concentrated within the pedosphere—occurs throughout the whole mining area and in the area covered with waste dumps.

The combined impacts of direct and indirect deterioration caused by open-cast mining is a profound interference in the entire structure and function of ecosystems. In less intensive cases structural and functional changes take place only inside the ecosystems. Ecological transformations do, however, occur in the entire mining area even outside the respective ecosystems. This in turn affects the natural landscape of the area and the socio-economic sphere of the environment, i.e. the human ecosphere.

Quantification of the impact of mining in the area under consideration is evident from the following data, which show the areas of active mining carried out by units of the North Bohemia brown coal mining concern:

Year	1945	1950	1960	1970	1980
ha	5300	6380	11 600	13 620	15 580

In 1980 the mining concern used the available land in the manner shown in Table 14.2. Of the total area not included in the process, 878.37 ha was reserved for various types of construction.

14.4 RECLAMATION AND REHABILITATION OF AREAS DEVASTATED BY MINING

Legislative theory and practice in the Czechslovak Socialist Republic uses the term 'reclamation' for restoration of the ecological and utility values of areas devastated by the exploitation of mineral raw materials.

To comply with the relevant laws, reclamation of areas devastated by mining in the area under consideration is fully secured by the North Bohemia brown coal mining concern. In practice it is divided into two spheres:

(1) the technical stage of reclamation is realized by the mining activities of the individual mines;
(2) the biotechnical stage of reclamation is secured by a system of special reclamation units.

14.4.1 Technical Stage of Reclamation

This stage covers the measures taken before and during mining with the aim of creating optimal conditions for the later biotechnical stage of reclamation. The measures are therefore purposeful and cover, for example, the location of waste dumps in the landscape, the removal of overburden rocks and the construction of waste banks. The possibilities are influenced by the geological conditions, by the system of exploitation of the mineral raw materials, by the physical and technical parameters of the mining area, and by the planned future use of the area.

14.4.1.1 Concept of Brown Coal Mining

The choice of the most suitable mining system is influenced mainly by the coal deposited in the coal seams, the availability of efficient equipment and technology, the demand for brown coal, the physical and technical state of the mining area, and demands made by the principles of technical reclamation—including requirements for the use of the landscape following the discontinuation of mining.

Evaluating this situation with regard to current and future technical possibilities, we may assume that practically the whole seam is exploitable using open-cast methods. The practical orientation of mining corresponds to this situation. New mines are constructed and old mines reconstructed into technological units which are equipped with rail-track excavators, conveyor belts and stowers.

The current state of the basin is characterized by the large number of operating mines whose planned orientation is only being implemented in the second half of this century. The mining is significantly affected by the extremely complex physical and technical structure of the area—especially the intensity and density of settlements and production units—which makes it necessary to operate in constricted spaces.

Modern principles of technical reclamation were introduced at the design stage of mines opened in the 1960s. In other cases they are being introduced with greater difficulty in mines that have been in operation for a long time.

14.4.1.2 Siting of Waste Dumps in the Landscape

The siting of waste dumps significantly influences not only their reclamation, but also the physical and technical arrangement of the whole area. In the North Bohemia brown coal basin the principle is adopted of using landfills whenever possible, rather than the establishment of waste dumps and tips. The implementation of this principle in practice is influenced by the specific situation of the basin, which ensues from the multitude of operating mines that have not respected the principle. Also, mining to ever-greater depths necessitates increasing the handling area of the mines, increasing the removal of overburden rocks, and building a large number of high tips outside the actual mining area. The situation is further complicated by the construction of the whole exploited area, which has an extraordinarily high density of settlements, a technical infrastructure, and which is also used for intensive industrial production and for agricultural activities. Under these circumstances the principles of waste dump management are observed only with the greatest of difficulty.

For example, in the project design of one of the waste dumps of the V. I. Lenin mine in Komořany, three alternatives were considered: the Všehrdy waste dump, the Havraňská waste dump and enlargement of the Velebudín waste dump. The criteria applied were the following:

(1) the coefficient of use of the area of the waste dump;
(2) the total taking up of agricultural land;
(3) losses in agricultural production in terms of yields and gross value in Czechoslovak crowns per annum;
(4) the extent of topsoil stripping in terms of volume and cost.

Data provided in Table 14.3 show that the most suitable alternative was enlargement of the Velebudice waste dump, and this was done. This dump has an almost square foundation, the smallest ratio of slope yardage deposit, a high utilization of area and the lowest losses in agricultural soil. Compared with the alternative of the Havraňská waste dump, 919 ha of highly fertile agricultural plots will be saved, with an annual yield of 2,400 t of wheat or 29 000 t of sugar-beet.

14.4.1.3 Selective Removal of Overburden Rocks

In cases where the overburden varies in quality (i.e. where it is composed of rocks whose suitability for reclamation purposes differs), the selective removal of rocks is one of the basic preconditions for the reclamation of waste dumps. The length of the reclamation cycle, the intensity of the reclamation process and its final effect mainly depend on the mineralogical and petrographical composition of the overburden rocks which in the process of removal

Table 14.3 Data considered in choosing the site of a waste dump at the V. I. Lenin mine in Komořany

Criteria	Všehrdy	Havraňská	Velebudice
Total area (ha)	1 622	1 988	1 380
Content of dump area ($\times 10^6$ m^3)	912 260	850 790	819 000
Coefficient of usage of dump area (m^3/m^2)	56.2	42.8	59.3
Area of agricultural land covered by dumps (ha)	1 479	1 889	970
Loss of agricultural production:			
wheat (t/year)	3 697	4 723	2 425
sugar-beet (t/year)	44 370	56 670	29 100
Loss of agricultural production (crowns per year)	8 874 000	11 334 000	5 920 000
Topsoil stripping ($\times 10^6$ m^3)	5 179	6 611	3 397
Cost of topsoil stripping ($\times 10^6$ crowns)	119 727	152 824	78 532

have been placed on the surface of the waste dump. The properties of the rocks on the surface of the dump affect not only the course of pedogenetic changes, but also the course of the deformation processes on the dumps, a new hydrogeological and hydropedological regime of dumps and subsequent changes in other ecological factors.

In the North Bohemia brown coal basin the overburden rocks and earths were classified in 1959 with regard to their suitability for reclamation, as part of the first Master Plan for Basin Reclamation. This classification was carried out as part of a complementary evaluation of the results of a geological survey and of complex analyses of samples of rocks taken from all rock formations uncovered at that time at the faces of brown coal quarries.

All overburden rocks were divided into five categories:

1st category: rocks and earths very suitable as soil formation substrates for agricultural reclamation

2nd category: rocks and earths for soil formation substrates for agricultural reclamation

3rd category: rocks and earths suitable for forest reclamation

4th category: rocks and earths suitable for land improvement and for green zones

5th category: rocks and earths not suited to reclamation.

Substrates of the first category in the North Bohemia brown coal basin include top humus profiles with proper or degraded chernozem, loess which has a

5–25% content of $CaCO_3$ and a mild alkaline reaction. They have very favourable physical, chemical and mineralogical properties, and are highly suitable for the formation by reclamation of good agricultural soils.

Overburden rocks included in the second category comprise, in the top humus profiles, brown earths and brown soils, and in the lower layers, loess and slope clays. In the North Bohemia brown coal basin the overwhelming majority of overburden rocks are Miocene loam sediments whose properties are significantly influenced by the proportion of koalinite, illite and montmorillonite. They may be further divided into three groups: loams in the vicinity of coal seams which often contain a considerable amount of coal, grey loams and yellow and brown loams.

In the North Bohemia brown coal basin the third and fourth categories of overburden rocks include grey KMI and IKM loams whose scaled aggregates strongly resist the weathering process, contain a large amount of available nutrients (with the exception of phosphorus and nitrogen), have a neutral reaction, favourable sorption capacity with a high saturation of the sorption complex, and are favourable from the physical point of view. In the aggregate state they are also good for the formation of agricultural soils with gleisation tendency. Much better results on these loams have been achieved by forestry reclamation, during which the scaled aggregates do not disintegrate so quickly.

In areas with a prevalence of kaolinite it is best to plant green zones or forests. Substrates of the third category therefore include MIK, KI and IK loams and gravel sands. Compared with the grey KMI and IKM loams, they show a smaller resistance of flaky aggregates, they are susceptible to peptization, they have a lower sorption capacity and usually a lower content of organic substances. Their reactions are neutral. Their suitability for forest reclamation decreases with the increasing content of kaolinite.

Overburden rocks should be placed on the waste dump in natural stratigraphic sequence, and the surface should be formed of the pedogenetically most valuable earths. In this respect the situation in the North Bohemia brown coal basin is very complicated: there is a very low proportion of Quarternary result of the coherent structure with no aggregation of the loam particles. They also include Tertiary and Quarternary gravels and gravel sands which, at the foothills of the Krušné Hory Mountains, pass into layers of boulders and sandy gravels.

Rocks classified into the fourth and fifth categories for reclamation include sands, clays, loams and sands with a coal admixture, slack coal and oxyhumolites. This is a group characterized by a great differentiation in quality.

Overburden rocks should be placed on the waste dump in natural stratigraphic sequence, and the surface should be formed of the pedogenetically most valuable earths. In this respect the situation in the North Bohemia brown coal basin is very complicated: there is a very low proportion of Quarternary rocks, which are unevenly distributed, and there are phytotoxically acting Tertiary rocks of the coal seam strata series. Another difficulty is the development of the basin-orientated large-scale mining.

Regardless of these factors the selective stripping and removal of rocks should be orientated to:

(1) the use of all earth from the top humus profiles;
(2) the use of loess, loess clays and slope debris;
(3) preservation of the greatest amount possible of overburden rocks and earth which can be used for land improvement (such as oxyhumolites, montmorillitic loams and peats);
(4) differentiation of the Tertiary overburden loams (of which the grey loams with a higher content of illite and montmorillitite characterized by flakiness usually have better pedogenetic properties than yellow loams and brown loams, which have a high content of kaolinite, are more weathered and have a coherent structure);
(5) preventing the placing on top of the dump of Tertiary rocks of the coal seam strata series when they contain a coal admixture with a high content of iron sulphides, oxyhumolites, possibly sandy rocks with an extremely high content of SiO_2, and minimal mineralogical ability to release plant nutrients.

14.4.1.4 Construction of Waste Dumps

The geographical and climatic conditions of the basin allow for intensive agricultural use of the landscape. In the Czechoslovak Socialist Republic the demands on the supply of food are increasing at a time when the area of agricultural land is constantly decreasing. In the North Bohemia industrial and highly urbanized conurbation, the quality of agricultural land as well as the quality of the environment have greatly deteriorated. The destruction of the water regime is part of this deterioration. The complex negative impacts of industrial activities on the landscape in the basin are the main cause of the degradation of this area as the natural base for settlement and for recreation.

In the open landscape the prevalent role will be taken by agricultural reclamation, in the area surrounding settlements recreation will be given priority, and in the area of industrial centres priority will be given to clearance and rehabilitation. To these trends must be subordinated any views on the anthropogeomorphological formation of the exploited parts of the landscape, and on the basic geometry of the waste dumps. However, the basic aim of the technical stage of reclamation of dumps is the attainment of a desirable, stable, geometry in relation to the chosen method of reclamation.

In the North Bohemia brown coal basin, with its prevalence of brown coal seams and therefore loam rocks on waste dumps, deflation and water erosion have but little effect, but solifluxion and landslides are a danger. The requirement is that the body of the dump should be in a steady state not only during construction but also during reclamation and subsequent use. Since space is limited, there is a tendency towards the building of large numbers of high landfills, and their general gradients are mostly within the range of 1:7 to 1:12, which corresponds to a slope gradient of 8°08′ to 4°46′, i.e. a gradient suitable for agriculture. This approach leads to very satisfactory stability and makes it possible to realize a good number of reclamation projects for agriculture.

14.4.2 Biotechnical Stage of Reclamation

The biotechnical part of the reclamation cycle is mainly concerned with the organization and management of reclamation—and the optimization of methods in relation to an integrated development of the whole region.

14.4.2.1 Organization of Reclamation Projects in the Basin

All reclamation projects in the basin are planned, managed and financed by a special unit of the general management of the North Bohemia Brown Coal Mining Concern in Most, which secures their preparation and realization through functionally subordinate working units (Figure 14.7) on a commercial basis.

The reclamation units of the individual enterprises of the concern secure the preparation, realization and implementation of all reclamation projects within their own unit in compliance with the Master Plan of Reclamation Projects.

The basin concern project unit Báňské projekty, k.ú.o. Teplice, has its own specialized reclamation unit which, for the general management of the concern, secures the processing and specification of the Master Plan and for the individual enterprises of the concern secures all project documentation of all reclamation projects. Reclamation specialists are involved at the project design stage.

The realization of the biotechnical stage of the reclamation is carried out by Báňské stavby, k.p. Most, which is a unit of the North Bohemia Brown Coal Mining Concern. Drilling and reclamation unit O4 in Odek provides such work for all concern enterprises in the basin. The reclamation department of this unit has personnel, organization, technical equipment and funds for all types of biotechnical reclamation projects.

14.4.2.2 Reclamation Methods

Reclamation of individual devastated areas must be implemented with regard to the character and trend of development of the whole area, as already

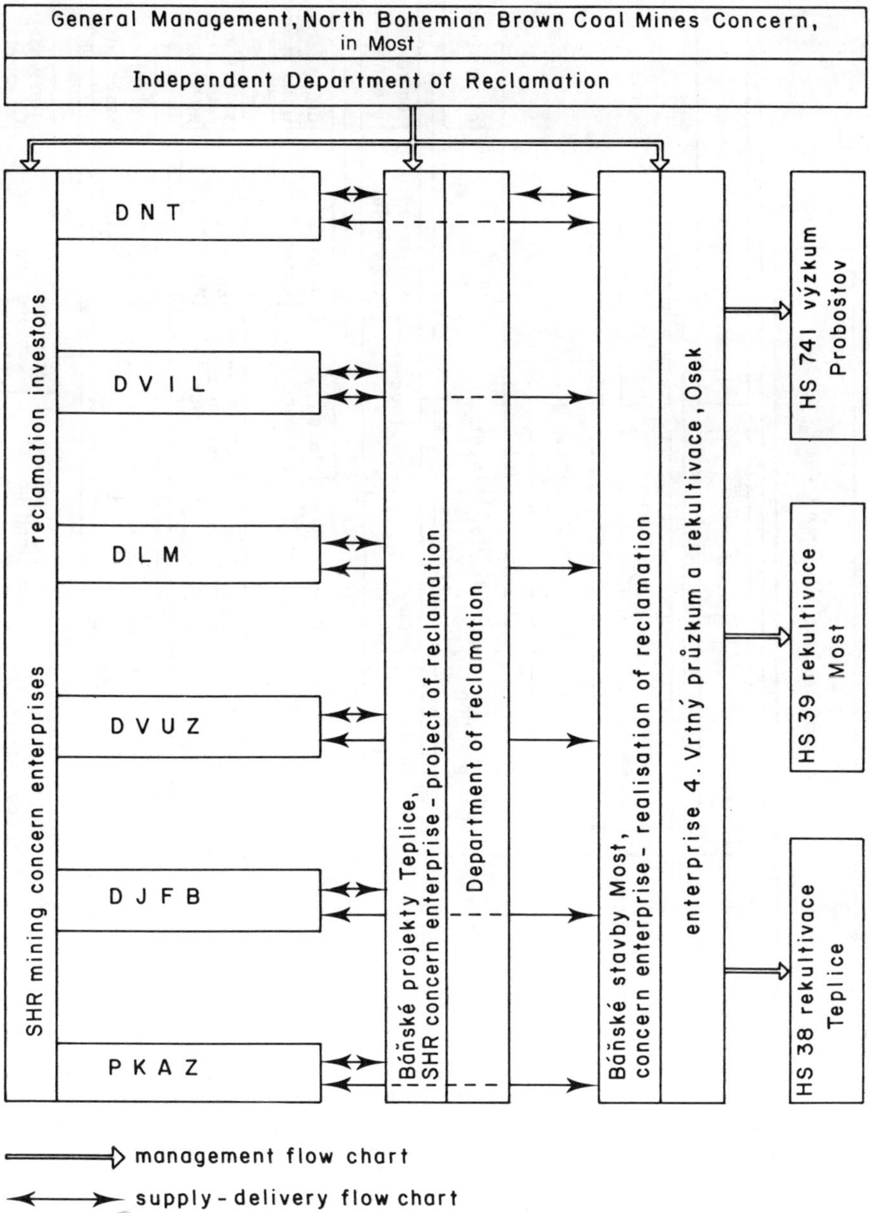

Figure 14.7 The management structure for reclamation projects

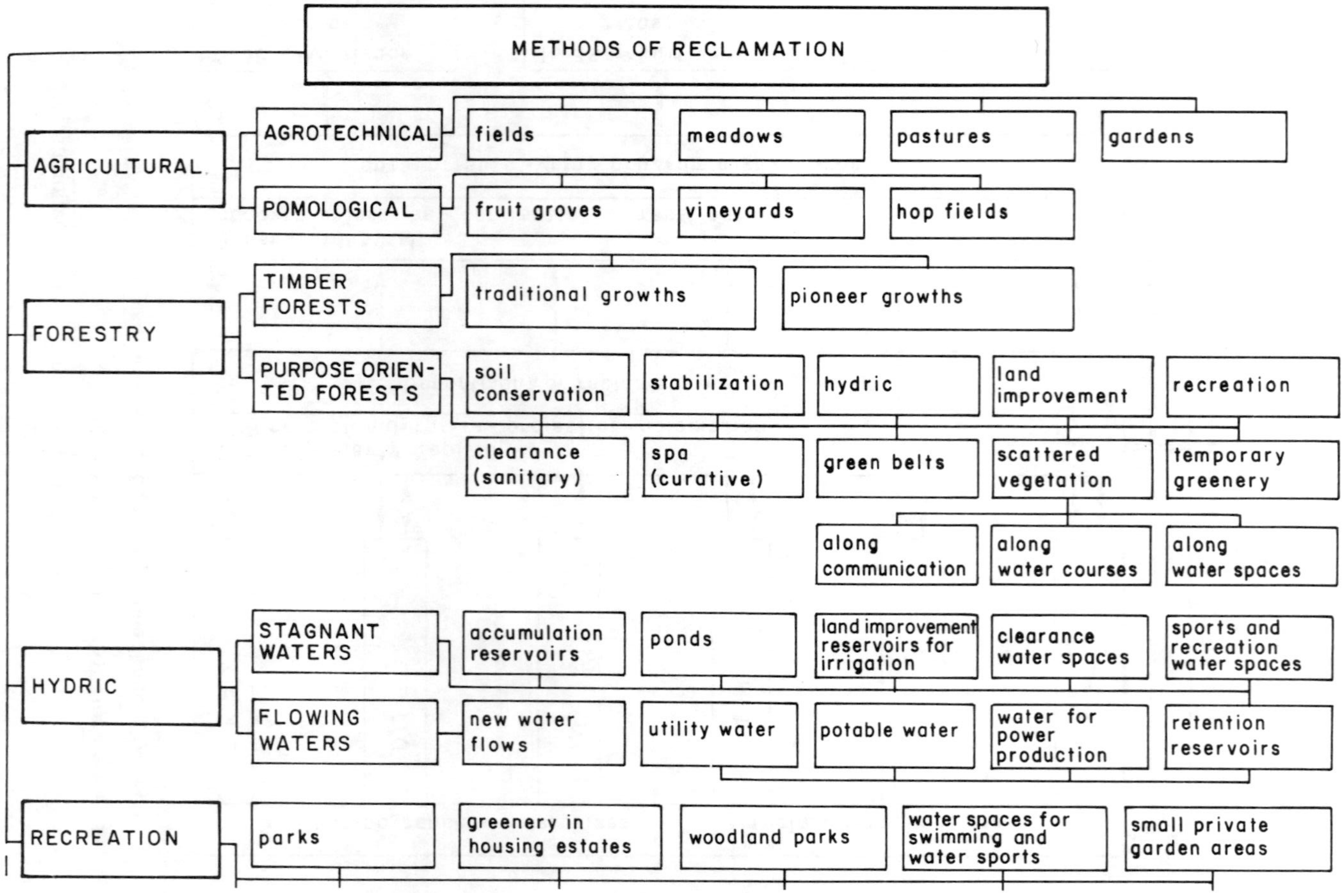
METHODS OF RECLAMATION
AGRICULTURAL
AGROTECHNICAL
fields
meadows
pastures
gardens
POMOLOGICAL
fruit groves
vineyards
hop fields
FORESTRY
TIMBER FORESTS
traditional growths
pioneer growths
PURPOSE ORIENTED FORESTS
soil conservation
stabilization
hydric
land improvement
recreation
clearance (sanitary)
spa (curative)
green belts
scattered vegetation
temporary greenery
along communication
along water courses
along water spaces
HYDRIC
STAGNANT WATERS
accumulation reservoirs
ponds
land improvement reservoirs for irrigation
clearance water spaces
sports and recreation water spaces
FLOWING WATERS
new water flows
utility water
potable water
water for power production
retention reservoirs
RECREATION
parks
greenery in housing estates
woodland parks
water spaces for swimming and water sports
small private garden areas

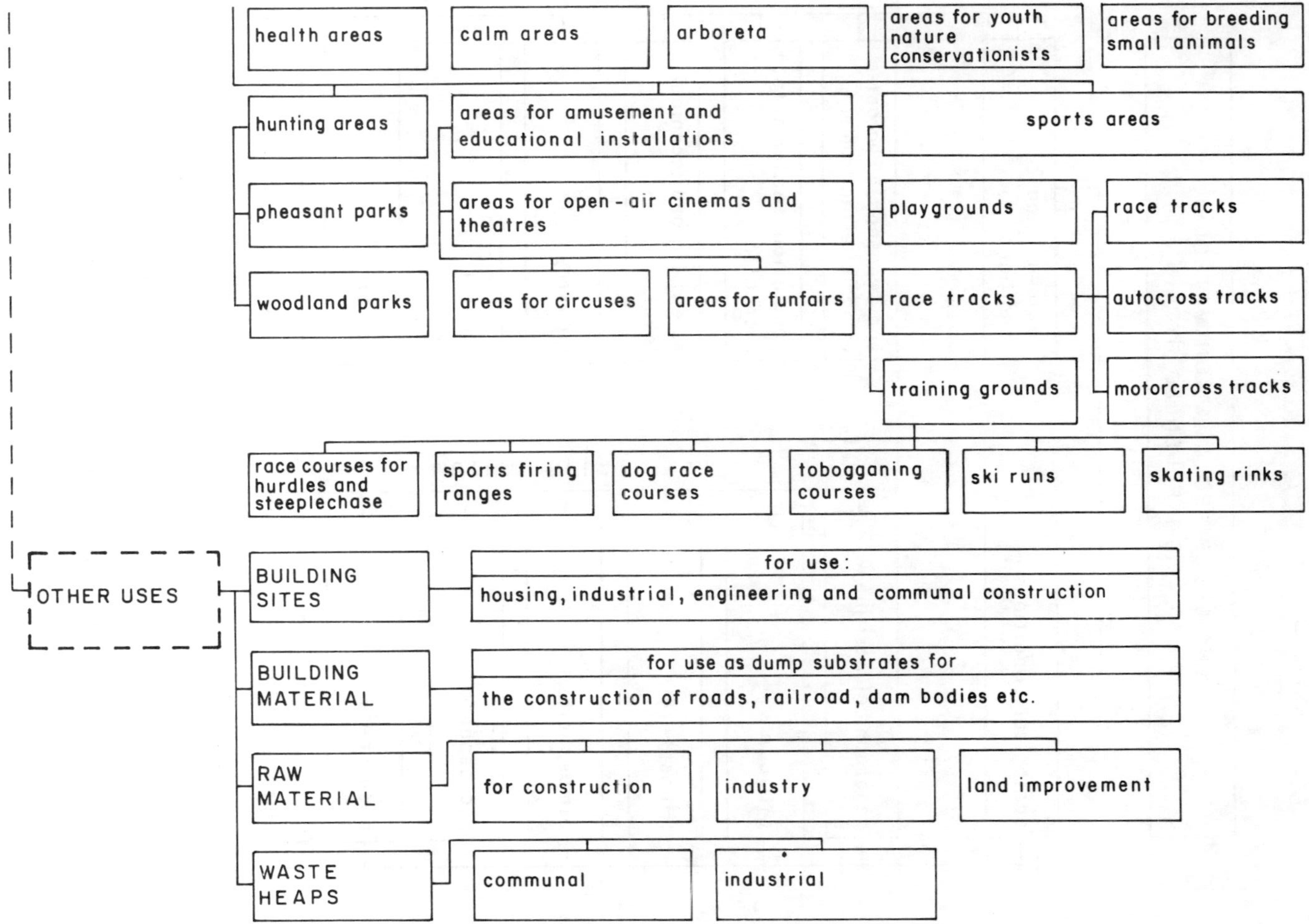

Figure 14.8 General types of reclamation

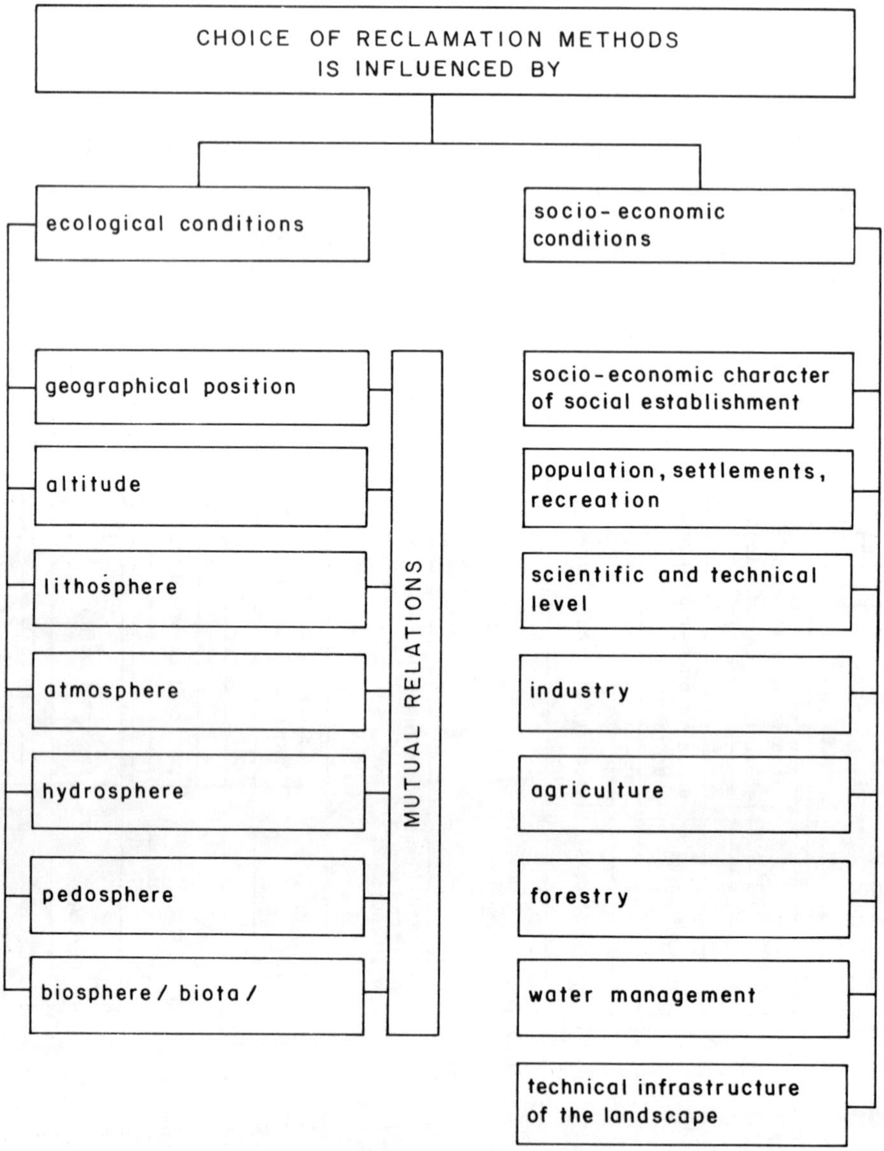

Figure 14.9 Factors affecting the choice of reclamation

discussed. A new landscape with proportionate representation of agricultural soils, forests, woodland parks, parks and water areas meets not only the ecological but also the socio-economic interests of society.

Assessed from the ecological point of view, the reclamation principles applied in the basin must together lead to a state of inner ecological balance. From the socio-economic point of view, projects are orientated mainly at establishing a suitable environment and at increasing food production. In ecological terms reclamation mainly includes trees and hydric plants which are highly effective from the environmental point of view. Agricultural reclamation—i.e. agro-technical and pomological—is significant for its ecological and economic effectiveness (Figure 14.8).

The optimization of reclamation methods in the basin is based on a set of ecological and socio-economic conditions (Figure 14.9).

In the ecological factors (Figure 14.10) it is necessary to take into consideration the basic fact that the whole area of the basin is, with regard to its geographical position, altitude and the surrounding orographic entities, a very fertile area, advanced geonomically and as regards forestry. The very warm and semi-arid character of the climate makes it necessary to economize with moisture. In the localization of reclamation projects it is basically possible to implement all agricultural and forestry methods, whose ecological and economic effectiveness is magnified by the planned formation of new hydric conditions which may be influenced by new water management methods of reclamation. The limitations usually arise from inclinations and the topography of the area, which in turn are linked with resistance to water erosion and the extent to which mechanization may be applied. The topical nature of this problem is brought out by the necessity of establishing large numbers of waste dumps made up of a great amount of overburden rock with a considerable content of slope debris.

Ecological motivations are affected by many socio-economic factors, such as demographic conditions, the density of population, the density of settlement, the level of the technogenic deterioration of the landscape of the basin as a biotope, and the ensuing intensity of environmental control and the societal aspects of the necessity of providing food and recreational spaces for the population. The landscape of the basin used to be characterized by a dense network of small and medium-sized settlements, but these are currently being dispersed as a result of the development of coal mining. The housing stock is being concentrated in several towns, around which reclamation projects are implemented with a primarily aesthetic and recreational effect (trees, bushes and water spaces). Agricultural reclamation is given priority mainly in the open landscape.

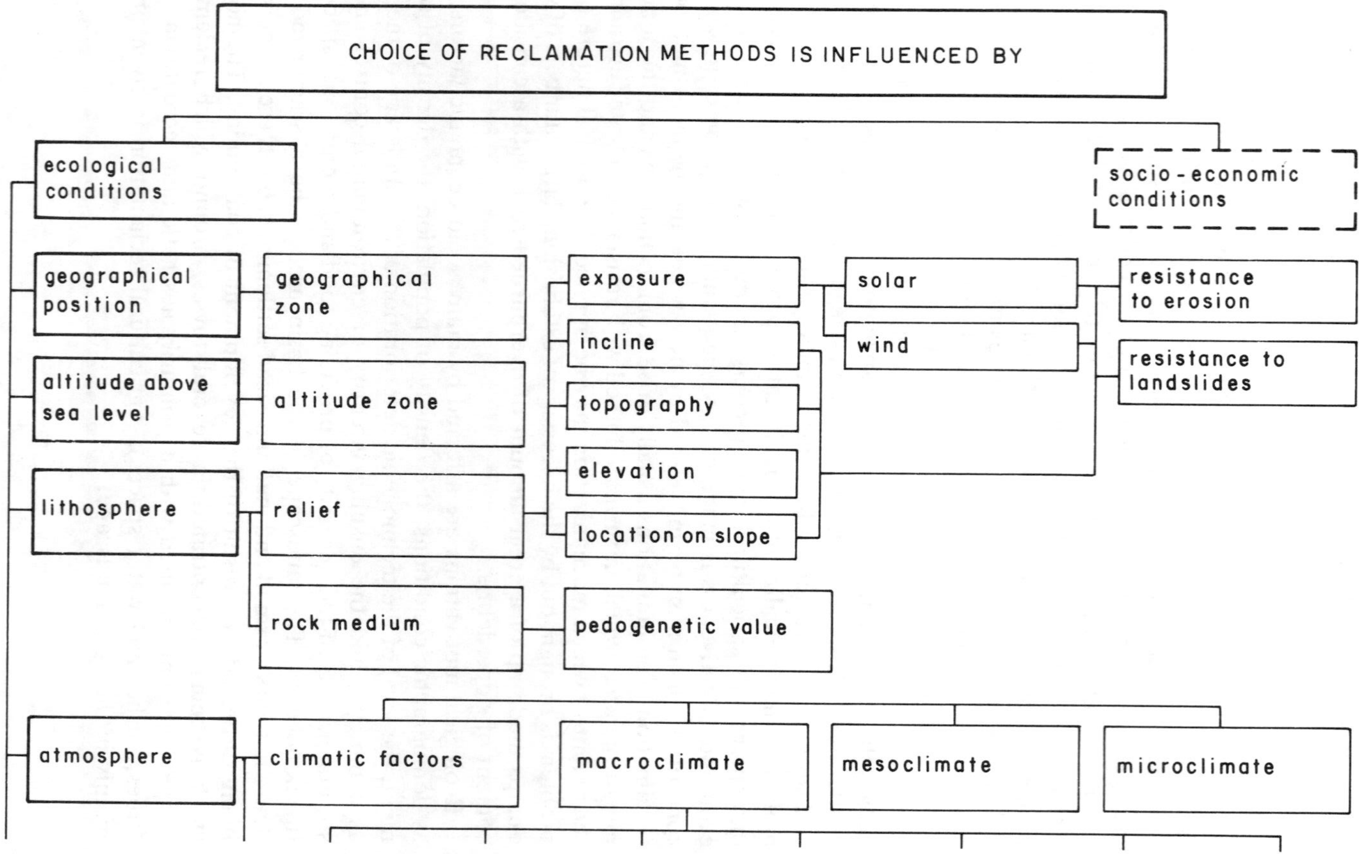
CHOICE OF RECLAMATION METHODS IS INFLUENCED BY
ecological conditions
socio-economic conditions
geographical position
geographical zone
altitude above sea level
altitude zone
lithosphere
relief
rock medium
exposure
incline
topography
elevation
location on slope
pedogenetic value
solar
wind
resistance to erosion
resistance to landslides
atmosphere
climatic factors
macroclimate
mesoclimate
microclimate

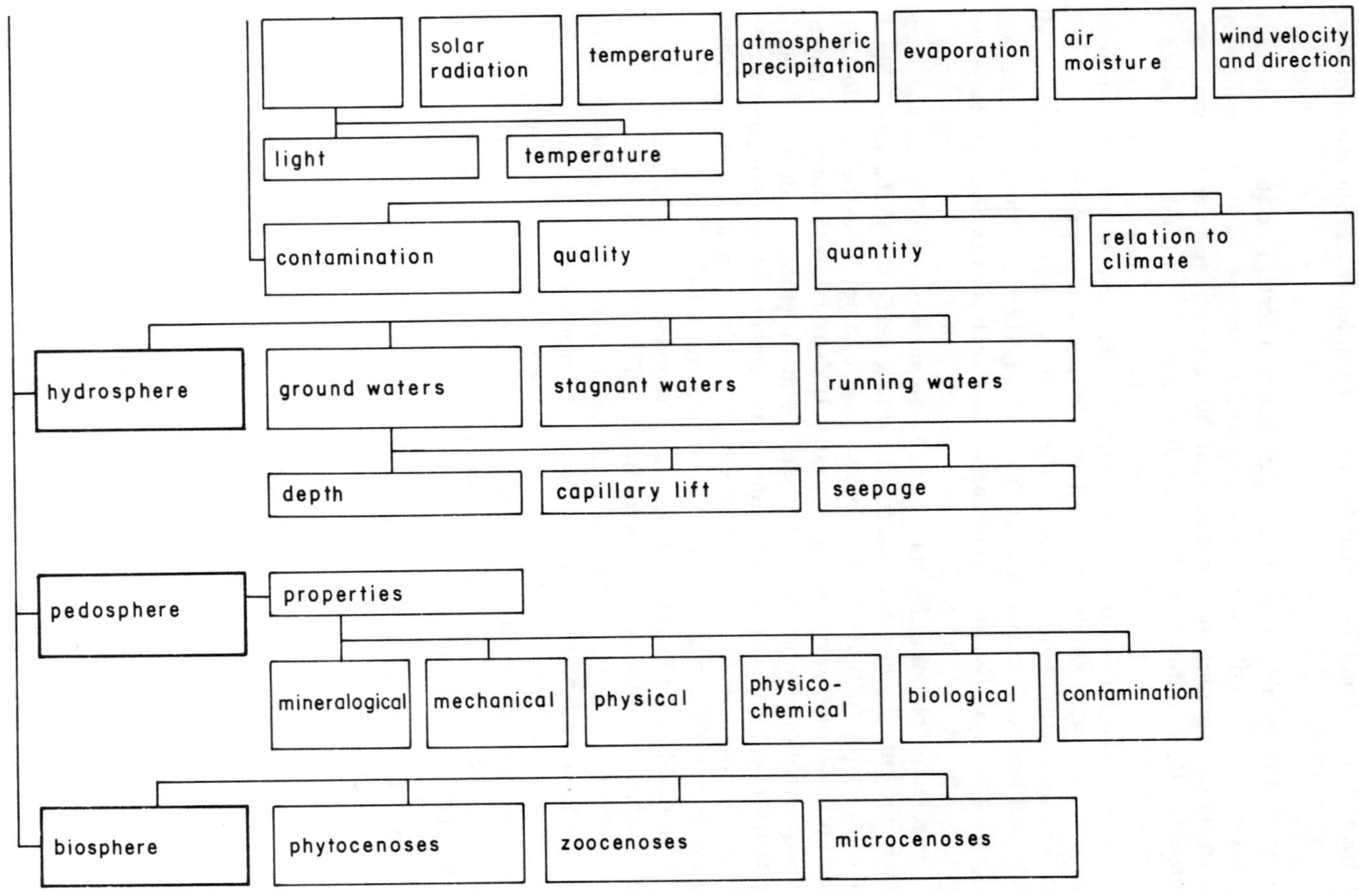

Figure 14.10 Detailed analysis of the ecological factors affecting the choice of reclamation

14.4.2.3 Regional and Technical Methods of Implementing the Reclamation Concept

So far each of man's activities in nature has to a certain extent been marked by one-sidedness, influenced by limited and subjective views. As civilization expands, the amount and intensity of human activities on the landscape increases. Reclamation of the landscape must be solved by applying a system-integrated approach which incorporates all natural and economic activities in the landscape. These demands are best met by a system of regional planning which most effectively leads to optimization of the development of the landscape and to the socially effective integration of all ecological and socio-economic activities; this is therefore the most effective long-term method. Regional planning in the Czechoslovak Socialist Republic benefits from being part of the state planning system.

The regional planning for the North Bohemia brown coal basin considers mainly the future functional use of the whole territory of the basin by all branches of production—mainly brown coal mining, industrial production, power production, agriculture and forestry. It also deals with the development of the population: employment, the settlement structure, the development of the tertiary sector, transport and water management. Next to safeguarding an acceptable measure of emission levels, the regional plan considers reclamation as being the basic factor of clearance and rehabilitation.

The initial Master Plan was processed in the years 1958–59 and projected the reclamation of the whole basin for the period 1960–80. It was modified several times and became the basic strategic directive for implementation of the individual reclamation projects by the enterprises of the concern.

In the period 1961–78 the Master Plan was modified seven times. The objective nature of the initial plan is brought out by the fact that throughout the whole period the initial distribution of reclamation projects was preserved, i.e.

agricultural	46.4–50.9%
forest	52.5–41.4%
water management	1.1–7.7%

The larger proportion of agricultural reclamation projects shows the immense pressure that has been brought to bear to offset the loss of agricultural plots by mining, and the growth of water spaces is the result of efforts made to create an area for short-term recreation in the basin itself.

The subsequent stage of the Master Plan was for the period 1981–90. The basic problem was the rapid take-up of plots for mining, with the limited possibilities for including devastated areas in reclamation projects. Table 14.4 shows how the plan was expected to work out. The new reclamation cycle in the time period comprises all devastated areas which no longer serve mining.

Table 14.4 Planned development of the land in the basin

Year	Take-up of agricultural land for mining (ha)	Newly-launched reclamation projects (ha)	Planned completion of reclamation projects (ha)
1981	389	406	255
1982	413	375	259
1983	472	277	325
1984	469	281	296
1985	391	124	435
1986–90	1556	1339	1208
1981–90	3691	2804	2779

In the period 1981–90 reclamation projects operated by the North Bohemia Brown Coal Mining Concern will cover a total area of about 2800 ha, made up as shown in Table 14.5. The Master Plan assumed that the reclamation cycle will be completed as shown in the last column of Table 14.4. These projects will cover the areas shown in Table 14.6.

The orientation of reclamation to intensive land use, namely to agriculture, and the increase in the selective stripping of topsoil and loess, are the reasons for thc increasing cost trends, which are evident from Table 14.7.

Table 14.5 Reclamation projects in the period 1981–90

	(ha)	(%)
Agricultural land	1771	63.4
Forest land	644	23.0
Parks	87	3.1
Water spaces	152	5.4
Miscellaneous	143	5.1

Table 14.6 Completion of reclamation projects in the period 1981–90

	(ha)	(%)
Agricultural land	898	32.3
Forest and parks	1622	58.4
Water spaces	191	6.9
Miscellaneous	68	2.4
Totals	2779	1 00.0

Table 14.7 Cost trends for reclamation (in thousands of Crowns)

Year	Topsoil and loess	Reclamation	Totals
1981	94 768	122 264	217 032
1982	95 505	103 327	195 831
1983	72 217	88 458	160 675
1984	85 135	157 303	242 438
1985	69 918	166 910	236 828
1986–90	337 429	897 009	1 234 438
1981–90	751 970	1 535 273	2 287 243

14.4.2.4 Technology of Reclamation Projects

In the implementation of agricultural, forestry, water management and recreational reclamation projects certain technologies are being applied, which include:

(1) *technical methods*—levelling, the bringing in of fertile and potentially fertile make-up earths, a system of land improvement measures, the building of road communications, hydrotechnical and hydro-amelioration work, technical stabilization of slopes;
(2) *biotechnical methods*—erosion control measures, and a system of agricultural, forestry and grove and park reclamation (the methods have developed from extensive to intensive projects).

In the first years extensive methods of reclamation were used. The whole set of measures was usually based on the necessity to subordinate the project to the unfavourable and atypical site of the devastated plots. The aim of the reclamation was to plant vegetation on the devastated site—using mainly pioneer and ecologically adaptable plants with a low economic effect. This in practical terms meant the least earthwork. Agricultural reclamation was restricted to planting without the use of make-up ground and without basic land improvement, without intensification of hydro-amelioration measures, mostly by the direct cultivation of the waste dump.

The establishment of suitable prerequisites for reclamation during the technical (mining) stage, and growing demands on the quality of the reclamation, led to intensification of the technology. Greater emphasis was placed on the stage of improving the ecological factors of waste dump sites in compliance with the set target. The improved system of waste dump construction was followed up by a more intensive earth-work plan, the bringing in of fertile soils for land improvement, and water-course training. Measures were taken to improve the mineralogical and hydropedological structure of the surface layers of the dumps. All sites were made accessible.

Forests are planted on waste dumps with gradients not exceeding 1:5 or 1:6, which makes mechanization possible. Such trees will increasingly be exploited for timber. Forest cultures are fertilized and intensive care is devoted to the site before it is handed over to its users.

In agricultural reclamation projects higher demands are placed on area, compactness and regularity of the shape of the plots, and on the possibility of using mechanized tillage. In the preparatory stage increasing attention is devoted to securing the pedogenetic quality of the surface layers of waste dump rocks, namely their mineralogical and hydropedological properties. The number of agricultural reclamation projects is increasing on made-up ground. The make-up consists of the top humus profile, clayey rocks of the Quaternary, and various rocks which are used for land improvement (such as bentonites, tuffitic loams, marl, peats, oxyhumolites). Greater attention is being devoted to the intensity of plant nutrition: mineral organic carbonic fertilizers manufactured by the V. I. Lenin mines in Komořany from waste sludges from brown coal treatment plants have been used successfully. Almost every year the mines selectively strip and remove greater quantities of humus earths, significantly contributing to the intensification of the soil-forming process. Hydro-amelioration is becoming a frequent method in agricultural reclamation projects, including not only drainage but also irrigation. The original 2–3 year cultivation process has been extended to 8–12 years, and agricultural reclamation is now increasingly subordinated to the final target of attaining an anthropogenic soil profile with the highest potential fertility.

Technologically demanding and costly reclamations of areas designated for recreational use were only launched in the second half of the reclamation period. Woodland parks are being equipped with facilities for intensive use by the population, mainly by means of a dense system of communications. The building of recreational water reservoirs with a controlled water regime is very demanding. These reservoirs are usually set in woodland parks or parks which are planted around them. The planting of parks is very demanding because it usually requires terrain levelling and other treatment of the relief of a considerable part of the waste dump, with considerable quantities of fertile make-up soil, a wide variety of seeds and seedlings of trees and bushes. The concept is that of a combination of free grassy areas with solitaires and groups of trees and bushes resembling the English park. The areas usually have a dense network of paths, and recreational facilities corresponding to the possibilities of the park.

14.5 PRACTICAL RESULTS OF RECLAMATION

Reclamation projects have been implemented in the basin for three decades now. The initial stage, 1950–60, was marked by the search for suitable

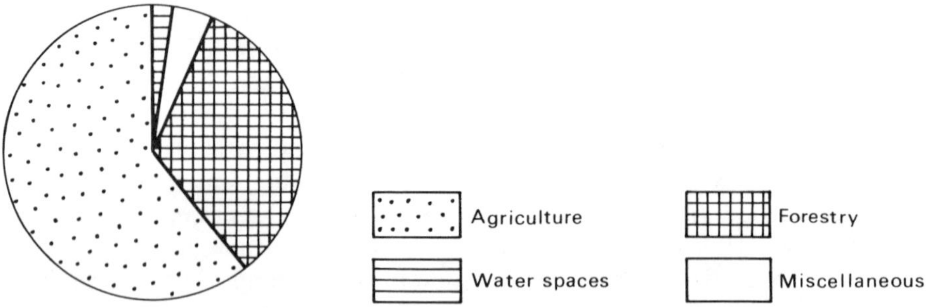

Figure 14.11 Completed reclamation projects handed over to users by the SHD Concern up to 1 January 1981

technologies in the field of forestry and agricultural reclamation, the organization of reclamation projects in the basin, and methodical specification of the whole system of reclamation ranging from the compilation of legal regulations and prescriptions to practical implementation. The first Master Plan of Recultivation was launched in 1960; this introduced the concept of regional and landscape reclamation, and the concept of regional reclamation became the long-term strategic basis for practical strategy.

Figures 14.11–14.16 summarize the 30 years' development and the current status of reclamation in the basin.

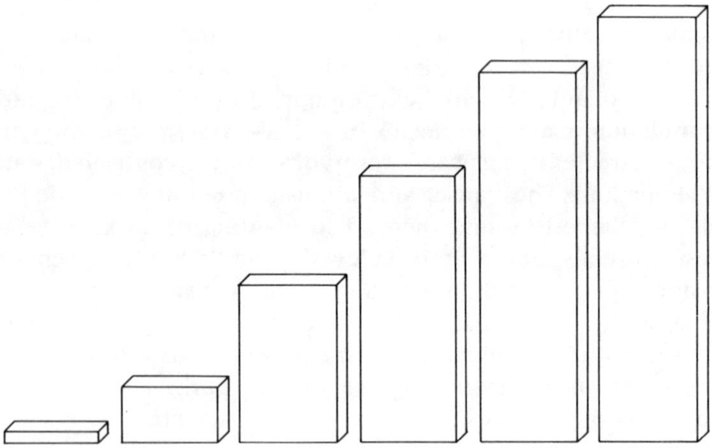

in ha	173	567	1643	2640	3888	4139.47
year	1955	1960	1965	1970	1975	1980

Figure 14.12 Development of reclamation projects by the SHD Concern

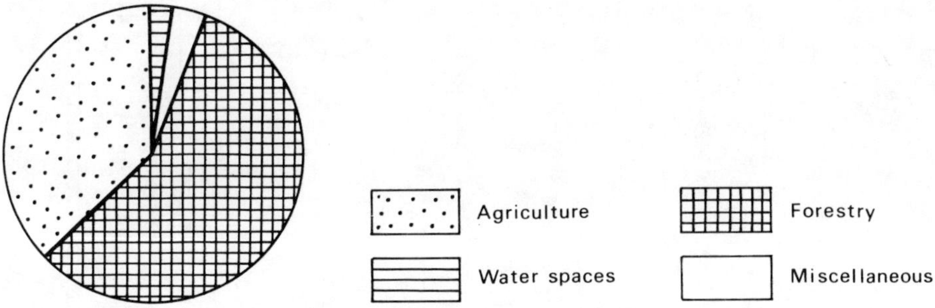

Figure 14.13 Current reclamation projects by the SHD Concern up to 1 January 1981

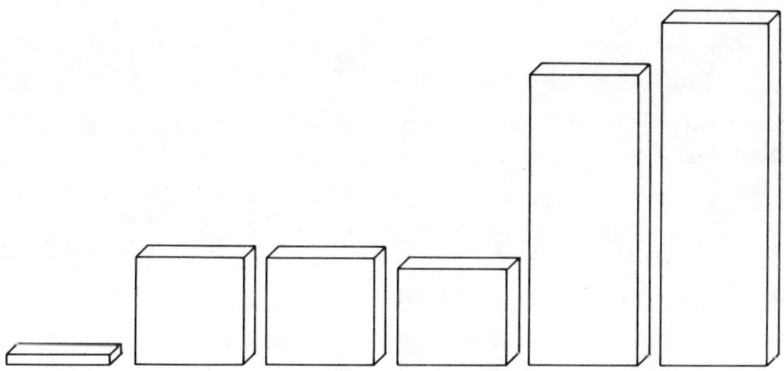

in $10^6\ m^3$	0.02	0.48	0.50	0.41	1.32	1.54
year	1956	1960	1965	1970	1975	1980

Figure 14.14 Development of volume of selective removal of topsoil by the SHD Concern

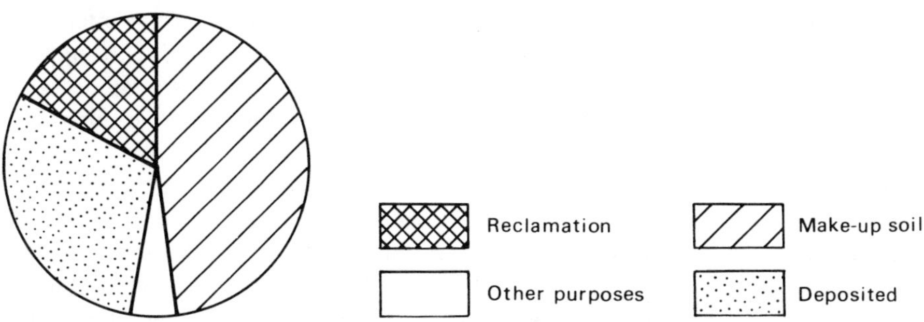

Figure 14.15 Survey of the use of topsoil selectively removed by the SHD Concern, 1956–80

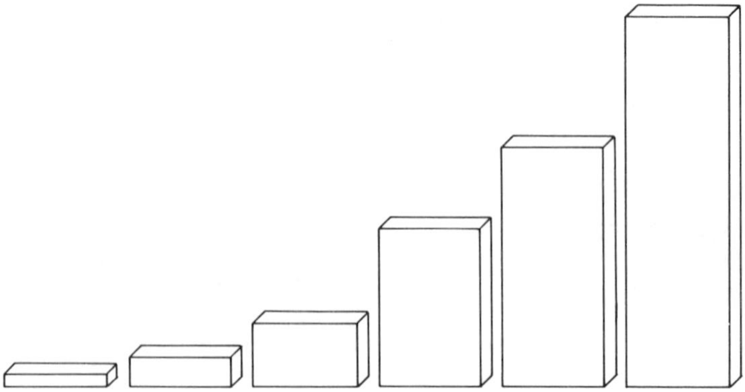

in 10^6 crowns	5.0	11.7	25.0	74.7	112.4	175.0
year	1955	1960	1965	1970	1975	1980

Figure 14.16 Development of financial volumes of reclamation by the SHD Concern

14.6 DYNAMIC BALANCE OF THE LANDSCAPE IN THE BASIN

Before settlement the landscape was characterized by an abundance of riparian woodlands, water surfaces, swamps and water-courses. Medieval colonization changed the area by drying out the swamps and water spaces and by felling the forests into an area with a significant prevalence of agricultural cultures. This state culminated in the mid-nineteenth century when the opening of mines began on a large scale.

Hand in hand with the development of open-cast mining the mining industry in the area intensified, and as a result the following period saw an increase in industrialization and urbanization, with the accompanying process of denaturation. Only since 1950 has there been planned management of the devastated areas. Table 14.8 shows that after exploitation of the basin the area will again be used for agricultural purposes. Compared with the mainly agricultural use which preceded the period of mining, there will be a higher proportion of forests, woodland and parks (especially in the environs of cities), and water spaces, for recreational purposes.

The data in Table 14.8 for the years 1960 and 1980 show that the greatest losses to mining were inflicted on agricultural soils (5.6%); of this, arable soils lost 4.3%. Other areas, which include mining facilities, increased over this period by 4.1%. These data favourably reflect the reclamation work done in this period, as without it the loss of agricultural soil would have been 7.5%. The area under forest is being preserved only thanks to reclamation work; otherwise it would have recorded a loss of 0.9%.

Table 14.8 Dynamic balance of the area of the North Bohemia brown coal basin landscape[a]

Year[b]	Total area (ha)	Arable land	Hop fields	Vine-yards	Gardens	Fruit growers	Meadows	Pastures	Agricultural soil	Forest soil	Ponds	Other water surfaces	Built-up areas	Other areas[c]
1850	105 539 (100%)								83 376 (79%)	10 554 (10%)	2 111 (2%)	4 222 (4%)	4 222 (4%)	1 055 (1%)
1960	105 539 (100%)	61 411 (58.2%)	1 166 (1.1%)	– –	2 087 (2.0%)	953 (0.9%)	3 220 (3.0%)	3 816 (3.6%)	72 653 (68.8%)	4 785 (4.6%)	152 (0.1%)	2 035 (1.9%)	2 288 (2.2%)	23 626 (22.4%)
1980	105 539 (100%)	56 918 (53.9%)	1 268 (1.2%)	46 (0.05%)	1 787 (1.7%)	1 027 (1.0%)	2 440 (2.3%)	3 127 (3.0%)	66 613 (63.2%)	4 812 (4.5%)	139 (0.1%)	3 452 (3.3%)	2 539 (2.4%)	27 984 (26.5%)
2050	105 539 (100%)							(60%)	63 323 (24%)	25 329 (2%)	2 111 (8%)	8 443 (5%)	5 277 (1%)	1 055

[a]The inventory of cultures was made only in those cadastral areas in which the consequences of brown coal mining have, are or will be felt.
[b]Figures for the years 1850 and 2050 are qualified estimates; figures for 1960 and 1980 correspond accurately to the records of real estate property kept by the relevant respective units of geodesy and cartography.
[c]The category of 'mining' is classified into 'other areas'.

14.7 PROPOSALS AND RECOMMENDATIONS

14.7.1 Criteria

A set of criteria that could be applied in studying and assessing changes in the structure, dynamics and functions of land use in the area, and the impacts of these changes on quantifiable factors of the environment, were set out in detail in Figures 14.2–14.6 and Table 14.3.

14.7.2 Suggestions for Research

Ecological considerations should be introduced more systematically into technical development and technology. Research should cover this very important area, particularly with regard to simultaneous optimization of:

(1) the development of the mine;
(2) the siting of waste dumps in the landscape;
(3) the shaping of waste dumps;
(4) the technology of removal of sub-grade rocks in relation to the principles of selective stripping, transport and deposition of fertile, potentially fertile and valuable overburden rocks and earths.

There should be an integrated and systems approach to research on the socio-ecological optimization of the reclamation of the area, taking into account all ecological, socio-economic and political considerations.

Since the work considered would cover a large area (of the order of 1000 km^2) it would not be possible to use traditional methods. It would be useful to organize remote geological prospecting.

The results of this suggested research should effectively contribute to the controlled creation of good environmental conditions during the period of mining, which is important for the stabilization of the population and labour in the area. This would in turn be positively projected into the creation of conditions suitable for the optimization of reclamation in the basin.

14.7.3 Suggestions for Minimizing Negative Environmental Impacts and Optimizing Reclamation

There should be continuing research on measures to eliminate excessive dust and gaseous industrial emissions from power plants and chemical industries.

During the project design stage, and then during actual mining, there should be means to implement systematically all necessary measures for technical (mining) reclamation, namely:

(1) the selective stripping and removal of fertile, potentially fertile and valuable overburden rocks and earths;

(2) the planned construction of waste dumps observing the set future use of the area.

The above measures, again, are necessary to create good conditions for the effective realization of all the biotechnical stages of reclamation.

The concept of reclamation should be optimized so as to be integrated with all other socio-economic activities in the area of the North Bohemia brown coal basin. To achieve this it will be necessary to:

(1) speedily complete and approve plans for brown coal exploitation, with a time horizon for breaking out the coal;
(2) determine the regional prognosis for the basin, in terms of the final reconstruction and use of the whole area for mining, industrial production, agriculture, forestry, and water management, and the infrastructure, dwellings, sports and recreation spaces.

Land Transformation in Agriculture
Edited by M. G. Wolman and F. G. A. Fournier

CHAPTER 15
Changes and Consequences Related to the Transformation of Small Villages into Rural Cities in Czechoslovakia

M. GOTTLIEB AND J. PAVLŮ

15.1 INTRODUCTION

Our attention will focus first of all on studying interlinkages—relationships between fields and their arrangement in the landscape, on the one hand, and the agricultural infrastructure on the other. The agricultural infrastructure will be conceived as the system of agricultural settlement, both in terms of the overall system of settlement, including higher settlement structures (small, medium-sized and big towns and cities), and in terms of the elementary units of the agricultural structure (the farm and farm buildings, such as stables, granaries, etc.). We are well aware that this infrastructural concept inclines to sociological methods of investigation. In this case study it will, however, also include the essential aspects of notions and concepts such as the man-made environment, the material environment or anthropo-cenoses—terms mainly used by ecologists.

The relationships between field plots and their arrangement in the landscape will be examined historically (i.e. from the past through the present into the future), as shown in plans and hypotheses. In our analysis of this relationship we proceed from the basic premise that the function of an arbitrary object in society reflects a sovereign form of interrelationship between the components of the unified process of social development, when changes in one are induced by changes in the others. On the basis of our own specific concept of anthropo-ecology (Hadač *et al.*, 1977; Blažek *et al.*, 1974; Gadač *et al.*, 1978; Gottlieb, 1976) we shall be looking at these relationships not from the point of view of the balanced circulation of matter and energy, but from a point of view that includes the problems of changes in the informative values of cultural meanings of objects in settlements and in the landscape over longer time intervals. Generally, we are thus concerned with harmonizing and balancing natural cenoses and agro-cenoses with an agricultural infrastructure, not only with regard to the harmony of matter and energy exchange but also on the psychological and aesthetic plane—which is usually expressed by the term 'a beautiful landscape', each part of which has its own 'genius loci'.

15.2 SPECIFIC FEATURES OF AGRICULTURAL SETTLEMENTS IN THE CZECH REGIONS

Throughout its historical development the land has been characterized by a great density of settlements—especially settlements of less than a few hundred inhabitants—and thus by small distances between them. In the Czech lands the average distance between settlements is 2.76 km. This distance increases in the easterly direction (i.e. in the direction of Slovakia), so that in western Slovakia it reaches 3.2 km and in eastern Slovakia 3.9 km. The average separation in Slovakia is given at 3.53 km.

The individual villages formed explicit morphological types given by the spatial relations between the dwellings, the farm houses and fields (Figure 15.1). The villages could be compact, with a common, where the dwellings are concentrated around the central village 'square' with the church. In these cases the dwellings are relatively close to one another, separated by small gardens. The fields are not adjacent to the dwellings but are mostly scattered and located in different parts of the village environs, and approached by an arrangement of paths. Alternatively the villages could be open and linear, and established along roads. In these cases the distances between the dwellings are much larger than in the previous types. There are also 'run-based' villages where the distances between the dwellings again differ. The variety of forms arises not only from the morphological variety of the landscape, but also from cultural and historical tradition. Thus, for instance, in the border areas of eastern Bohemia the villages were mostly established in the eighteenth century by colonization, and they are carbon copies of the villages in the native lands of the colonizers (i.e. in the German regions of Thuringia and Westphalia) (Sommer, 1836; Laurentius Wintera, 1894; Šimák, 1936).

In the individual types of landscape specific vernacular architecture was characteristic not only of dwellings but also of farm houses and outbuildings. In the ponded area of southern Bohemia, where the inhabitants have the sky above their head and its reflection at their feet, there originated the village Baroque house which, in its own original way, transformed the Baroque style in the cities. We find transformed elements of Alpine architecture only in the Šumava area—the south-western border region of Bohemia—and nowhere else. We find the Frankish court only in north-western Bohemia. According to

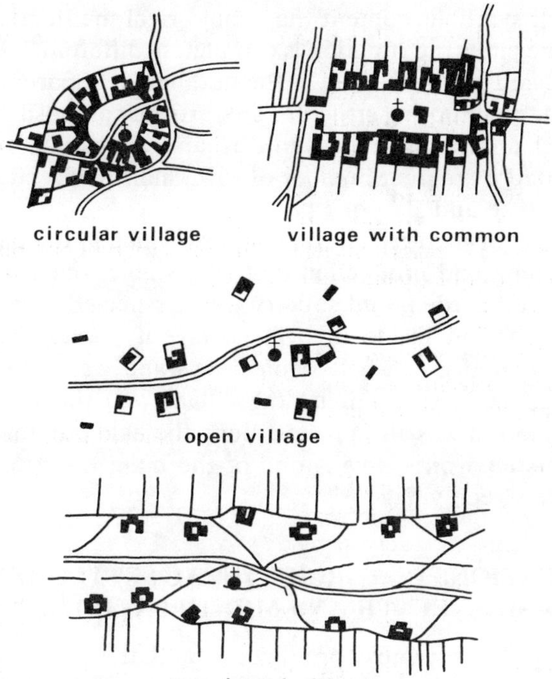

Figure 15.1 Examples of village types in Czechoslovakia

the results of the latest research on the development of vernacular architecture in agricultural settlements, there are about twenty-five basic types of peasant and village houses, each with a number of sub-types. If we take the layout of the houses as the main criterion, there are about fifty types (Mencl, 1980).

The current state of vernacular architecture in the villages is the result of the historical stratification of values from various periods from the middle ages to the mid-nineteenth century. The basic characteristic of vernacular culture in general was the trend to preserve and conserve the old, not to remove it.

The dominant silhouette in the Czech village was, and very often still is, the church tower—most frequently the spire of the Baroque church which rises above the surrounding trees. This domination of the sacral symbol in the rural settlements in no way corresponds to the significance of religion to the rural population in the past century—with the exception of certain areas which are traditionally religious, the importance of religion has waned, especially in the last 50 years. It is therefore interesting that no social force, no social institution, in the whole pre-socialist period created a material structure with such a significant symbolic content that would rival the dominant position of the church. In the cities the situation is completely different. There the growth of the economic and political might of the burghers is apparent in the building of the town hall; its imposing structure and artistic composition heralded the end of the church monopoly in the whole urban composition of the town. This fact is documented by a great number of town halls in Czech towns from the Gothic, Renaissance and Baroque periods.

An integral part of the agricultural complex which in the past was a spatial unit of the dwelling and production unit (the stable, the barn, the granary) was vegetation: the garden, and solitary trees, especially the popular linden (lime) tree. Vegetation inside the village layout passed into purposefully planted and protected green-belts along ways and water-courses, and then into the broader areas surrounding the village and thence into the whole landscape. Vegetation was also planted along the field boundaries, etc., in the village and in its environs—this is one of the main traditional traits of the landscape of the Czech village.

15.3 MODIFICATIONS IN AGRICULTURAL SETTLEMENT BETWEEN THE MID-NINETEENTH AND MID-TWENTIETH CENTURIES

The development of capitalist production, and the rapid development of certain branches of industry, made the Czech region at the end of the nineteenth century (then part of the Austro-Hungarian Empire) the Empire's industrially most highly advanced region. This fact has key importance, as it not only significantly influenced the development of agricultural settlement at the end of the century, but also the basic relationship between the development of industry and agriculture in the whole capitalist epoch. Moreover, this relationship became the basic factor in the changes connected with the beginning of the socialization of agricultural production and the long-term reconstruction of the whole settlement network (Table 15.1). Without understanding this historical fact it is not possible to understand certain 'parameters' of current agricultural large-scale production and the system of settlement that goes with it.

The upsurge of industrial production was already being reflected in agricultural settlement by the end of the nineteenth century. At the time the first decline in the population is recorded of small villages with fewer than 200–500 inhabitants. Thus began the process of the depopulation of the rural areas, which became significant around 1930 and then again after 1945 (Table 15.2). This started the process of concentration. The number of small settlements decreased; this means that they became the integral components of higher settlement units, and because of their closeness they often became architecturally, visually interlinked. They concurrently lost their agricultural function and acquired other functions (residential, recreational, industrial) within the framework of the higher category settlement.

Table 15.1 Numbers of people working in the national economy and in agriculture

Year	Numbers working in the national economy (millions)	Numbers working in agriculture (millions)	(%)
1948	5.545	2.239	40.4
1950	5.577	2.058	36.9
1953	5.683	1.858	32.7
1955	5.956	1.933	32.5
1960	6.063	1.468	24.2
1965	6.477	1.262	19.5
1968	6.794	1.207	17.8
1969	6.919	1.192	17.2
1970	7.033	1.183	16.8
1971	7.115	1.167	16.4
1972	7.179	1.097	15.3
1973	7.254	1.066	14.7
1974	7.357	1.058	14.4
1975	7.435	1.033	13.9
1976	7.476	1.029	13.8
1977	7.539	1.006	13.3

Source: Federal Office of Statistics, *Statistical Yearbook of the Czechoslovak Socialist Republic*, 1978, pp. 22–23, SNTL, ALFA, Prague–Bratislava.

The mid-nineteenth century is the dividing line which marks the end of individuality in the construction of agricultural buildings, from building materials which were provided by nature. Since then standard structural forms, easily reproducible, have become common. This trend was especially strong in the vicinity of large cities in the 1920s and 1930s. The social differentiation which took place in the villages, the crystallization of certain social classes (i.e. large landowners, medium landowners and small farmers and hired farm hands), was reflected in the architecture of the agricultural buildings: the estate, the farm, and large and small cottages. Industrial development in the Czech regions resulted in the gradual appearance of a non-agricultural population in the Czech villages in the early twentieth century.

In the mid-twentieth century the non-agricultural element was strongly represented in almost all villages. This meant that in the final stage of capitalist development the Czech village was no longer the exclusive dwelling and production location of the agricultural population. Socially it had become heterogeneous, with a population of farmers, part-time farmers (people with a small acreage of land who leave the village in the winter season to work in industry and building), craftsmen and transport workers (railwaymen), etc. This social heterogeneity of the village corresponds to spatial heterogeneity:

Table 15.2 Number of localities by size in Czechoslovakia in the period 1921–80[a]

Population range	1921	1930	1950	1960	1970	1980
0–2000	13 991	14 210	13 994	10 916	9 566	6 441
	(93.7%)	(93.2%)	(94.5%)	(91.3%)	(90.2%)	(85.8%)
2000–5000	704	786	583	757	716	669
	(4.8%)	(5.2%)	(3.9%)	(6.3%)	(6.8%)	(8.9%)
5000–10 000	149	153	133	171	173	191
	(1.0%)	(1.0%)	(0.9%)	(1.4%)	(1.6%)	(2.6%)
10 000–20 000	56	63	54	72	82	114
	(0.4%)	(0.4%)	(0.4%)	(0.6%)	(0.8%)	(1.5%)
20 000–50 000	19	28	28	33	44	58
	(0.1%)	(0.1%)	(0.2%)	(0.3%)	(0.4%)	(0.8%)
>50 000	6	7	11	14	19	30
	(–%)	(0.1%)	(0.1%)	(0.1%)	(0.2%)	(0.4%)
Totals	14 925	15 247	14 803	11 963	10 600	7 503

[a]According to the Federal Office of Statistics, localities with a population of fewer than 5000 are considered rural. This means that in 1980 there were 7110 such localities, or 94.7% of the total.

Sources: Vladimir Srb (1967) *Demographic Handbook*, Svoboda, Phraha. Alexander Keviczký (1966) *Survey of Large Localities in Slovakia with Special Regard to the Use of their Building Stocks within the Long-term Concept of Settlement*, Research Institute of Agricultural Construction, Faculty of Civil Engineering, Bratislava. *Development of the Society in the Czechoslovak Socialist Republic* (1975) (results of the 1970 census), Federal Office of Statistics, Prague. *Preliminary Results of the 1980 Census* (1981) Federal Office of Statistics, Prague.

Table 15.3 Construction of family houses in the period 1961–79 (see Note 1)

Year	Total numbers of flats	Numbers of family houses in localities with a population of:									Total family house	Total flats in family houses	Flats in family houses (%)
		<500	500–1000	1000–2000	2000–5000	5000–10 000	10 000–20 000	20 000–50 000	50 000–100 000	>100 000			
1961	86 032	5 635	6 513	7 469	6 046	1 951	600	499	157	404	29 274	30 890	35.9
1962	85 221	3 412	5 172	6 090	5 717	1 864	1 237	566	194	357	24 609	26 821	31.0
1963	82 189	2 521	4 468	5 664	5 417	1 660	984	527	201	334	21 776	23 098	28.1
1964	77 301	2 312	3 983	4 992	4 731	1 388	783	446	198	276	19 109	20 289	26.2
1965	77 818	2 196	3 779	4 607	4 334	1 357	713	603	150	276	18 015	19 195	24.7
1966	75 526	1 971	3 628	4 529	4 519	1 325	790	632	231	304	17 929	19 172	25.4
1967	79 297	1 854	3 491	4 596	4 715	1 545	1 015	612	223	248	18 299	19 835	25.0
1968	86 571	1 766	3 167	4 190	4 877	1 601	1 108	551	254	274	17 788	19 695	22.8
1969	85 656	1 787	3 303	4 000	4 807	1 725	1 199	800	250	422	18 101	21 282	24.8
1970	107 653	1 948	4 568	5 100	5 974	2 256	1 839	1 085	347	599	23 716	26 755	24.9
1971	107 380	2 281	4 102	4 869	5 058	1 946	1 789	1 174	324	584	22 127	29 551	27.5
1972	115 559	2 439	4 103	4 701	5 438	2 208	1 992	1 305	326	671	23 183	32 150	27.8
1973	118 594	2 371	3 721	4 146	4 964	2 013	1 778	1 366	344	948	21 651	32 038	27.0
1974	128 988	2 542	3 680	4 543	5 134	2 093	1 840	1 563	348	976	22 719	34 713	26.9
1975	144 678	2 857	3 864	4 475	5 348	2 279	1 945	1 547	444	1 829	24 588	38 579	26.7
1976	132 451	2 701	4 243	5 066	6 244	2 654	2 098	1 586	580	1 184	26 356	39 589	29.9
1977	134 820	2 757	4 097	5 105	6 224	2 372	2 357	1 875	583	1 511	26 881	40 107	29.7
1978	129 330	2 741	4 329	5 211	5 853	2 533	2 124	1 594	514	1 509	26 408	37 982	29.4
1979	122 741	2 178	3 546	4 885	5 901	2 787	2 307	1 742	882	1 386	25 594	35 666	29.1

there is no spatial segregation among the rural population. The large or small cottage of the medium or small farmer stands next to the small cottage of the industrial worker who commutes to work in the district town, or next to the house of the railwayman. The situation was different in Slovakia which, during the pre-war Republic, was an agricultural appendage of the Czech regions. There, especially in the most fertile areas such as the Rye Island on the Danube, there existed villages with thousands of inhabitants, purely agricultural.

The vegetation in the individual villages developed, too. In areas with intensive development of capitalist large-scale type agriculture, on large fields with mono-cultures (wheat or sugar-beet), vegetation was radically removed and the landscape was transformed into a cultural steppe. In the village itself the state of vegetation declined rapidly.

15.4 REVOLUTIONARY CHANGES TAKING PLACE IN SOCIALIST AGRICULTURE

The development of agriculture on a socialist basis has led to revolutionary changes in the whole system of the rural landscape, including the system of rural settlements. The setting of the areas around the villages—the agricultural landscape—has in the past 30 years (1950–80) changed its appearance, and has taken on large-scale dimensions with the new organization of the soil stock (large runs of fields which comply with highly efficient large-scale technologies). In the village the traditional linkage of the dwelling area with the outbuildings (stables, granaries, barns) in one unit has ceased to exist on large farms and estates. The concentration of agricultural large-scale production into a smaller number of larger production or service units enforces their location on the boundary of the built-up area of the village, or at a certain distance from it. To start with, the locations of these production and service units were determined on purely production and economic criteria, regardless of the ecological consequences and of aesthetic factors.

Great changes in the built-up parts of villages in the past 30 years have resulted from modernization of former farm buildings, such as barns, stables, stys and sheds; many have been converted into dwelling areas or into storage rooms for fruit, tools, etc., or into garages. In addition the villages have been overpowered by a wave of new construction. (We shall discuss the consequences of this construction later in this study.) For example, of the total number of flats built in the Czechoslovak Socialist Republic in the period 1976–80, flats built in village houses make up around 30% (Table 15.3).

A completely new phenomenon, which was not reckoned with in the initial planning of the socialist village, was the use of rural areas for recreation by city dwellers while living in their own second homes. The city dwellers no longer come to the villages, as they used to in the 1920s and 1930s, as

Table 15.4 Populations in villages in Czechoslovakia in the period 1921–80[a]

Population range	1921	1930	1950	1960	1970	1980
0–2000	7 398 000	7 363 000	6 026 259	5 740 120	5 399 356	4 182 203
	(56.9%)	(52.6%)	(48.8%)	(41.8%)	(37.6%)	(27.4%)
2000–5000	2 087 000	2 369 000	1 759 824	2 292 611	2 138 264	2 011 155
	(16.0%)	(16.9%)	(14.3%)	(16.7%)	(14.9%)	(13.2%)
5000–10 000	1 012 000	1 460 000	974 380	1 208 154	1 218 636	1 320 340
	(7.8%)	(7.5%)	(7.9%)	(8.8%)	(8.5%)	(8.6%)
10 000–20 000	774 000	831 000	795 292	989 220	1 124 484	1 651 097
	(5.9%)	(5.9%)	(6.4%)	(7.2%)	(7.8%)	(10.8%)
20 000–50 000	542 000	775 000	714 186	1 020 626	1 300 513	1 730 101
	(4.2%)	(5.6%)	(5.8%)	(7.4%)	(9.1%)	(11.3%)
>50 000	1 190 000	1 614 000	2 068 509	2 494 846	3 171 112	4 381 903
	(9.2%)	(11.5%)	(16.8%)	(18.1%)	(22.1%)	(28.7%)
Totals	13 003 000	13 998 000	12 338 450	13 745 577	14 352 365	15 276 799

[a]According to the Federal Office of Statistics, localities with a population of fewer than 5000 are considered rural. This means that in 1980, the total population of these villages was 6 193 358, or 40.5% of the total population.

Sources: as for Table 15.2.

holiday-makers spending their holidays in rented rooms: they now come en masse to their *own* houses which they have converted from what used to be village cottages, mills, granaries, etc., purchased from former farmers. Small village localities began to provide recreational possibilities to all social and age categories of the town population on a very large scale indeed. The recreational function of the rural areas was strengthened, which certainly bore its consequences. We shall say a few words about the social and economic processes that were taking place.

In Czechoslovakia, as in other industrially advanced countries, this century has been marked by a depopulation of the rural areas. In 1930, 67% of the total population of the Czech regions lived in localities with fewer than 5000 inhabitants; by 1970 their numbers had declined to 47%. In Slovakia this process was somewhat slower; the number of inhabitants in such localities had declined from 77% in 1930 to 65% by 1970.

According to the preliminary results of the October 1980 census, only 40% of the total population of the Czechoslovak Socialist Republic then lived in localities with a population of fewer than 5000. This basic shift in the demographic structure of the villages was most marked in the smallest localities (Table 15.4). The decrease of the population results from the departure of the younger members and the retirement of the aged. It appeared that these localities would soon disappear altogether. On the other hand, vernacular architecture was in very good condition in all these small settlements, and they had a thick growth of lush and old trees and other vegetation and formed oases of calm, with a healthy, unpolluted atmosphere. So, 25–30 years ago, when the first waves of a demand from the municipal population for recreation in the rural areas was heard, this abandoned building stock was used for the recreation of the urban population, who could purchase the houses and make them their private property (Table 15.5). The demand, however, was not controlled purposefully and sensibly: the open landscape was inundated with chaotic groupings of individual cottages, forming agglomerations of 'secondary' settlements whose existence has many

Table 15.5 Numbers of individually owned recreational buildings in Bohemia and in the Czechoslovak Socialist Republic

Year	Cottages	Converted agricultural buildings
1923	23 000	0
1955	45 000	0
1965	84 000	8 500
1972	132 000	24 500
1980	Data from last census not yet processed	

negative consequences for the whole rural area. Construction of these cottage agglomerations threatened to devastate the most beautiful nature areas and landscapes, and so measures had to be taken to control and check the process of construction of cottage colonies. Special legislation enacted in 1971–74 is based *inter alia* on a system of protective ecological criteria.

Official statistics record the continuous decline of the permanent population in small rural localities, but they do not record the opposite process, i.e. the number of urban inhabitants who spend short or long periods in these settlements. The settlements are pulsating organisms which fill up over the weekends and in the summer holidays. In mountain villages where there are good conditions for winter sports, especially skiing, city dwellers travel to their mountain cottages regularly. It is therefore not illogical to consider these settlements as points of relatively high 'potential urbanization'. Moreover, the urban population which comes regularly is one of the most important vectors of the spread of city behaviour—the city 'life-style'. This is often manifested in very subtle ways: sociologists have noted that in these settlements where the urban 'second-home' population is strongly represented, the trend is more evident towards keeping the village neat and tidy.

The 'second home' is an organic component of the plan for the establishment of a new settlement system whose main trends of development were formulated 25 years ago and will be discussed later in this study.

A major revolutionary measure which affected the very substance of the development of rural settlement was planning. The earlier random development of rural and other settlements was replaced by planning, which was increasingly based on a broad spectrum of scientific information and knowledge. One of the targets of the socialist programme is the elimination of antagonisms between the town and the village.

15.5 THE CHANGING VILLAGE CONCEPT IN THE CZECHOSLOVAK SOCIALIST REPUBLIC

Within the framework of this study it is not possible to present a complete survey of the problems related to the substantial differences between the towns and the villages. We do, however, consider it necessary to mention some factors that have influenced development in an undesirable way.

At one time it was thought that the villages could be made more like the towns by providing similar buildings. Housing in the villages was therefore oriented to the construction of multi-storeyed apartment houses. These, however, were totally unsuitable for the way of life in the villages (with the lack of small storage areas for fruit, potatoes and fuel, and the lack of any space for keeping small animals such as chickens, geese and ducks). What is more, the flats in these houses lacked a large kitchen serving as a living area, and had many other aspects unsuitable for the villagers.

The construction of these apartment houses was mostly only possible on the boundaries of villages—on arable land—and their scale, mass and general appearance often disturbed the whole panorama of the village, the architectural and rural character of the whole landscape. This type of housing construction was therefore considerably restricted, in favour of various types of small family houses. In some cases the practitioners went so far as to demand that the vegetation inside the village setting be replaced by something like city parks. It was thought that country people would not need to keep poultry or grow vegetables, because everything would be supplied by shops, as in the towns. This trend was later described as 'bulldozerism', whose proponents vehemently advocated the speedy elimination and liquidation of small villages and localities, by which operation they wanted to acquire agricultural land and to win the absolute social and political domination of settlements of the municipal type. The opposition between the town and the village concepts was to have been resolved by the liquidation of small localities and the municipalization of bigger villages. Today these views, which were advocated 25–30 years ago, are considered as having been rather sombre.

We need to consider first the very small average distances between settlements, the social heterogeneity of the rural settlements, the total absence of spatial segregation between the farming population and the other social and professional groups living in the village settlement, etc. Among the socialist countries, the Czechoslovak Socialist Republic and the German Democratic Republic have the highest population densities. Czechoslovakia also has one of the highest capital-construction rates per square kilometre, and an accumulation of activities and functions. Like the modern woman, the Czechoslovak landscape continuously faces the problem of how to harmonize often highly contradictory activities and yet remain pretty and attractive. Moreover, Czechoslovakia has one of the lowest acreage rates per inhabitant in Europe (0.42 ha), of which 75% is arable land. The amount of agricultural land is continuously decreasing, raw material resources are limited, and there is a lack of utility and drinking water. All these factors exert pressure on land management.

Czechoslovakia and the GDR are foremost among the socialist countries in the elimination of differences between the towns and the villages (Linhart, 1980). This is manifest in many ways. The earlier economic and social differences between industrially advanced areas and areas that are mainly agrarian (southern Bohemia, the Czech-Moravian Highlands, and certain parts of the less industrially developed Slovakia), have been levelled. It is also very significant that small and medium-sized towns have experienced sweeping development, while excessive growth has been prevented of the three major cities Prague, Brno and Bratislava; depopulation of rural areas has also been limited (Table 15.4).

Table 15.6 Average wages (in Crowns) of workers in the socialist sector of the national economy of the Czechoslovak Socialist Republic (see Note 2)

Category	1955	1960	1965	1970	1975	1977
Workers in industry	1300	1450	1590	1980	2360	2520
Workers on state farms	895	1110	1315	1830	2250	2410
Workers in the building industry (including project design work)	1410	1550	1720	2235	2615	2760

There have been structural changes in village family life. The number of family members who remain in agriculture is decreasing and the social and professional spectrum of the other members of the family is broadening—they work in industry, the services, transport, etc. This may be interpreted as another channel through which city life 'flows' into the villages.

Working and social living conditions of workers in agriculture and in other production sectors have become more similar. A considerable number of workers in agriculture now have a five-day working week like workers in industry, and some do shift work. Their social and health security, and that of their families and children, do not differ from workers in other sectors. They have the same rights to paid holidays and are given the same opportunities to spend their holidays in recreational centres and in spas. The average wages of workers on state farms are close to those of workers in industry (Table 15.6). The pattern of ownership of consumer durable goods is also similar (Table 15.7).

The statistics must be evaluated taking into account the total cost of living of families living in urban and rural areas. Costs of certain important indicators—such as expenditure on food—are much more favourable for the villagers. This has resulted in the concentration of considerable financial means in the rural areas, which has allowed not only the basic reconstruction of the housing stock but also the construction of new buildings of various

Table 15.7 Numbers of certain consumer durable goods in Czech households, according to Social groups, at the end of 1978 (see Note 2)

Household	Refrigerator	TV sets	Radios	Motor cars	Second homes
Workers	93.6	98.8	132.3	49.7	8.7
Cooperative farmers	95.8	102.3	127.1	61.0	2.0
Other employees	92.6	95.6	132.9	52.6	15.7

types. In the solution of housing problems the rural areas are still well in advance of the big cities, and this gap is widening.

15.6 FROM PLANNING THE DEVELOPMENT OF THE SOCIALIST VILLAGE TO PLANNING THE DEVELOPMENT OF THE SYSTEM OF SETTLEMENTS

The village 'space' has undergone a number of stages during the development of socialist villages. It is not the purpose of this study to make a detailed analysis of each of these stages. When comparing the past with the present we may point to several general differences in the planning of rural settlements. Earlier planning processes proceeded from the one-sided and linear theoretical postulate of the removal of differences between the towns and the villages, conceived in the sense that the villages would be transformed into towns, that the towns would not undergo any major *qualitative* changes in arrangement, functions or composition, but only rapid *quantitative* changes.

Reconstruction of the villages was not approached in a systematic manner (a term not even used 25–30 years ago); it was always only related to this or that settlement as an individual unit, and the reconstruction only concerned a certain part, such as the road system. This meant that the plans were not concerned with the whole of a village space, nor were they concerned with the development of the rural settlement in linkage with a higher-level settlement system. Regional plans were only devoted to built-up areas, without any continuity with the other systems in the landscape. Planning was thus oriented to the technical and economic parameters of the individual settlements, little attention being devoted to social laws, even less to psychological and aesthetic considerations, and none at all to ecology.

Almost 20 years ago the concept of a new type of settlement system was started. We have said already that somewhat later, in the period 1971–74, the whole system of 'secondary settlement' of summer cottages became its integral component. Its basic theoretical postulate was that the elimination of substantial differences between the towns and the villages was attainable only within the framework of individual agglomerations of settlements, i.e. a new settlement structure which overcomes the traditional concept of the town as well as the traditional concept of the village and of the village space in general. The new settlement structure of the Czechoslovak Socialist Republic was therefore based on 19 explicit conurbations:

(1) Prague, Brno, Ostrava, Bratislava and their environs;
(2) conurbations in which the dominant role is played by several close towns, such as Pardubice and Hradec Králové;
(3) strings of towns, such as Chomutov, Klášterec, Kadaň, Most and Teplice in the North Bohemian brown coal basin.

The development of another 35 secondary agglomerations of settlements is envisaged. They will be a decentralized analogy of today's compact town. The individual parts (quarters) of a town have their administrative, commercial, industrial, recreational and residential functions, and a similar division of functions should be formed inside each agglomeration, even though the relations between the individual parts will be looser. Large-scale agriculture will form part of these agglomerations.

The harmonization of all aspects of the development of each part, of all functions within the agglomeration (i.e. production, social activities, etc.), with ecology will not be an easy matter. A very grave limiting factor will be the small acreage of arable land per inhabitant, which limits the use of land for any construction; and so the cores of these agglomerations will take on the character of 'diffused nebulae'. In the residential parts a high density of settlement will have to be enforced.

Thus, further development of settlements traditionally known as villages will become part of a new system. The development of socialist villages will in the future be solved within a certain settlement agglomeration.

From the very beginning the newest concept of settlement proceeded from a three-tier settlement structure which developed gradually with the development of society, production and the settlements themselves. Its main principle—the three-tier character of the settlement system—was preserved.[3]

The district and regional centres are obvious, but a decision as to what is a local centre and how it is distinguished from other rural settlements which do not have the character of a centre remains in the fore of the attention of administrative bodies and urban planners. This is understandable, because it is here that the changes in the historical settlement structure are most acutely felt. They result from the development of agricultural production and the concentrated build-up of farms and other agricultural installations, such as services, joint supra-enterprise storage areas, silos, etc., and from the development of the 'recreational phenomenon' which has changed the character of many settlements by underlining their recreational character.

A categorization was made of rural settlements with regard to all the factors mentioned, and taking into account the further development of agricultural construction (especially the expansion of existing enterprises) in compliance with the functions of the respective installations. Localities were classified into centres of local importance and localities that do not have the character of a centre. Large-scale animal farms with a great concentration of manpower should be located in the local centres. In the first period of the organization of agricultural production the unified farmers cooperative was linked to one locality. In the present period of the merger of cooperatives this has been overcome. Farmers cooperatives today operate on cadasters comprising two or more localities. There are state farms with more than 50 000 ha which cover the whole district. These facts alone are bound to lead to changes in the functions of the existing settlement structure.

As a result of cooperation and integration processes, some settlements become centres of agricultural production and development and gain in importance, while others lose their importance when development is transferred to another locality. The recreational function of many localities strengthens; this mainly happens in the vicinity of industrial towns and conurbations and in localities outside industrial areas, located in protected nature conservation areas and national parks.

We thus see that some former agricultural settlements grow into rural towns while other villages lose their agricultural importance and gain importance as residential or recreational areas.

This significant functional differentiation of the existing rural settlements within the framework of settlement agglomerations will also determine the approach to their further development. In settlements with a prevalently dwelling function, the following should be given priority:

(1) the development of concentrated housing construction, based on new structural and architectural principles, unobtrusively incorporated into the existing area and surrounding landscape;
(2) the strengthening of communications with the centres of job opportunities and centres of social life;
(3) the building of local leisure centres.

The last point is related to a completely new problem area. It appears that activities which have so far been made possible by existing or newly built cultural houses and centres are not adequate. The view that agricultural work in the open air is so healthy that farm workers have no need for other physical exercise or sport no longer holds. These facts will have to be taken into account in conceiving the content of the activities planned and organized in these cultural centres. It has become evident that they will have to have sports areas and swiming pools.

In settlements with a mainly recreational function, the main effort will be towards doing everything to prevent the emigration of the former inhabitants. It is becoming more and more evident that when settlements have a significantly mono-functional recreational character, outside the weekends they attract antisocial elements and criminal activities. It is also very important to keep the construction of recreational houses and cottages in these villages within certain limits (pressure in this direction is very strong). Public amenities must be built adequately.

In summary, then, we can say that in the socialist development programme the villages have been influenced by many factors and impacts of which we believe the most important are the following:

(1) the transition from small-scale to large-scale agricultural production, and with it the industrialization of agricultural production, the influence of

technical development not only on productivity but on the whole way of thinking;

(2) the rise in the standard of living of workers in agriculture;

(3) the growth of the technical intelligentsia, higher demands on management, and the scientific development of agriculture;

(4) the growth of the size of agricultural enterprises and farms, integration and cooperation processes and their impacts and changes in the settlement structure of rural settlements;

(5) the construction of public amenities, shops, etc., in view of the municipalization of the village whose inhabitants were no longer able to grow and breed their own food;

(6) the high mobility of the rural population, be it commuting to work or for other purposes;

(7) the recreational use of free cottages and small estates in rural areas, by their conversion into recreational cottages;

(8) the construction of cottage colonies in recreational areas, and the construction of public amenities and services.

15.7 OPEN QUESTIONS

15.7.1 Theory of Urbanism

Continuous reconstruction of socialist villages is impossible without clarification of the wide range of possible future intentions for the landscape. One has to visualize the future development of agricultural production and its orientation—mainly the further stages of concentration and specialization, the acreage of large runs, the communcations for farm machines of future generations, land improvement measures, etc. Other questions to be considered will be the mining of raw materials—especially mines and quarries of local character, the construction of new sand mines, stone quarries, etc.—the development of road and rail transport, long-distance power supply systems and other limiting factors. Here we shall side with those specialists (mainly urban planners of rural settlements and sociologists oriented to rural problems) who emphasize that this problem area, be it very important, is only part of the question. We are also concerned with the unresolved problem of the overall composition of the settlement structure, including its dominant features. Past historical periods show the typical 'silhouette' of the Czech village with the church dominant, the adjacent parish houses engulfed by trees and flowers, passing into scattered vegetation in the adjoining landscape beyond the gardens. This silhouette had its semantic importance: it was the expression of the power of the Catholic church. Regardless of the waning power of the church and of religion in general, the church remained the natural and logical centre of the layout of the village and of the arrangement of the whole landscape.

What *should* the silhouette of the socialist village incorporated in the landscape be like? The erection, next to the church, of vast silos, haylage batteries and similar structures has led to very unsatisfactory results. Radical changes in the scale of individual and group structures, in volume, height, layout and area, and their incorporation into the settlement and in the landscape, and their role in the formation of the silhouette of the settlement and panorama, are all problems that are currently being discussed. The exchange of views has so far not yielded any conclusions. There always remain the basic questions: what should the present generation build to stand next to the value represented by the church? and what should be the dominant features of the larger landscape to personify the meaning and aims of a socialist society?

15.7.2 Management

Hitherto the attention devoted to the reconstruction of the spaces of rural settlements has been very small in proportion to its significance. Of the total number of architects and urban planners, only a very negligible minority are devoting their attention to this problem. With few exceptions there are no project design institutes to design village dwellings, and the departments of construction at district national committee level are inundated with routine work and are not oriented to the solution of fundamental questions (Vidláková, 1979; Vidláková *et al.*, 1979). The personnel of these departments include a large number of new graduates and people without adequate qualifications.

As a result the inhabitants of the villages, with adequate financial means in hand, began reconstructing their old houses or building new ones according to old stereotypes and prejudices. Nobody showed them new housing models, particularly the advantages of living in terraced houses which, in certain combinations, could positively influence the outlook of a new village. Most of them saw their ideal in the suburban villa. It is understandable from the sociological point of view why they believed that this was the form that represented all that was new: it pointed to a new way of life, whereas the picturesque vernacular architecture, with its small windows and low ceilings, was for them something of the past, something to be got rid of very quickly. Especially in Slovakia the original log-cabins were, in the period of a rising standard of living, identified with the period of poverty and therefore replaced with 'villas'. The location of villas and family houses along the communications links, which used to be an advantage, now became very uncomfortable owing to the dense traffic (heavy lorries, and the intensification of public transport). This type of housing construction represented, from the point of view of the whole rural space, a disproportionate diffusion of the settlement into the landscape, while in the centre of the existing villages and localities there remained unused buildings which lost their raison d'être (such as granaries, barns or little-used gardens). This purely extensive form of development has in recent years been halted, but the consequences of this spontaneous process will bear their negative impacts for a long time to come.

15.7.3 Questions of an Ecological Nature

The best savings of soil and energy will, it appears, be the two decisive factors influencing the construction of both production units and dwellings. The whole sphere of legislation (new laws on soil) will enhance trends to save soil and to increase its biological potential. In place of the diffusion of settlements—their sprawl into the countryside—it will be necessary to enforce a more concentrated concept of construction, with a more purposeful use of space inside the village area, and the use of modern forms of construction, such as terraced and semi-detached houses. This trend should be enforced even more vehemently in the case of agricultural production units, to seek optimal solutions. The ecological problem area thus gains importance.

In those places where the traditional systems of built-up areas are being replaced by new housing constructions such as multi-storey complexes, it is necessary to ensure that combinations of styles do not result in incongruence and chaos. This has often happened in the past owing to the purely technical approach to housing construction. An important factor is the aesthetic harmony of the broad range of traditional materials which used to be used in the construction of village buildings with the increasing variety of new building materials. The arguments voiced by ecologists should not be overlooked, pointing to certain chemical properties of new materials which may have negative impacts on various life processes (e.g. low chemical stability, mutagenic characteristics, undesirable electric fields, excessive water absorption).

Current trends prove that the linkage between agriculture and industry will deepen, that new agricultural construction will take place especially in those places where there will be a direct connection with the use of industrial wastes, waste heat and energy. Let us give one example. When, ten years ago, a project was launched for a transit gas pipeline to run across Czechoslovakia, a small circle of specialists began discussing the possible uses of the large amounts of waste heat which were generated by the operation of turbo-sets in compressor stations along the route. Two variants were considered: to use this heat for heating towns and villages, or to use it in agriculture. Calculations showed that the latter would be more suitable and more advantageous. The pipeline avoids large towns and new housing estates, and to transfer the heat would be economical only to a distance of 4–5 km (the costs of transfer to larger distances rising very rapidly). The distribution of small amounts of this heat into individual village family houses also proved unfavourable. Under an agreement concluded between the Federal Ministry of Fuels and Power and the three ministries of agriculture (Federal, Czech and Slovak), a broadly conceived plan was drawn up for the use of waste heat from the individual compressor stations for the needs of agriculture. The plan was to be implemented mainly in the years of the 7th five-year plan (1981–85), and possibly in later years. Close to these compressor stations installations and complexes were to be built for a wide variety of purposes, such as hothouses, fodder driers, hop driers, fisheries, duck farms, broiler plants, central storage areas with a controlled atmosphere for seed samples for a genetic bank, etc. The concept has the form of a closed circuit. Hot water from the compressor and pumping station will not only heat the agricultural units, but will return to the compressor where it will be recycled for heating. We are convinced that in the coming period these principles will find broader application, mainly because various types of industrial wastes discharged into water media contain considerable volumes of wastes which pollute the water and which are in fact a valuable raw material for agricultural production. This mainly concerns the food industry itself: sugar mills, distilleries, slaughter-houses, and the amounts of proteins they discharge as waste.

Finally we should like to mention one completely new problem which arises with the problem of ecological pathways in the landscape: what role will the rural settlements play in the system of these pathways? It should be said at this point that Czech zoology has a long tradition in the research of animal life in big cities. Czech zoologists have proved that with advancing urbanization the composition of animal species in the big cities changes, that some species are replaced by others, but that it would be a mistake to believe that big cities repudiate the whole animal world. The outstanding Czech zoologist, Prof. Julius Komárek, wrote a book more than 40 years ago devoted to animals in a big city: the book has the appropriate title *The Unknown Face of Prague* (1941) and is dedicated to all lovers of the city and to all those who decide its fate—especially to the urban planners and architects—calling on them to remember that 'people are not the only inhabitants of Prague'.

Today we could in an analogous way speak of the unknown face of all settlements, small and large. Scientists have not yet succeeded in describing all manifestations of the 'urbanization' of animals, for whom the radical transformation of the whole agricultural landscape made human settlements and their immediate environs the most dangerous refuge. Prof. Komárek writes that even in small rural settlements the situation arose when 'new and previously unobserved animals appear which had previously lived outside human settlements and had been known for their extreme shyness. It is of special interest that this process is not concluded and that we may expect the arrival of further imigrants', and he adds that the phenomenon has not yet been clarified by scientists. In stating this, Prof. Komárek depicted the general trend of synathropization which was dramatically accelerated by the total reconstruction of agricultural production and by the introduction of new scales in the landscape. We are thus following up on the ideas of this leading Czech scientist and humanitarian of Schweitzerian provenance. It will be necessary to seek an answer to the question: what should a human settlement look like, considering that it is not inhabited solely by people? This problem, logically following up on the concept of ecological pathways, is still *in statu nascendi*. The answer to the question forms part of a general reply to the question of how can one rationally cultivate the surrounding world. People are beginning to realize, albeit with difficulties, what the presence in their settlements of animals, plants and nature means for them.

Possible forecasts of future changes may be based on correction based on experience, on a systems analysis approach including simulation, on the existence of loading limits, on the reversibility or irreversibility of observed processes.

Forecasting future changes in the whole rural space (landscape and settlements) is more and more difficult, owing to revolutionary discoveries and technical progress, on the one hand, and energy and ecological barriers on the other. It has become a general phenomenon that processes of development are becoming entangled in an ever greater mesh of contradictory trends, which are often the result of forecasts that did not reveal certain important phenomena which took place in the process of development.

We believe that most plans and forecasts do not show enough respect for ecological 'parameters' which enter all events surrounding man. It may be expected that in a short period of time new ecological knowledge will influence a wide range of man's activities, and thus many production technologies and the entire economic development of society. This fully applies to agricultural production, and with it the related system of rural settlement. No concept of the development of the village designed so far has adequately applied the ecological aspects which science has discovered. This is to a certain extent understandable, because we are concerned with a relatively young problem area. In the rural areas anthropo-ecology has yet to be applied.

Prognoses of the development of the whole network of settlements in the rural landscape up to the year 2000 must be very cautious, and it may be better to specify development in very short intervals (of approximately 5 years) so that new knowledge may be taken into account. This point is exemplified by the waste heat from the gas pipeline mentioned earlier, which has initiated the construction of a whole range of agricultural production units.

Current experience shows that, contrary to previous radical concepts, older settlements will remain preserved in view of the recreational demands that were not correctly estimated before. Small localities and non-prospective settlements will therefore continue to be used for recreation, and their recreational function is likely to be enhanced. It may be envisaged that the concept of the settlement system with the local settlement centre will be strengthened in localities with a permanent economic function by the construction of public amenities and new housing, mainly in the form of family houses (in compliance with knowledge about the healthy environment). It will probably not be possible to exclude new housing from those settlements that do not serve as local centres but which have a singularly good natural environment.

It may also be expected that the broader environs of big cities will in the near future be exposed to unprecedented urbanization pressure. This phenomenon was recorded in the big cities of the advanced western countries 20–25 years ago. This trend will be caused not only by the strengthening of the 'recreational phenomenon', but also by the quality of the environment and of housing in big cities which is not expected to undergo any radical improvement in the near future. From this, and from the desire of the inhabitants of large towns and big cities to spend their weekends in natural surroundings, will stem the increased importance of individual recreation, which will in turn exert increasing pressure and demand on new spaces for the construction of private cottages, recreational houses in the villages and the construction of public and enterprise recreational projects. It may be expected that strengthening pressure on the recreational use of the village and the landscape, together with limited space in the landscape, will give rise to other, higher forms of recreational use and facilities.

In large integrated centres of agricultural, livestock or plant production, or in centres of cooperative associations of agricultural services, there will logically be a further concentration of new housing construction and of public amenities. This assumes the accumulation of certain branches of the food industry which will create the nucleus of small rural cities.

All these speculations are obviously conditioned by a wide range of pre-conditions, including a sufficient supply of fuels and energy. It is, moreover, not yet clear to what extent agriculture will be limited by ecological laws, principles and barriers which may, on the basis of factors which are as yet little known or unknown to us, either confirm the current development or necessitate a different development. The current development of large-scale agricultural production is already coming into conflict with certain ecological principles, although not everyone who deals with such production is aware of this. It will, however, be essential to put large-scale agricultural production and ecological laws into harmony within a relatively short period. The manner and orientation of this process will also specify the answer to the question of the future development of rural settlements as part of our landscape.

15.8 REFERENCES

Blažek, B., Petz, J., and Stoklasa, J. (1974) Anthropo-ecological decision making. *Lund Studies in Geography*, Series B: *Human Geography*, **40**, 7–31.

Gadač, E., Bureš, O., and Pavlů, J. (1978) Někotoryje osnovnyje ponjatija ekologii landšafta. *Acad. Nauk SSSR, Serija Geografičeskaja*, **1**, 135–144.

Gottlieb, M. (1976) Notes on the concept of the 'ecology of man'. *Bull. Ústavu krajinné ekologie ČSAV, Praha*, **1**, (1), 1–65.

Hadač, E., *et al*. (1977) Complex interdisciplinary investigation of landscape. *Landscape Planning*, **4**, 333–348.

Komárek, J. (1941) *The Unknown Face of Prague*. Čin, Prague.

Laurentius Wintera, P. (1894) Geschichtsbild der Stadt und des deutschen Schulbezirkes Braunnau. In: *Braunauer Heimatskunde*.

Linhart, J. (1980) Dynamics of the class structure and changes in the life style of Czechoslovakia. In: *Sociologický časopis*, **XVI**, (3), 273–293.

Mencl, V. (1980) *Vernacular Architecture in Czechoslovakia*. Academia, Prague, 561–585.

Šimák, J. V. (1936) Počátky Broumova a Broumovska. *Česky časopsis historicky*, **XLII**, sesit 3–4.

Sommer, J. G. (1836) *Das Königreich Böhmen statistisch–topographisch dargestellt*, IV Bild, Prague.

Vidláková, O. (1979) Landscape conservation in the activities of national committees. *Správní pravo*, **XII**, (6), 361–373.

Vidláková, O., Obrdlíková, J., Pazderová, M., and Gottlieb, M. (1979) *Empirical Study of State Administration Oriented to Landscape Conservation and its Individual Components Selected by National Committees*. Ústav krajinné ekologie, Prague.

15.9 NOTES IN THE TEXT

(1) Villages with fewer than 5000 inhabitants are considered as being rural localities. Therefore all family houses built in such localities are considered as being rural, even though it is apparent that the size of the village is not always the criterion which will allow us to judge whether we are actually concerned with a rural village or with a village of municipal character. Next to family houses the unified farmers cooperatives and state farms build multi-storey houses. This is not, however, separately recorded by the Federal Office of Statistics and is thus included in cooperative or enterprise housing construction. We can therefore not give the data separately. The sources of the data in the table are the annual surveys of capital construction published by the Federal Office of Statistics (Department of Investments), Prague.

(2) Tables 15.6 and 15.7 were processed on the basis of data published in the *Statistical Yearbook* of the Czechoslovak Socialist Republic, Prague, SNTL, for the individual years.

(3) The new three-stage concept of the development of settlements, the first draft of which was developed after 1960, starts from a three-tier settlement structure:

(i) The lowest tier in the hierarchy of settlement centres are *centres of local importance*, i.e. selected rural settlements primarily serving workers in agricultural production and workers in the services of these settlements. They are usually located in the natural centre of the catchment. They satisfy to a certain extent the economic, cultural and social needs of the inhabitants of the respective centre and of the whole catchment. The minimum prospective size of such a centre is a locality with a population of 1500, but the optimum population will be 3000–4000. In industrial agglomerations, industrial production may be respresented in the locality.

(ii) The middle link in the three-tier system are *centres of district importance*. The population of such localities should reach approximately 15 000 inhabitants. Mostly these will be district and former district towns. They will serve as the catchment areas for centres of local importance within a radius of 12–16 km. The population of the whole catchment should be around 50 000 inhabitants. These centres will form the 'skeleton' of the whole settlement structure, in which will be concentrated a considerable proportion of production and social installations and facilities forming the required municipal standard.

(iii) The highest link in the three-tier system are the *regional centres*, with the highest-level public amenities, usually linked to the territorial administrative organization of the state (regional and former regional capitals). They are significant production and social centres with large transport junctions. Their populations should not prospectively be less than 50 000.

All *other* settlements belong to the extensive and varied category of non-centres. They are divided into three groups:

(a) settlements of relatively permanent importance fulfilling a certain special or explicit economic function of permanent character in the settlement structure;

(b) settlements of temporary importance whose economic function is justified but which cannot be determined as permanent;

(c) settlement with no contemporary significant function and no envisaged future function.

On the basis of Working Instructions for the elaboration of the Draft of the Long-Term Development of Settlement in the Czechoslovak Socialist Republic in 1967, the Regional National Committees in the Czech Region drew up a draft selection of centres of local and district importance. In the Czech Socialist Republic they chose 922 centres of local importance and 163 centres of district importance. These figures do not include centres in the Slovak Socialist Republic.

Index